AF688983

FUNDAMENTALS OF BIOFUELS ENGINEERING AND TECHNOLOGY

Kruger Brentt
Publishers

FUNDAMENTALS OF BIOFUELS ENGINEERING AND TECHNOLOGY

VOLUME-1

Diego Rossit
Sebastian Ubal
Ioan Gergely

Kruger Brentt
Publishers
2025

Kruger Brentt Publishers UK. LTD.

Company Number 9728962

Regd. Office: 68 St Margarets Road, Edgware, Middlesex HA8 9UU

© 2025 AUTHORS

ISBN: 978-1-78715-318-9 (Vol. 1)

Disclaimer:

Every possible effort has been made to ensure that the information contained in this book is accurate at the time of going to press, and the publisher and author cannot accept responsibility for any errors or omissions, however caused. No responsibility for loss or damage occasioned to any person acting, or refraining from action, as a result of the material in this publication can be accepted by the editor, the publisher or the author. The Publisher is not associated with any product or vendor mentioned in the book. The contents of this work are intended to further general scientific research, understanding and discussion only. Readers should consult with a specialist where appropriate.

Every effort has been made to trace the owners of copyright material used in this book, if any. The author and the publisher will be grateful for any omission brought to their notice for acknowledgement in the future editions of the book.

All Rights reserved under International Copyright Conventions. No part of this publication may be reproduced, stored in a retrieval system, or transmitted in any form or by any means, electronic, mechanical, photocopying, recording or otherwise without the prior written consent of the publisher and the copyright owner.

For information on all our publications visit our website at http://krugerbrentt.com/

PREFACE

"Fundamentals of Biofuels Engineering and Technology" offers a comprehensive exploration into the principles, processes, and technologies that underpin the production, utilization, and sustainability of biofuels. This book serves as an essential resource for students, researchers, engineers, policymakers, and industry professionals interested in understanding and advancing the field of biofuels engineering.

Biofuels represent a promising renewable energy source that can help mitigate the environmental impacts of fossil fuels and contribute to energy security and sustainability. From first-generation biofuels such as biodiesel and ethanol to advanced biofuels derived from lignocellulosic biomass and algae, biofuels offer a diverse range of options for reducing greenhouse gas emissions and dependence on finite fossil resources.

As editors of this volume, we have endeavored to compile a diverse array of perspectives, methodologies, and case studies from leading experts and practitioners in the field of biofuels engineering and technology. Through a blend of theoretical foundations, practical insights, and real-world examples, this book covers a wide range of topics, including biomass feedstocks, biofuel production processes, biorefinery concepts, biofuel characterization, and sustainability assessment.

Our overarching goal in presenting this book is to provide readers with a comprehensive understanding of the principles and technologies involved in the production and utilization of biofuels. By offering insights into the latest advancements, best practices, and emerging trends in biofuels engineering, we aim to inspire innovation, collaboration, and excellence in the development of sustainable bioenergy solutions.

We extend our sincere gratitude to all the contributors who have generously shared their expertise, experiences, and insights in the field of biofuels engineering and technology. It is our hope that this book will serve as a valuable reference and guide for students, researchers, engineers, policymakers, and industry professionals as they work to unlock the full potential of biofuels as a clean, renewable, and sustainable energy source for the future.

Diego Rossit

Sebastian Ubal

Ioan Gergely

CONTENTS

PREFACE V

VOLUME-1

1. THE IMPACT OF THERMAL PRETREATMENT PROCESS ON BIO-FUEL CONVERSION 1

1. Introduction: An Overview 1

2. Methodology 1

3. Experimental findings and their analysis 4

4. CONCISE SUMMARY 19

References 19

2. AN OVERVIEW OF THE BIOENERGIES CHALLENGE 21

1. Introduction: An Overview 21

2. Plant biofuels 32

3. Microdiesel 39

4. Biohydrogen 40

5. The future of transport technology 42

6. Options for grid contributions 45

7. Management and the ability to maintain long-term viability 47

8. Concise Summary 51

References 51

3. ENHANCING THE QUALITY OF BIOGAS BY PRESSURE SWING ADSORPTION 63

1. Introduction: An Overview 63

2. Carbon dioxide reduction technologies for biogas 64

3. Principles of Pressure Swing Adsorption (PSA) for the purification of biogas 66

4. Emerging markets and advancements in PSA technology 74

5. Concise Summary 78

References 78

4. UTILIZATION OF RAPESEED STRAIGHT VEGETABLE OIL AS FUEL MANUFACTURED IN SMALL-SCALE OPERATIONS 83

1. Introduction: An Overview 83

2. Production of rapeseed 83

3. *Processing of rapeseed* 85
4. *Use of rapeseed oil as fuel* 86
5. *Use of rapeseed cake for animal feeding* 88
6. *Proposed model for crop cultivation and agricultural utilization* 89
7. *Environmental and economic analyses* 91
8. *Concise Summary* 95
References 96

5. NANOTECH BIOFUELS AND FUEL ADDITIVES 99

1. *Introduction: An Overview* 99
2. *The future of oil refining [1] and the shift of the oil industry to other energy sources [2]* 100
3. *Challenges related to liquid biofuels [3]* 100
4. *The vastness of nanotechnology* 102
5. *Converting algae into biofuels* 103
6. *The application of nanotechnology to landfill facilities[13]* 104
7. *The utilization of nanotechnology for the conversion of biomass into biofuels.* 104
8. *Nanotech liquid additives* 105
9. *Public apprehensions over nanotechnology encompass issues of security, health, and the environment* 106
10. *Concise Summary* 108
References 108

6. UTILIZING BIORESOURCES FOR ADVANCED (THIRD-GENERATION) BIOFUELS 111

1. *Introduction: An Overview* 111
2. *First-generation biofuels* 111
3. *Second-generation biofuels* 115
4. *Third-generation biofuels* 120
5. *Concise Summary* 127
References 128

7. CORN-BASED FUEL ETHANOL COPRODUCTS: PRODUCTION AND USE 137

1. *Introduction: An Overview* 137
2. *The study aims to achieve specific goals and objectives* 140
3. *Methods of production* 140
4. *Significance of coproducts* 141
5. *Evolution of coproducts* 149
6. *Concise Summary* 154
References 154

8. BIOREFINERY PROCESSES FOR THE CONVERSION OF BIOMASS INTO LIQUID FUEL 161

1. *Introduction: An Overview* 161
2. *Concept of Biorefinery* 161
3. *Production of bioethanol from lignocellulosic biomass* 162

4. *Production of bio-oil from lignocellulosic biomass and biomass with high moisture content* 166
5. *Conclusion* 176
References 176

9. UTILIZATION OF DIGESTERS' WASTE FOR BIOGAS PRODUCTION 183

1. *Introduction: An Overview* 183
2. *Significant issues* 185
3. *Information* 186
References 209

10. PRODUCTION AND ASSESSMENT OF BIODIESEL 213

1. *Introduction: An Overview* 213
2. *Production of biodiesel* 214
3. *Assessment of Biodiesel Quality* 219
4. *Standards for biodiesel* 221
5. *Concise Summary* 228
References 228

11. BIOBUTANOL PRODUCTION AND UTILIZATION PERSPECTIVES 233

1. *Introduction: An Overview* 233
2. *Theoretical framework* 233
3. *Biobutanol fermentation experience at ICT Prague* 238
4. *Biobutanol's application in road transportation* 244
5. *Concise Summary* 249
References 250

12. ADVANCING THE PROGRESS OF ALGAL BIOFUELS THROUGH THE ESTABLISHMENT OF A GENETIC FRAMEWORK 257

1. *Introduction: An Overview* 257
2. *Utilizing microalgal biomass for the generation of biofuels.* 259
3. *Advancing the creation of selectable markers for Dunaliella salina* 267
4. *Concise Summary* 275
References 275

13. RHEOLOGICAL ANALYSIS OF BIO-OILS PRODUCED BY PILOT SCALE MICROWAVE ASSISTED PYROLYSIS 281

1. *Introduction: An Overview* 281
2. *Materials and methods* 284
3. *Results and discussion* 287
4. *Concise Summary* 294
References 294

14. SIMULTANEOUS PRODUCTION OF BIOETHANOL AND POWER 299

1. *Introduction: An Overview* 299
2. *Energy equilibrium in the process of traditional bioethanol manufacturing* 300

3. Self-heat recuperation technology and self-heat recuperative processes 301
4. Integration with biomass gasification 309
5. Energy balance for self-heat recuperative bioethanol production 309
6. Concise Summary 311
References 311

15.CONVERSION OF NON-HOMOGENEOUS BIOMASS TO ULTRACLEAN SYNGAS AND CATALYTIC CONVERSION TO ETHANOL 315

1. Introduction: An Overview 315
2. Synthesis gas (syngas) production by gasification 316
3. Syngas conversion into methanol 323
4. Synthesis of Ethanol 327
5. Concise Summary 330
References 331

VOLUME-2

16. INNOVATIVE APPROACHES TO BIODIESEL MANUFACTURING 335

1. Introduction: An Overview 335
2. Supercritical process 338
3. Microwave-assisted procedure 344
4. Ultrasound assisted process 349
5. Concise Summary 358
References 359

17. CATALYTIC HYDROTREATMENT EMPLOYED FOR OF STABILIZING PYROLYSIS OIL 369

1. Introduction: An Overview 369
2. Investigation of the hydrotreatment of fast pyrolysis oil using catalysts 371
3. Investigations on the catalytic hydrotreatment of fast pyrolysis oils 373
4. Proposed paths for reactions and their potential consequences 381
5. Enhanced catalyst compositions for the efficient catalytic hydrotreatment of fast pyrolysis oil 384
6. Potential applications of enhanced oils and essential features of the product 387
7. Conclusions 388
References 388

18. BIOMASS FEEDSTOCK PRE-PROCESSING-PART-1: PRE-TREATMENT 393

1. Introduction: An Overview 393
2. Characterization of Lignocellulosic biomass 394
3. Pre-treatment of lignocellulosic biomass 397
4. Reduction of particle size and its impact on physical qualities 405
5. Concise Summary 411
References 412

19. BIOMASS FEEDSTOCK- PRE-PROCESSING- PART-II: DENSIFICATION 421

1. Introduction: An Overview 421

2. Lignocellulosic composition and higher heating values 423

3. Lab-scale pelleting of agricultural biomass 425

4. Biomass Compression characteristics 430

5. Pilot-scale pelleting of agricultural biomass 434

6. Concise Summary 440

References 441

20. PERFORMANCES OF ENZYMATIC GLUCOSE/O_2 BIOFUEL CELLS 447

1. Introduction: An Overview 447

2. Enzymatic electrodes for glucose/O_2 biofuel cells 448

3. Abiotic catalysts for glucose/O_2 biofuel cells 455

4. Design of glucose/O_2 biofuel cells 462

5. Concise Summary 465

References 465

21. THE SIGNIFICANCE OF BIOPHYSICS IN BIOFUEL RESEARCH: A QUANTITATIVE APPROACH TO BIO-ENGINEERING 473

1. Introduction: An Ovbverview 473

2. ThermoGravimetric Analysis (TGA) and Differential Scanning Calorimetry (DSC) 473

3. Cellulose crystallinity measurement 475

4. Pyro-GC/MS and GC/MS 476

5. Plant stem mechanical strength testing 478

6. Analysis of cell walls using vibrational spectroscopy and microscopy 480

7. Cell wall analysis by 2D and 3D electron microscopy 484

8. Analysis of biomass using acanning and transmission electron microscopy 487

9. Analyzing microbial populations using sophisticated electron microscopy techniques 487

10. Concise Summary 489

References 489

22. INVESTIGATION OF THE RATE OF DECOMPOSITION OF PALM OIL WASTE 499

1. Introduction: An Overview 499

2. Materials and methods 500

3. Results and discussions 502

4. Concise Summary 509

References 509

23. THE COLOMBIAN EXPERIENCE WITH BIOFUELS AND ACHIEVING ENERGY SELF-SUFFICIENCY 513

1. Introduction: An Overview 513

2. Energy replacement and consumption 514

3. Precedents 521

4. The most common raw materials 528
5. Technical considerations 531
6. Concise Summary 535
References 537

24. MICROFLUIDIC BIOFUEL CELL UTILIZING ENZYMES FOR MICROPOWER GENERATION 539

1. Introduction: An Overview 539
2. Miniature BFCs 540
3. Microfabricated devices 541
4. Concise Summary 553
References 553

25. ENERGY PATHWAYS RESULTING FROM THE BLUE TOWER PROCESS 557

1. Introduction: An Overview 557
2. Blue Tower technology 559
3. Concept of the biomass Life Cycle Assessment 565
4. Future application of bio-fuel 569
5. Concise Summary 577
References 577

26. ADVANCEMENT OF BIOETHANOL DEVELOPMENT 581

1. Introduction: An Overview 581
2. Raw materials 583
3. Production processes 586
4. Environmental issues 599
5. Conclusions 601
References 601

27. THE IMPACT OF A VARIETY OF FRIED DISHES ON CERTAIN CHARACTERISTICS OF UTILIZED PLANT OILS AS PRIMARY RESOURCES FOR THE CREATION OF AN ALTERNATIVE TO DIESEL FUEL 607

1. Introduction: An Overview 607
2. Assessment of usability of post-frying edible oils as a raw material for production of diesel fuel substitute 608
3. Result of investigation 610
4. Concise Summary 618
References 619

28. ADVANCEMENTS IN MINIATURIZED ENZYMATIC BIOFUEL CELLS 621

1. Introduction: An Overview 621
2. Miniature biofuel cells 628
3. C-MEMS 3D architecture electrodes 630
4. Surface functionalization of C-MEMS 632
5. Simulation of C-MEMS based EBFCs 634
6. Concise Summary 638
References 639

29. PROCESSING LIGNOCELLULOSIC BIOMASS USING FEEDSTOCK IMPREGNATION FOR BIOREFINING. SWIFT AND CONSECUTIVE APPLICATION OF STEAM TREATMENT 645

1. Introduction: An Overview 645

2. Biomass 647

3. Fractionation 653

4. Hydrolysis 662

5. Concise Summary 666

References 667

30. PRODUCTION OF BIOMETHANOL FROM FORAGE GRASSES, TREES AND CROP RESIDUES 673

1. Introduction: An Overview 673

2. Gasification technology and the test plants 674

3. Concise Summary 687

References 688

INDEX 691

CHAPTER-1

THE IMPACT OF THERMAL PRETREATMENT PROCESS ON BIO-FUEL CONVERSION

1. INTRODUCTION: AN OVERVIEW

In Russia, the utilization of biofuels is employed for various reasons due to the exploitation of local low-grade fuels. To build advanced combustion and gas-generating facilities that operate on low-grade fuel, it is necessary to understand the mechanisms of burning and gasification in moderate low-temperature combustion modes. It is widely recognized that the quantity and quality of the fuel particle surface available for interaction with the oxidizing agent greatly influences the rate of thermochemical conversion. When conditions are the same, the fuel with the biggest response surface will burn down at the fastest rate. The qualitative characteristics of the internal surface that developed before to entering the furnace (reactor) may differ from those that are now evolving within the furnace. When the above fuels are gasified in autothermal mode, the resulting gas contains a high concentration of fully burned products ($CO2$ and $H2O$) and hydrocarbons, but has low chemical efficiency. In order to enhance efficiency, it is imperative to incorporate allothermal conditions and optimize heat recirculation. A series of tests and calculation-based estimates were conducted to investigate marginal allothermal conditions, sometimes known as "ideal gasification," and explore methods for controlling them.

2. METHODOLOGY

The experiments were conducted on "model fuels" that exhibited significant variations in their thermal and kinetic characteristics. The study focused on examining the impact of pyrolysis kinetics and material density on the creation of coke-ash residue reaction structure. This was done by using low-ash high-reactivity biofuels with medium density (wood) and high density (date seeds), as well as their treated products (charcoal). The study utilized charcoal, a result of pyrolysis that acts as an oxidizing agent, to examine several gasification processes under varied blow circumstances in gas generators. The attributes of fuels are provided in Table 1.

Conditions of porous structure formation and porosity during preheating (devolatilization) were studied Using biomass particles with a uniform size of 10 mm in diameter. The particles were subjected to heating using two methods: rapid heating by placing them in a preheated muffle furnace at a specific temperature (ranging from 100 to 800 °C with an accuracy of 20°C), and gradual heating in conjunction with muffle heating while limiting the supply of oxidizing agent. This enabled the simulation of authentic thermal conditions, such as rapid heating (e.g., particle pyrolysis in a fluid flow or fluidized bed carbonizer) and progressive heating (when the particle enters a cold fluidized bed and warms up slowly with the bed). The porosity was evaluated after the cooling process using mercury porometry to determine the volume and size of pores with a diameter greater than 5.7 nm. Additionally, the specific surface area was determined using nitrogen adsorption to measure the surface area of pores with a diameter greater than 0.3 nm.

Table 1. Model fuels characteristics. NA – not available.

Parameter	Charcoal	Wood (pine)	Wood pellet	Date seed
Original particle				
Moisture of fuel as receivedWar, %	1.4	8	10	4
Ash (dry basis),Ad, %	0.9	1	2	0.97
Volatile contentVdaf, %	15	88	87	85
Low heat valueQdaf, MJ/kg	31.5	18.1	17.5	18.9
Apparent density of fuel as received ϱ, kg/m^3	380	520	1200	1150
Porosity Π, %	75	65	20	25
Specific surface area S0, m^2/g	8.6	1.0	2.0	0.01
Coke-ash residue after pyrolysis (fast heating/slow heating)				
Ash content A, %	NA / 1.5	NA / 3	NA / 6	NA / 2
Volatile content Vdaf, %	NA / 1	NA / 1	NA / 1	NA / 1
Apparent density ϱ, kg/m^3	280/320	230/260	NA / 360	200/620
Porosity, Π, %	80/77	85/83	NA / 70	87/60
Specific surface areaS0, m^2/g	NA /29.2	454/366	NA / 436	NA /9.1

The kinetics of conversion in combustion mode was investigated on individual particles with an equivalent diameter (dp) ranging from 3 to 75 mm. The range of diameters investigated aligns with the values presented in (Tillman D.A., 2000) for the permissible use of biofuels in both individual and co-combustion (gasification) processes. A test sample, positioned at the center of the thermocouple junction (Ch-A type), was introduced into the muffle furnace that had been prepared to the desired temperature (ranging from 100 to 800 °C with an accuracy of 20 °C). The experiments were conducted using an air flow rate ranging from 0 to 3.5 m3/h, with an upstream flow velocity of 0 to 0.5 m/s under normal conditions. The average effective burning velocity for coke-ash residue was determined by calculating the weight loss of the residue per unit surface area of an equivalent sphere (based on its initial size) during the burning process: j = ΔM / (Δτcar. F). The burn-out time (Δτcar) of the coke-ash residue (CAR) was determined using thermograms (fig. 1). It is defined as the time gap between points C and D. The measurement of the A"–B segment was overlooked. Individual particle burning, combustion in fluidized bed, and combustion in flame can be evaluated using similar criteria in certain respects. In both flame and fluidized bed systems, the fuel particles are widely separated from each other and are typically treated as separate entities. The heat-mass exchange intensity of particles burning in an inert medium in a fluidized bed is similar to the intensity of individual particles. The application of experimental data on individual particle burning is justified for calculating and evaluating the thermochemical pretreatment of large-size particles in furnaces with dense beds. This is because the heat-mass exchange processes in the large size elements are the same as those for individual particles, as stated in the problem. Consequently, the empirical data regarding the combustion of individual particles are commonly employed in computations to evaluate the thermochemical preparation of sizable particles in furnaces of diverse kinds.

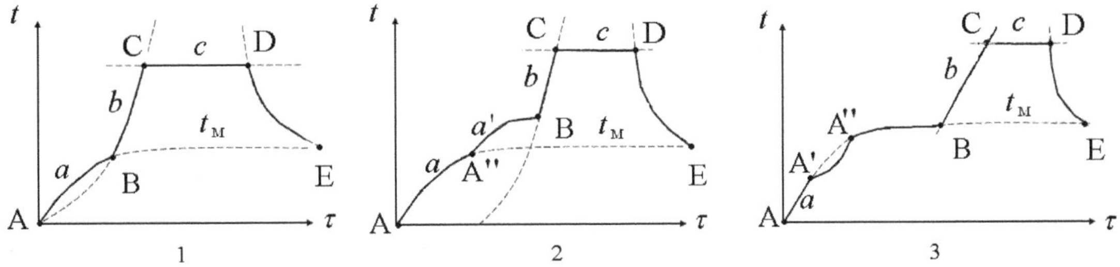

Figure 1. Schematic view of fuels thermograms attm= 400 °C; Moments when: A – the particle enters the furnace, A' – endothermic reaction starts to dominate, A" – process returns back to curve of inert matter heating curve, B – intense oxidation (self ignition) of coke ash residue begins, C – quasi-stationary burning of coke-ash residue begins, D – coke-ash residue has burnt out, E – ash cooled down to muffle temperature. Processes: a – heating by inert matter curve, a' – heating due to pyrolysis gases burning; b – sef-heating of coke-ash residue, c – quasi-stationary process of coke-ash residue burning; temperature in particle centert,°C; time since the moment the particle entered the muffleτ, s

Experimental data was compared with other researches' data on thermal pretreatment of low-grade fuel particles in the air showed inTable 2.

Table 2. Experimental data on thermal pretreatment of particles in the air (IP – individual particle, FB – fluidized bed)

No	Material (method)	Particle size, mm	Environment temperature, °C	Speed of blowing, m/s	Red
1	Charcoal (IP)	3-80	250-1200	0-0.5	11-43
2	Wood (IP)	3-80	250-1200	0-0.5	11-43
3	Pellet (IP)	13	250-1200	0	0
4	Date seed (IP)	11	100-1200	0	0
5	Pellet (IP)	13	600-1000	0.18	10-24
6	Charcoal (IP)	3-5	280-335	0	0
7	Brown coal (FB)	2.5-5.15	800-950	0.23-0.46	2-11
8	Brown coal (IP)	0.1-1.2	850	0.01	0.004-0.05
9	Brown coal (IP)	0.1-1.0	950-1200	0.02-0.03	0.008-0.13
10	Antracite (IP)	0.1-1.0	950-1200	0.02-0.03	0.008-0.13
11	Antracite (FB)	2-9	750-950	0.54	8-40
12	Antracite (IP)	15	1000-1500	0.27-1.0	18-65
13	Fossil coal (FB)	2-10	800	0.25	4-20
14	Carbon (IP)	5.5-8.5	850-1450	0.01-10	0.5-700
15	Graphite (IP)	15	800	0.6	69
16	Electrode C (IP)	15, 25	1300	0.02	1.3-2.2

The study investigated the kinetics of conversion under gasification conditions using a plant that included a quartz retort with an inner diameter of 37 mm and a length of 650 mm. The retort was located in a cylindrical muffle furnace (Nel = 2.5 kW, Tmax = 1250°C) and was equipped with an air blower (Qmax = 6 m3/h, Hmax = 0.6 m), an electric heater, a steam generator, a rotameter, a set of thermocouples, a carbon dioxide cylinder, and a thermocouple polling and temperature recording system. The gas chromatograph was used to determine the combustible gas components (CO, H_2, CH4), while the effluent gas composition was used to determine the air flow coefficient. The studies were conducted in a dense bed, ensuring the precise fulfillment of fuel thermochemical pretreatment. This procedure was carried out in a stratified manner, following certain temperature and concentration requirements, without any flow disruptions or fluid dynamics issues. The process of gasification was founded on the principle of downdraft. Particles with starting diameters ranging from 3 to 20 mm were placed in a retort equipped with a tube welded to its bottom for the purpose of releasing gas and sample for analysis. The fuel bed was subjected to heating in a muffle furnace, reaching temperatures ranging from 600 to 1000°C, with the specific temperature being determined by the experiment. The gasifying agent, which could be air, air and water vapor, water vapor, or carbon dioxide, was introduced into the bed at various depths through the furnace tuyere. The fluid was heated by an electric heater to a temperature range of 700-750°C. The experiment was deemed complete when the levels of CO and H2 in the gas decreased by less than 1% of the volume.

3. EXPERIMENTAL FINDINGS AND THEIR ANALYSIS

3.1. Pyrolysis for fuel structural transformation

Pyrolysis induces substantial alterations in the physical and chemical characteristics of fuel particles. Measurements indicated a 50% decrease in the density of bio-fuel particles within a specific temperature range. The particle undergoes a negligible reduction in size, barely exceeding 30% of its original dimensions. Due to the volume change being less than 20% of the starting amount, and the significant decrease in density, the porosity of particles rises. As a result of the opening of pores, oxygen is able to reach a previously inaccessible surface (see figure 2).

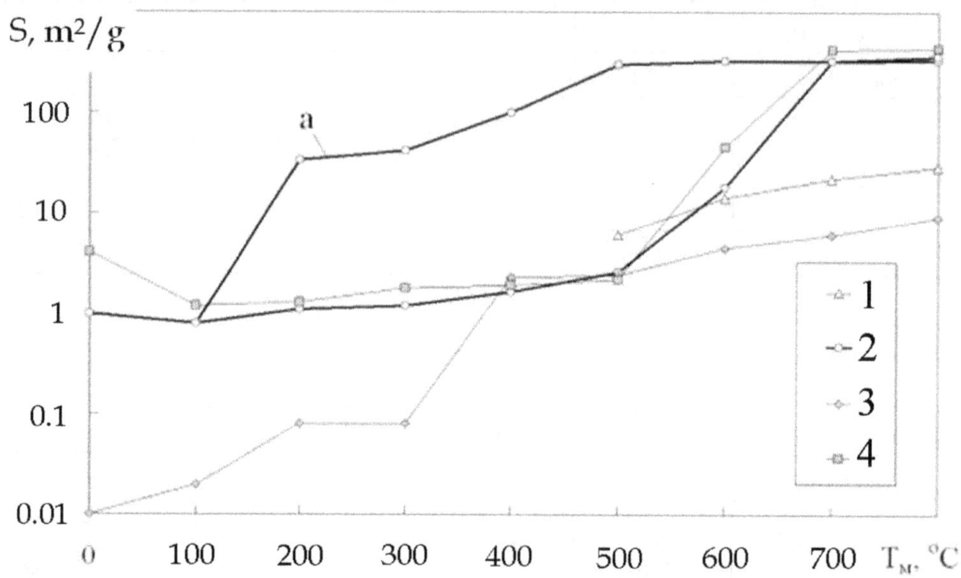

Figure 2. Specific surface area of fuels during heating. Symbols: 1 – charcoal; 2 – wood chips, 3 – date seeds, 4 – wood pellet; a – fast heating, others – slow heating

The specific surface area of wood chip in the temperature range of 200 to 700°C is highly influenced by the pace of heating. Specifically, when the wood chip is heated rapidly, its specific surface area can be one or two orders of magnitude higher compared to when it is heated slowly. The fuel porosity curves (fig. 3) of the test samples showed three distinct peaks in the mezzo- and macro-porosity range (5–50000 nm) at 5–50, 100–3000, and 10000–50000 nm. The first nanolevel exhibits a porosity that is characteristic of compact fuel particles. The average diameter of holes in most coals typically falls between 4 and 10 nm, although in certain cases it may be as large as 30-40 nm. These highly-burned activated coals are artificial materials designed to have pores that can absorb big molecules, such as organic colors. When a particle is introduced into the furnace, the volume of pores in this category will significantly expand as a result of the thermal decomposition of organic compounds. This decomposition is triggered by an increase in temperature and is suppressed by an increase in pressure, which acts as a natural regulator of gas production within the pores of the material.

The nuclei responsible for pore development at the nanoscale are composed of thinner holes and fractures. It is inside these structures that the detachment of gaseous fragments from the splitting macro molecule of coal material takes place. The initial size of these pores is about equivalent to the diameter of a gas molecule, which ranges from 0.4 to 1.0 nanometers. The cavity degasification process is slowed down by molecular repulsion forces that impede the passage through narrow points in extremely small cracks. Overcoming these forces requires a significant amount of energy (activation), resulting in a shift in the dispersion phase state. Flexible materials, such as wood, can create colloidal structures in a swelled state that are resistant to both contraction and additional expansion. Structures resembling those found in metals or interstitial compounds can be created in solid materials, such as cokes. The rate of gas diffusion from these pores is contingent upon the activation energy and temperature. The gas molecules move within the furnace for a duration that is suitable for the size of the coal macro molecule. Throughout the whole particle burning time, gas emissions occur from many ultra micropores into bigger pores that function as collectors. The presence of porosity in the system leads to a notable increase in flow resistance, causing the pressure within the pores to rise during the early stage of destruction, reaching saturation pressure. This, in turn, creates a positive flow in the larger holes, which hampers the entry of external gas into the particle pores. Concurrently, the particle may experience mechanical stress, leading to rupture. During the damping stage of the destruction process, as the intra-pore pressure decreases, the gaseous products of pyrolysis can react with an external oxidizing agent. This reaction occurs not only on the surface of the particle, but also inside it, creating favorable conditions for homogeneous intra-pore burning. Once the degasification process is finished, the nanolevel pores establish a kind of free molecule diffusion, known as Knudsen diffusion, which is accompanied by convective Stefan's flow. According to various estimates, the extent to which porous space (with a specific surface area Sp) participates in the reaction is not significantly influenced by particle size for particles ranging from 10 to 1000 µm. At a temperature of 600 °C, the participation of oxygen in the porous space is within the range of Sp / Scar < 0.1 for fast heating cokes and Sp / Scar < 0.03 for slow heating cokes. Pores with a diameter range of 0.1–3 µm are found in the region where the transition from Knudsen mode to normal diffusion takes place. They enhance oxidant accessibility and enable greater participation in reactions due to their increased volume. Within the pores of the third peak, when the diameter is greater than 10 µm, diffusion occurs in a manner that is comparable to diffusion in open space. These pores make up a small portion of the interior surface and their impact on the burnout rate is considered to be minor. Nevertheless, their function is highly significant since they have the ability to transport reagent to the interconnected pores of both the first and second peaks. The data obtained indicate that wood particles and pellets have a low-porous structure, with a specific surface area (S0) of less than 2 m2/g. Charcoal, on the other hand, has a mesoporous pattern, with a S0 value of less than 8.6 m2/g. Lastly, the seed has a dense microporous structure, with a S0 value of less than 0.01 m2/g. The initial state of specific surfaces can vary greatly, however this disparity tends to diminish when the surfaces undergo heat treatment. The increase is cubic for seed (up to 9 m2/g), quadratic for wood and its products (pellet) (400 m2/g), and negligible for charcoal (three times).

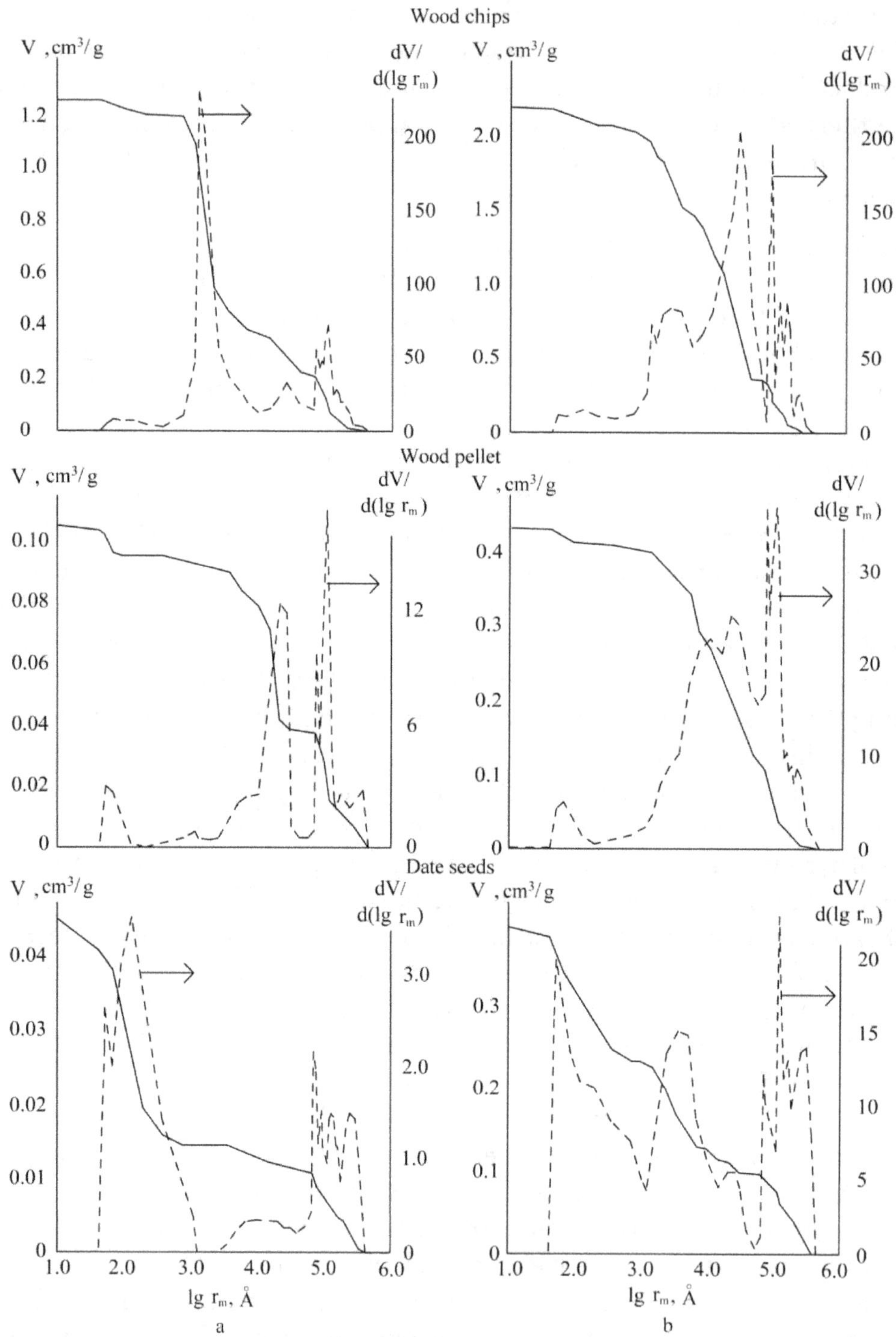

Figure 3. Porosity curves for test fuels; pyrolysis temperature: a) 20°C, b) 800°C; rm– medium (average) radius of pores

The seed, which is used for biofuels and has the maximum natural density, falls between wood and fossil coal in terms of density. The initial peak pores, which make up more than 65% of the volume, are dominant. The third peak pores (25%) follow them. The seed pores have a total volume of 0.045 cm3/g, which is 2.7 times smaller than the volume of the pellet (0.12 cm3/g) and 27 times less than the volume of the wood (1.22 cm3/g). The specific surface area of seeds is two orders of magnitude lower than that of the wood. Following heat treatment, the seed's pore volume expanded by a factor of 10. Additionally, a second peak emerged alongside the existing first and third peaks. The height of these two peaks also rose by one order, and they remained consistent with each other. In biofuels composed of natural density materials such as wood, the predominant pores are of the second kind, whereas the pores of the first type have not been observed, and the volume of the third type is negligible. The thermal treatment of wood leads to a modest doubling of the total pore volume, while also causing the formation of bigger pores. The second peak's height decreased by a factor of three, while the third peak's height climbed by a factor of three. During the pelletization process, the wood undergoes intense mechanical processing such as crushing and pressing, resulting in significant changes to its structure. This leads to the formation of larger pores and a substantial reduction in the overall volume of these pores, with a tenfold decrease compared to the original volume. The pore distribution in the pellet following heat treatment remains essentially the same as the initial distribution, but the overall volume has expanded by a factor of 4 and the peak heights have doubled. When comparing the porosity curves of different fuels, it is observed that subjecting biofuels with varying initial structures to thermal treatment will reduce the differences and create a uniform pore structure for all fuels. As a result, the burn out rates of their coke residues by volume become similar.

3.2. Thermal consequences of pyrolysis

Thermo-gravimetric analysis (TGA) was employed to study the precise changes in mass transfer and conversion of small-sized fuel particles (Bi < 0.1) under controlled temperature conditions. The analysis was conducted in a thermally inert medium with restricted oxygen access, which prevented the particles from overheating and prematurely burning out. The mass transfer correlation used in the analysis had a specific value and sign that corresponded to the thermal effect. The experiments were conducted at installation Q1500D in Hungary, following the conventional protocol in an air medium. The ground fuel sample weighed 100 mg, with an inert medium charge of 400 mg. The temperature was increased at a rate of 0.3 K/s, reaching a final temperature of 1000°C. The samples consisted of wood particles, seeds, charcoal, and soot obtained from the ash box of a pilot downdraft gas production. These samples were produced through quick and slow thermal treatment using the procedure described above. Figure 4 displays the thermograms. Given that the samples were completely devoid of moisture, the primary factor contributing to the reduction in weight was the combined influence of coke-ash residue pyrolysis and oxidation. The overheating value and the sign of thermal impact were caused by exothermic processes involving the oxidation of volatiles generated by coke-ash residue, with the exception of the early stage. The controlled combustion of carbon from ground charcoal and coke-ash residue of biofuels begins at a moderate temperature over 350°C. This process is characterized by the emergence of a distinct temperature peak at 500°C for charcoal and wood particles. After reaching its highest point, the combustion of charcoal becomes consistent and concludes after some exposure at a temperature of 1000°C. The temperature profile for wood particles during overheating closely resembles the curve observed for charcoal, but in a condensed form. The pattern for seeds is much different from what was said earlier. In this scenario, there is no excessive heating occurring in the area where volatile emissions are released (a phenomenon that is less intense compared to when wood particles are involved). Consequently, these chemicals exhibit characteristics similar to chemically inactive compounds. The temperature of the seed sample starts to surpass the surrounding temperature only when it reaches a value above 370°C. Nevertheless, it demonstrates its distinct characteristics within this particular range as well. The burnout curve for seeds exhibits a low and prolonged (truncated) peak, with slightly higher levels of overheating in the steady burning region. Therefore, pyrolysis can be characterized as a prolonged process that occurs simultaneously with the heterogeneous oxidation of coke-ash residue, often occurring until temperatures reach roughly 750°C.

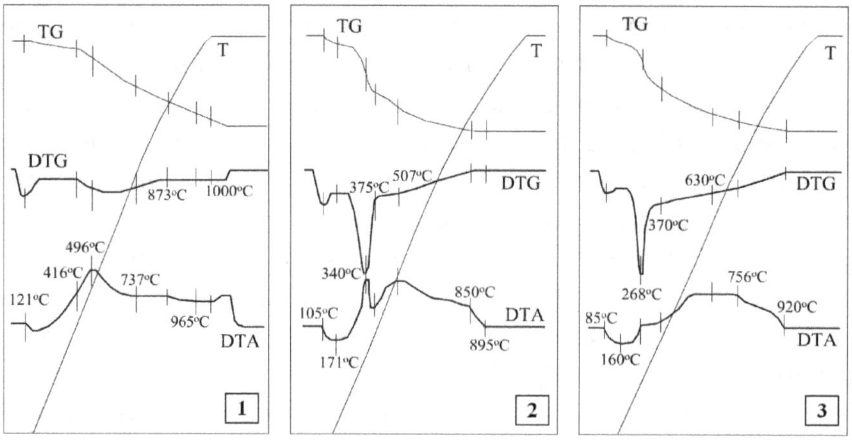

Figure 4. Fuel thermograms: T – temperature in thermo gravimeter chamber, TG – sample weight, DTG – rate of sample weight loss, DTA – thermal effect; numbers: 1 – charcoal; 2 – wood particle, 3 – date seed

Following the initial thermal treatment, which involved subjecting the material to thermal shocks for 15-20 minutes at temperatures of 400, 600, and 800°C, the following observations were made: Wood subjected to thermal treatment at a heating rate of 200 K/20 min maintains the peak of volatile emission at 342°C. However, the initial preheating peak, known as the "gas" peak, disappears. Additionally, the height of the second peak, which corresponds to the burning of coke-ash residue, increases. However, this peak appears at a higher temperature range, specifically 585°C. Furthermore, the conversion process, indicated by weight loss, is completed earlier at 706°C instead of the original temperature of 850°C. Therefore, the initial heating of fossil fuel enhances its reactivity. After subjecting charcoal to thermal treatment under thermal shock conditions for 15-20 minutes at temperatures of 400, 600, and 800°C, it was observed that increasing the treatment temperature caused a shift in the occurrence of the coke-ash residue peak (from 462°C to 496°C). Additionally, the heating value and conversion rate decreased, while the process time and final temperature increased. Furthermore, the fuel partially ceased to burn. Therefore, during thermal treatment at 400 K / 20 min, the combustion of coke-ash residue occurs at the same time as the plant reaches its maximum temperature of 1000°C. However, during thermal treatment at 800 K / 15 min, the burning process is incomplete, with 9% of carbon emaining unburned, and it takes much longer for the furnace to reach its maximum temperature. Charcoal that has undergone thermal treatment using the "heating simultaneously with furnace" method is distinguished by a reduced burn out rate and a higher percentage of unburned carbon (14%) at equivalent final temperature values. The behavior of volatile breakdown products in solid phase settling in the ash box of a gas generator (soot) is similar. The percentage of inert components in these goods, under specified sampling conditions, exceeds 50%. As a result, the burn-out process is completed earlier, corresponding to a temperature of 800°C.

3.3. Ignition of biofuel particles

At temperatures between 250-500°C, when the particle enters the furnace and before pyrolysis begins, the particles are heated by the muffle irradiation as an inert compound (curve an in figure 1). In the case of charcoal, the coke-ash residue will ignite at point B, resulting in a rapid increase in temperature (self-heating) following an exponential curve (curve b). The burning process will then continue under quasi-stationary conditions (curve c). Charcoal possesses the least amount of volatile substances and has a predominantly prepared reaction surface (with 75% porosity and a specific surface area of 8.6 m2/g). During the process of pyrolysis, the properties of charcoal undergo little changes (with only a slight increase in porosity up to 77-80% and specific surface area up to 29.2 m2/g). Additionally, self-heating begins at a temperature of 250°C. Quasi-stationary combustion initiates at an earlier stage compared to the combustion of wood and seeds. The onset of self-heating in charcoal occurs around 50 seconds and 400 seconds earlier than in wood and seed, respectively (as indicated by the time difference between B points). When wood has porosity similar to that of coal (65%) and a specific surface area one order of magnitude smaller (1 m^2/g), heating the particle after reaching the temperature at which severe pyrolysis begins (particle temperature tp 275°C) follows the curve a' (up to tp 420°C > tm). The analysis indicates that at a temperature of 400°C, the volatile combustion heat, which accounts for 60% of the overall heat flux, is of significant importance. The role of bio fuel volatiles combustion in heating up their solid residue is highly significant. At a temperature of 800°C, the primary source of heat transfer is the irradiation from the muffle, which accounts for approximately 60% of the total heat flow. The emission and combustion of volatile substances strongly obstruct the oxygen's reach to the residual coke-ash. This phenomenon has been extensively discussed in other sources. Additionally, the heating of the residue is less intense compared to charcoal. Once the majority of volatile substances have been consumed (at point B), the remaining coke-ash residue of the wood particle is heated along curve b. The intensity of this heating is similar to that of charcoal self-heating, but it reaches a higher temperature. Eventually, the residue burns off 1.5-2.0 times faster. It is important to mention that the porosity and specific surface of wood coke-ash residue are greater than those of coal. Regarding wood, the sources of warming are heat fluxes emitted by a muffle furnace, exothermic reactions during pyrolysis, and the burning of volatiles. The magnitude and proportion of these flows are contingent upon the temperature of the muffle and particles. The preheating step and combustion of solid residue at 400°C for wood are in sync, with a ratio of around 1:1. As the temperature in the muffler increases, the ratio of the duration of solid residue burnout will also grow. At a temperature of 800°C, this ratio will be 1:4. The heat emission intensity for volatile combustion and coke oxidation is equal, with a ratio of 1:1, measured in watts per square meter (W/m2). As the temperature rises, the amount of heat emitted from volatile combustion will increase dramatically compared to coke oxidation, approximately at a ratio of 4 to 1. The initial features of the seed's inner surface exhibit significantly lower values compared to other biofuels, with a porosity of 25% and a specific surface area of 0.01 m2/g. The seed is heated at a slower rate on segment A'–A" compared to the "model" inert compound due to variations in thermal capacity and the presence of endothermic effects observed throughout the investigation. Volatiles are released in large quantities as boiling liquid tar fractions ("cuts") in the pores and on the surface of coke-ash residue, to the right of point A, similar to coking coals. During decomposition, a portion of the substance forms soot on the surface of the porous coke-ash residue, while another portion is released as a thick smoke that does not ignite. The smoke has a pale tint. During the time interval of roughly 250–300 seconds on segment A"–B, the temperature at the center of the seed aligns with the temperature within the muffle. This interval is formally classified as a type of well-established induction phase, as described by Pomerantsev in 1973. The presence of tar fractions partially covering the reaction surface and the lower initial porosity (as shown in Table 1) cause a significant delay in the start of self-warming in seeds compared to wood and charcoal self-heating. At instant B, the liquid phase on the surface disappears, allowing the particle to be exposed to air. This leads to the ignition and self-heating of the coke ash residue. In this instance, the rate at which the temperature increases is over 10 times slower than that of wood and charcoal. This is likely because of incomplete pyrolysis during the preceding stage. The self-heating seed coke-ash residue exhibits the highest porosity, resulting in the highest overheating temperature and the maximum rate of burning in the end. The ratio of the time of thermal pretreatment and solid residue burning at 400 °C for seeds is 3:1. In this scenario, the preparation stage for seeds lasts significantly longer than that for wood. When the temperature inside the muffle hits 800 °C, the ratio of the relative solid residue burn out duration increases, comparable to the ratio for wood, and becomes 1:4. The correlation between the intensity of heat radiation during the stages of volatile combustion and the oxidation of seed coke is 0.3:1 (volatile:coke) at a temperature of 400 °C and 4:1 at a temperature of 800 °C. At muffle temperatures exceeding 500 °C, the differences between the warming and igniting of various biofuels become less noticeable. The delay in seed warming reduces in relation to the charcoal particle, eventually reaching zero. Additionally, the overheating that occurs during self-heating and the burning rate of the coke-ash residue in the main segment become more closely aligned. The significance of qualitatively diverse pyrolysis situations for various biofuels, which yield similar quantitative results in terms of burning intensity, is rather low. This correlation is consistent with a widely recognized high-temperature experiment. The processes of wood particle and seed igniting exhibit significant qualitative distinctions and both deviate from the process of charcoal ignition. Upon visual inspection, it is observed that when a wood particle is placed in a muffle at a temperature of 800 °C, the release of volatiles occurs at a slow rate. This results in the production of a "faint" burning layer near the surface of the particle, accompanied by a slightly flickering brief flame directly above it, which is consistent with the particle's diameter. The discharge of volatile substances from seeds is characterized by an intense and explosive nature, with a continuous burst of burning substance that can travel a distance of up to 3-4 times the diameter of the seed (Fig. 5).

Figure 5. Process of inflammation: a) wood chip, b) date seed; 1-4 – stages of inflammation from time of placing particle into the muffle (1) separated by 20 s.

3.4. Study of the rate at which coke-ash residue is converted under burn out conditions

Visual observations indicate that wood chips with low ash content burn completely within an ash enclosure under natural convection conditions. The burning carbon nucleus gradually decreases in size, while the remaining ash enclosure, which is soft, maintains the original shape of the particle without changing in size. This phenomenon can also be observed in a very turbulent fluidized bed, specifically when burning high-ash flotation tailings that contain stiff mineral enclosures (Belyaev A.A., 2009). The combustion of wood and seed coke-ash residues in the low-temperature range occurs when most of the volatile substances are released at a particle temperature of 350–400 °C. However, for charcoal with a low volatile content, the combustion begins at a lower temperature of 250–300°C. The burning of coke-ash residue at temperatures near to the values mentioned above, in an intra-kinetic mode, results in a notable increase in the temperature at the center of the particles compared to the surrounding environment. The temperature at the center of the particles is around 360°C for seed, 250°C for wood, and 300°C for charcoal (6,a). At high temperatures when diffusive resistance is dominant, the difference in temperature between the center of the particle and the surrounding environment is insignificant. For example, at 800 °C, the temperature difference is 145 °C for pellets and seeds, 105 °C for wood, and 85 °C for charcoal. At 1200 °C, the temperature difference is 75 °C for wood and 40 °C for charcoal. The transfer from kinetic to intra-diffusion conditions is particularly prominent for charcoal, as its combustion is not intensified by the significant release of volatile substances and instead involves the complete burning of coke residue. The burning rate curve exhibits a distinct inflection point at an ambient temperature of 400 °C (tp = 600 °C). The steep section of the curve corresponds to the kinetic mode, while the flat segment corresponds to the diffusion mode (see figure 6,a). At a high temperature range, the particle center experiences the highest level of overheating at tm = 800 °C for a seed with the highest porosity after pyrolysis. On the other hand, charcoal, which has the lowest porosity and reactivity, exhibits the least amount of overheating when stationary burning conditions are reached, compared to all other fuels that were examined. The temperature range for the overheating of wood particles is bounded by these two values. The burnout rate ratio of the analyzed fuels is directly related to the ratio of porosity of the coke-ash residue in the examined fuels, specifically at point B on the thermal curves. This relationship is comparable to the overheating relations described in table 1, which shows the porosity after fast heating. Figure 6b illustrates the relationship between the rate of burning and overheating of a fragment and the blow rate in the high temperature region. Air speed fluctuations ranging from 0 to 0.5 m/s lead to a roughly twofold increase in the combustion rate and cause the examined fuels to overheat. Despite having an incredibly little amount of ash, when the wood particle is not moving, it burns within the ash envelope. The section of the particle that contains carbon decreases, but the ash enclosure actually maintains the form and size of the original particle. When the air flow reaches a speed of 0.1 m/s or higher, it forcefully removes the ash envelope, exposing the coke-ash residue. This behavior is identical to what happens with anthracite, charcoal, and electrode coal particles during burning. The rate at which charcoal burns out is lower than that of biofuels' coke-ash residue across all the blow speeds that were tested. The fuel thermochemical conversion mode that resembles the examined mode without blowing is commonly observed in fluidized bed gasifiers and complete combustion furnaces. This occurs when the blowing process is interrupted for some reason, and a low-intensity residue burn-out continues for several hours, with a fire bed surface intensity (qR) of less than 0.3 MW/m^2. When modes with a low speed of about 0.5 m/s are examined under normal conditions, they represent a critical scenario for a conventional FB furnace. These modes do not result in the loss of bed stability and retain a fire bed surface intensity of qR 4 MW/m^2. The scientists utilized experimental results to design a state-of-the-art gas generator that burns wood fuel. Figure 6(c) displays the relationship between burning rates and particle temperature. The experimental data points are situated within the "segment" defined by the boundaries of the kinetic and diffusion modes. The combustion rate of wood under high-temperature conditions at a temperature of 900 °C and with no air flow is lower than the predicted limit for diffusion mode by approximately 5-15%, and lower than the combustion rate of charcoal by 10-30%.

The data we have collected on charcoal at temperatures ranging from 300 to 1200 °C are closely aligned with the predicted curve for diffusion mode. Additionally, the charcoal points obtained at temperatures between 280 and 335°C (Khitrin, 1955), where the oxygen content in the heating medium is lower and there is burn out without flame-forming self-heating, are more closely aligned with the kinetic curve. By overlaying experimental data points for coal onto a single curve, it becomes evident that the particle temperature is the crucial variable to consider in computations. Selecting the appropriate reference temperature is crucial for analyzing the experimental data in Arrhenius coordinates, as it directly influences the gradient of the curve. It is crucial to select the particle temperature as the reference temperature when dealing with medium temperatures ranging from 300 to 700°C, as this is the range where substantial overheating occurs. The combustion of brown coal particles with a diameter of 3 mm in a fluidized bed occurs through intra-diffusion within a temperature range of 400-900 °C. The burning rate of brown coal coke-ash residue is approximately half that of wood coke-ash residue with a particle diameter of 3 mm, because to its reduced reactivity. The transition to intra kinetic mode occurs at a temperature that is 150-200 °C higher. The diffusion behavior of biofuel combustion is studied across the complete range of particle sizes, dp = 3–75 mm, at ambient temperatures above 500 °C. An analysis of many investigations reveals that microscopic particles with a diameter smaller than one millimeter burn in a kinetic manner within the temperature range of 800 to 950°C. Based on figure 8, the observed overheating of coal dust is lower than the calculated values (line 1). The calculation using current combustion rates (line 2) closely approximates the actual values of overheating.

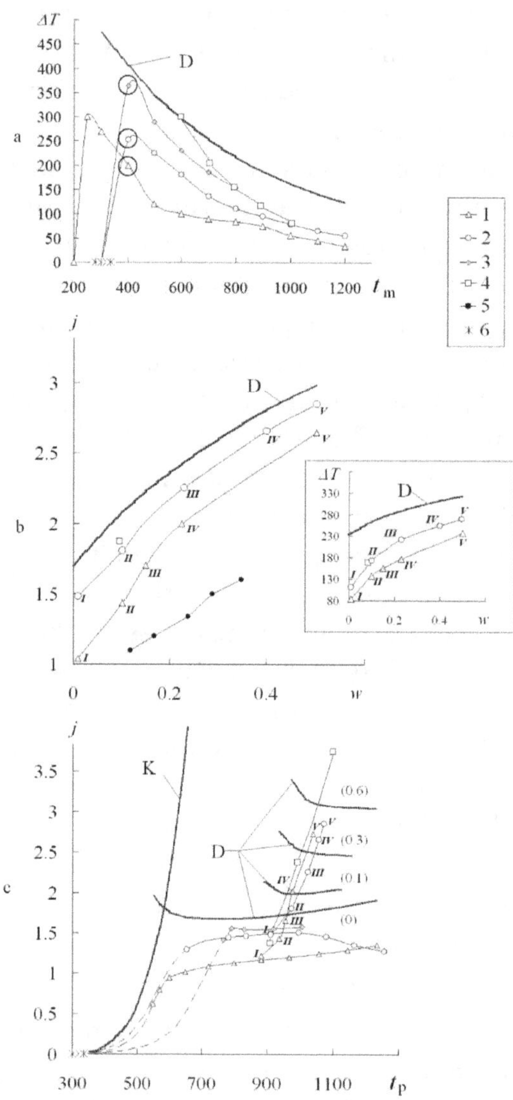

Figure 6. Curves: a) particle center overheating, °C vs muffle temperature, °C, with blow speed w = 0, m/s, b) coke-ash residue burning rate j, g/(m2s) vs. blow speed w, m/s, at tm= 800–900°C,) coke-ash residue burning rate j, g/(m2s) vs. particle temperature tp, °C (numbers in brackets indicate blow speed, m/s); designations are pertable 2; the fragment shows particle center overheating ΔT, °C vs. blow speed; roman numbers are numbers of experiments; D – diffusive mode, K – kinetic mode

Fig. 7. shows the particle burning rate j, g/(m^2s) vs. reverse (1000 / Tp, 1000 / C) and normal (tp, °C) temperatures of particle.

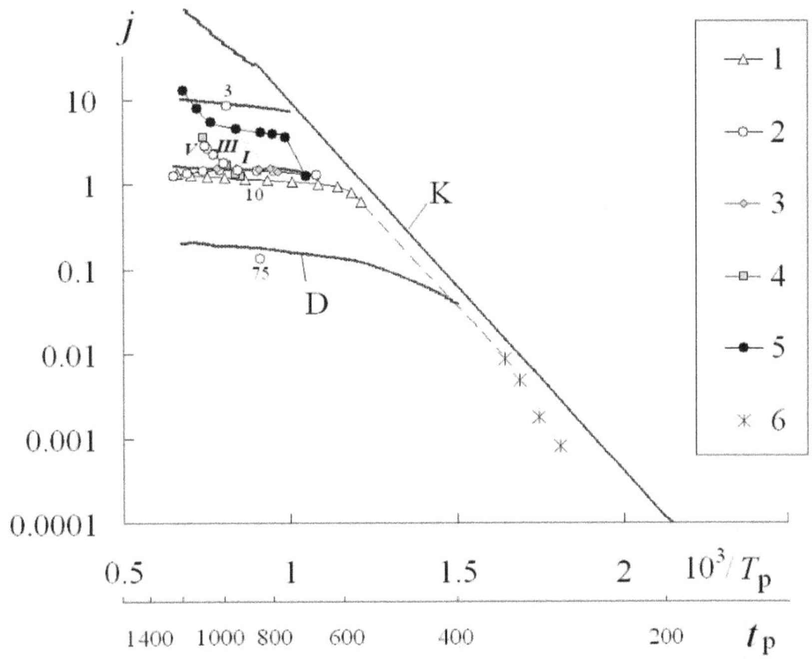

Figure 7. Particle burning rate j, g/(m2s) vs. reverse (1000 / Tp, 1000 / K) and normal (tp, oC) temperatures of particle; numbers in brackets designate particle diameter, dp, mm; symbols are pertable 2; roman numbers designate burning with blow, the same as infig. 6

As the particle size increases, its specific surface area per unit volume falls and the boundary layer becomes larger. This leads to a reduction in the emission of specific heat, an increase in overheating and burning rate, and the particle moves closer to isothermal burning. At a temperature of 800°C, the diameter of the object decreases from 10 to 75 mm, resulting in a decrease in overheating from 120–150°C to 30°C. Additionally, the burning rate decreases from 2 to 0.2 g/(m2s).

In the range of dp= 10–75 mm calculation with accuracy of 15% is approximated by dependence $\Delta T=1130\ d_p^{-0.84}$, and $j=23 \cdot d_p^{-1}$, g/(m²s). Diameter dpis expressed in mm.

The observed results for FB were acquired using a fluidization rate significantly greater than the rate of natural convection. The flow structure of FB (fluidized bed) systems is highly distinctive, as indicated by the Sherwood criterion, which approaches 1.0 for an inert medium at rest. At Reynolds numbers (Re) of 80 and above, the combustion of particles in a fluidized bed (FB) and in a medium without inert particles happens at similar rates. However, at lower Reynolds numbers, the combustion in a fluidized bed is less intense compared to the combustion in a medium without inert particles, and the overheating is approximately half as significant. The estimated range of overheating (field 3 in fig. 8) based on the diameter of inert substance particles (ranging from 0.4 to 1.5 mm) is in good agreement with the overall trend observed for individual fuel particles across a wide range of diameters. However, it is significantly lower than the actual data reported by V.I. Babiy (Babiy, 1986) for dust burning. The overheating of pellets in FB (point 5) is more pronounced compared to wood particles of similar size in spontaneous convection. This could be attributed to the higher density of the pellets (Palchonok, 2002).

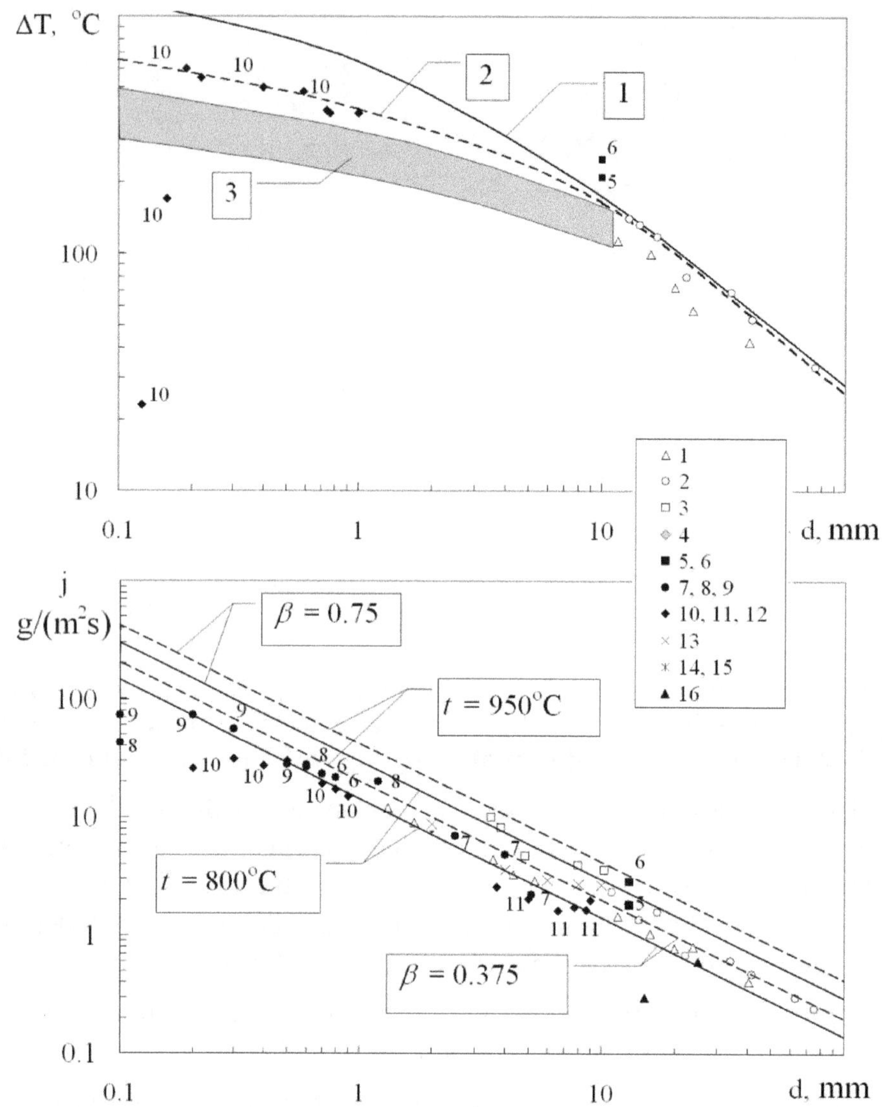

Figure 8. Overheating and burning rate vs. particle size: 1 – calculated, 2 – ditto for actual burning rates, 3 – calculation for FB with inert material diameter from 0.4 to 1.5 mm. Numbers are perTable 2

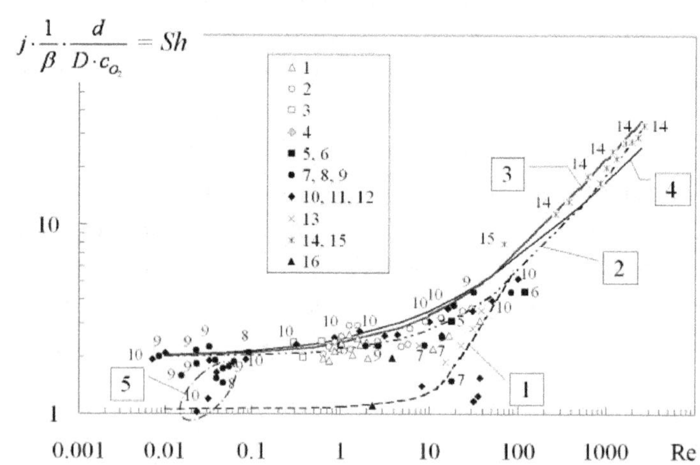

Figure 9. Dimensionless rate of burning vs. Re number: 1 – calculation for FB, 2, 3, 4 – calculation data; 5 – V.I. Babiy points at particle key temperature; numbers are perTable 2

Fig. 9 Illustrates the overall correlations between the dimensionless burning rate and Reynolds number within the temperature range of 400–1500 °C, considering particle diameter (dp) ranging from 0.1–80 mm, and blow rate (w) ranging from 0.01–50 m/s. The data collected by V.I. Babiy on dusts can be well represented by his curve, provided that the medium temperature is considered as the crucial temperature. When using particle temperature as the major temperature reference (domain 5 in fig. 9), the calculations show that the points are positioned below the curve. At these temperatures, the combustion of dusts happens in kinetic mode. The experimental findings obtained by L.I. Khitrin closely align with his curve.

3.5. Gasification mode kinetics of coke-ash residue conversion

The gasification process was conducted using intermittent charging. The composition of the gas altered as the bed burned out: initially, the combustible content increased, but as it reached its maximum value at a specific bed depth, it began to drop (see figure 10, d). The coal bed depth in the retort at which the concentrations of H2 and CO started to decline was found to be approximately 6 to 7 times the initial size (average equivalent diameters) of the fuel, which aligns with existing reference data. When the bed depth is between 1 and 3 times the particle size, the carbon monoxide (CO) concentration decreases to 9-10%, and the hydrogen (H2) content becomes very close to nil, within the range of instrument error. The range of 1 to 3 sizes complied with the available data on the height of the burning zone. The study investigated the influence of air flow rate on the composition of charcoal conversion gas for two types of blowing (air and steam-air) at the maximum concentration point (for a bed depth of 6-7 sizes). This is illustrated in Figure 10, parts a and b. A series of trials was undertaken with a progressive rise in the intensity of each subsequent blow. The increased rate of air flow and subsequent elevation in bed temperature resulted in a corresponding increase in the concentration of carbon monoxide in the gas.

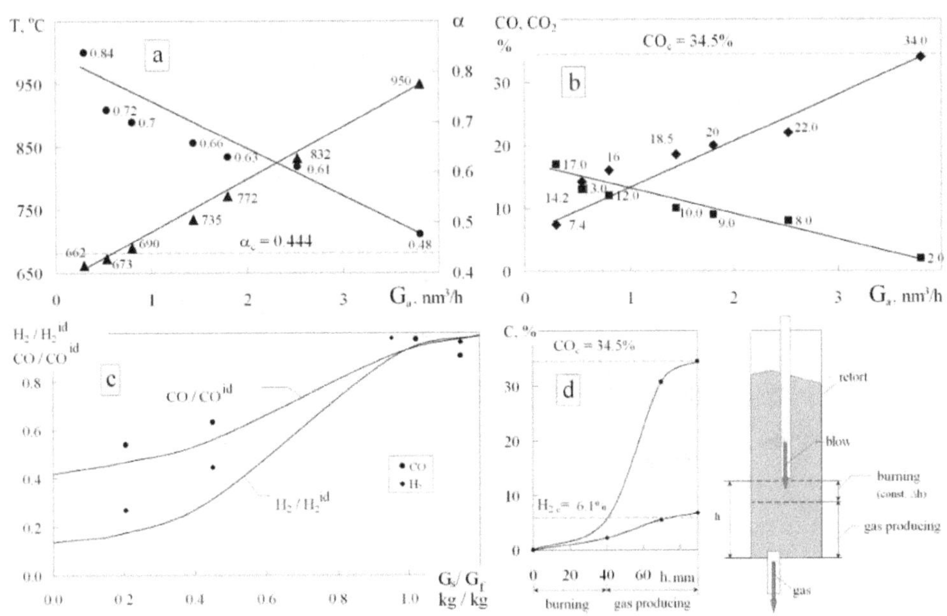

Figure 10. Relationship between gasification parameters; air flow rate effect on: a) temperature in reduction zone (T) and air flow rate coefficient (α); b) CO and CO2content in gas; COcand αc– maximum CO content in gas and blow-limiting factor for air gasification of pure carbon; c) H2and CO relative concentration in producer gas depending on steam content in blow: s – steam, f – fuel, id – ideal regime; lines represent calculation of gas equilibrium composition; d) gas composition vs. bed depth in retort

Gasification reactions did not occur at temperatures below 700°C, and the gas composition was consistent with pyrolysis. The content of CO reached the projected limit for pure carbon when air was blown (COc = 34.5%), and the coefficient of blow, determined by the composition of the gas emitted, was approaching αc. The maximum concentration limits and minimal blow rate coefficient were achieved when the blow rate was 4 m3/h (the speed of the blow in the bed, computed for an empty cross-section, was 1.4 m/s) and the bed temperature was 950°C. The results elucidate the phenomenon of fast combustion. It is hypothesized that a greater blow rate in the combustion process leads to a decrease in the solid-carbon based reduction of CO_2 to CO beyond the combustion zone due to insufficient time for reduction. Once a specific rate is reached, the reduction process ceases and the ratio of CO and CO_2 in the gas is determined by the ratio set during the burning zone (primary CO and CO_2). Experiments involving activated charcoal demonstrated that when the blow rate exceeds 0.3 m/s, the carbon monoxide (CO) content is 34.5%, whereas the carbon dioxide (CO_2) content is 2%. The study results provide a more concise explanation of the "rapid gasification" phenomenon, specifically that gasification conditions can be reached at $\alpha = \alpha$c. The findings validate the feasibility of generating gas devoid of three-atom gases (CO_2 and H_2O) and methane. It also demonstrates a method for manufacturing such gas in a compact environment, by maintaining an optimal depth and temperature of at least 900-950°C. Figure 10c presents the outcomes of tests on allothermal steam-air gasification in a dimensionless format. The gas concentration is decreased to the level found in an optimum air-steam conversion of char coal, with a carbon monoxide (CO) concentration of 43% and a hydrogen (H2) concentration of 26%. The steam content (Gs / Gf, kilogram steam/kg fuel) was systematically increased in each subsequent trial, while maintaining a constant air supply at α = 0.7 for dry blow ("semi-gas" mode). In this instance, the gas underwent enrichment with combustible elements through the reaction $C+H_2O=CO+H_2$. No significant progress in steam generation was reported. The steam supply was at a rate of 1-1.2 kilogram of steam per kg of fuel, with the moisture content in terms of fuel as received being 60%. The experiment showed that the CO and H_2 content in the steam-air mixture was near to the optimal concentration of 35-42% and 22-25% respectively. The chemical efficiency for converting steam and air was 100%.

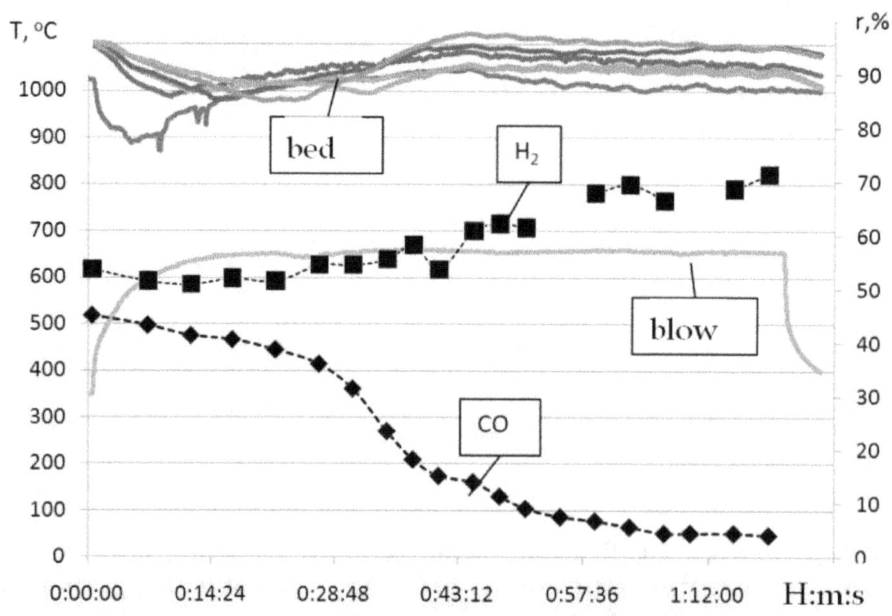

Figure 11. Steam allothermal conversion

Fig. 11 The text illustrates the alteration in gas composition as the bed depth changes during the conversion of allothermal steam (water). Figure 12 depicts the conversion of carbon dioxide from char

coal in a bed operating at temperatures ranging from 950 to 1000°C, with a blow temperature of 600 to 750°C.

The gas composition for steam conversion at a temperature of 1100°C corresponds to 97.5% of the stoichiometric composition, with carbon monoxide (CO) accounting for 42.6%. The hydrogen yield is 54.8% (H_2 = 54.8%), and the chemical efficiency of steam conversion is 1.26 (ηchem = 1.26). The ratio of carbon monoxide (CO) content to the stoichiometric value for carbon dioxide (CO_2) conversion has reached 0.8, while the chemical efficiency of carbon dioxide conversion, denoted as ηchem, is 1.35. A linear relationship between bed height and CO concentration was observed for both carbon dioxide and steam blow at bed depths below 70 diameters. The hydrogen content had no correlation with bed height and consistently increased over the duration of the experiment. The allothermal approach used in the experiment effectively decomposed all the water vapor introduced into the bed by utilizing muffle heat. The trials successfully produced marginal conditions for allothermal gasification, involving the use of steam-water, steam, and carbon dioxide blow. In the two latter situations, the homogeneous gas-water shift reaction $CO + H_2O = CO_2 + H_2$ was carried out to increase the H^2 concentration beyond the stoichiometric value. This was achieved through a heterogeneous reaction and the steam conversion of charcoal.

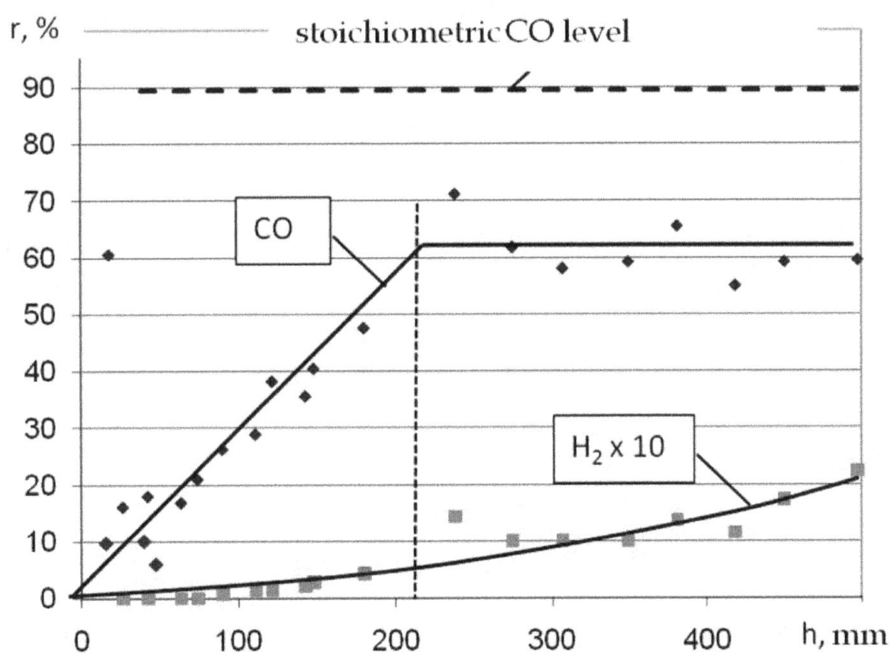

Figure 12. Carbon dioxide allothermal conversion

3.6 Utilizing heat recovery (recirculation) in sophisticated combined cycle gas turbine (CCGT) systems with thermochemical conversion of low-grade fuels.

The temperature limits and ranges of mass concentration for combustible elements and oxidizing agents have been established. These conditions enable the low-grade fuel gasification process to occur with greater efficiency, resulting in the conversion of tars into useful products without any residue. Fuels can be categorized into two groups: Group "A" consists of fuels that can be converted into an ideal mixture ($CO + H_2$) through their own internal chemical source of heat, specifically in auto-thermal modes. Group "B" includes fuels that require an external heat power source (allo- and auto-thermal modes) for ideal conversion. In the case of auto-thermal processes, the resulting gas is not ideal and contains complete combustion products (CO_2 and H_2O). Wood, peat, and certain types of high-moisture brown coals are classified as belonging to the "B" category. In the "A" range, ideal gasification is defined by a consistent heating value of gas and a consistent chemical efficiency of converting fuel into syngas. Additionally, highly metamorphosed fuels necessitate the addition of a significant amount of water vapor, while brown coals primarily rely on fuel hydrogen as the main source of molecular hydrogen. The allothermal conversion process enables the achievement of the highest level of chemical efficiency enhancement. This is demonstrated by the results of allothermal steam-air, steam, and carbon dioxide gasification, as shown in figure 13, points 8, 8, and 8. Nevertheless, the increase in conversion efficiency resulting from external sources is seldom utilized (p.p. 4, 7'). This can be attributed to the significant investments required and the limited efficacy of production.

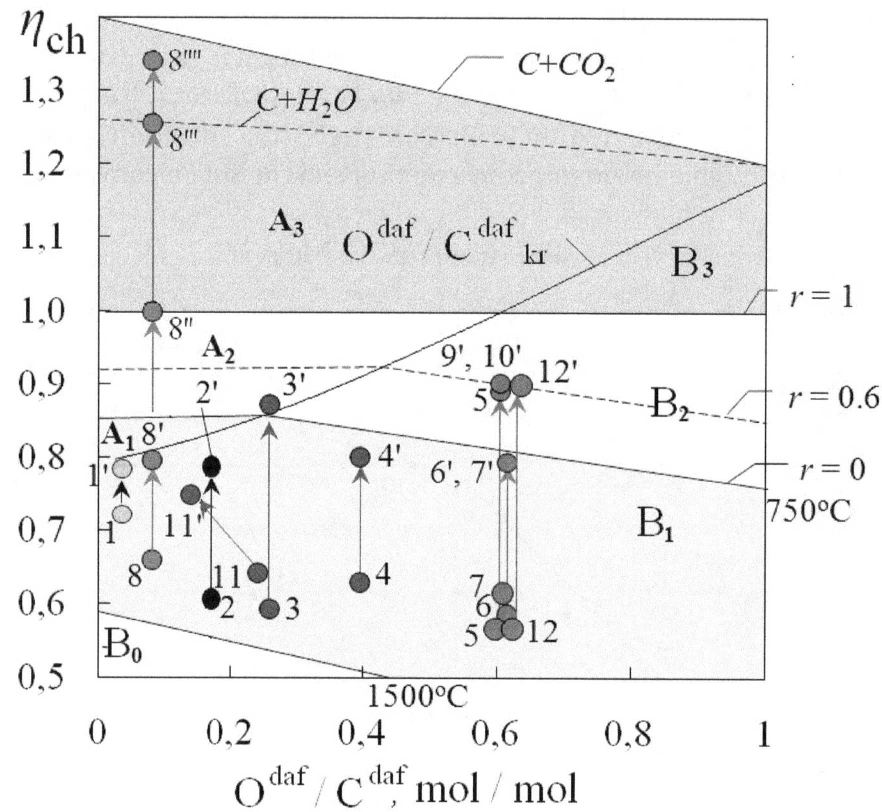

Figure 13. Chemical efficiency of air thermochemical conversion:

1 – fuel oil gasification without regeneration in flow-type gasifier (by VNII NP); 1 – ditto, calculation with regeneration; 2 – first-generation dense-bed gasifier operating on black coal; 2 – tuyere gas producer (JSC VTI) operating on black coal; 3 – gas producer by MHI without regeneration (estimated); 3 – gas generator MHI (expected upgrading of recirculation system); 4 – first-generation dense bed gas producer operating on peat; 4 – allothermal pyrolyzer for peat OIVTAN; 5, 6, 7, 8 – first-generation dense bed gas generator; 5 – first-generation gas producer with superadiabatic heating by IPKhF RAN (estimated); 6 – upgraded downdraft producer by UrFU with regeneration; 7 – plasma gas generator, IEE RAN; 8 – three-zone downdraft producer by UrFU, 8 – with steam and air blow, 8 – with steam blow at 1000°C to the third zone, 8 – with carbon dioxide blow at 1000°C to the third zone; 9 – mini-CHP gasifier Viking, Denmark (estimated); 10 – mini-CHP gasifier, UrFU (estimated); 11 – brown coal gasification (estimation); 11 – joint gasification of brown coal and natural gas, VNII NP; 12, 12 – downdraft producer by ISEM with different blow temperatures

Improving heat recirculation within the gas generator or "gas generator-gas consumer" system has significant potential to increase chemical efficiency. This is apparent when examining the patterns of progress in the advancement of industrial coal gasifiers with high capacity (from the Winkler gasifier to the HTW plant), as well as air gasification plants, particularly those that operate using low-quality fuels and biomass. In gasifiers operating on biomass with steam and air, without recirculation, a significant amount of fuel (up to 30%) is required to achieve complete combustion. The temperature in these gasifiers is low, resulting in the presence of a considerable amount of intermediate gasification products such as tars, CO_2, and H2O in the gas. As a result, the chemical efficiency of these gasifiers does not exceed 60%. These reactors are incapable of producing gas that possesses nearly perfect qualities, even in theory. The lack of prospects for the development of simple-procedure low-efficiency air reactors is evident.

4. CONCISE SUMMARY

Improving heat recirculation within the gas generator or "gas generator-gas consumer" system has significant potential to increase chemical efficiency. This is apparent when examining the patterns of progress in the advancement of industrial coal gasifiers with high capacity (from the Winkler gasifier to the HTW plant), as well as air gasification plants, particularly those that operate using low-quality fuels and biomass. In gasifiers operating on biomass with steam and air, without recirculation, a significant amount of fuel (up to 30%) is required to achieve complete combustion. The temperature in these gasifiers is low, resulting in the presence of a considerable amount of intermediate gasification products such as tars, CO_2, and H2O in the gas. As a result, the chemical efficiency of these gasifiers does not exceed 60%. These reactors are incapable of producing gas that possesses nearly perfect qualities, even in theory. The lack of prospects for the development of simple-procedure low-efficiency air reactors is evident.

REFERENCES

1. Babiy V. I. Kuvaev Yu. F.1986 Burning of fuel dust and modeling of coal torch. Energo AtomIzdat, Russia, Moscow.
2. Belyaev A. A. 2009 Teploenergetika. Jan. 2009, No.1, 918.
3. KhitrinL. I.1955 Physics of burning and explosion. AS USSR, Russia, Moscow.
4. Palchonok G.2002 Combustion characteristics of wood pellets. Proceedings of the First World Pellets Conference. Stockholm, Sept. 2002.
5. Pomerantsev V.V. 1978 Self-burning and Explosions of Fossil Fuels Dust. Energiya, Russia, Leningrad.
6. Tillman D.A. 2000 Biomass co-firing: the technology, the experience, the combustion consequences. Biomass and Bioenergy. 2000, No.19,365384

CHAPTER-2

AN OVERVIEW OF THE BIOENERGIES CHALLENGE

1. INTRODUCTION: AN OVERVIEW

The exponential increase in the cost of crude oil, the depletion of oil reserves, heightened security apprehensions, and a growing acknowledgment of the ecological consequences of fossil fuels have sparked significant enthusiasm for biofuels as a viable substitute for energy. The transportation revolution that took place at the start of the previous century resulted in Western economies becoming reliant on the burning of hydrocarbon fuels. T.A. Edison's discovery of the electric light bulb in 1883 paved the way for the establishment of the first energy distribution facility in Roselle, New Jersey. This landmark development ultimately led to the creation of the electric grid worldwide. The electric grid has become the central neurological system of our civilization due to its high conversion rate (usually >90%) and the ease of converting electricity between constant voltage and vice versa, as well as mechanical conversion. In the 20th century, humanity established a framework where electrical applications are utilized across all sectors of society. This revolution is currently establishing a comprehensive network of ubiquitous computing and is getting humanity ready for the age of nanotechnologies and robotics. Nevertheless, the electric grid exhibits a precarious reliance on the accessibility of carbon-based resources, such as coal and natural gas (Song, 2006). Currently, human society has the task of discovering sustainable and ecologically sound energy sources to meet the needs of our power grid and economic expansion. Fossil fuels will remain the most economically efficient energy source for the next 20 years. Nevertheless, despite their elevated expenses, alternative energies have commenced to be thoroughly examined. The objective of this work is to offer a comprehensive overview of this matter. Fossil fuels are thought to have formed over millions of years by biochemical and geochemical processes that transformed organic molecules on the earth's surface. Geological carbon storage, which involves storing carbon in the form of fossil fuels, can be seen as an alternative method to the carbon reduction/oxidation cycle happening on the Earth's surface. Multiple investigations have demonstrated that coal is created through the biochemical breakdown and geochemical transformation of higher-plant components that were once produced during photosynthesis. Crude oil exhibits distinct compounds, such as phytane, which serve as evidence of its origin from plants (Johnston et al., 2007). Crude oil consists of a variety of hydrocarbons, which span from light gasses (such as C1-C5) to heavy leftovers (Fialkov et al., 2008). The hydrocarbons are separated during distillation into three primary products: naphtha, middle distillate, and a residual fraction. Naphtha, with a boiling range of 90-190ºC, is mostly utilized for the production of motor gasoline, namely in the range of C3 to C12 hydrocarbons. It constitutes roughly 20% of the whole volume of crude oil. The intermediate distillate can be divided into two distinct categories: kerosene-range products, which are lighter, and diesel-range products, which are heavier. The light-end middle distillates, which make up approximately 25-30% of the total crude oil, have a boiling range of 150-260ºC. These distillates are utilized in the production of solvents, kerosene (C8-C16), commercial and military jet fuels (C3-C10), and light diesel fuel (diesel fuel No. 1, C8-C22). The middle distillates, which make up 25-30% of the total crude oil and have a boiling range of 190-400ºC, are refined to generate diesel fuel No. 2 (C10-C25) and heating oils (15-20%) (Kaplan et al., 1997). The proportion of lubricating oil with a carbon chain length greater than C18 is estimated to be around 5-7% of the whole crude oil volume. Aside from the alkane fraction, there is also a portion of polycyclic aromatic hydrocarbons (PAH) that becomes more significant in the composition of distillates as we move from kerosene to heating oil. In addition to greenhouse gases (GHG) and their impact on climate change, polycyclic aromatic hydrocarbons (PAH) are the primary cause of significant environmental problems due to oil spills. PAHs exhibit a high level of resistance to microbial decomposition (bioremediation) in anaerobic circumstances, resulting in a lengthy process of removing them from affected areas. PAHs are a matter for worry when it comes to human health since they have the potential to adhere to DNA, leading to harmful genetic changes and ultimately the development of cancer.

The global daily oil consumption stands at around 79 million barrels (bl), with each barrel equivalent to 159 liters. The transportation sector accounts for 50% of oil consumption, which is equivalent to 20% of total energy consumption. Additionally, it experiences an annual average growth rate above 2%. The global energy demand is projected to increase by around 50%-60% in the next two decades, reaching 112 million barrels per day (Song, 2006). The majority of the readily accessible and inexpensive oil, sometimes referred to as "cheap oil," has already been depleted from the earth and sold at prices ranging from US$ 20-30 a barrel. Consequently, fossil fuel prices have experienced an upward trend in recent decades. The price of crude oil rose from US$ 18/bl in 1990 to US$ >100/bl in 2008 (Tan et al., 2008).

Fuel prices will probably continue to increase in the future because of the following factors:

- Fossil fuel resources are limited;
- There is a lack of balance between supply and demand;
- The demand for fossil fuel is rising rapidly; and
- Geopolitical instability and international conflicts are increasing (Rout et al., 2008).

Renewable energy derived from various biofuels is becoming economically competitive when fossil fuel prices reach US$ 45 per barrel. Biodiesel fuel typically exceeds US$ 0.5 per liter (compared to US$ 0.35 per liter for fossil diesel fuel, as stated in Zhang et al., 2003). Consequently, the adoption of biodiesel has experienced significant growth in the past two decades. The volume of liquid has increased significantly from almost nothing in 1995 to over 20 billion liters in 2009, both in the United States and Brazil. Brazil and the United States (US) possess the most extensive biofuel initiatives globally, while the European Union (EU) holds the third position in terms of scale. Biofuels have the potential to reduce greenhouse gas emissions, promote energy self-sufficiency, and create more job prospects. Due to their compatibility with current technologies, they can help reduce resistance to change and facilitate a seamless transition to alternative technologies and laws (Shinnar & Citro, 2006). Agricultural and forestry byproducts are now being widely used as biofuels in several nations. The primary advantages of biofuels are as follows: (i) they can be derived from readily available biomass sources; (ii) they can be seamlessly integrated into the carbon dioxide (CO2) combustion cycle; (iii) they are environmentally compatible; and (iv) they promote environmental sustainability. For instance, biodiesel fuel derived from vegetable oils or animal fats is compatible with conventional diesel engines, making it a viable substitute for fossil-based diesel fuel.

The emissions of sulfate generated by the combustion of biodiesel are minimal, and the overall impact of CO_2 from biofuels, taking into account their complete life cycle (including cultivation, oil production, and conversion to biodiesel), can be relatively low. The rate of pollutant emission during the whole lifespan of biofuels is similar to that of fossil diesel fuel. However, due to its renewable nature, the combustion of biofuels does not lead to the accumulation of CO_2 in the atmosphere (Agarwal, 2007). Plants utilize solar energy to convert atmospheric CO_2 and water into organic carbon and hydrogen, effectively storing energy. Subsequently, the organic molecules undergo decomposition during the plant's demise, leading to the release of carbon back into the atmosphere in the form of CO_2. When cultivating a crop for fuel, a portion of the plant's biomass is utilized to generate energy directly, and the CO_2 that was initially absorbed by the plant is released back into the atmosphere through the combustion process. This CO_2 is considered "renewable" since it is a component of the overall CO_2 participating in the natural cycle. Nevertheless, the production of biofuel necessitates a specific quantity of energy. This intricate production process requires a substantial quantity of energy for the cultivation, collecting, and processing of the required biomass. The term "well-to-wheel" (WTW) refers to the total energy consumption involved in the production and transportation of a fuel, as defined by Gnansounou et al. (2009).

The WTW analysis is often divided into the following five stages:

- Feedstock production,
- Feedstock transport,
- Fuel production,
- Fuel distribution and
- Vehicle use.

These stages can be divided even further into "well-to-tank" (WTT) and "tank-to-wheel" (TTW) processes. According to Agarwal (2007), demonstrates that the pollution rates associated with the rise in energy consumption of crude oil, natural gas, biomass, and wind power are 1.3, 1.8, 7.1, and 45 times lower, respectively, than the pollution rate of coal. However, biofuels, including ethanol and biodiesel, have a similar trend as crude oil, with a comparable slope. Nevertheless, the intercept is comparatively lower, indicating that the advantage of biofuels over fossil diesel fuel stems from the recycling of the former through photosynthesis. Advanced technologies such as synthetic fuels derived from biomass gasification or wind generation primarily rely on renewable energy sources for the conversion processes and produce little greenhouse gas emissions.

The carbon dioxide (CO_2) balance

Carbon cycling is not present when fossil fuels are used since the carbon dioxide (CO_2) that is emitted does not go back to the fossil fuel sources. The atmospheric CO_2 concentration has reached a substantial level, estimated to be approximately 780 billion metric tons as of 2002 (Song, 2006). CO_2 concentrations are quantified using infrared absorption spectroscopy. Due to data heterogeneity, estimates can range from 400 to 780 Gt, representing a potential doubling of the simple estimate. Nevertheless, all measurements align with a consistent pattern of continuous growth in CO_2 levels. The current concentration stands at around 390 parts per million (ppm), a significant increase from the 310 ppm recorded in 1955 (Song, 2006). During this time, the atmospheric temperature has risen by around 0.6–0.74 °C, leading to various impacts on the environment, climate, ecosystems, and human populations. As an illustration, the duration of growing seasons is extending, resulting in a sevenfold increase in the accumulation of carbon in terrestrial biomass from the 1980s to the 1990s. Aside from the carbon sink on land, which helps to slow down the increase of CO_2, we must also take into account the carbon sink in the ocean. The oceanic carbon sink is able to store twice as much human-caused carbon as the land-based sink (Ridgwell & Zeebe, 2005). Collectively, these sinks capture and store over half of the approximately 24 gigatons of carbon dioxide released into the atmosphere each year as a result of human activities. While it is important to carefully consider the variability among estimates, it can be concluded that phototropic carbon sinks play a crucial role in reducing atmospheric CO_2 concentrations and mitigating the increase in global temperatures projected for the 21st century (Reay, 2007). Nevertheless, there exists a constraint on the reduction of CO_2 through phototropic carbon sinks. Increasing atmospheric CO_2 concentrations will lead to a decrease in oceanic pH levels, causing significant alterations in ocean chemistry. This is likely to reduce the efficiency of carbon recycling and, along with a decrease in precipitation, may cause a shift from oceans acting as a storage for CO_2 to being a source of CO_2 by the year 2050. Warmer soils will indeed emit a greater amount of carbon, and forests that experience heightened drought conditions will ultimately deteriorate. Furthermore, alterations in land use could result in the release of 40 gigatons (Gt) of carbon dioxide (CO_2) that is currently stored inside 50 Gt of terrestrial biomass. According to Miles and Kapos (2008), deforestation poses the greatest risk to the terrestrial carbon sink as woody biomass is the most effective biological system for storing carbon. Approximately 4 billion hectares (ha), which is equivalent to 30% of the earth's land surface, are covered by the world's forests. In 2005, a total of 3.5 gigameters cubed (Gm^3) of wood was extracted from 434 Gm3 of forests. Approximately 60% of this quantity was utilized for industrial applications, while the remaining portion was employed as fuel. Terrestrial forests can be classified as either primary, accounting for 36% of the total, or modified, accounting for 53%. The extent of primary forest has been gradually diminishing at an approximate rate of 6 million hectares per year since the 1990s. The deforestation rate is particularly elevated in Brazil and Indonesia, with these two nations being accountable for the annual loss of 4.9 million hectares of forests. Forest depletion is more prevalent in low-income countries, particularly those located in the tropics. Conversely, higher-income countries have successfully halted the decline of their forests, as evidenced by the findings of Kirilenko and Sedjo in 2007. The conversion of rainforests, peatlands, savannas, or grasslands into areas for biofuel production and food crops in Brazil, Southeast Asia, and the United States results in a "carbon debt" due to the release of 17 to 420 times more CO_2 than the annual greenhouse gas (GHG) reductions that these biofuels would achieve by replacing fossil fuels. On the other hand, biofuels made from waste biomass or biomass cultivated on low-quality land using perennial plants do not generate significant carbon emissions (Fargione et al., 2008; Searchinger et al., 2008). Tropical deforestation is responsible for over 25% of human-caused carbon emissions, as well as the depletion of biodiversity and environmental services. The Amazon jungle in Brazil has been deforested at a rate of approximately 25,000 square kilometers every year since the 1990s. To put this into perspective, this is equivalent to the size of Belgium, which is approximately 30,000 square kilometers. According to Kindermann et al. (2008), implementing a program that reduces deforestation by 10% from 2005 to 2030 might result in a yearly decrease of 0.3-0.6 gigatonnes (Gt) of carbon dioxide (CO_2) emissions. The anticipated cost of this program would be around US$ 0.4-1.7 billion each year, totaling over a period of 30 years. A decrease of 50% in deforestation between 2005 and 2030 has the potential to result in a reduction of 1.5-2.7 gigatons of carbon dioxide emissions annually. This effort would require an investment of US$ 17.2-28.0 billion per year. According to the World Bank, illicit logging was approximated to cause developing nations to lose revenue of US$ 15 billion per year (Agrawal et al., 2008). This implies that using stricter measures to combat illegal logging would offset the financial losses resulting from the decrease in deforestation. Currently, it is believed that safeguarding and improving the terrestrial carbon sink is the most effective method to counteract harmful changes in the climate (Reay, 2007). It is crucial to prioritize the global monitoring of each country's greenhouse gas

(GHG) emissions, as stated by Yang and Sirianni (2010). The CarbonTracker is a comprehensive CO2 measurement system that evaluates the overall CO_2 exchange between the land biosphere and the atmosphere. It is designed to be beneficial for both the scientific community and policymakers (Peters et al., 2007). For the purpose of achieving successful reduction of greenhouse gas (GHG) emissions, which are on the verge of becoming unsustainable without intervention, governments must prioritize the establishment of consistent and harmonized regulations. For instance, the United Nations and other international organizations aim to elevate the GDP per capita in developing nations to match the level attained by industrialized countries, such as US$ 9,000 per capita. Nevertheless, implementing such political measures would lead to a fivefold rise in annual worldwide CO_2 emissions, as documented by Tomkiewicz in 2006. This outcome can be attributed to the following factors: (i) When the energy consumption per capita is below the saturation threshold of US$ 25,000, there is a positive association between purchasing power and energy consumption per capita. (ii) The average GDP per capita worldwide is US$ 5,210 according to Tomkiewicz (2006). Therefore, to offset the rise in CO2 emissions, one or more of the following actions must be taken: (i) expanding the global forested area, (ii) adopting a new technical approach such transitioning to solar energy, or (iii) reducing the size of the human population. Global warming, caused by the escalated utilization of fossil fuels, aligns with predator-prey equations (Lonngren & Bai, 2008). The solutions to these equations exhibit periodicity in time. If humans persist in following the present patterns of fossil fuel usage, as we have done since the onset of the industrial revolution, we can expect significant variations in the size of the human population in the future. Due to the limited nature of the earth and fossil fuel supplies, the global-warming situation will eventually result in a decline in both the human population and the availability of fossil fuels. In the worst-case scenario, this might lead to the extinction of humanity. One potential disastrous consequence of climate change is the forecast that by 2050, if CO_2 emissions are not significantly decreased, over one billion people could face a shortage of drinking water (Parry et al., 2008).

1.3 Greenhouse gasses and how to control them

The alteration in atmospheric trace gases, such as SO_2, O_3, CO, CO_2, CH_4, NO_2, and CFCs, has a significant role in influencing climate change (Hopkin, 2007). Climate change, in turn, can cause alterations in agricultural circumstances that may have negative socioeconomic consequences (Howden et al., 2007). Trace gases in the atmosphere exhibit pronounced absorption bands in the infrared (IR) region and interact with IR radiation emitted by both the Earth's surface and the atmosphere. This has a direct impact on the thermal composition of the atmosphere and leads to the phenomenon of the greenhouse effect. Gaseous compounds like NH_3, SO_2, and their derivatives exhibit short lives of just a few days. However, they can exert substantial impacts on the atmosphere (Begum, 2005). Currently, N2O and CH_4 emissions are the primary factors driving the problem. While carbon dioxide (CO_2) is the predominant greenhouse gas in terms of quantity, it is important to also take into account other greenhouse gases. Nitrous oxide (N_2O) is the primary contributor to greenhouse gas emissions in agriculture, with substantial amounts being produced from cultivated fields, especially when fertilizers are used intensively (Snyder et al., 2009; Ceschia et al., 2010; Mander et al., 2010). Due to its high global warming potential, with N2O being approximately 300 times more potent than CO_2, even little amounts of N2O can have substantial effects on the total greenhouse gas balance (Cherubini, 2010). The utilization of the carbon sequestration abilities of the terrestrial biosphere has been acknowledged as a potentially effective and cost-efficient method to counterbalance carbon emissions (Dorian et al., 2006). Various models have been studied for this goal (Werner et al., 2010). Nevertheless, the capacity of terrestrial carbon sequestration to fulfill more than 25% of the necessary global CO_2 emissions reductions by 2050 has been deemed inadequate. Considering that carbon sinks are the most effective option now available, an emissions credit system has been implemented to offer CO_2 emitters (businesses or countries) a way to fulfill their carbon responsibility resulting from their carbon emissions into the atmosphere. The emitter fulfills their responsibility by temporarily "storing" the same amount of carbon in a terrestrial carbon sink, such as a forest, for a fee (Sedjo & Marland, 2003). This notion involves the utilization of the "willing-to-pay" principle in the global economic market. In essence, the privilege to release CO_2 (in the form of a carbon credit) is offset by cultivating biomass that will absorb and store an equal quantity of carbon. The marketing of carbon credits is designed to incentivize activities that lead to the long-term sequestration of CO_2 in a non-gaseous state. In essence, a carbon fee has been suggested to be imposed on industrial nations based on their greenhouse gas emissions. This fee would be paid to developing countries, who can then utilize the funds for carbon reduction initiatives, such as creating or preserving forest sinks (Jones, 2010). The Kyoto Protocol currently allows for the official recognition and endorsement of various activities, including revegetation, forest management, cropland management, grazing-land management, and carbon sequestration in deep crustal layers (such as oil fields and deep saline aquifers), which can be traded as carbon credits. Additionally, efforts are being made to incorporate principles of fairness and equity in proposals and policy approaches aimed at preventing deforestation. These efforts are being pursued through negotiations on Reducing Emissions from Deforestation and forest Degradation (REDD).

Scientists have predicted that the natural process of absorbing and storing carbon dioxide (CO_2) in living organisms, known as the biological sink, may potentially capture and store a total of 100 gigatonnes (Gt) of CO_2 over the next 50-100 years. The majority of this CO_2 storage is expected to occur in forest ecosystems. This indicates that 10-20% of the projected net emissions from fossil fuels until 2050 will be captured (Sedjo & Marland, 2003). Nevertheless, carbon stored in the terrestrial biosphere may not be permanent. Forests can be exploited for timber, which can be utilized to create perishable goods, or they may be removed for alternative objectives. Wildfires have the ability to liberate significant quantities of carbon that was previously stored. Farmers may revert to agricultural methods that result in the release of carbon that has been previously sequestered. Terrestrial carbon sequestration can be seen as a temporary interruption in the release of carbon from fossil fuels into the atmosphere. Nevertheless, there should be a rise in economic incentives to promote the long-term storage of carbon. Whenever there are financial incentives for carbon-sequestration services, one would anticipate a greater occurrence of sequestration compared to situations where no payments are made (Johnston & Holloway, 2007; Tollefson, 2008). Carbon sequestration in living forests can be carried out in unproductive lands that are unsuitable for agriculture or intense forestry and are also compatible with the objective of conserving biodiversity across vast areas. Conversely, in order to be financially sustainable, intensive crops used for biofuels typically require land that is more fertile and capable of producing higher yields. Intensive biofuel crops have the potential to create competition with food production and the less-productive lands that currently support a significant portion of the earth's biodiversity (Huston & Marland, 2003; Miles & Kapos, 2008). For instance, this occurrence has been documented in Brazil, Indonesia, and Malaysia, where livestock, soybean, sugarcane, and palm oil may vie with existing forests (Darussalam, 2007; Laurance, 2007; Malhi et al., 2008; Stone, 2007; Venter et al., 2008). This competition in Indonesia has devastating repercussions for biodiversity. In order to address this issue, it is necessary for the Kyoto protocol and its later iterations to incorporate "wildlife credits" with carbon credits. This will help to maintain wildlife populations and their ability to mitigate the impact of human activities, as proposed by Lovelock and Margulis in 1974. This approach would offer the benefit of acknowledging the essential functions performed by ecosystem services and initiating their inclusion in calculations (Mäler et al., 2008). Additional financial incentives are required to serve as a balancing influence against the economic forces that drive deforestation

(Jones, 2010). The new accord, referred to as the "Bali Roadmap", seeks to prolong the Kyoto Protocol beyond 2012. It includes instructions for offering compensation to countries who possess rainforests in return for their management of deforestation and environmental deterioration. Compensation for such matters can be administered either via international carbon markets or through voluntary funds. These recommendations could potentially alter the equilibrium of economic market forces that currently support deforestation by generating billions of dollars to finance the ecosystem services offered by rainforests. Nevertheless, in order to achieve desired results, it is imperative that they possess extraordinary skills in strategic planning, flawless execution, and consistent long-term commitment. The new proposal seeks to decrease European Union CO_2 emissions by 30% by 2020, contingent upon the achievement of a worldwide climate agreement in international discussions. In the event that such an agreement is not reached, the reduction target will be lowered to 20%. The European Union is also strategizing to safeguard its economy against carbon "dumpers" by utilizing leverage to compel enterprises who import goods from countries with high pollution levels to purchase emissions permits (Barnet, 2008).

Usually, carbon-credit compensation monies are utilized in underdeveloped nations to build new permanent plantations, such as those for rubber trees or oil palm. An obstacle arises when the renter unlawfully burns the original forest, gains the carbon-credit funds, and then creates a new plantation that will never be as effective in capturing carbon as the original forest, thereby undermining the actual objective of carbon sequestration. Environmental crimes in certain areas may evade detection and get inadequate punishment. Scarlat and Dallemand (2010) suggest that implementing certification programs, which include international approaches and global monitoring, is crucial for ensuring the environmental sustainability of biofuels and controlling the production process. Although facing challenges, the carbon-credit market achieved stability by 2007 and is not anticipated to experience any further decline (Haag, 2007). The upcoming years give a distinctive chance, possibly the final one, to uphold the durability of biodiversity and ecosystem services (Malhi et al., 2008; García-Montero et al., 2010; Hagerman et al., 2010).

1.4 What are the reasons for using biofuels?

When compared to a collection of solar cells, plants are far less efficient in converting the sun's energy. The energy storage through photosynthesis is roughly 2 watts (W) per square meter (m^2). The key distinction between plants and solar cells lies in their cost-effectiveness, with plants being far more affordable. These plants have the capacity to thrive with a moderate amount of water, nutrients, and CO_2, which they convert into stable organic molecules. While no fuel technology is flawless, the current dilemma of greenhouse gas emissions and the worry about oil availability highlight the importance of expanding the variety of fuel sources. This diversification not only makes logical sense, but also allows mankind to gain valuable knowledge and experience in the process (Haug et al., 2011). Governments have been compelled to explore alternative sources of energy due to the ongoing rise in energy demands. The purpose is to streamline the shift from an economy reliant on fossil fuels to one that relies on renewable energy sources. Multiple low-emission scenarios have shown that the objectives of the Kyoto Protocol cannot be met without assigning a significant role to biofuels in the global energy market by 2050. There are several reasons why biofuels are suitable for this transition: The advantages of these technologies include their simplicity, their ability to be produced using well-established agricultural methods, their potential to mitigate climate warming without requiring a complete overhaul of the current energy system, their compatibility with existing engines for transportation purposes (including aviation turbofans), their ability to foster global cooperation through shared regulations, their availability as a readily accessible energy source with high public acceptance, their more evenly distributed availability compared to fossil fuels and nuclear resources, and their potential to generate economic benefits for rural areas, such as job creation.

1.5 Biofuels

Biofuel is a word that encompasses liquid, gas, and solid fuels mostly derived from biomass. The development of biofuels may raise concerns regarding security, environmental impact, trade, and socioeconomic issues associated with the rural sector. Biofuels encompass a range of substances such as bioethanol, biomethanol, vegetable oils, biodiesel, biogas, bio-synthetic gas (bio-syngas), bio-oil, bio-char (charcoal produced from biomass pyrolysis), Fischer-Tropsch liquids, and biohydrogen. Biogas and bio-oil are the main outputs, and their initial processing mainly involves the collection of raw materials. The majority of conventional biofuels, such as ethanol derived from corn, wheat, or sugar beets, and biodiesel made from oil seeds, are sourced from traditional agricultural food crops that necessitate fertile agricultural land for cultivation. The biofuel industry is projected to have significant growth in the 21st century (Demirbas, 2008a).

Biogas

There are two fundamental methods for converting solid biomass into either liquid or gaseous biofuels. One approach is to utilize microbiological fermentation to convert the polysaccharides into alcohols like bioethanol or biobutanol (Gavrilescu & Chisti, 2005). Another option is to convert raw plant biomass into biomethane using anaerobic fermentation (Demirbas & Balat, 2006). A major limitation of the anaerobic system is its sluggishness, primarily due to the limited energy available to the organisms. Hence, the production of methane is constrained, and the sustainability of this technology is contingent upon specific circumstances (Asam et al., 2011). Nevertheless, the implementation of a modest facility to convert agricultural and human wastes into methane can significantly impact the quality of life in small communities (Arthur et al., 2011; Parker, 2002). The second step seeks to thermally transform the entire biomass into a synthesis gas with a high calorific value, also known as syngas, which consists of hydrogen (H_2) and carbon monoxide (CO). This process is followed by the creation of different types of liquid and gaseous fuels (Tijmensen et al., 2002). Syngas generation presents a promising opportunity for the large-scale conversion and usage of CO_2. The process of converting CO_2 into CH_4 has been well examined and documented in the literature (Song & Pan, 2004). The utilization of renewable energy sources in the catalytic reduction of CO_2 to produce methanol (or even CH_4) has the potential to serve as a feasible substitute for limited or costly fossil fuels. Only a fraction of the entire biomass is utilized for the production of biodiesel derived from plant oils and bioalcohol obtained from sugar. Researchers are now working on advanced methods to convert biomass into syngas, which can be further transformed into fuels or chemicals using the Fischer-Tropsch (FT) process (Baker & Keisler, 2011; Fagernäs et al., 2010).

$$CH0.8 + 0.7O_2 \rightarrow CO + 0.4\,H_2O \quad (1)$$

$$CH\,0.8 + H_2\,O \rightarrow CO + 1.4H_2 \quad (2)$$

$$CO + H_2O \rightarrow CO + H_2 \quad (3)$$

$$(2n+2)H_2 + nCO \rightarrow CnH_{2n+2} + nH_2O \quad (4)$$

Given that coal inputs have a H/C ratio of 0.8 to 1, the entire FT process can be succinctly described as follows. Equation (1) represents the process of partially oxidizing coal with oxygen. Equation (2) is the result of the interaction between water and carbon monoxide, known as "steam reforming". As a result, the H/CO ratio is enhanced by transferring oxygen from the molecular water to CO, which generates an extra hydrogen and carbon dioxide according to equation (3). Once the sulfur and carbon dioxide impurities have been eliminated, the syngas undergoes a reaction on a catalyst in the FT reactor to generate top-notch, environmentally-friendly fuels, as described by formula (4) (Greyvenstein et al., 2008). Biomass exhibits higher reactivity compared to coal and is typically subjected to gasification processes at temperatures ranging from 550 ºC to 1,500 ºC and pressures ranging from 4 to 30 bars (Damartzis & Zabaniotou 2011; Leibold et al., 2008; Steinberg, 2006). Usually, biomass is combusted in an electrically heated furnace comprised of multiple-tube units that can be individually heated to temperatures as high as 1,350 °C (Theis et al., 2006). Alternatively, fossil fuel or biomass can be converted into hydrogen plasma. The electric arc in hydrogen plasma generates a temperature that is extremely high, reaching around 1,500 ºC. As a result, this method achieves a conversion rate close to 100% in producing hydrogen and CO gas (Steinberg, 2006). The process of FT synthesis produces intermediate substances that can be used to create synthetic fuels (Liu et al., 2007). The thermal efficiency of power and hydrogen production using hydrogen plasma and carbon fuel cells ranges from 87% to 92%, depending on the fuel type and biomass feedstock. This system is over two times more efficient than a traditional coal-burning steam plant that produces power with an efficiency of approximately 38%. Furthermore, the combination of hydrogen plasma and carbon fuel-cell technology enables a significant decrease of 75% in carbon dioxide emissions per unit of power (Steinberg, 2006).

The technique is currently generating significant interest due to the fact that FT creates mainly linear hydrocarbon chains. Sasol, Petro S.A., and Shell have successfully implemented Fischer-Tropsch (FT) technology on a commercial level. This technology is mostly used to create transportation fuels and chemicals, using coal or natural gas as the raw materials. This fuel alternative offers numerous significant benefits. First, the FT process can produce hydrocarbons of different lengths (typically <C15, Liu et al., 2007) from any carbonaceous feedstocks; these hydrocarbons can then be refined to easily transportable liquid fuels. Secondly, because of their functional similarities to conventional refinery products, the synthetic fuels (synfuels) produced by the FT process (i) can be handled by existing transportation systems; (ii) can be stored in refueling infrastructure for petroleum products; (ii) are largely compatible with current vehicles; and (iv) can be blended with current petroleum fuels (Tijmensen et al., 2002). Thirdly, synfuels are of high quality (this is especially true for FT diesel), have a very high cetane number and are free of sulfur, nitrogen, aromatics, and other contaminants typically found in petroleum products. The principal drawbacks of the FT process are that the capital cost of FT-conversion plants is relatively high and that the energy efficiency for the production of FT liquids by conventional techniques is lower than the energy efficiency for the production of alternative fuels.

1.52 Bio-oil

Bio-oils are viscous, dark-colored liquids that are generated through the process of biomass pyrolysis. Biomass is commonly derived from municipal garbage or agricultural and forestry by-products (Demirbas, 2007). Pyrolysis is a highly effective process for converting biomass, with an efficiency rate of over 70%. The liquid has a density of around 1,200 kg/m3, which is greater than that of fuel oils and considerably greater than that of the original biomass. Researchers are now studying the gasification of bio-oil using pure oxygen and the subsequent conversion of the resulting syngas into synthetic fuel by the FT process. However, this method does not seem to be financially appealing, according to Demirbas and Balat (2006).

1.5.3 Plant oils

There is a growing interest in utilizing virgin plant oils as a source of fuel for diesel engines. Over 2,000 oily plant species have been found, thriving in various climates and latitudes. Over 350 plant species have been identified as potential sources of oil for diesel engines (Goering et al., 1982). The composition of plant oils consists primarily of 98% triglycerides, along with minor quantities of mono- and diglycerides. There are primarily two categories of vegetable oils: those with a predominant amount of fatty acids in C_{12} (such as palm oil) and those with a predominant amount of fatty acids in C_{18}. Utilizing plant oils directly (or in combination with fossil fuels) has typically been deemed unsatisfactory or impractical for both direct and indirect diesel engines. Evident issues encompass the elevated viscosity (Ramadhas et al., 2005), acidic composition, presence of free fatty acids, carbon deposition tendency, lubricating-oil thickening propensity, and gum production due to oxidative polymerization during storage and burning. The viscosity of the blend of vegetable oils and fossil diesel fuel can be significantly modified. According to the EN 14214 guidelines, the highest allowed proportion of vegetable oils in a blend is B30, which means a maximum of 30% plant oil mixed with fossil diesel, measured by volume (Abollé et al., 2008). Pyrolysis can decrease the oil viscosity by breaking down the big triglycerides, resulting in an alternative fuel for diesel engines (Lima et al., 2004). Utilizing plant oils in blends also substantially raises their cloud points, hence restricting their application to nations with appropriate climates.

1.5.4 Bioalcohol

Due to the energy crisis and climate change, there is a pressing need for a substantial immediate supply of biofuels. Bioethanol and biodiesel are often regarded as the most suitable options for meeting these requirements, and thus are sometimes referred to as the initial generation of biofuels. Ethanol can be derived from several crops such as sugarcane, sugar beets, maize, barley, potatoes, cassava, and mahua (Baker & Keisler 2011; Kremer & Fachetti 2000). Flexible-fuel engines have been engineered to combust hydrous ethanol/gasoline mixes in any proportion, including pure ethanol. These engines electronically manage the automatic modification of combustion parameters based on the required oxygen level for the fuel in the tank (Marris, 2006). Gasohol, often known as ethanol-gasoline blend, is a mixture of ethanol and gasoline. Ethanol is synthesized through the process of fermenting a sugar slurry, which is commonly derived from sugar or grain crops. The fermentation process, facilitated by yeast, results in a solution with an ethanol concentration of roughly 12%. The yeast invertase enzyme facilitates the process of breaking down sucrose into glucose and fructose through hydrolysis. Afterwards, yeast zymase transforms both glucose and fructose into ethanol. Through the process of distillation, the alcohol can be condensed to yield ethanol with a maximum concentration of 96% (hydrous ethanol). Ethanol, being a polar solvent, has distinct chemical properties compared to hydrocarbon fuels, which are non-polar solvents. Consequently, the act of combining ethanol with hydrocarbon fuels has certain distinct difficulties. The issues encompass several factors: (i) increased fuel volatility when using low levels of ethanol/gasoline blends, (ii) the need for higher octane ratings, (iii) a rise in dissolved water content in motor gasoline, which leads to fuel mix heterogeneity and engine corrosion, and (iv) a greater solvency requirement. Akzo Nobel Surface Chemistry and the Lubrizol Corporation have collaborated to create an inexpensive additive that enables the blending of ethanol and diesel fuel, resulting in a fuel that is both stable and transparent (Lü et al., 2004). The name of this fuel is "Dieshol". Biomethanol can be synthesized from biomass by utilizing bio-syngas derived from the steam-reforming process of biomass. Recovering biomethanol from biomass is significantly more straightforward compared to recovering bioethanol. Nevertheless, the current economic feasibility of sustainable methanol manufacturing systems is lacking. The conversion of biomass into methanol is a chemically complex process that requires a significant investment of financial resources. Thus, given the present circumstances, methanol production exclusively relies on the utilization of waste biomass, such as wood or municipal trash.

1.5.5 Biodiesel

Biodiesel possesses the benefit of being compatible with all diesel engines without necessitating any alterations. The production of this substance occurs through the conversion of renewable oils, which are derived from many sources such as plants, algae, bacteria, and fungi. First-generation biodiesel is produced by a two-stage process. Firstly, raw materials, usually oilseeds, are crushed in specialized mills to extract the oils. Secondly, the extracted oil is then converted into biodiesel. Free fatty acids (FFA) or triglycerides undergo a chemical reaction with short-chain alcohols (such as methanol or ethanol) in the presence of a catalyst, resulting in the formation of alkyl-esters. The process of converting FFA to alkyl-esters is referred to as esterification, while the process of converting triglycerides is known as transesterification. Fatty acid methyl esters are partially derived from biological sources, as the methanol used in their production is typically derived from fossil methane, which is obtained from natural gas. Alternatively, the production of biodiesel can involve substituting methanol with ethanol, which leads to the formation of fatty acid ethyl-esters. If the ethanol is derived from living organisms, then the resulting product is entirely biological. The goal of the transesterification process is to decrease the oil's viscosity, which is achieved at a lesser cost compared to the pyrolysis method used in bio-oil processing. As per the EU regulations for alternative diesel fuels, the alkyl-esters content in biodiesel must be equal to or greater than 96.5% by weight.

1.5.6 The concept of biofuels can be classified into four distinct generations

The initial iteration of biofuels proved the technical viability of energy crops, but it became evident that there is no universal solution that can address all circumstances (Venturi & Venturi, 2003). Furthermore, the production of first-generation biofuels is hindered by factors that are in direct opposition to the principles of biofuel, such as the devastation of tropical rainforests (Kleiner, 2008). Tropical rainforests are the most effective carbon sinks on the planet. Thus, if biofuels contribute to the loss of ecosystems, it can be inferred that the carbon balance of biofuels is negative. This factor restricts the practicality of initial biofuels. Additionally, this implies that the range of raw materials used for biofuel production will need to be expanded in the future. Second-generation bioethanol is specifically designed to address this difficulty. Second-generation bioethanol will be derived from lignocellulosic biomass, which is a more appropriate and sustainable source of renewable energy (Frondel & Peters, 2007; Tan et al., 2008; Tilman et al., 2007). Lignocellulose is derived from abundant cellulosic biomass that is widely distributed worldwide. Nevertheless, the inexpensive conversion of lignocellulose into bioethanol remains a difficult task. Genetic modification of plants is a potential technology that raises concerns in society. In order to ensure that biomass ethanol becomes a significant source of renewable energy, it is essential to implement energy regulations, regardless of how technology may evolve in the future (Tan et al., 2008). According to Brennan and Owende (2010), algae and cyanobacteria are more efficient than higher plants at capturing solar energy. They are expected to outperform first- and second-generation biofuels in terms of energy capture per unit of surface area. Algae are currently employed in experimental CO2-sequestration systems to mitigate emissions in certain traditional power stations that rely on fossil fuels. The process is referred to as carbon dioxide filtering. Regrettably, algae necessitate financial resources for investing in reactors capable of cultivating them, hence presenting a promising prospect for advancing this technique in CO2 filtering. From that perspective, algae can be considered a fuel of the third generation. Novel techniques and advanced technology are currently being devised to produce second- (e.g., synfuels, Baker & Keisler, 2011) and third-generation biodiesel fuels. These advancements will necessitate a revision of the prevailing definition of biofuels commonly employed in governmental laws (Lois, 2007). Additionally, one can also imagine the investigation of photosynthetic systems with the purpose of producing biohydrogen and bioelectricity. These biofuels are considered fourth-generation (Gressel, 2008). It is anticipated that the creation of efficient fourth-generation biofuels would not occur until the latter part of the 21st century.

2. PLANT BIOFUELS

2.1 Bioethanol

The process of alcohol fermentation has been recognized for millennia. For many years, the industry has been conducting ethanol distillation as a means to regulate sugar prices in the global market. Ethanol is commonly derived from the isomerose (high-fructose syrup) of grain crops like maize or wheat, as well as from sugar crops such as sugar beet or sugarcane. Sugar beet is the preferred crop in Europe. This holds particularly true in nations like the UK, France, Holland, Belgium, and Germany, where it is exceptionally efficient. A single hectare of this crop can provide 5.5 tons of ethanol, whereas 1 hectare of wheat only generates 2.5 tons of ethanol (Demirbas & Balat, 2006). It is necessary to compare these values with the ethanol output from sugarcane, which amounts to 7.5 t in Brazil (Bourne, 2007). The United States of America utilizes maize as a source for ethanol production, while India employs sugarcane, China utilizes sweet potatoes, and Canada relies on wheat. China, Austria, Sweden, New Zealand, and Ghana are now developing their biofuels infrastructure using wood-based feedstocks (Herrera, 2006). The sugarcane production area in Brazil occupies 8 million hectares, which is approximately 0.94% of Brazil's total land area of 850 million hectares. Sugarcane yields an energy return on investment (EROI) of eight, meaning that the energy produced from sugarcane is eight times more than the energy input required to generate it. One metric ton of sugarcane used for ethanol production results in a reduction of 220.5 units of CO2 emissions compared to the emissions produced by burning fossil fuels. Therefore, if the rainforest is preserved instead of being cleared for sugarcane cultivation, the production of ethanol from Brazilian sugarcane results in a reduction in greenhouse gas emissions equivalent to 25.8 million metric tons of CO2 per year (Marris, 2006; Walter et al., 2010). Since 2009, the government has issued a decree to prohibit sugarcane growing in the Amazon, Pantanal, and Alto Rio Paraguai regions in order to protect these ecosystems. Approximately 47% of Brazil's transport fuel consumption in 2009 was attributed to ethanol. The "Flex" automobile fleet has the capability to utilize either ethanol or gasoline as fuel, with the ability to use 100% of either option (Orellana & Neto, 2006). Indeed, ethanol provides a 20% to 30% lower mileage per liter compared to gasoline, and individuals adjust the mixture based on the optimal balance between fuel efficiency and cost (Marris, 2006). Brazil's current ethanol export capacity is at approximately 8 gigaliters (Gl). The primary export destinations include the United States, the European Union, Japan, and Central America. Based on conservative predictions, the land area dedicated to sugarcane cultivation in Brazil is projected to expand from 8 to 11 million hectares by 2015. The government has imposed a restriction stating that the total area available for growing sugarcane cannot exceed 64 million hectares, which is equivalent to 18.5% of the country's land. Brazil is currently the sole country capable of maintaining the burgeoning global ethanol market in the near future. In order to achieve long-term success in the market, countries like Australia, Columbia, Guatemala, India, Mexico, and Thailand will have to enhance their export activities (Orellana & Neto, 2006). Ethanol manufacturing in Brazil commenced in 1973. During that period, the country relied significantly on foreign crude oil, with around 80% of its oil supply being imported. In 1975, the program PROALCOOL was initiated (Goldemberg et al., 2004). It provided financial assistance in the form of subsidies and low-interest loans to bioethanol manufacturers with the aim of expanding their production capacity. The government implemented a program of price dumping to stimulate the use of gasohol. According to Pousa et al. (2007), the ethanol concentration in regular gasoline has increased from 5% to 25% as mandated by law.

2.1.1 Production of bioethanol using lignocellulosic biomass

Plant structural polysaccharides are the most plentiful sources of renewable carbon in the biosphere. Annually, around 1,011 metric tons of these polymers are produced, which is comparable to an energy content of 640 gigatons of oil (Proctor et al., 2005). Some examples of non-food plant species used for bioenergy generation are Sorghum halepense, Arundo donax, Phalaris arundinacea (Raghu et al., 2006), poplar, switchgrass (Panicum virgatum), the hybrid grass Miscanthus x giganteus, and big bluestem. These species possess energetic, economic, and environmental benefits compared to first-generation biofuel crops (Hill et al., 2006; Havlík et al., 2010). Switchgrass, for instance, has a net energy production of 60 Giga Joule per hectare per year (GJ/ha/yr) (Schmer et al., 2008). The estimated amount of fuel that may be obtained from cellulosic biomass production on land is 135 gigajoules per hectare per year, which is somewhat greater than the yield from corn (85 gigajoules per hectare per year) or soybean biodiesel (18 gigajoules per hectare per year). Perennial and indigenous species that are well-suited for growing on marginal soils are the most suitable forms of specialized biofuel crops. Eucalyptus is a rapidly growing woody tree that is used for biomass production in tropical and Mediterranean nations. High-yielding species of willow (Salix nigra), Miscanthus (a rhizomatous grass that produces up to 26 t of dry matter per hectare per year), and poplar are readily accessible in wet and temperate nations. These energy crops have a lower requirement for chemicals and energy compared to traditional crop production. They can also grow on less fertile terrain, which prevents them from competing with food crops. In an Ireland-based context, the use of Miscanthus and willow for generating heat and electricity might result in a reduction of up to 5.2% of 2004 greenhouse gas (GHG) emissions, while occupying just 4.6% of the overall agricultural area (Styles & Jones, 2008). According to estimates, lignocellulosic biomass has the potential to provide 70-100 exajoules (1 exajoule = 1,000,000,000 gigajoules) of energy by the year 2020 (Gielen et al., 2002). Poplar is a suitable option for short rotations of approximately 5 years. Poplar exhibits extensive seed and pollen dispersal, surpassing other crops in this regard. This dispersal occurs over multiple years prior to harvesting, and poplar possesses numerous wild relatives with which it can engage in outcrossing. Furthermore, poplar can be propagated asexually, enabling the utilization of low-lignin variants by multiplying infertile specimens. Poplar biotechnology has been the prevailing field for a number of years, and its genome has been fully sequenced. Trees have a lignocellulosic energy-conversion factor of 16, which is higher than corn (1-1.5) and sugarcane (8-10). Additionally, trees can be cultivated on marginal soils, minimizing conflict with food crops. The global wood consumption currently stands at 3.4 gigameters cubed per year (Gm3/yr) and is expected to experience a significant rise due to the manufacturing of ethanol from biomass. Establishing high-yield plantations is crucial in order to meet the growing need for wood (Fenning et al., 2008). Sweden and Austria have always used the residues of the forest sector to generate liquid and solid fuel, which has been essential for small towns, schools, buses, ski resorts, and factories (Herrera, 2006). Plant breeding can benefit from the application of biotechnology and systems biology. A significant number of plant species utilized for bioenergy generation are in a state ranging from wild to semi-domesticated. Utilizing molecular techniques can enhance the pace of domestication and increase productivity (Chen & Dixon, 2007). Several potential genes associated with domesticated features have been found through a comparative analysis of the genomes of poplar, rice, and Arabidopsis, focusing on gene function and expression at a wide scale. The examined genes were associated with the production of cellulose and hemicellulose, as well as with several morphological growth traits, including as height, branch number, and stem thickness (Ragauskas et al., 2006; Chapple et al., 2007; Sticklen, 2008). Researchers have developed transgenic plants that express mutant alleles or use RNA interference (RNAi) to silence native genes. This has resulted in obtaining cell-wall components that can be more readily turned into ethanol (Chen & Dixon, 2007; Himmel et al., 2007). Some options that can be used to reduce the cost of converting cell walls to ethanol include reducing lignin levels and increasing cellulose components in cell walls, or overexpressing cellulases in plant cells (Sticklen, 2008). Nevertheless, the potential impact of lignin interference on biomass production needs to be thoroughly assessed, since it may result in unintended consequences such as heightened vulnerability to fungal infections. The investigation of low-lignin transgenic trees has been conducted due to the comparatively high resistance of lignin to enzymatic breakdown (Herrera, 2006). The inhibition of p-coumaroyl-CoA 3'-hydroxylase using RNA interference in hybrid poplars exhibited a strong positive correlation with the decrease in lignin concentration. The suppressed lines of poplars showed an increase of approximately 13.5% in cell-wall carbohydrates compared to wild-type poplars, as observed by Coleman et al. in 2008. The bioconversion of lignocellulosic biomass is mostly achieved by the process of lignocellulose pretreatment followed by enzymatic hydrolysis (Sanderson, 2006). The selection of pretreatment determines the most suitable combination of enzymes to be utilized and the specific composition of the resulting sugar mixture. Ultimately, the sugars undergo fermentation using ethanol-producing microbes including yeasts, Zymomonas mobilis, Escherichia coli, or Pichia stipitis (Fischer et al., 2008).

2.2 Biodiesel

2.2.1 Biodiesel production procedure

The primary constituents of plant oils consist of fatty acids and their derivatives, including mono-, di-, and triacylglycerides. Plant oils consist predominantly of tri-acylglycerides, accounting for 95% of their composition. Glycerides are esters that result from the condensation of fatty acids with the tri-alcohol glycerol (propanetriol). Triacylglycerides can be formed by fixing three fatty acids onto a glycerol molecule, while diacylglycerides are formed by fixing two fatty acids, and monoglycerides are formed by fixing only one fatty acid. Indeed, the fatty acids have the potential to be either identical or distinct. As mentioned in the introduction, biodiesel can be acquired through the processes of esterification or transesterification. Esterification is the chemical reaction in which a fatty acid combines with a mono-alcohol to produce an ester. Acids act as catalysts in the esterification reaction. Esterification is a frequently employed procedure in the production of biodiesel to remove free fatty acids (FFAs) from low-grade oil that has a high concentration of acidic compounds. Transesterification, also known as alcoholysis, is a chemical reaction where one alcohol replaces another alcohol in an ester, resembling the process of hydrolysis. This procedure has been extensively employed to decrease the thickness of triglycerides. The transesterification reaction is denoted by the generic equation (5).

$$RCOOR' + R''OH \rightarrow RCOOR'' + R'OH \quad (5)$$

The reaction proceeds in a step-by-step manner, involving the synthesis of di- and monoglycerides as intermediate products (Canakci et al., 2006). In theory, transesterification necessitates three alcohol molecules for every triglyceride molecule. Nevertheless, an abundance of alcohol is required due to the reversibility of the three intermediate processes (Marchetti et al., 2007; Om Tapanes et al., 2008). Following the completion of the reaction, the phase containing a high concentration of glycerol is separated from the layer containing the ester using the process of decantation or centrifugation. The ester phase obtained (crude biodiesel) contains impurities such as methanol, glycerides, soaps, catalysts, or glycerol that need to be purified in order to meet the requirements of the European Standard EN 14214.

Biodiesel manufacturing can utilize many technologies, such as chemical or enzyme catalysis and supercritical alcohol treatment (Demirbas, 2008b). EN 14214 mandates the evaluation of 25 factors in order to verify the quality of biodiesel. Conventional biodiesel manufacturing methods involve the use of potent alkalis or acids as chemical catalysts in transesterification and esterification processes. The significant energy consumption of these processes, along with the lack of precise reaction selectivity commonly observed in physicochemical synthesis, provides a valid reason for the further investigation into enzymatic catalysis. Furthermore, a further purification process is necessary to exclude glycerol, water, and other impurities from alkyl-esters. Base catalysis has significantly higher reaction rates compared to acid catalysis. Due to its low cost and excellent kinetics, NaOH has become the catalyst of choice in the industry. Nevertheless, the reaction might give rise to the formation of soap and emulsion, which can complicate the process of purification.

2.2.2 Non-consumable raw materials for the manufacturing of biofuels

Rapeseed oil currently accounts for around 84% of global biodiesel output. The remaining sections consist of 13% sunflower oil, 1% palm oil, and 2% soybean oil and other oils (Gui et al., 2008). The majority of biodiesel, over 95%, is currently derived from food oils. In order to address this unfavorable circumstance, there is a growing trend in the production of biodiesel from non-edible oils and waste cooking oil (WCO). Non-consumable oils has the benefit of not conflicting with consumable oils in the food industry. Recycled cooking oil is considered a byproduct and, as a result, it is less expensive than fresh plant oil. The acid-catalyzed method necessitates a greater initial investment due to the need for stainless-steel reactors and methanol-distillation columns. However, this is balanced off by the advantage of having cheap feedstock cost (Zhang et al., 2003). Utilizing WCO esters for recycling is a sophisticated method, as waste oils are forbidden for animal feed, pose risks to the environment and human health, and interfere with the functioning of wastewater treatment plants, leading to increased expenses for maintenance and water purification. The manufacture of biodiesel from waste cooking oil (WCO) is currently limited in scale, although it is seeing a global growth trend. The United States and China are prominent users of WCO, with annual consumption rates of 10 and 4.5 million metric tons, respectively. The EU, Canada, Malaysia, Taiwan, and Japan collectively manufacture around 0.5-1 million metric tons per year, as reported by Gui et al. in 2008. The potential utilization of waste cooking oil (WCO) as a principal feedstock for biodiesel fuel is significant because it would alleviate the majority of difficulties related to the conflict between food and biodiesel crops for land (Bindraban et al., 2009; Odling-Smee 2007). Through the process of converting edible oils into biodiesel fuel, food resources are effectively transformed into motor fuels. The manufacturing of biodiesel fuels on a wide scale using edible oils is thought to potentially disrupt the global food supply and demand market. However, there are differing opinions on this matter (Ajanovic, 2010). Nevertheless, there are no obstacles to use edible oils initially for culinary purposes and subsequently for the production of biodiesel fuel.

2.2.3 Global sources of biofuel feedstocks

The United States is confronted with the urgent need to address the reduction of greenhouse gas (GHG) emissions by 17.7% in order to mitigate the possible harms caused by climate change, such as coastal changes and the spread of tropical illnesses (Lokey, 2007). Due to the electrical sector's contribution of 40% to total greenhouse gas (GHG) emissions, it is advisable to invest in cost-competitive renewable energy sources including wind, geothermal, and hydroelectricity. The United States possesses abundant solar resources, which offer numerous unexplored opportunities for generating renewable energy (Flavin et al., 2006). The Biomass Program of the US Department of Energy, initiated in 2000, proposed a target of 5% utilization of biofuels by 2010, 15% by 2017, and 30% by 2050. According to a prediction by Szulczyk et al. (2010), the use of ethanol in transportation is expected to reach approximately 50% of gasoline use by 2030. At now, maize and other grains, such as sorghum, serve as the main raw materials for ethanol manufacturing in the United States. Maize is classified as a low-efficiency biofuel crop due to its high input requirements, significant topsoil erosion (10 times faster than sustainable), and other adverse effects (Donner & Kucharik, 2008; Laurance, 2007; Sanderson, 2006; Scharlemann & Laurance, 2008). Despite consuming 40 Ml of ethanol per day, maize falls short in terms of efficiency and has detrimental impacts. In contrast, soybean biodiesel necessitates reduced inputs. Nevertheless, both of these biofuels are incapable of replacing fossil fuel without causing a significant impact on food supplies. According to Hill et al. (2006), if all corn and soybean production in the US were used for biofuels, it would only be able to meet 12% of the gasoline demand and 6% of the diesel demand. Nevertheless, the combination of agricultural, municipal, and forest wastes has the potential to supply 1 Gt of dry matter each year in a sustainable manner, and it should be used in addition to other biofuel crops (Vogt et al., 2008). The proposal suggests converting an area of land ranging from 3.1 to 21.3 million hectares into biomass production, as stated by Schmer et al. (2008). Algal biodiesel is being incorporated into a renewable-energy park, as documented by Singh & Gu (2010) and Subhadra (2010).

Brazilian bioethanol yields more than 90% reduction in greenhouse gas emissions, according to Hill et al. (2006). Besides the PROALCOOL initiative, the Brazilian government established the PRO-ÓLEO program in 1980 with the aim of achieving a 30% blend of vegetable oils or derivatives in diesel fuel, with complete substitution in the future. Regrettably, with the decline in the price of crude oil on the global market in 1986, this initiative was discontinued and was not reinstated until 2002. Brazil possesses a wide range of plant-oil feedstocks due to its rich biodiversity and diverse climate and soil conditions. These feedstocks mostly include soybean, sunflower, coconut, castor bean, cottonseed, oil palm, physic nut, and babassu (Nass et al., 2007). In 2010, Brazil marked the opening of the Embrapa Agroenergia research center to facilitate the incorporation of oil derived from these feedstocks into the biodiesel production network. The National Program of Production and Use of Biodiesel (PNPB) was initiated in 2004 with the aim of determining the economic feasibility of biodiesel production while promoting social and regional growth. The present annual diesel use in Brazil is roughly 40 billion liters, whereas the potential market for biodiesel is currently 800 million liters and is expected to reach 2 billion liters by 2013. Furthermore, B5 has been compulsory since 2010. The auction prices have fluctuated between US$ 0.3 and 0.8 per liter, depending on the manufacturing location (Barros et al., 2006). According to a study by Moreira and Goldemberg in 1999, an investment of US$ 5 billion was made in bioenergy between 1975 and 1999. This investment led to the creation of 700,000 new employment and a savings of US$ 43 billion in gasoline imports. According to Johnston and Holloway (2007), the anticipated job creation rate for biodiesel production is 1.16 jobs per million liters of annual production. Nevertheless, it is anticipated that the current tendency towards business centralization will lead to a decrease in this rate (Hall et al., 2009). Petrobras is now utilizing a processing capacity of 425,000 tons to refine a blend of plant oil and crude oil, which is referred to as "H-Bio". The country possesses over 90 million hectares of land suitable for cultivating oleaginous crops, primarily located in Mato Grosso (southwest), Goiás, Tocantins, Minas Gerais (central), Bahia Piauí, and Maranhão (northeast). This vast expanse of land benefits from a tropical environment.

The European Union comprises a population of 454 million individuals, which represents approximately 7% of the global population and is 50% larger than the population of the United States (Solomon & Banerjee, 2006). The EU is committed to a gradual transition towards a hydrogen-based economy. The primary sources of hydrogen anticipated for usage in the period of 2020-2050 are renewable energy sources and, potentially, enhanced nuclear power (Adamson, 2004). Nevertheless, the EU anticipates that hydrogen production from fossil fuels, together with carbon sequestration, will continue to be significant in the distant future, alongside renewable energy and nuclear power. Due to their renewable nature, biodiesel and bioethanol in the EU have been estimated to provide GHG savings of 15-70% compared to fossil fuels. Frondel and Peters (2007) discovered that the energy and greenhouse gas balances of rapeseed biodiesel are unequivocally favorable. The EU now relies on bioethanol derived from sugar beets or wheat, as well as biodiesel made from rapeseed, as its primary means of achieving its biofuel production goals. Due to the expansion of area dedicated to biofuel production, there is currently a competition between biofuel crops and food crops (Odling-Smee, 2007), which is anticipated to have significant impacts on the economy. According to Malça and Freire (2011), the estimated usage of fossil diesel fuel in Europe is around 210 Gigaliters (Gl), while the consumption of biodiesel is approximately 9.6 Gl. The European Union (EU) cultivates more than 2 million hectares (Mha) of rapeseed (about 1 gigaliter) and sunflower (approximately 0.6 kiloliters per hectare) for biofuel production, as reported by Fischer et al. (2010). This indicates a significant reliance on imported biofuels in order to meet the proposed target of B5.75. Due to the greater energy capacity of synfuel derived from biomass and the limitations on the amount of arable land, second-generation biofuels are expected to become competitive in the biofuel industry in the near future (Fischer et al., 2010; Havlík et al., 2010). The cost of biodiesel that complies with the EU quality standard (EN 14214) is roughly €730 per metric ton. The profit gained from selling biodiesel from abroad on the EU market can be calculated by deducting the biodiesel export value from the EU market price. The export value encompasses both the expenses of production and the costs associated with exporting the goods. The production costs consist of the expenses incurred in the production of plant oil or animal fat, as well as the costs associated with biodiesel processing, excluding the value of by-products such as glycerol. The exportation costs encompass expenses such as scaling, insurance, taxes, and administrative fees. The figures for these costs can be found in Johnston & Holloway's publication from 2007. The price of biodiesel, at US$ 0.88 per liter, was 45% greater than the price of fossil diesel fuel during the same period in 2006. While this price serves as a useful reference point, the cost of biodiesel in the EU market is subject to rapid fluctuations influenced by various factors including domestic production levels, fossil diesel prices, agricultural productivity, and legislative measures. Identical regulations will be enforced in the developing markets of China. Malaysia, Indonesia, Argentina, the US, and Brazil are the top five countries that exhibit the most favorable mix of large volumes and cheap production costs, as determined based on their volume and profitability. Together, these nations contribute to more than 80% of the overall biodiesel production. Only about 2% of the total production of vegetable oil is now used for biodiesel, while the majority is mostly allocated for food purposes. Although India has not achieved the same level of ethanol production as Brazil, it holds the distinction of being the world's largest producer of sugar. Indian ethanol is now mixed with gasoline at a rate of 5% in nine states of India. To fully comply with the directive, an additional 500 Ml of ethanol would be required. The overall demand for ethanol is around 4.6 gigaliters (Subramanian et al., 2005). The nation consumes over 44 million metric tons (Mt) of fossil diesel fuel, which is three times greater than its consumption of gasoline. This significant amount of fuel is primarily utilized for transportation purposes. Given that India relies on imports for over 70% of its fuel, which amounts to almost 111 million metric tons, any form of renewable energy is highly valued. India has successfully created a market for biodiesel blends containing 10% of biodiesel (Kumar & Sharma, 2008). India, being a country that imports more edible oils than it exports, prioritizes the use of non-edible oils derived from plants like physic nut, karanja, neem, mahua, and simarouba. Physic nut and karanja are the top two contenders on the Indian plant list for biodiesel generation.

Out of the total land area of 306 million hectares, 173 million hectares are now being used for farming. The remaining land is categorized as either degraded agricultural land or non-arable barren terrain.

Approximately 40% (80-100 Mha) of the land area has been degraded due to inadequate land utilization and population pressures over an extended period of time. These unused or unproductive regions are being examined as potential sites for restoration using physic nuts, as suggested by Kumar and Sharma in 2008. Approximately 80,000 out of India's 600,000 villages now lack access to fuel or power due to insufficient fuel supply, which makes it impractical to establish a comprehensive distribution network. Physic nuts have the potential to provide the direct production of oil inside villages, hence enabling the development of local economies (Fairless, 2007). This also pertains to emerging regions of Brazil and Africa. Furthermore, conventional motorcycles equipped with internal combustion engines of 100 cm3 have been modified to operate using hydrogen as fuel. The motorcycles have demonstrated an efficiency above 50 kilometers per charge. This achievement is highly significant due to the fact that motorbikes and scooters account for 70% of privately owned vehicles in India. There are ongoing efforts to modify small automobiles and buses to run on hydrogen. This transition is expected to be facilitated by the increasing presence of electric and compressed natural gas (CNG) vehicles in and around New Delhi (Solomon & Banerjee, 2006).

China has a smaller per capita arable land area compared to the global average. Consequently, the majority of edible oils are brought in from other countries, and it is estimated that the demand for edible oils in 2010 will be 13.5 million metric tons. China's substantial population necessitates the urgent implementation of sustainable energy sources. Due to limited arable land, China is investigating options for the development of second- and third-generation biofuels (Meng et al., 2008). China, as a sizable emerging nation, possesses extensive areas of degraded land and requires substantial amounts of renewable energy to support its quickly expanding economy and the associated need for sustainable development. The energy yield of biomass cultivated on depleted soil is approximately equivalent to that of ethanol derived from traditional maize cultivated on fertile land. Biofuel derived from biomass is significantly more cost-effective compared to conventional biofuels like corn ethanol or soybean biodiesel. The estimated annual potential energy production from biomass is 6,350,971 terajoules (TJ/yr), as stated by Zhou et al. (2008). It is widely acknowledged that the value of biomass in China's energy portfolio would inevitably expand. Biodiesel derived from microalgae might be cultivated efficiently over the 18,000 km Chinese coastline, making use of the abundant availability of seawater (Song et al., 2008). Marine microalgae production necessitates the utilization of unoccupied desert terrain, seawater, carbon dioxide, and sunlight. China's extensive mudflats and salty soils offer significant opportunities for the development of biodiesel production using marine microalgae. The sales of electric bicycles and scooters in China have shown a substantial growth over the past decade, reaching an annual total of over 1 million units. The increase in demand for electric bicycles and scooters in Beijing and Shanghai, as well as other major cities, has been driven by the prohibition of gasoline-fueled alternatives. These prohibitions were implemented due to growing apprehensions about pollution (Solomon & Banerjee, 2006). Due to this factor, China has emerged as one of the most significant prospective markets for hydrogen fuel cells in the transportation industry. The prevalence of recurrent droughts in numerous Asian nations has posed a significant challenge in emulating Brazil's triumph in cultivating sugarcane, as this crop need a copious water source. Thailand and Indonesia are harnessing the potential of palm oil. Japan has emerged as a key actor in the global advancement of a hydrogen-based economy due to its desire to maintain its status as a leading high-tech superpower. According to Japanese estimates, the potential for hydrogen generation from renewable energy in Japan is 210 GNm^3/yr. (Nm^3 is the volume of gas in cubic meters at 0 ºC and one atmosphere). This amount is four times greater than the projected need for hydrogen in 2030. Nevertheless, hydrogen derived from renewable sources is projected to account for just around 15% of the total hydrogen consumption by 2030. According to Solomon and Banerjee (2006), using methanol or gasoline for fuel cell propulsion through on-board reforming is considered the most feasible method in the near future. However, the ultimate objective is to transition to using pure hydrogen.

3. MICRODIESEL

Oleaginous microorganisms are microorganisms that have an oil content greater than 20%. The manufacture of biodiesel from microbial lipids, often known as single-cell oil or microdiesel, has garnered significant global interest. While there are different types of microorganisms that may store oils, such as microalgae, bacillus, fungi, and yeast, not all of them are ideal for the manufacture of biodiesel (Demirbas, 2010). The majority of bacteria typically do not possess the ability to make oil effectively. Actinomycetes are an exception because they can produce large quantities of fatty acids (up to approximately 70% of their dry weight) from simple carbon sources like glucose, even when their growth is restricted. These fatty acids are stored inside the cells as triglycerides (Alvarez & Steinbuchel, 2002).

Cryptococcus curvatus, the most effective oily yeast, can store over 60% of lipids when cultivated in settings with little nitrogen. The lipids are typically retained as triglycerides containing around 44% saturated fatty acids, a composition comparable to that of numerous plant seed oils. Rhodotorula glutinis has been employed for the purification of wastewater in the production of monosodium-glutamate. Monosodium-glutamate wastewater serves as an inexpensive fermentation medium for the synthesis of biodiesel using lipids derived from R. glutinis. In order to achieve optimal efficiency, the fermentation process requires an additional source of glucose to maintain the appropriate C:N:P ratio of 1:2.4:0.005. After 72 hours of culture, this technique results in a lipid production equivalent to 20% of the biomass. Additionally, the rate of oil transesterification is 92% (Xue et al., 2008). Furthermore, R. glutinis has the ability to utilize different carbon sources such as dextrose, xylose, and glycerol. It can collect varying percentages of triglycerides, specifically 16% from dextrose, 12% from xylose, 25% from glycerol, 10% from a combination of dextrose and xylose, 21% from a combination of xylose and glycerol, and 34% from a combination of dextrose and glycerol. The study by Easterling et al. (2008) revealed that the rate of accumulation of unsaturated fatty acids in R. glutinis is influenced by the carbon source. Specifically, when grown on xylose and glycerol, the accumulation rates were found to be 25% and 53% respectively. The results suggest that the utilization of R. glutinis can enhance the worth of various by-products, such as glycerol. Nevertheless, the elevated quantities of unsaturated fatty acids may necessitate an additional saturation process in order to comply with biodiesel regulations. Cyanobacteria are photosynthetic prokaryotes that are gram-negative and can be grown in water with varying levels of salinity, from freshwater to severe circumstances. Their capacity to generate a diverse array of lipids, oils, sugars, and useful bioactive chemicals has led to suggestions of incorporating them into wastewater treatment procedures (Markou & Georgakakis, 2011). Their replication period is 3.5 hours during the logarithmic phase of cell proliferation (Chisti, 2007). By harnessing light energy, they have the capability to transform carbon substrates into oil, which has a fatty acid composition resembling that of plants. This conversion process occurs at a rate of 20-40% of dry biomass, as reported by Meng et al. in 2008. Microalgae, despite their ability to store high amounts of lipids, necessitate greater surface areas and longer fermentation periods compared to bacteria. The microalgae market yields around 5,000 metric tons of dry biomass per year and generates roughly US$ 1.25 billion annually (Pulz & Gross, 2004).

Eukaryotic diatoms, green algae, and brown algae obtained from oceans and lakes generally contain lipid levels ranging from 20% to 50% of their dry mass (Brennan & Owende, 2010). Microalgae can contain exceptionally large amounts of lipids. Botryococcus, for example, can have a hydrocarbon content that surpasses 80% of its dry weight. By contrast, the quantities of lipids in dry biomass plant oil typically range from 15% to 40% (Spolaore et al., 2006).

There are almost 300 different types of algae, with diatoms (such as Amphora, Cymbella, and Nitzschia) and green algae (especially Chlorella) being the most appropriate for producing biodiesel. Triglycerides, which are high in C16 and C18, make up over 80% of the oil found in nearly all microalgae (Meng et al., 2008). The process of lipid formation in oleaginous bacteria is initiated either by nitrogen depletion or when there is an excess of carbon (Ratledge 2002).

Chlorella protothecoides has the ability to acquire lipids at a rate of 55% through heterotrophic growth when CO2 filtration is used. Significant amounts of microalgal oil have been effectively obtained from these heterotrophic cells by the process of n-hexane extraction. The microdiesel produced through acidic transesterification of heterotrophic microalgal oil is similar in quality to fossil diesel and has the potential to be a strong competitor to traditional biodiesel. This is due to the higher efficiency of photosynthesis, larger biomass production, and faster growth rates of microalgae compared to plants (Song et al., 2008).

Microalgal oils, as mentioned earlier, contain a higher amount of polyunsaturated fatty acids with four or more double bonds compared to most plant oils (Belarbi et al., 2000). This renders them vulnerable to oxidation during storage and diminishes their appropriateness for commercial biodiesel (Chisti, 2007). Nevertheless, fatty acids containing more than four double bonds can be readily reduced through partial catalytic hydrogenation (Dijkstra, 2006). Modifying the lipid content of microalgal oils poses significant problems, particularly in altering the degree of fatty acid unsaturation and adjusting the length of fatty acids. Enzymes that primarily bind to the cell membrane govern these characteristics, making their study more challenging (Certik & Shimizu, 1999). Presently, the majority of genetic alterations that have sought to enhance metabolic pathways have been conducted on oleaginous microbes. The fundamental reason for this is their capacity to store large quantities of lipids within their cells, their rapid development rates, and their similarity in oil composition to plants (Kalscheuer et al., 2006a, 2006b). Microalgae are commonly employed for the capture and reutilization of CO2 through a process known as "CO2 filtration" (Haag, 2007). They can decrease CO2 emissions by 82% on sunny days and by 50% on gloomy days (Vunjak-Novakovic et al., 2005). The utilization of microdiesel allows for a more refined approach to carbon recycling compared to carbon storage (CCS) in depleted oil fields or aquifers. The projected storage capacity of CCS is between 2,000-11,000 Gt CO2. However, these aquifers are not uniformly distributed worldwide (Schiermeier et al., 2008). Furthermore, CCS does not generate any revenue from the stored CO_2 and instead incurs an additional expense during the entire operation. On the other hand, algae transform carbon dioxide (CO_2) into oil. Consequently, the energy stored in the CO_2 can be reintroduced into the power plant following filtration by the algae and conversion into microdiesel. One proposed method for improving fish productivity is to stimulate the growth of phytoplankton by injecting CO_2 into selected areas of the ocean (Markels & Barber, 2001). Nevertheless, the practice of ocean fertilization has faced significant opposition due to its potential to ultimately disrupt the local environment (Bertram, 2010; Glibert et al., 2008).

4. BIOHYDROGEN

Electricity and hydrogen are the primary alternative energy sources being studied for transportation. Hydrogen energy has been of interest for about two centuries due to its practical applications, making it far from a new concept. According to Solomon and Banerjee (2006), only Iceland and Brazil are considering renewable-energy feedstocks as the primary or exclusive source of hydrogen in the future. Many analysts consider fuel-cell vehicles (FCVs) fuelled by hydrogen to be an imperative and the sole feasible solution for the future of transportation (Cropper et al., 2004). Unlike crude oil or natural gas, there are no reserves of molecular H_2 on Earth. Thus, it is more accurate to classify H_2 as an energy transporter, similar to electricity, rather than an energy source (Song, 2006). H_2 can be obtained from conventional fuel sources like natural gas, methanol, or gasoline. However, the most optimal and sustainable approach is to generate H_2 from water, utilizing heat from solar sources and extracting O_2 from the atmosphere. Currently, the primary technique of producing hydrogen involves removing carbon from fossil fuels. However, in the future, it will be feasible to generate hydrogen by alternate means such as water photolysis using semiconductors (Khaselev & Turner, 1998) or ocean thermal-energy conversion (Avery, 2002). These methods are currently in the phase of research and development and are not yet prepared for implementation in industrial settings. The process of generating hydrogen from biomass involves a series of sequential chemical reactions. The reformation of fuels in the water-gas shift reaction involves two subsequent steps: a carbon monoxide purification step and the elimination of carbon dioxide. Biomass can undergo thermal conversion by either gasification or pyrolysis. The primary gaseous byproducts produced from biomass can be represented by equations (6), (7), and (8) (Kikuchi, 2006).

$$\text{pyrolysis of biomass} \rightarrow H_2 + CO_2 + CO + \text{hydrocarbon gases} \quad (6)$$

$$\text{catalytic steam reforming of biomass} \rightarrow H_2 + CO_2 + CO \quad (7)$$

$$\text{gasification of biomass} \rightarrow H_2 + CO_2 + CO_2 + N \quad (8)$$

Hydrogen from organic wastes has generally been produced through equations (9), (10) and (11).

$$\text{solid waste} \rightarrow CO + H_2 \quad (9)$$

$$\text{biomass} + H_2\,O_2 + \text{Air} \rightarrow H_2 + CO_2 \quad (10)$$

$$\text{cellulose} + H_2\,O_2 + \text{Air} \rightarrow H + CO + CH_4 \quad (11)$$

Over time, the techniques employed for hydrogen production are anticipated to be tailored to the unique geographic area. According to Gummer and Head (2003), it is anticipated that the process of steam reforming methane and electrolysis would be incorporated, particularly in regions with abundant hydropower, such as Brazil, Canada, and Scandinavia. Once hydrogen becomes a ubiquitous energy source, it is expected to be distributed via pipelines. Previous systems, such as the established H2-distribution network in Germany that has been in operation for over 50 years, and the intercontinental liquid-hydrogen transport chain, have proven that leak rates of less than 0.1% can be attained in industrial settings (Schultz et al., 2003). Nevertheless, a significant concern related to the hydrogen paradigm is its status as the smallest atom, which makes leakage seemingly inevitable. It is necessary to acknowledge the potential for a substantial release of H_2 into the stratosphere. The reaction between hydrogen and ozone is anticipated to occur according to the equation $H_2 + O_3 \rightarrow H_2O + O_2$. This mechanism, as assessed by Kikuchi in 2006, poses a significant risk as a catalyst for ozone depletion. Alternatively, hydrogen can be generated from alternative fuels like as ethanol, biodiesel, gasoline, or synfuel through the use of onboard reformers, also known as hydrogen fuel processors. This method is highly favorable due to the ability to create synfuel from nearby feedstocks using the Fischer-Tropsch process. Furthermore, it can be easily transported and dispersed using current technologies and infrastructures (Agrawal et al., 2007; Takeshita & Yamaji, 2008). This consideration is equally relevant to biofuels. Furthermore, the viability of vehicles equipped with onboard reformers has already been demonstrated. The significance of synfuel is anticipated to escalate swiftly due to the abundance of natural gas reserves, also known as "stranded" gas, in remote areas. These reserves are deemed insufficient for liquefied natural gas (LNG) or pipeline initiatives.

The biological generation of hydrogen (or biohydrogen) provides a wide range of approaches for generating hydrogen, including direct biophotolysis, indirect biophotolysis, photo-fermentation and dark-fermentation (Lin et al., 2010). Biological hydrogen production processes are found to be more environmentally friendly and less energy intensive as compared to thermochemical and electrochemical processes. There are three types of microorganisms that produce hydrogen, namely cyanobacteria, anaerobic bacteria, and fermentative bacteria (Demirbas, 2008a). The process of photosynthetic H_2 generation from water is a biological mechanism that can transform sunlight into valuable, stored chemical energy. Hydrogen production is a characteristic exhibited by several phototrophic organisms. The list of hydrogen producers encompasses several hundred species from various genera of both prokaryotes and eukaryotes. The synthesis of H2 in green algae is facilitated by an enzyme called reversible hydrogenase. This enzyme is capable of catalyzing the oxidation of ferredoxin even without the presence of ATP, as observed by Beer et al. in 2009. The enzyme is susceptible to oxidation; nevertheless, resistant allozymes are being chosen (Seibert et al., 2001). Glucose has been utilized to produce hydrogen by the action of NADP+-dependent enzymes, namely glucose-6 phosphate dehydrogenase (G6PDH), 6-phosphogluconate dehydrogenase (6PGDH), and hydrogenase (Heyer & Woodward, 2001).

Carbon monoxide (CO) can undergo metabolism by many naturally existing microorganisms in the presence of water to generate H_2 and CO_2 according to equation (12), which represents the "water-gas shift" process, even at normal temperatures.

$$CO + H_2O \leftrightarrow CO_2 + H_2 \tag{12}$$

The bacteria Rubrivivax gelatinosus (Wolfrum & Watt, 2001) has been employed in the utilization of the biological water-gas shift reaction for the conversion of syngas derived from biomass.

Nitrogenases have the ability to generate hydrogen, however this process necessitates a significant amount of energy. Nevertheless, the nitrogenase reaction is fundamentally irreversible, hence enabling hydrogen pressurization. Rhodopseudomonas palustris is capable of catalyzing the nitrogenase reaction through the utilization of light as an energy source (Wall, 2004).

5. THE FUTURE OF TRANSPORT TECHNOLOGY

5.1 Fuel cells

The fuel cell serves as the main component in hydrogen-powered cars, responsible for turning the energy from fuel into electricity by mobilizing protons. Fuel cells operate without any mechanical parts, producing just water and low-voltage electricity through the use of hydrogen and oxygen. These machines function quietly and have an efficiency of 60%, which is higher than the 45% efficiency of internal combustion engines (ICE). Laboratory investigations indicate that fuel cells possess a minimum theoretical efficiency of 85%. Fuel cells, when combined with an electric motor that has an efficiency of 80%, might potentially be twice as efficient as utilizing hydrogen directly in an internal combustion engine (Ross, 2006).

Due to the issues surrounding the security and cost of hydrogen infrastructure for distribution and storage, ethanol is presently the most practical choice for fuel cells. Ethanol can undergo onboard steam reforming to be turned into hydrogen, or it can be employed more conveniently as a proton donor in certain fuel-cell technologies (Lamy et al., 2004). The process of ethanol-based steam reforming is carried out according to equation (13) as described by Velu et al. (2005).

$$C_2H_5OH + 3H_2O \rightarrow 2CO_2 + 6H_2 \tag{13}$$

Deluga et al. (2004) described an onboard system for hydrogen production by auto-thermal reforming from ethanol. Following this system, ethanol and ethanol-water mixtures were converted directly into H2by catalytic oxidation with ~100% selectivity and >95% conversion and with a residence time on rhodium catalysts of <10 milliseconds. This process has great potential for low-cost H_2 generation in fuel cells for small portable applications in which liquid-fuel storage is essential and in which systems must be small, simple, and robust.

The glycerol biofuel cell is an additional method for extracting energy from basic organic compounds (Arechederra et al., 2007). A biofuel cell closely resembles a conventional proton exchange membrane (PEM) fuel cell. Biofuel cells utilize biological molecules, such as enzymes, instead of precious metals, as catalysts to facilitate the reactions. Arechederra et al. (2007) successfully attached two oxidoreductase enzymes (pyrroloquinoline quinine-dependent alcohol dehydrogenase and pyrroloquinoline quinine-dependent aldehyde dehydrogenase) to a carbon anode surface. They then carried out a series of reactions to convert glycerol into mesoxalic acid, utilizing 86% of the energy present in glycerol. The bioanodes achieved power densities of up to 1.21 mW/cm2 by utilizing glycerol at concentrations as high as 99%. Due to the lack of swelling of the Nafion membrane in the presence of glycerol, it is anticipated that the longevity of the biofuel cell will surpass that of the current technology.

Formula 1 has joined the competition to enhance and improve environmentally friendly technologies. Since 2009, Formula 1 has implemented new regulations that require racing teams to harness and utilize the energy dissipated while braking in order to move the car forward (Trabesinger, 2007). The process responsible for this is referred to as a "kinetic-energy recovery system" (KERS), more often known as "regenerative braking". In a hybrid car equipped with both internal combustion and electric motors, the batteries can be recharged either by the internal combustion engine (ICE) or through regenerative braking. The accumulated electrical energy is subsequently utilized to propel the vehicle at lower velocities, particularly in urban traffic conditions, where the internal combustion engine's effectiveness is diminished due to frequent halting and restarting.

Fuel cells remain prohibitively expensive, with a current price of around US$ 4,000 per kilowatt (kW), making them 100 times more than internal combustion engines (ICEs). The current generations of vehicles require the replacement of fuel-cell stacks around 4-5 times over their lifespan. Therefore, the cost of 4-5 fuel-cell units should be compared to alternative internal combustion engines (ICEs) (Marcinkoski et al., 2008; Sorensen, 2007).Therefore, to be competitive with ICEs, the technology must reach the threshold of US$ 30/kW. To address this situation, Honda is selling its first prototype fuel-cell car under a leasing contract in California. BMW has been a pioneer of fuel-cell technology and produced its first hydrogen-car prototype in the 1960s (Hissel et al., 2004). Its current vehicle uses liquid hydrogen with autonomy of up to 386 km. The Ford Motor Company has set a new land-speed record for a fuel-cell powered car (334 km/h).

Despite these pilot experiments, it is likely that urban buses will be among the first large scale commercial applications for fuel cells. This is due to the fact that urban buses are highly visible to the public, contribute significantly to air and noise pollution in urban areas, have few size limitations and are fueled via a centralized infrastructure. Folkesson et al. (2003) reported the following: (i) the net efficiency of a Scania bus powered by a hybrid PEM fuel-cell system was approximately 40%; (ii) the fuel consumption of the hybrid bus was between 42 and 48% lower than that of a standard ICE Scania bus; and (iii) regenerative braking saved up to 28% energy. The bus prototype was fitted with a 50 kW fuel cell and powered by compressed ambient air and compressed hydrogen stored on the roof. All fossil fuel choices generate significant quantities of greenhouse gas (GHG) emissions. Ethanol and hydrogen possess the capacity to substantially diminish greenhouse gas emissions. Nevertheless, the utilization of ethanol and hydrogen will heavily rely on the specific methods of their generation. Certain hydrogen alternatives yield greater greenhouse gas emissions compared to internal combustion engines fueled by gasoline. The automotive options that will be competitive in the next two decades are those that employ advanced internal combustion engines, including hybrid vehicles that operate on eco-friendly gasoline or diesel fuel. At present, vehicles that utilize hydrogen and onboard carbon fuel reforming are still less environmentally efficient compared to gasoline internal combustion engines (ICEs). The significant energy consumption required for hydrogen production is expected to influence the distribution of hydrogen-powered vehicles across different regions. These cars are likely to be better suited for places that have abundant access to solar, wind, or hydro-electric power sources, as suggested by Eugenia Corria et al. (2006).

5.2 Energy storage

A variant of the hybrid car is the 'plug-in hybrid', which has the capability to be linked to the electric grid. The energy cost reductions over the whole charging and discharging cycle of an onboard battery, used to power an electric motor in an electric-hybrid (e-hybrid) car, amount to 80%. According to Romm (2006), this value is around four times greater than the savings achieved by fuel-cell automobiles powered by hydrogen produced by electrolysis, and it is 30% larger than the savings achieved by gasoline-powered cars. These cars enable the substitution of a significant proportion of fuel consumption and emissions from the exhaust. When power is generated from sources that do not emit CO2, e-hybrids can significantly decrease their overall greenhouse gas emissions.

The electrical storage system is a crucial component of the e-hybrid car since its power capacity and lifespan significantly determine the expenses of the entire system (Bitsche & Gutmann, 2004).

Emerging bio-based energy-management processes have the potential to make a substantial impact in the near future. Electricity can be generated by the process of whole-microorganism fermentation. Microorganisms belonging to the Geobacteraceae family have the ability to transport electrons directly onto electrodes, namely those containing Fe(III). This electron transfer process has been documented in studies conducted by Bond et al. (2002) and Bond & Lovley (2003). Nevertheless, the scope of electron donors that these organisms can utilize is restricted to uncomplicated organic acids. In contrast, Rhodoferax ferrireducens can efficiently oxidize glucose and other sugars, such as fructose and xylose, and transfer electrons to graphite electrodes in a quantifiable manner. The sugar is metabolized in the anode chamber. One molecule of glucose undergoes oxidation to create carbon dioxide (CO_2), hydrogen ions (H^+), and 24 electrons with an efficiency of approximately 83%. The reaction generates a persistent and stable electric current that continues even after refreshing the glucose-containing liquid in the anode chamber. The anode media may be modified to replenish this microbial fuel cell. The charge/discharge cycle does not exhibit significant capacity fading, but only demonstrates minimal capacity losses during open circuits and extended idle periods (Chaudhuri & Lovley, 2003).Another bacterium that is able to transfer electrons to solid metal oxides isShewanella oneidensisMR-1. In addition, to their remarkable anaerobic versatility, analyses of the genome sequences of Shewanellae species suggest that they can use a broad range of carbon substrates; this creates possibilities for their application in biofuel production (Fredrickson, 2008). Production and storage of electricity are expected to evolve quickly within the new paradigm of emerging bioelectronics (Willner, 2002).

Sol-gels have been demonstrated to be usable for the entrapment of membrane-bound proteins in a physiologically active form and have been proven to be capable of maintaining protein activity over periods of months or more (Luo et al., 2005). Using a membrane- associated F0F1-ATP synthase, Luo et al. (2005) showed that the photo-induced proton gradient can be used to 'store' light energy as ATP. This has the advantage of eliminating passive leakage of ions across the membrane. In addition, ATP can be used for direct powering of motor proteins for the conversion of chemical energy to mechanical energy (Browne & Feringa, 2006). Nano power plants based on the rotation of magnetic bead propellers mounted on F0F1-ATP synthase rotors that are fed by ATP to induce electric current in microarrays of nanostators are now being designed and are in the research and development stage of construction (Soong et al., 2000; Yasuda et al., 2001).

6. OPTIONS FOR GRID CONTRIBUTIONS

Electricity is the fundamental basis of contemporary civilizations, but, over 1.6 billion individuals still lack access to the electrical grid. The majority of this people resides in South Asia and sub-Saharan Africa. If significant governmental incentives are not implemented, approximately 1.4 billion individuals, or for 18% of the global population, will remain without access to electricity by 2030, despite the worldwide economic growth and advancements in energy technologies (Dorian et al., 2006).

The global mean annual power consumption ranges from 2 to 4 terawatts (TW). The price of electricity generated from fossil fuels currently falls within the range of US$ 0.02–0.05 per kilowatt-hour, which includes the costs of storing and distributing the electricity (Lewis & Nocera, 2006). To provide a comparison, the available alternatives for power generation that do not involve biological processes are as follows. (i) The majority of the world's nuclear power plants, which are light-water reactors, generate electricity at a cost of US$ 0.025-0.07/kW/. However, there is no agreement on how to address the issue of nuclear waste that has accumulated in these power plants over the past 50 years (Schiermeier et al., 2008). (ii) Hydroelectric energy sources possess a generating capacity of 800 GW, which is tenfold greater than the combined power of geothermal, solar, and wind energy sources. Currently, hydroelectric sources contribute approximately one-fifth of the global electricity consumption. The annual running costs for such sources range from US$ 0.03 to US$ 0.10 per kilowatt-hour (kW/h), making them cost-competitive with coal and gas. According to Schiermeier et al. (2008), there is potential to increase the amount of energy generated from hydroelectric sources, as only around 30% of the global hydropower capacity is now being utilized. (iii) Wind turbines have the capacity to generate 1,500 kilowatts at a cost of US$ 0.05-0.09 per kilowatt-hour, which makes wind energy economically competitive with coal. Additionally, wind power has the potential to supply up to 20% of the electricity in the system. The European Union has the potential to fulfill 25% of its present electricity demands by harnessing wind power in less than 5% of the North Sea. The EU is making significant investments in this option. (iv) The cost of exploiting and utilizing the most favorable geothermal locations is expected to be around US$ 0.05 per kilowatt-hour. Therefore, 70 gigawatts (GW) of the worldwide heat flow is considered usable for practical purposes. Nevertheless, due to the substantial amount of investment needed, the utilization of geothermal power is not currently a top priority, except in areas with notable volcanic activity (Schiermeier et al., 2008). (iv) The cost of commercial photo-voltaic (PV) electricity is approximately US$ 0.25-0.30/kW/h, which is still tenfold higher than the present price of electricity on the grid.

The existing PV technology is limited to a maximum efficiency of 31% due to theoretical constraints. By converting high-energy photons into electricity instead of heat, it is conceivable to achieve a conversion efficiency above 31%. By utilizing this technology, it is possible to achieve a conversion efficiency of above 60% (Lewis, 2007). Furthermore, the lack of an affordable and efficient means of storing solar-generated electricity is a significant challenge. At now, the most economical approach to capturing, converting, and storing solar energy is through solar thermal technology, which can be as little as US$ 0.10-0.15 per kilowatt-hour for generating electricity. This necessitates the concentration of solar energy for the production of syngas or synfuel (Lewis & Nocera, 2006), or alternatively, the absorption of solar energy by heat-transfer fluids capable of withstanding high temperatures (>427 ºC), resulting in the generation of electricity through steam production (refer to Shinnar & Citro, 2006). Solar power is one of the most promising technologies that does not produce carbon emissions, according to Schiermeier et al. (2008). The Earth annually receives an estimated 100,000 terawatts (TW) of solar energy. Specific regions in the

Sahara Desert, the Gobi Desert in central Asia, the Atacama in Peru, and the Great Basin in the US provide favorable conditions for harnessing solar energy and converting it into power. Solar energy harnessed in an area less than one-tenth the size of the Sahara might meet the global energy demands. Roof surfaces of both residential and commercial buildings are currently being utilized in many nations to enable individuals to sell their own photovoltaic (PV) electricity to the power grid. This practice helps them save significant annual expenses. This sophisticated approach has the potential to be applied to various energy production systems.

The capital expenditures for biomass plants are comparable to those of fossil fuel facilities. The cost of power can be as low as US$ 0.02 per kilowatt-hour when biomass is combusted alongside coal in a typical power station. The costs for a co-generation plant rise to US$ 0.04-0.09/kW/h, but, the utilization of waste heat significantly enhances the efficiency of the process. The first challenge faced by new biomass power plants is the identification of a dependable and concentrated feedstock that is accessible within the immediate vicinity. The production of biomass is constrained by the availability of land surface, the efficiency of photosynthesis, and the water supply. The projected biomass potential is approximately 5 terawatts (Schiermeier et al., 2008). Photosynthesis exhibits a rather low level of efficiency when taking into account the fact that switchgrass, a rapidly growing crop, stores energy in biomass at an average rate of less than 1 watt per square meter per year. Considering that the typical amount of solar radiation received is 200-300 W/m2, it is worth noting that the average annual efficiency of converting and storing energy in the most rapidly developing crops is less than 0.5% (Lewis 2007; Lewis & Nocera, 2006). Genetic engineering has the potential to enhance photosynthetic efficiency (Ragauskas et al., 2006). Another potential issue associated with biomass production is the possibility of a significant rise in water use, maybe by a factor of two to three orders of magnitude. This is a significant factor to address because the essential needs of humans and the creation of power are increasingly vying for water resources (King et al., 2008).

The potential accessibility of wind (Pryor & Barthelmie, 2010), solar, and biomass energy fluctuates based on both temporal and geographical factors. The variance in resources is influenced by individual features such as wind and solar regimes, as well as geographic factors like land use and land cover. Additionally, techno-economic factors such as scale and labor costs, as well as institutional elements like policy regimes and legislation, also contribute to this variation (de Vries et al., 2007). The energy potential per year in a specific region must be combined with the corresponding geographical units within that region. The model presented by de Vries et al. (2007) demonstrated the following: (i) Solar energy is predominantly available in Northern Africa, South Africa, the Middle East, India, and Australia. (ii) Wind energy is primarily concentrated in temperate zones like Chile, Scandinavia, Canada, and the USA. (iii) Biomass can be produced on extensive areas of abandoned agricultural land commonly found in the USA, Europe, the Former Soviet Union (FSU), Brazil, China, as well as on grasslands and savannas in other regions. Renewable energy sources in regions such as India, China, Central America, South Africa, and equatorial Africa are readily accessible at rates below US$ 0.1/kW/h. These places also have a significant existing or anticipated demand for electricity. Utilizing a mix of wind, biomass, and solar energy sources (Eugenia Corria et al., 2006) has the potential to generate cost advantages in transportation and storage systems. Regions having significant ratios of solar, wind, and biomass potential compared to their present electricity demand include Canada (mostly wind), several regions in Africa (solar photovoltaic and wind), the Former Soviet Union (wind and biomass), the Middle East (solar photovoltaic), and Oceania (all energy sources). In regions such as Southeast Asia and Japan, there is a notable disparity between the supply and demand for power generated from solar, wind, and biomass sources. Europe and South Asia exhibit ratios about equal to one. The mentioned potentials are contingent upon several characteristics, and their realization will be contingent upon future land-use regulations (de Vries et al., 2007; Miles & Kapos, 2008).

7. MANAGEMENT AND THE ABILITY TO MAINTAIN LONG-TERM VIABILITY

Adam Smith's concept that an individual's pursuit of self-interest often leads to the betterment of society, even without intending to do so, and Karl Marx's vision of a society where the unrestricted growth of each person is necessary for the unrestricted growth of all, are both constrained by a clear limitation. The world has a limited size. When a particular group of individuals prioritizes their own interests, it adversely affects the interests of others (Vertès et al., 2006). This rationale was employed to build the model of Western economies. The philosophical underpinning of this theory is a mathematical framework derived from energy-conservation equations developed by von Helmholtz in 1847. In this framework, economic variables are substituted for physical variables in an arbitrary manner.

The consequences of this model are as follows:

- The market is a closed circular flux between production and consumption, without inflows or outflows;
- Natural resources are located in a domain that is separate from that of the closed market system;
- The costs of environmental destruction because of economic activities must be considered as unrelated to the closed market system (or at least they cannot be included in the price-formation processes of that system);
- The natural resources that are used by the market system are endless and those that are limited in quantity can be substituted by others that are endless; and
- Biophysical limits to the increase of the market system simply do not exist (Nadeau, 2006). This model is obsolete and is based on hypotheses that have no grounding in scientific bases. Sustainable economic solutions to global warming and environmental destruction are impossible to establish under the logic of this model.

Consequently, the United States alone has achieved a level of oil consumption in the transportation sector that is close to 14 million barrels per day (Mbl/day), resulting in the emission of around 0.53 gigatons of carbon per year (Gt C/yr). According to Agrawal et al. (2007), the present worldwide carbon emissions from fossil fuel consumption are estimated to be 7 gigatons of carbon per year. It is projected that this number will increase to approximately 14 gigatons of carbon per year by 2050. By 2050, it is projected that worldwide energy consumption could range from 30 to 60 terawatts (TW). Given the projected increase in global population to 8 billion by 2030, it is crucial to significantly increase energy consumption in order to sustain economic growth. China, boasting a population of 1.3 billion and a rapidly expanding economy, has surpassed Japan to claim the position of the world's second-largest oil user, trailing only behind the United States. Currently, the Asian superpower has the title of being the largest producer and user of coal (Tollefson, 2008). Additionally, they have made public their plans to build 24-32 new nuclear reactors by the year 2020 (Dorian et al., 2006). Based on the existing trajectory, it is projected that the global energy sector would require an approximate expenditure of $16 trillion within the next 30 years in order to sustain and enhance its energy production. Approximately 65% of the investment will be allocated to the generation, transmission, and distribution of electricity, while the oil and gas sectors will receive around 20% of the global energy investment. Experts predict that the global oil production peak is unlikely to happen for at least another 30-40 years. To provide a clear understanding of global oil requirements, the anticipated increase in oil consumption is expected to go from approximately 80 million barrels per day at present to more than 120 million barrels per day by the year 2030. The OPEC nations are presently functioning at almost maximum capacity, resulting in oil prices surging to US$ 120 per barrel in August 2008. Undoubtedly, it is imperative for the global community to discover and implement more effective methods of energy management. According to Rafaj and Kypreos (2007), it is argued that the increasing global demand for oil may be met by a mix of pricing and technical incentives. Due to the escalating challenges in accessing the remaining oil reserves, the ongoing surge in oil prices is expected to promote the emergence of alternative energy sources (Herrera, 2006). The world should make a transition from oil to renewable energy before depleting its oil supplies (Dorian et al., 2006). Currently, solar, wind, and geothermal energy sources, excluding biofuels, contribute a mere 3% to the global energy demand (Dorian et al., 2006). There should be a global rise in the implementation of solutions that utilize various energy sources, and they should be integrated into the existing electricity grid. Palm oil and sugarcane are now the two most productive sources of renewable biodiesel and bioethanol, respectively, in terms of plant bioenergy production per hectare. Their cultivation is experiencing a steady rise, but the ongoing expansion of their output is not viable and will not address the exponentially growing energy needs. The production of palm oil results in an average yield of around 5,000 liters per hectare. The highest bioethanol production obtained from sugarcane in Brazil is 7,500 liters per hectare. The majority of the energy required for cultivating sugarcane and transforming it into ethanol is obtained through the combustion of its byproducts, such as bagasse. According to Bourne (2007), the production of sugarcane ethanol results in the recovery of around 8 units of energy for every unit of fossil energy consumed. The energy recovery rates from other biofuel crops typically fall below 5%. EU biofuel crops exhibit significantly lower productivity compared to palm oil and sugarcane. Consequently, the implementation of B5 enforcement would need allocating approximately 13% of the arable land in the EU25 for biofuel production. The current condition is significantly unsustainable, being approximately five times less sustainable than desired. When it comes to the environmental impact, ethanol derived from maize, for instance, incurs expenses due to the substantial use of nitrogen fertilizer and the significant erosion of topsoil during cultivation. Annually, pesticides, herbicides, and fertilizers from the corn fields seep into the groundwater. River pollution facilitates eutrophication, the proliferation of algal blooms, and the formation of oxygen-depleted areas known as 'dead zones'. Moreover, the importation of ethanol by industrialized nations has the potential to cause heightened ecological devastation in developing countries. This is due to the clearing of indigenous natural habitats to make way for energy crops, as supported by studies conducted by Gui et al. (2008), Marris (2006), and Thomas (2007).

There is a prevailing belief that first-generation biofuels have reached a point of saturation due to the scarcity of arable lands. Brazil possesses surplus area suitable for cultivating sugarcane and physic nut, whereas India is actively encouraging the development of physic nut on its vast wastelands. Nevertheless, the progress made in the advancement of these fuels has already proven to be successful, as they have shown that motor technology utilizing ethanol or biodiesel is both viable and capable of powering public transportation, at the very least. Fortunately, biofuels derived from biomass in the second generation have supplementary prospects. Lignocellulose has a cheaper feedstock cost compared to agricultural crops, which now make up to 70% of the overall production expenses for first-generation bioethanol. Despite their current higher price, synfuel derived from biomass sources such as poplar, willow, and reed grass may become more cost-effective in the near future compared to fuel derived from sugar beets, wheat, and rapeseed sources (Wesseler, 2007; Styles & Jones, 2008). Biomass fuels present a viable alternative for the European Union to achieve its objective of increasing energy output from renewable sources. However, this objective was not achieved by 2010 as originally anticipated (Fischer et al., 2010;

Havlík et al., 2010). According to Frondel and Peters (2007), the European CO2 emissions-trading system, which involves buying carbon credits, is more cost effective than its biodiesel program. This is because the system allows for the purchase of units of CO2 sequestration in tropical climates, where the rate of CO2 fixation is substantially higher compared to temperate regions. Third-generation biofuels have also joined the competition for fuel sustainability. Sugarcane normally produces approximately 75 metric tons of biomass per hectare, but microalgae can produce twice as much biomass per hectare (Brennan & Owende, 2010; Chisti, 2007, 2008). Assuming a productivity of 150 metric tons per hectare and an average dry-weight oil content of 30%, the oil yield per hectare would be approximately 123 cubic meters over 90% of the year (equivalent to 98.4 cubic meters per hectare). In order to meet the demand of 0.53 gigameters cubed (Gm3) of biodiesel for powering transport vehicles in the United States, it would be necessary to cultivate microalgae over an approximate area of 5.4 million hectares, which represents 3% of the total land area of the US. The cost of producing algal biomass in a facility with an annual capacity of 100 tons has been estimated to be around US$ 3,000 per ton. The viability of extracting oil from microalgal biomass has been proven (Belarbi et al., 2000; Sánchez Mirón et al., 2003). Additionally, most of the leftover biomass following oil extraction may be reused by anaerobic digestion to generate biogas.

The primary obstacles to the widespread production of microalgae are mostly economic in nature and are closely linked to the financial investments needed for algae farming. An approach to address this issue would involve enhancing oil productivity by the application of genetic and metabolic engineering techniques (León-Bañares et al., 2004; Mathews & Wang, 2009). The spread of algal technology through CO2 filtration is anticipated because to its immediate integration potential into power plant management systems. This technology is anticipated to gradually disseminate as experience accumulates. Approximately 50% of global oil usage is allocated to the transportation industry, which is also responsible for 32% of greenhouse gas (GHG) emissions. The energy conversion efficiency for work in the transportation sector is often lower compared to large-scale power plants. The objective is to enhance this efficiency from the existing range of 15-35% to 60-80% (Song, 2006). Regrettably, it is improbable that advanced transportation technologies, such as hydrogen fuel cell vehicles and alternative fuels like gas-to-liquids, coal-to-liquids, and biodiesels, will have a substantial impact on the conventional transportation fuel industry before 2030, except in specific regions. The proliferation of fourth-generation technology, such as hybrids and fuel cell cars, may impede the growth of oil usage for transportation purposes in the future. Nevertheless, the essential advancements in technology will not happen unless there are extraordinary policy measures taken globally to encourage the use and integration of new technologies into daily life (Doniger et al., 2006; Haug et al., 2011; Michel 2009). Presently, the global count of hybrids stands at around 500,000, while the number of sophisticated clean-diesel engines reaches approximately 30 million. Hybrid cars are experiencing an increase in popularity in the United States and Japan, whereas advanced clean-diesel engines are primarily found in Europe (Dorian et al., 2006). Auto-mobility is a complex and dynamic system that relies on the interaction of automobiles, drivers, roads, fuel supply, and other items and technologies. Auto-mobility has been intricately linked to modern social life. Nevertheless, this form of transportation is not essential or unavoidable in society (Urry, 2008). A total of one billion vehicles were manufactured throughout the previous century. According to Hawken et al. (2002), the number of people traveling by car worldwide is expected to quadruple from 1990 to 2050. Currently, global inhabitants collectively travel a total distance of 23 gigakilometers every year. The use of automobiles requires individuals to deal with the limitations in time and space that it creates (Mills et al., 2010). Thankfully, several projects that were initiated 35 years ago are now being implemented, specifically the incorporation of automobile and bicycle rentals into public transportation networks, as seen in certain European towns. In a world without cars, there will be significant alterations in lifestyles, city architecture, cognitive processes, and social customs. Enhancing active transport, such as walking and bicycling, will lead to significant carbon reductions and improve public health. Urban areas necessitate secure and enjoyable surroundings for active transportation, as well as convenient access to public transportation. Transportation can have negative impacts on health, such as traffic injuries, lack of physical activity (which contributes to the high cost of obesity in the USA, estimated at roughly US$ 139 billion per year), urban air pollution, disputes connected to energy use, and environmental degradation. Urban air pollution is responsible for 750,000 deaths annually, with 530,000 of those occurring in Asia (Woodcock et al., 2007). Due to the scarcity of energy resources, there is a contention that the world will need to transition towards virtual forms of travel, such as internet browsing, virtual sensory experiences, and video conferences, in order to minimize physical travel (Moriarty & Honnery, 2007). The predicament described above is a consequence of considering people solely within social circumstances, without taking into account the essential environmental perspective (Thomas, 2007). The concept of environmental crime is now in its nascent stage and has not yet been widely implemented. Rational human communities should consider the necessary land area to adequately support both human people and wildlife. Failing to do so will result in escalating global devastation (Urry, 2008) and jeopardize the prospects of human civilization. The Gaia principle (Lovelock & Margulis, 1974) was formulated based on these arguments. This principle asserts that the planet Earth should be regarded as a whole entity, and as a result, the destruction of one ecosystem might have an impact on all other ecosystems. The recognition of the importance of ecosystems has emerged relatively recently (Costanza et al., 1997). The issue of human integration with the environment has gained attention in society primarily due to the imminent danger of global warming and its potentially catastrophic consequences (Stern, 2006). The authors (Mäler et al., 2008) have also suggested a conversation about the economic measurement of ecosystem services, considering sustainable development. The idea of "willingness-to-pay" (WTP) has been recently established. This approach enables the quantification of personal choice in monetary terms to prevent any adverse consequences. Its objective is to assess the need for enhanced environmental conditions. WTP quantifies the amount individuals are willing to spend in order to enhance their own or others' quality of life. The aggregate Willingness to Pay (WTP) of all persons represents the monetary value that a collective of individuals is willing to contribute in order to preserve their environment in an unaltered condition. For instance, the routes taken by harmful compounds are traced from where they are released to the locations where they cause damage, together with the corresponding expenses of repairing the damage. Incorporating external costs into the total cost of energy production allows for the calculation of the true cost of an activity and provides an effective policy tool for mitigating the adverse effects of energy use (Nast et al., 2007). The method of combining production costs with external expenses to calculate a total specific cost is used as a comparison measure to assess the economic and environmental performance of different energy options and technologies (Rafaj & Kypreos, 2007). The scenarios presented in this new cost-accounting technique demonstrate the potential for the spread of innovative technologies and the transition to alternative fuels in the power generation system. By adopting this approach, renewable energy enhance their competitiveness and significantly reduce the reliance of the electrical sector on fossil fuels. In addition, there will be a reduction of 70-85% in emissions of SO_2 and NOx by the year 2030. While the analysis suggests that advanced technologies equipped with emission controls and carbon sequestration will experience substantial cost reductions and eventually become economically viable, it is essential to have policies in place to support the adoption of these technologies in electricity markets, particularly during their initial phase of market entry. This model pertains to policy initiatives aimed at promoting technical advancement through investments in research and development. These efforts support the evolution of carbon-free technologies along their required learning curves (Haug et al., 2011; Rafaj & Kypreos, 2007).

8. CONCISE SUMMARY

It is now necessary to combine the fields of technology and social sciences in order to discover a path towards achieving environmental and economic sustainability on our planet. In the absence of a resolution, economic expansion will come at the expense of the number of the human population (Urry, 2008). Due to the ongoing rise in fossil fuel prices, research on renewable energy sources has become economically feasible. It is evident that renewable energy technologies have reached a critical point where there is no possibility of reverting back. There are a minimum of five regional blocs, including the USA, the EU, China, Brazil, and India, that have a keen interest in reducing their reliance on fossil fuels. There is no apparent benefit to stopping this process by using aggressive tactics such as reducing oil prices and flooding the market. Biotechnology is closely linked to agricultural processes, which are likewise backed by governments due to geostrategic concerns. Furthermore, the evident progression of climate change will soon transcend individual interests and become a universal issue for all of humanity. Biofuels and sources of bioenergy will undergo a series of rapid technological advancements and innovations before reaching their ultimate states. Bioethanol derived from biomass is anticipated to overtake bioethanol derived from sweet crops. The gradual substitution of plant biodiesel with synfuel derived from biomass and solar energy is also recommended. Biotechnology is anticipated to play a larger role in the manufacture of microdiesel fuel, the genetic modification of plants and microorganisms, and the use of enzymes in nanotechnology. The incorporation of renewable energies into the electricity grid is in its early stages, however it is already advancing swiftly. In order to ensure that technological advancements and developments have a positive impact, it is crucial to implement energy management and urbanization policies to prevent excessive energy waste that could undermine their benefits. Emerging ideas, such as willingness-to-pay, carbon credits, and external costs, are currently being included when calculating the life cycles of energy. The scope of this toolkit will broaden as government rules become more extensive. It should encompass essential principles like "biodiversity credits" and the establishment of a "minimal territorial unit" for living organisms to ensure the long-term viability of both wildlife and people. Biodiversity serves as a reservoir of nanostructures and nanomachines. It would be unwise to destroy it without careful thought, considering that it took three billion years to create and humanity is only just beginning to explore it. Due to energy conservation regulations, upcoming cars will operate on a combination of gasoline combustion and electricity. These choices have the potential to decrease the usage of fossil fuels and the release of greenhouse gases by 30 to 50%, without the need for significant modifications to vehicles. Furthermore, they will facilitate the connection to the electricity grid, resulting in substantial savings on electricity use at an extra expense. These purported plug-in hybrids are expected to achieve a travel distance that is three to four times greater per kilowatt-hour (kW/h) compared to conventional automobiles. These advanced hybrids should ideally possess flexibility and the ability to operate on mixtures of bio/fossil fuels as well as gas (Romm, 2006). In the first half of this century, there will be a shift from fossil fuels to a non-carbon-based global economy. This transition will have a significant impact on the future civilization experienced by upcoming generations (Dorian et al., 2006).

REFERENCES

Abollé, A., Kouakou, L. & Planche, H. 2008. The viscosity of diesel oil and mixtures with straight vegetable oils: Palm, cabbage palm, cotton, groundnut, copra and sunflower.Biomass Bioenergy33:1116-1121.

Adamson, K.A. 2004. Hydrogen from renewable resources - the hundred year commitment. Energ. Policy32:1142–1231.

Agarwal, A.K. 2007. Biofuels (alcohols and biodiesel) applications as fuels for internal combustion engines. Prog. Energ. Combust. Sci. 33: 233-271.

Agrawal, R., Singh, N.R., Ribeiro, F.H. & Delgass, W.N. 2007. Sustainable fuel for the transportation sector. Proc. Natl. Acad. Sci. USA104:4828-4833.

Agrawal,A.,Chhatre,A.&Hardin,R.2008.ChangingGovernanceoftheWorld'sForests.Science320:1460-1462.

Ajanovic, A. 2011. Biofuels versus food production: Does biofuels production increase food prices?Energy, doi:10.1016/j.energy.2010.05.019

Alvarez, H.M. & Steinbuchel, A. 2002. Triacylglycerols in prokaryotic microorganisms.Appl. Microbiol. Biot. 60:367-376.

Arechederra, R.L., Treu, B.L. & Minteer, S.D. 2007. Development of glycerol/O_2biofuel cell. J. Power Sources173:156-161.

Arthur, R., Baidoo, M.F. & Antwi, E. 2011. Biogas as a potential renewable energy source: A Ghanaian case study.Renew. Energ. 36:1510-1516.

Asam, Z.-u.-Z. 2011. How can we improve biomethane production per unit of feedstock in biogas plants?Appl. Energ., doi:10.1016/j.apenergy.2010.12.036

Avery, W. 2002. Ocean-thermal energy conversion. In: Encyclopedia of Physical Science and Technology, edited by Meyers, R.A., 123- 160. San Diego: 3rd ed. Academic Press.

Baker, E. & Keisler, J.M. 2011. Cellulosic biofuels: Expert views on prospects for advancement. Energy36:595-605.

Barnet, A. 2008. Europe's 2020 vision.Nat. Rep. Clim. Change2:36.

Barros, G.S.A.C., Silva, A.P., Ponchio, L.A., Alves, L.A., Osaki, M. & Cenamo, M. 2006. Custos de produção de biodiesel no Brasil.Revista de Política Agrícola3:36-50.

Beer, L.L., Boyd, E.S., Peters, J.W. & Posewitz, M.C. 2009. Engineeringalgaefor biohydrogen and biofuel production.Curr. Opin. Biotech. 20:264–271.

Begum, Z.N. 2005. Modeling of chemical tracer transport in the atmospheric environment and its impact on the global climate.J. Quant. Spectrosc. Radiat. Transfer95:423-427.

Belarbi, E.-H., Molina Grima, E. & Chisti, Y. 2000. A process for high yield and scaleable recovery of high purity eicosapentaenoic acid esters from microalgaeand fish oil.Enzyme Microb. Tech.26:516-529.

Bertram, C. 2010. Ocean iron fertilization in the context of the Kyoto protocol and the post- Kyoto process. Energ. Policy38:1130–1139

Bindraban, P.S., Bulte, E.H. & Conijn, S.G. 2009. Can large-scale biofuels production be sustainable by 2020?Agr. Syst. 101:197–199

Bitsche, O. & Gutmann, G. 2004. Systems for hybrid cars.J. Power Sources127:8–15.

Bond, D.R., Holmes, D.E., Tender, L.M. & Lovley, D.R. 2002. Electrode-reducing microorganisms that harvest energy from marine sediments.Science295:483-485.

Bond, D.R. & Loveley, D.R. 2003. Electricity production byGeobacter sulfurreducensattached to electrodes. Appl. Environ. Microb. 69:1548-1555.

Bourne, J.K.Jr. 2007 Biofuels: green dreams.National Geographic Magazine(October), 41–59.

Brennan, L. & Owende, P. 2010. Biofuels from microalgae- A review of technologies for production, processing, and extractions of biofuels and co-products.Renew. Sustain. Energ. Rev. 14:557-577

Browne, W.R. & Feringa, B.L. 2006. Making molecular machines work.Nature Nanotechnology1:25-35.

Canakci, M., Erdil, A. & Arcaklioglu, E. 2006. Performance and exhaust emissions of a biodiesel engine. Appl. Energ. 83, 594-605.

Certik, M. & Shimizu, S. 1999. Biosynthesis in microorganisms being used for single cell oil production. Biochimie87:1-14.

Ceschia, E. et al. 2010. Management effects on net ecosystem carbon and GHG budgets at European crop sites.Agricult. Ecosys. Environ. 139:363–383.

Chapple, C., Ladisch, M. & Meilan, R. 2007. Loosening lignin's grip on biofuel production. Nat. Biotechnol. 25:746-748.

Chaudhuri, S.K. & Lovley, D.R. 2003. Electricity generation by direct oxidation of glucose in mediatorless microbial fuel cells.Nat. Biotechnol. 21:1229-1232.

Chen, F. & Dixon, R.A. 2007. Lignin modification improves fermentable sugar yields for biofuel production. Nat. Biotechnol. 25:759-761.

Cherubini, F. 2010. GHG balances of bioenergy systems - Overview of key steps in the production chain and methodological concerns.Renew. Energ.35:1565–1573.

Chisti, Y. 2007. Biodiesel from microalgae.Biotechnol. Adv. 25:294-306.

Coleman, H.D., Park, J.-Y., Nair, R., Chapple, C. & Mansfield, S.D. 2008. RNAi-mediated suppression of p-coumaroyl-CoA 3'-hydroxylase in hybrid poplar impacts lignin deposition and soluble secondary metabolism.Proc. Natl. Acad. Sci. USA105:4501- 4506.

Costanza, R. et al. 1997. The value of the world's ecosystem services and natural capital. Nature387, 253-260.

Cropper, M., Geiger, S. & Jollie, D. 2004. Fuel cells: a survey of current developments.J. Power Sources131:57–61.

Damartzis, T. & Zabaniotou, A. 2011. Thermochemical conversion of biomass to second generation biofuels through integrated process design-A review.Renew. Sustain. Energ. Rev. 15:366–378.

Darussalam, B. 2007. Last-Gasp effort to save Borneo's tropical rainforests.Science317:192.

Deluga, G.A., Salge, J.R., Schmidt, L.D. & Verykios, X.E. 2004. Renewable hydrogen from ethanol by autothermal reforming.Science13:993-997.

Demirbas, M.F. & Balat, M. 2006. Recent advances on the production and utilization trends of bio-fuels: A global perspective.Energ. Convers. Manage.47:2371-2381.

Demirbas, A. 2007. The influence of temperature on the yields of compounds existing in bio- oils obtaining from biomass samples via pyrolysis.Fuel process. Technol.88:591-597.

Demirbas, A. 2008a. Biofuels sources, biofuel policy, biofuel economy and global biofuel projections.Energ. Convers. Manage.49:2106-2116.

Demirbas, A. 2008b. Comparison of transesterification methods for production of biodiesel from vegetable oils and fats.Energ. Convers. Manage.49:125-130.

Demirbas A. 2010. Use ofalgaeas biofuel sources.Energ. Convers. Manage. 51:2738–2749.

de Vries, B.J.M., van Vuuren, D.P. & Hoogwijk, M.M. 2007. Renewable energy sources: Their global potential for the first-half of the 21st century at a global level: An integrated approach.Energ. Policy35:2590-2610.

Dijkstra, A.J. 2006. Revisiting the formation of trans isomers during partial hydrogenation of triacylglycerol oils.Eur. J. Lipid Sci. Tech. 108:249-264.

Doniger, D.D., Herzog, A.V. & Lashof, D.A. 2006. An ambitious, centrist approach to global warming legislation.Science314:764-765.

Donner, S.D. & Kucharik, C.J. 2008. Corn-based ethanol production compromises goal of reducing nitrogen export by the Mississippi River.Proc. Natl. Acad. Sci. USA 105:4513-4518.

Dorian, J.P., Franssen, H.T. & Simbeck, D.R. 2006. Global challenges in energy.Energ. Policy34:1984- 1991.

Easterling, E.R., French, W.T., Hernandez, R. & Licha, M. 2008. The effect of glycerol as a sole and secondary substrate on the growth and fatty acid composition ofRhodotorula glutinis.Bioresource Technol. 100:356-61

Eugenia Corria, M., Melian Cobas, V. & Silva Lora, E. 2006. Perspectives of Stirling engines use for distributed generation in Brazil.Energ. Policy34:3402-3408.

Fagernäs, L., Brammer, J., Wilén, C., Lauer, M.& Verhoeff, F. 2010 Drying of biomass for second generation synfuel production.Biomass Bioenergy34:1267-1277.

Fairless, D. 2007. The little shrub that could - maybe.Nature449:652-655.

Fargione, J., Hill, J., Tilman, D., Polasky, S. & Hawthorne, P. 2008. Land clearing and the biofuel carbon debt.Science319:1235-1238.

Fenning, T.M., Walter, C. & Gartland, K.M.A. 2008. Forest biotech and climate change.Nat. Biotechnol. 26:615-616.

Fialkov, A.B., Gordin, A. & Amirav, A. 2008. Hydrocarbons and fuels analyses with the supersonic gas chromatography mass spectrometry - The novel concept of isomer abundance analysis.J. Chromatogr. A 1195:127-135.

Fischer, C.R., Klein-Marcuschamer, D. & Stephanopoulos, G. 2008. Selection and optimization of microbial hosts for biofuel production.Metab. Eng.10:295–304.

Fischer, G., Prieler, S., van Velthuizen, H., Lensink, S.M., Londo, M. & de Wit, M. 2010. Biofuel production potentials in Europe: Sustainable use of cultivated land and pastures. Part I: Land productivity potentials.Biomass Bioenergy34:159 - 172.

Flavin, C. 2006. American Energy: The Renewable Path to Energy Security. 40 p. Washington DC: Worldwatch Institute.

Folkesson, A., Andersson, C., Alvfors, P., Alaküla, M. & Overgaard, L. 2003. Real life testing of a Hybrid PEM Fuel Cell Bus.J. Power Sources118:349-357.

Fredrickson, J.K. et al. 2008. Towards environmental systems biology ofShewanella.Nat. Rev. Microbiol.6:592-603.

Frondel, M. & Peters, J. 2007. Biodiesel: a new Oildorado?Energ. Policy35:1675-84.

García-Montero, L.G., López, E., Monzón, A. & Pastor, I.O. 2010. Environmental screening tools for assessment of infrastructure plans based on biodiversity preservation and global warming (PEIT, Spain).Environ. Impact Assess. Rev.30:58–168.

Gavrilescu, M. & Chisti, Y. 2005. Biotechnology - a sustainable alternative for chemical industry.Biotechnol. Adv. 23:471-499.

Gielen, D.J., Fujino, J., Hashimoto, S. & Moriguchi, Y. 2002. Biomass strategies for climate policies?Climate Policy2:319–333.

Glibert, P.M. et al. 2008. Ocean urea fertilization for carbon credits poses high ecological risks.Mar. Pollut. Bull. 56:1049-1056.

Gnansounou, E., Dauriat, A., Villegas, J. & Panichelli, L. 2009. Life cycle assessment of biofuels: Energy and greenhouse gas balances.Bioresource Technol. 100:4919–4930.

Goering, E., Schwab, W., Daugherty, J., Pryde, H. & Heakin, J. 1982. Fuel properties of eleven vegetable oils. Transactions ASABE25:1472-1483.

Goldemberg, J., Coelho, S.T. & Lucon, O. 2004. How adequate policies can push renewables. Energ. Policy32:1141-1146.

Gressel, J. 2008. Transgenics are imperative for biofuel crops.Plant Sci. 174:246-263.

Greyvenstein, R., Correia, M. & Kriel, W. 2008. South Africa's opportunity to maximise the role of nuclear power in a global hydrogen economy.Nucl. Eng. Des.238:3031-3040.

Gui, M.M., Lee, K.T. & Bhatia, S. 2008. Feasibility of edible oil vs. non-edible oil vs. waste edible oil as biodiesel feedstock.Energy33:1646-1653.

Gummer, J.H. & Head, C.R. 2003. Hydrogen, hydropower and world poverty.Int. J. Hydropower Dams5:67-71.

Haag, A.L. 2007.Algaebloom again.Nature447:520-521.

Hagerman, S., Dowlatabadi, H., Satterfield, T. & McDaniels, T. 2010. Expert views on biodiversity conservation in an era of climate change.Global Environ. Change 20:192–207.

Hall, J., Matos, S., Severino, L. & Beltrão, N. 2009. Brazilian biofuels and social exclusion: established and concentrated ethanol versus emerging and dispersed biodiesel.J. Clean. Prod. 17:S77–S85.

Haug, C., Huitema, D. & Wenzler, I. 2011. Learning through games? Evaluating the learning effect of a policy exercise on European climate policy.Technol. Forecast. Soc. Change, doi:10.1016/j.techfore.2010.12.001

Havlík, P. et al. 2010. Global land-use implications of first and second generation biofuel targets.Energ. Policy, doi:10.1016/j.enpol.2010.03.030

Hawken, P., Lovins, A. & Lovins, L.H. 2002. Natural Capitalism: Creating the Next Industrial Revolution. London: Earthscan. Accessed February 2, 2010. http://www.natcap.org/.

Herrera, S. 2006. Bonkers about biofuels.Nat. Biotechnol. 24:755-760.

Heyer, N.I. & Woodward, J. 2001. Efficient hydrogen production using enzymes of the pentose phosphate cycle. Proceedings of the 2001 DOE Hydrogen Program Review, NREL/CP-570-30535. Accessed February 2, 2010. http://www1.eere.energy.gov/hydrogenandfuelcells/pdfs/30535g.pdf.

Hill, J., Nelson, E., Tilman, D., Polasky, S. & Tiffany D. 2006. Environmental, economic, and energetic costs and benefits of biodiesel and ethanol biofuels.Proc. Natl. Acad. Sci. USA103:11206-11210.

Himmel, M.E., Ding, S.-Y., Johnson, D.K., Adney, W.S., Nimlos, M.R., Brady, J.W. & Foust, T.D. 2007. Biomass Recalcitrance: Engineering Plants and Enzymes for Biofuels Production.Science315:804-807.

Hissel, D., Péra, M.C. & Kauffmann, J.M. 2004. Diagnosis of automotive fuel cell power generators.J. Power Sources128:239-246.

Hopkin, M. 2007. Climate sceptics switch focus to economics.Nature445:582-583.

Howden, S.M., Soussana, J.-F., Tubiello, F.N., Chhetri, N., Dunlop, M. & Meinke, H. 2007. Adapting agriculture to climate change.Proc. Natl. Acad. Sci. USA104:19691-19696.

Huston, M.A. & Marland, G. 2003. Carbon management and biodiversity.J. Environ. Manage. 67:77-86.

Johnston, M. & Holloway, T. 2007. A global comparison of national biodiesel production potentials.Environ. Sci. Technol41:7967-7973.

Johnston, C.D., Bastow, T.P. & Innes, N.L. 2007. The use of biodegradation signatures and biomarkers to differentiate spills of petroleum hydrocarbon liquids in the subsurface and estimate natural mass loss.Eur. J. Soil Biol. 43:328-334.

Jones, D.B. 2010. Proposal for an international carbon fee.Energy for Sustainable Development 14:245–250.

Kalscheuer, R., Stoveken, T. & Steinbuchel, A. 2006a. Microdiesel:Escherichia coliengineered for fuel production.Microbiology152:2529-2536.

Kalscheuer, R., Stoveken, T., Luftmann, H., Malkus, U., Reichelt, R. & Steinbuchel, A. 2006b. Neutral lipid biosynthesis in engineeredEscherichia coli: Jojoba oil-like wax esters and fatty acid butyl esters. Appl. Environ. Microbiol. 72:1373-1379.

Kaplan, I.R., Galperin, Y., Lu, S.-T. & Lee, R.-P. 1997. Forensic environmental geochemistry: differenciation of fuel-types, their sources and release time.Org. Geochem. 27:289- 317.

Khaselev, O. & Turner, J. 1998. A monolithic photovolatic–photochemical device for hydrogen production via water splitting.Science280, 425-427.

Kikuchi, R. 2006. Penetration of hydrogen-based energy system and its potential for causing global environmental change: Scoping risk analysis based on life cycle thinking.Environ. Impact Asses. Rev. 26:206-218.

Kindermann, G., Obersteiner, M., Sohngen, B., Sathaye, J., Andrasko, K., Rametsteiner, E., Schlamadinger, B., Wunder, S. & Beach, R. 2008. Global cost estimates of reducing carbon emissions through avoided deforestation.Proc. Natl. Acad. Sci. USA105:10302-10307.

King, C.W., Holman, A.S. & Webber, M.E. 2008. Thirst for energy.Nat. Geosci.1:283-286.

Kirilenko, A.P. & Sedjo, R.A. 2007. Climate change impacts on forestry.Proc. Natl. Acad. Sci. USA104:19697-19702.

Kleiner, K. 2007. Civil aviation faces green challenge.Nature448:120-121.

Kleiner, K. 2008. The backlash against biofuels.Nat. Rep. Clim. Change2:9-11, doi:10.1038/climate.2007.71.

Kremer, F.G. & Fachetti, A. 2000 Alcohol as automotive fuel-Brazilian experience. SAE Technical Paper No. 2000-01-1965, doi:10.4271/2000-01-1965

Kumar, A. & Sharma, S. 2008. An evaluation of multipurpose oil seed crop for industrial uses (Jatropha curcasL.): A review.Ind. Crop. Prod. 28:1-10.

Lamy, C., Rousseau, S., Belgsir, E.M., Coutanceau, C. & Léger, JM. 2004. Recent progress in the direct ethanol fuel cell: development of new platinum–tin electrocatalysts.Electrochim. Acta49:3901-3908.

Laurance, W.F. 2007. Switch to corn promotes amazon deforestation.Science318:1721.

Leibold, H., Hornung, A. & Seifert, H. 2008. HTHP syngas cleaning concept of two stage biomass gasification for FT synthesis.Powder Technol. 180:265-270.

León-Bañares, R., González-Ballester, D., Galván, A. & Fernández, E. 2004. Transgenic microalgaeas green cellfactories.Trends Biotechnol. 22:45-52.

Lewis, N.S. & Nocera, D.G. 2006. Powering the planet: Chemical challenges in solar energy utilization.Proc. Natl. Acad. Sci. USA103:15729-15735.

Lewis, N.S. 2007. Toward cost-effective solar energy use.Science315:798-801.

Lima, D.G. et al. 2004. Diesel-like fuel obtained by pyrolysis of vegetable oils.J. Anal. Appl. Pyrol. 71:987-996.

Lin, C.-Y. et al. 2010. A pilot-scale high-rate biohydrogen production system with mixed microflora.Int. J. Hydrogen Energ., doi:10.1016/j.ijhydene.2010.07.115.

Liu, X., Li, X., Suehiro, Y. & Fujimoto, K. (2007). Elucidation of reaction network and effective control of carbon number distribution in the three phase Fischer-Tropsch synthesis.Appl. Catal. A-Gen.333:211-218.

Lois, E. 2007. Definition of biodiesel.Fuel86:1212-1213.

Lokey, E. 2007. How the next US president should slow global warming.Energ. Policy 35:5399-5402.

Lonngren, K.E. & Bai, E.-W. 2008. On the global warming problem due to carbon dioxide. Energ. Policy36:1567-1568.

Lovelock, J.E. & Margulis, L. 1974. Atmospheric homeostasis by and for the biosphere- The Gaia hypothesis. Tellus26:2-10.

Lü, X-c, L., Yang, J-g, Zhang, W-g, Z. & Huang, Z. 2004. Effect of cetane number improver on heat release rate and emissions of high speed diesel engine fueled with ethanol– diesel blend fuel.Fuel83:2013-2020.

Luo, T.-J.M., Soong, R., Lan, E., Dunn, B. & Montemagno, C. 2005. Photo-induced proton gradients and ATP biosynthesis produced by vesicles encapsulated in a silica matrix.Nat. Materials4:220–224.

Malça, J. & Freire, F. 2011. Life-cycle studies of biodiesel in Europe: A review addressing the variability of results and modeling issues.Renew. Sustain. Energ. Rev. 15:338-351.

Mäler, K.-G., Aniyar, S. & Jansson, Å. 2008. Accounting for ecosystem services as a way to understand the requirements for sustainable development.Proc. Natl. Acad. Sci. USA105:9501-9506.

Malhi, Y., Roberts, J.T., Betts, R.A., Killeen, T.J., Li, W. & Nobre, C.A. 2008. Climate change, deforestation, and the fate of the Amazon.Science319:169-172.

Mander, U. et al. 2010. Assessment of methane and nitrous oxide fluxes in rural landscapes. Landscape Urban Plan. 98:172–181.

Marchetti, J., Miguel, V. & Errazu, A. 2007. Possible methods for biodiesel production. Renew. Sustain. Energ. Rev. 11:1300-1311.

Marcinkoski, J., Kopasz, J.P. & Benjamin, T.G. 2008. Progress in the US DOE fuel cell subprogram efforts in polymer electrolyte fuel cells.Int. J. Hydrogen Energ.33:3894- 3902.

Markels, M. & Barber, R.T. 2001. Sequestration of carbon dioxide by ocean fertilization. Preprints - American Chemical Society, Division of Petroleum Chemistry 46:45-48.

Markou, G. & Georgakakis, D. 2011. Cultivation of filamentous cyanobacteria (blue-greenalgae) in agro-industrial wastes and wastewaters: A review.Appl. Energ., doi:10.1016/j.apenergy.2010.12.042

Marris, E. 2006. Drink the best and drive the rest.Nature444:670-672.

Mathews, J. & Wang, G. 2009. Metabolic pathway engineering for enhanced biohydrogen production.Int. J. Hydrogen Energ. 34:7404 – 7416.

Meng, X., Yang, J., Xu, X., Zhang, L., Nie, Q. & Xian, M. 2008. Biodiesel production from oleaginous microorganisms.Renew. Energ. 34:1-5

Michel, D. 2009. Foxes, hedgehogs, and greenhouse governance: Knowledge, uncertainty, and international policy-making in a warming World.Appl. Energ. 86:258–264.

Miles, L. & Kapos, V. 2008. Reducing greenhouse gas emissions from deforestation and forest degradation: global land-use implications.Science320:1454-1455.

Mills, G. et al. 2010. climate information for improved planning and management of mega cities (needs perspective).Procedia Environmental Sciences1:228–246

Moreira, J.R. & Goldemberg, J. 1999. The alcohol program.Energ. Policy27:229-245. Moriarty, P. & Honnery, D. 2007. The prospects for global green car mobility.J. Clean. Prod. 16:1717-1726.

Nadeau, R. 2006. The Environment Endgame: Mainstream Economics, Ecological Disaster, and Human Survival. 214 p. London: Rutgers University Press.

Nass, L.L., Pereira, P.A.A. & Ellis, D. 2007. Biofuels in Brazil: An Overview.Crop Sci. 47:2228-2237.

Nast, M., Langniss, O. & Leprich, U. 2007. Instruments to promote renewable energy in the German heat market-Renewable Heat Sources Act.Renew. Energ. 32:1127–1135.

Odling-Smee, L. 2007. Biofuels bandwagon hits a rut.Nature446:483.

Okereke, C. & Dooley, K. 2010. Principles of justice in proposals and policy approaches to avoided deforestation: Towards a post-Kyoto climate agreement.Global Environ. Change20:82–95.

Om Tapanes, N.C., Gomes Aranda, D.A., de Mesquita Carneiro, J.W. & Ceva Antunes, O.A. 2008. Transesterification ofJatropha curcasoil glycerides: Theoretical and experimental studies of biodiesel reaction.Fuel87:2286-2295.

Orellana, C. & Neto, R.B. 2006. Brazil and Japan give fuel to ethanol market.Nat. Biotechnol. 24:232.

Parker, J. 2002. Turning manure into gold.EMBO rep. 3:1114-1116.

Parry, M., Palutikof, J., Hanson, C. & Lowe, J. 2008. Squaring up to reality.Nature Reports Climate Change2, 68-70.

Peters, W. et al. 2007. An atmospheric perspective on North American carbon dioxide exchange: CarbonTracker.Proc. Natl. Acad. Sci. USA104:18925-18930.

Pousa, G.P.A.G., Santos, A.L.F. & Suarez, P.A.Z. 2007. History and policy of biodiesel in Brazil.Energ. Policy35:5393-5398.

Proctor, M.R., Taylor, E.J., Nurizzo, D., Turkenburg, J.P., Lloyd, R.M., Vardakou, M., Davies, G.J. & Gilbert, H.J. 2005. Tailored catalysts for plant cell-wall degradation: Redesigning the exo/endo preference ofCellvibrio japonicusarabinanase 43A.Proc. Natl. Acad. Sci. USA102:2697-2702.

Pryor, S.C. & Barthelmie, R.J. 2010. Climate change impacts on wind energy: A review. Renew. Sustain. Energ. Rev. 14:430–437.

Pulz, O. & Gross, W. 2004. Valuable products from biotechnology of microalgae.Appl. Microbiol. Biot. 65:635-48.

Rafaj, P. & Kypreos, S. 2007. Internalisation of external cost in the power generation sector: Analysis with global multi-regional MARKAL model.Energ. Policy35:828-843.

Ragauskas, A.J. et al. 2006. The path forward for biofuels and biomaterials.Science311:484- 489.

Raghu, S., Anderson, R.C., Daehler, C.C., Davis, A.S., Wiedenmann, R.N., Simberloff, D. & Mack, R.N. 2006. Adding biofuels to the invasive species fire?Science313:1742.

Ramadhas, A.S., Jayaraj, S. & Muraleedharan, C. 2005. Biodiesel production from high FFA rubber seed oil. Fuel84:335-340.

Ratledge, C. 2002. Regulation of lipid accumulation in oleaginous micro-organisms.Biochem. Soc. Trans. 30:1047-1050.

Reay, D. 2007. Spring-time for sinks.Nature446:727-728.

Ridgwell, A.T. & Zeebe, R.E. 2005. The role of the global carbonate cycle in the regulation and evolution of the Earth system.Earth Planet. Sci. Lett. 234:299-315.

Romm, J. 2006. The car and fuel of the future.Energ. Policy34:2609-2614.

Ross, D.K. 2006. Hydrogen storage: The major technological barrier to the development of hydrogen fuel cell cars.Vacuum80:1084-1089.

Rothengatter, W. 2010. Climate change and the contribution of transport: Basic facts and the role of aviation.Transportation ResearchPart D 15:5–13.

Rout, U.K., Akimoto, K., Sano, F., Oda, J., Homma, T. & Tomoda, T. 2008. Impact assessment of the increase in fossil fuel prices on the global energy system, with and without CO_2 concentration stabilization. Energ. Policy36:3477-3484.

Sánchez Mirón, A. Ceron Garcia, M.C., Contreras Gomez, A., Garcia Camacho, F., Molina Grima, E. & Chisti, Y. 2003. Shear stress tolerance and biochemical characterization ofPhaeodactylum tricornutumin quasi steady-state continuous culture in outdoor photobioreactors.Biochem. Eng. J. 16:287-297.

Sanderson, K. 2006. A field in ferment.Nature444:673-676.

Scarlat, N. & Dallemand, J.-F. 2010. Recent developments of biofuels/bioenergy sustainability certification: A global overview.Energ. Policy, doi:10.1016/ j.enpol.2010.12.039

Scharlemann, J.P.W. & Laurance, W.F. 2008. How green are biofuels?Science319:43-44. Schiermeier, Q. 2008. Europe spells out action plan for emissions targets.Nature451:504-505.

Schiermeier, Q., Tollefson, J., Scully, T., Witze, A. & Morton, O. 2008. Electricity without carbon. Nature454:816-823.

Schmer, M.R., Vogel, K.P., Mitchell, R.B. & Perrin, R.K. 2008. Net energy of cellulosic ethanol from switchgrass.Proc. Natl. Acad. Sci. USA105:464-469.

Schultz, M.G., Diehl, T., Brasseur, G.P. & Zittel, W. 2003. Air pollution and climate-forcing impacts of a global hydrogen economy.Science302:624-627.

Searchinger, T. et al. 2008. Use of U.S. croplands for biofuels increases greenhouse gases through emissions from land-use change.Science319 :1238-1240.

Sedjo, R.A. & Marland, G. 2003. Inter-trading permanent emissions credits and rented temporary carbon emissions offsets: some issues and alternatives.Climate Policy3:435-444.

Seibert, M., Forestier, M., Zhang, L. & Ghirardi, M. 2001. Development of an efficient algal H2-production system. Proceedings of the 2001 DOE Hydrogen Program Review NREL/CP-570-30535, Accessed February 2, 2010.http://www1.eere.energy.gov/hydrogenandfuelcells/pdfs/30535c.pdf.

Shinnar, R. & Citro, F. 2006. A road map to U.S. decarbonization.Science313:1243-1244. Simões, A.F., Kligerman, D.C., La Rovere, E.L., Maroun, M.R., Barata, M. & Obermaie, M. 2010. Enhancing adaptive capacity to climate change: The case of smallholder farmers in the Brazilian semi-arid region. Environmental Science & Policy13:801-808.

Singh, J. & Gu, S. 2010. Commercialization potential of microalgaefor biofuels production. Renew. Sustain. Energ. Rev. 14:2596–2610.

Snyder, C.S., Bruulsema, T.W., Jensen, T.L. & Fixen, P.E. 2009 Review of greenhouse gas emissions from crop production systems and fertilizer management effects.Agric. Ecosyst. Environ. 133:247–266.

Solomon, B.D. & Banerjee, A. 2006. A global survey of hydrogen energy research, development and policy. Energ. Policy34:781-792.

Song, C. 2006. Global challenges and strategies for control, conversion and utilization of CO_2 for sustainable development involving energy, catalysis, adsorption and chemical processing.Catalysis Today115:2-32.

Song, C.S. & Pan, W. 2004. Tri-reforming of methane: a novel concept for catalytic production of industrially useful synthesis gas with desired H/CO ratios.Catalysis Today98:463-484.

Song, D., Fu, J. & Shi, D. 2008. Exploitation of oil-bearing microalgaefor biodiesel.Chinese Journal of Biotechnology24:341-348.

Soong, R.K., Bachand, G.D., Neves, H.P., Olkhovets, A.G., Craighead, H.G. & Montemagno, C.D. 2000. Powering an inorganic nanodevice with a biomolecular motor.Science 290:1555-1558.

Sorensen, B. 2007. Assessing current vehicle performance and simulating the performance of hydrogen and hybrid cars.Int. J. Hydrogen Energ.32:1597-1604.

Spolaore, P., Joannis-Cassan, C., Duran, E., & Isambert, A. (2006). Commercial applications of microalgae.J. Biosci. Bioeng.101:87-96.

Steinberg, M. 2006. Conversion of fossil and biomass fuels to electric power and transportation fuels by high efficiency integrated plasma fuel cell (IPFC) energy cycle.Int. J. Hydrogen Energ. 31:405-411.

Stern, N. 2006. Review on the economics of climate change. Accessed February 2, 2010. http://www.hm-treasury.gov.uk/independent_reviews/stern_review_economics_climate_change/ stern_review_report.cfm.

Sticklen, M.B. 2008. Plant genetic engineering for biofuel production: towards affordable cellulosic ethanol.Nature Rev. Genet. 9:433-443.

Stone, R. 2007. Can palm oil plantations come clean?Science317:1491.

Styles, D. & Jones, M.B. 2008. Life-cycle environmental and economic impacts of energy- crop fuel-chains: an integrated assessment of potential GHG avoidance in Ireland.Environmental Science and Policy11:294-306.

Subhadra, BG. 2010 Sustainability of algal biofuel production using integrated renewable energy park (IREP) and algal biorefinery approach.Energ. Policy38:5892–5901.

Subramanian, K,A., Singal, S.K., Saxena, M. & Singhal, S. 2005. Utilization of liquid biofuels in automotive Diesel engines: an Indian perspective.Biomass Bioenergy29:65-72.

Sun, F. & Chen, H. 2008. Enhanced enzymatic hydrolysis of wheat straw by aqueous glycerol pretreatment. Bioresource Technol. 99:6156-6161.

Szulczyk, K.R., McCarl, B.A. & Cornforth, G. 2010. Market penetration of ethanol.Renew. Sustain. Energ. Rev. 14:394–403.

Takeshita, T. & Yamaji, K. 2008. Important roles of Fischer-Tropsch synfuels in the global energy future. Energ. Policy36:2773-2784.

Tan, K.T., Lee, K.T. & Mohamed, A.R. 2008. Role of energy policy in renewable energy accomplishment: The case of second-generation bioethanol.Energ. Policy36:3360- 3365.

Theis, M., Skrifvars, B.-J., Zevenhoven, M., Hupa, M. & Tran, H. 2006. Fouling tendency of ash resulting from burning mixtures of biofuels. Part 2: Deposit chemistry.Fuel85:1992-2001.

Thomas, C.D. 2007. A sixth mass extinction?Nature450:349.

Tijmensen, M.J.A., Faaij, A.P.C., Hamelinck, C.N. & Van Hardeveld, M.R.M. 2002. Exploration of the possibilities for production of Fischer Tropsch liquids and power via biomass gasification.Biomass Bioenergy23:129-152.

Tilman, D., Hill, J. & Lehman, C. 2007. Response to comment on "carbon-negative biofuels from low-input high-diversity grassland biomass".Science316:1567c.

Tollefson, J. 2008. Stoking the fire.Nature454:388-392.

Tomkiewicz, M. 2006. Global warming: science, money and self-preservation.Comptes rendus Chimie9 :172-179.

Trabesinger, A. 2007. Power games.Nature447:900-903.

UNFCCC 2002. Report of the Conference of the Parties on its Seventh Session, Marrakesh, 29 October to 10 November 2001, Document FCCC/CP/2001/13/add.1

Urry, J. 2008. Governance, flows, and the end of the car system?Global Environ. Change 18:343-349.

Velu, S., Suzuki, K., Vijayaraj, M., Barman, S. & Gopinath, C. 2005. In situ XPS investigations of Cu1_ xNixZnAl-mixed metal oxide catalysts used in the oxidative steam reforming of bio-ethanol.Applied CatalysisB: Environmental 55:287-299.

Venter, O., Meijaard, E. & Wilson, K. 2008. Strategies and alliances needed to protect forest from palm-oil industry.Nature451:16.

Venturi, P. & Venturi, G. 2003. Analysis of energy comparison for crops in European agricultural systems. Biomass Bioenergy25:235-255.

Vertès, A.A., Inui, M. & Yukawa, H. 2006. Implementing biofuels on a global scale.Nat. Biotechnol.24:761-764.

Vogt, K.A. et al. 2008. Bio-methanol: How energy choices in the western United States can help mitigate global climate change.Renew. Energ. 34:233-241.

Vunjak-Novakovic, D., Kim, Y., Wu, X., Berzin, I. & Merchuk, J.C. 2005. Air-lift bioreactors for algal growth on flue gas: mathematical modeling and pilot-plant studies.Ind. Eng. Chem. Res. 44:6154-6163.

Wall, D. 2004. Rain or shine – a phototroph that delivers.Nat. Biotechnol. 22:40-41.

Walter, A., Dolzan, P., Quilodrán, O., de Oliveira, J.G., da Silva, C., Piacente, F. & Segerstedt, A. 2010. Sustainability assessment of bio-ethanol production in Brazil considering land use change, GHG emissions and socio-economic aspects.Energ. Policy, doi:10.1016/j.enpol.2010.07.043

Werner, F., Taverna, R., Hofer, P., Thürig, E. & Kaufmann, E. 2010. National and global greenhouse gas dynamics of different forest management and wood use scenarios: a model-based assessment. Environmental Science & policy13:72 – 85.

Wesseler, J. 2007. Opportunities ('costs) matter: A comment on Pimentel and Patzek "Ethanol production using corn, switchgrass, and wood; biodiesel production using soybean and sunflower".Energ. Policy35:1414-1416.

Willner, I. 2002. Biomaterials for sensors, fuel cells, and circuitry.Science298:2407-2408.

Wolfrum, E.J. & Watt, A.S. 2001. Bioreactor design studies for a novel hydrogen-producing bacterium. Proceedings of the 2001 DOE Hydrogen Program Review, NREL/CP- 570-30535.

Woodcock, J., Banister, D., Edwards, P., Prentice, A.M. & Roberts, I. 2007. Energy and health 3, energy and transport.Lancet370:1078-1088.

Xue, F., Miao, J., Zhang, X., Luo, H. & Tan, T. 2008. Studies on lipid production byRhodotorula glutinisfermentation using monosodium glutamate wastewater as culture medium.Bioresource Technol. 99:5923-5927.

Yang, Z. & Sirianni, P. 2010. Balancing contemporary fairness and historical justice: A 'quasi- equitable' proposal for GHG mitigations.Energ. Econ. 32:1121–1130.

Yasuda, R., Noji, H., Yoshida, M., Kinosita, K. & Itoh, H. 2001. Resolution of distinct rotational substeps by submillisecond kinetic analysis of F1-ATPase.Nature410:898-904.

Yue, T.-X., Fan, Z.-M., Chen, C.-F., Sun, X.-F.& Li, B.-L. 2010. Surface modeling of global terrestrial ecosystems under three climate change scenarios.Ecol. Model., doi:10.1016/j.ecolmodel.2010.11.026

Zhang, Y., Dubé, M.A., McLean, D.D. & Kates, M. 2003. Biodiesel production from waste cooking oil: Process design and technological assessment.Bioresource Technol. 89:1- 16.

Zhou, X., Xiao, B., Ochieng, R.M. & Yang, J. 2008. Utilization of carbon-negative biofuels from low-input high-diversity grassland biomass for energy in China.Renew. Sustain. Energ. Rev. 13:479-485.

CHAPTER-3

ENHANCING THE QUALITY OF BIOGAS BY PRESSURE SWING ADSORPTION

1. INTRODUCTION: AN OVERVIEW

Biogas is a gaseous stream that is generated by the anaerobic breakdown of organic materials. Methane is the primary constituent of biogas, which is why it is regarded as a renewable energy and fuel source. The primary advantages of biogas stem from its globally distributed production and the environmental benefits of mitigating methane emissions by substituting bio-methane for fossil fuels. To effectively harness the energy derived from biogas, it is imperative to have control over its production. The generation of biogas from organic waste is an intricate process that encompasses numerous distinct bacterial species. The complete process of converting organic waste into biogas can be categorized into four distinct stages (Gavala et al., 2003; Demirbas et al., 2011):

- **Hydrolisis:** complex organic molecules are hydrolyzed into smaller units (sugars, amino-acids, alcohols, fatty acids, etc.
- **Acidogenese:** acidogenic bacteria further break down the molecules into volatile fatty acids, NH_3, H_2S and H_2.
- **Acetanogese:** the acetanogens transform the molecules into CO_2 , H2and mainly acetic acid.
- **Metanogese:** at the end of the process, the methanogenic archaea transform the H2and acetic acid molecules into a mixture of CO_2 , $CH_4$and water.

The production phases can be regulated in reactors, known as digesters, or they can occur naturally in landfills that can be enhanced to collect biogas efficiently. For more information, go visit http://www.epa.gov/lmop/. The digesters can function in mesophilic and thermophilic modes, where the biogas is produced at temperatures of 293-313 K and 323-333 K, respectively (Gavala et al., 2003). Biogas production in landfills primarily occurs under psychrophilic conditions, namely at temperatures ranging from 285 to 290 Kelvin (Monteiro et al., 2011). The primary components of biogas are methane, carbon dioxide, sulphur compounds (such as H2S and siloxanes), water, and minor impurities (such as 02, N2, ammonia, chlorine, fluorines, etc) (Wellinger, 2009; Pettersson and Wellinger, 2009, http://www.epa.gov/lmop/). The ultimate composition of biogas is subject to variation and is heavily influenced by the origin of the organic material (Pettersson and Wellinger, 2009). Landfills, waste-water treatment plants, manure fermentation, and fermentation of energy crops are the primary sources of biogas generation. Table 1 provides the composition of the biogas obtained from various sources. Table 1 includes the composition and properties of natural gas for the purpose of comparison. The methane concentration varies significantly depending on the type of molecules digested. For example, fat has a much higher bio-methane production compared to carbs. Cereal residues have a high biogas production, around 200 m3 CH4/ton, according to Pettersson and Wellinger (2009). This makes them a promising option for fermentation in farms worldwide.

Due to the high methane content in biogas streams, emitting it into the environment results in the wastage of a highly efficient hydrocarbon. Additionally, methane has a global warming potential that is 23 times greater than CO2. Therefore, the proper gathering and usage of the bio-methane included in biogas prevent significant methane emissions into the atmosphere and also convert this resource into a valuable and renewable source of energy. Bio-methane can be combusted and directly turned into electrical energy. Biogas utilization for electricity generation is growing globally.

Table 1. The average mix and qualities of natural gas and other biogas streams.

Gas	Biogas	Landfill gas	Natural gas
CH_4(%)	90-70	65-65	90
Hydrocarbons (%)	0	0	9
H_2(%)	0	0-3	0
CO_2 (%)	30-40	15-50	1
N_2(%)	~0.2	5-40	0.3
O_2(%)	0	0-5	0
H_2S (ppm)	0-4000	0-100	3
NH_3(ppm)	100	5	0
Heating value, kWh/Nm_3	6.5	4.4	11.0

Biogas can be utilized in a different way by converting it into purified bio-methane, which can then be fed into the natural gas grid or used as fuel for vehicles. The purification of biogas entails multiple stages, including the elimination of pollutants (mostly sulphur compounds), the extraction of water, and the upgrading process, which involves the large-scale removal of CO2. Typically, the initial stage in the purification of biogas involves the elimination of sulphur compounds. The sequence for separating H2O and CO2 is determined by the particular technology used for upgrading. The bio-methane produced must meet specific purity requirements, with a maximum allowable CO2 content ranging from 2-4%, depending on the regulations of each nation. Among all the purification procedures, the process of removing a large amount of CO2 is the most costly and will be the focus of this chapter.

2. CARBON DIOXIDE REDUCTION TECHNOLOGIES FOR BIOGAS

The process of upgrading biogas, specifically the removal of CO2, consumes approximately 3-6% of the energy contained in the biogas. This process can incur costs of up to €10 per gigajoule (GJ) for smaller volumes of biogas. However, studies have demonstrated that transporting enhanced biogas may yield higher energy efficiency compared to converting it into power at the location (Pettersson and Wellinger, 2009). Multiple commercial systems exist for the removal of CO2 from biogas streams. Currently, there is no definitive superior technology, and the cost of upgrading all technologies is rather uniform, dropping only when the plant capacity increases (Wellinger, 2009). The choice of the appropriate method for upgrading biogas is contingent upon factors such as gas flowrate, utility value, law, and so on. The primary biogas upgrading methods include water scrubbing, chemical scrubbing, membrane separation, and Pressure Swing Adsorption (PSA). This chapter will provide a concise explanation of the underlying phenomena involved in these processes. The primary objective of this chapter is to provide a comprehensive explanation of the PSA approach for biogas upgrading. It will include detailed examples of its use and discuss emerging research trends that can be utilized to economically manufacture bio-methane on a small scale, specifically for farm-oriented purposes.

2.1. Water scrubbing

Methane exhibits lower solubility in water compared to carbon dioxide. Water scrubbing systems utilize this process to eliminate CO_2 from biogas. The biogas is introduced into a column and undergoes a process of purification by being exposed to water that flows in the opposite direction from the top of the column. The column is typically filled with a substance to increase the surface area and facilitate the absorption of CO_2. The carbon dioxide (CO_2) is dissolved in the water and subsequently transported to a "regeneration column" where the CO2 is liberated. The water scrubbing process can be regenerated by increasing the temperature or reducing the pressure. CO_2 is used to eliminate H_2S in this technological process. Additionally, it is necessary to dehydrate the CH4 stream, which has been filtered to a level of up to 98%, as it exits the scrubber. The solubility of carbon dioxide (CO_2) in water significantly rises with lower temperatures. To minimize the energy required for pumping, it is necessary for the water to be accessible at lower temperatures. This strategy has been implemented in various nations characterized by frigid climates, such as Sweden, Switzerland, Germany, Austria, and others. While cooling water remains effective for large-scale facilities, it is not as practical for smaller applications. Currently, water scrubbing is the most often used method for enhancing the quality of biogas. Operational plants have processing capacities ranging from 80 to 10000 m^3/hour. The primary technology developers in this field include Malmberg (www.malmberg.se), Flotech Inc. (www.flotech.com), Rosroca (www.rosroca.de), DMT (www.dirkse-milieutechniek.com), and others. Some of these webpages contain a video that provides a detailed graphical explanation of the process.

2.2. Chemical scrubbing

Like water cleaning, it is feasible to utilize alternative chemicals for the absorption of CO2. The method also includes an absorption tower, where a chemical solvent is used to selectively absorb CO_2. Next, the saturated absorbent is subjected to heat in a regeneration tower, resulting in the release of carbon dioxide (CO_2). This method is extensively utilized for the purpose of cleaning expansive facilities within the natural gas business. The choice of solvent for this process is crucial as the "energy" of CO_2 absorption determines the overall energy consumption of the system. Amines, which are chemicals with a large capacity to absorb CO_2, are highly effective in purifying methane with low CO_2 content to achieve a very high level of purity. The CO_2 removal in biogas is a bulk removal procedure, which may result in greater energy penalties. Alternatively, when it comes to removing large amounts of CO_2 to achieve a CH4 purity between 97-98%, physical solvents are more energy efficient and need less energy consumption. Various physical absorbents include methanol, Selexol, Rectisol, Genosorb, Morphysorb, and others. There are now operational plants that have the capacity to handle biogas flowrates ranging from 55 to 13000 m^3/hour. Multiple businesses offer this technology (Pettersson and Wellinger, 2009).

2.3. Membranes

Gas cleaning using membranes is a widely adopted technology in the chemical industry. The membrane is a permeable substance that allows certain gases to pass through its structure. By utilizing a suitable material, it is feasible to achieve selectivity among the gases present in the mixture that needs to be separated. In this specific application, two distinct streams are acquired: a permeate gas consisting primarily of CO_2, water, and ammonia, and the retentate, which is a concentrated form of CH4. Hollow fibres comprised of various polymers are the materials most frequently used. Multiple firms offer this technology, with Air Liquide being the dominant company in this field (Air Liquide, 2011). During their procedure, the biogas is subjected to compression at a pressure of 16 bars. It is then directed to a two-stage membrane process, which allows for the extraction of methane with a purity above 90%. A PSA (Pressure Swing Adsorption) technique can be employed in a sequential manner to enhance the purity of CH4.

2.4. Pressure Swing Adsorption (PSA)

Pressure Swing Adsorption (PSA) is the second most commonly used method for enhancing the quality of biogas. Multiple businesses engage in the development and commercialization of this technology, including Carbotech (www.carbotech.de), Acrona (www.acrona-systems.com), Cirmac (www.cirmac.com), Gasrec (www.gasrec.co.uk), Xebec Inc (www.xebecinc.com), and Guild Associates (www.moleculargate.com). Currently, there are operational small-scale plants with a flowrate of 10 m^3/hour of biogas. However, it is worth noting that this technology can also be utilized for significantly larger flowrates, such as 10000 m^3/hour of biogas. In Pressure Swing Adsorption (PSA) operations, biogas is compressed to a pressure ranging from 4 to 10 bar. It is then introduced into a vessel, sometimes known as a column, where it comes into contact with an adsorbent material. This material selectively captures and retains CO_2. The adsorbent is a solid material that has pores and a large surface area. The primary adsorbents used in commercial operations are carbon molecular sieves (CMS), although activated carbons, zeolites, and other materials such as titanosilicates are also utilized. The purified methane (CH^4) is collected at the uppermost part of the column with minimal decrease in pressure. Once a given amount of time has passed, the adsorbent becomes completely filled with CO^2, and the column must be regenerated by decreasing the pressure (often to a vacuum for biogas enhancement). The adsorption of H_2S is often irreversible in the adsorbents, necessitating the implementation of a gas elimination step prior to the PSA. Alternatively, depending on the selected adsorbent, the moisture present in the biogas stream can be eliminated along with CO_2 in the same unit. Multi-column arrays are used to simulate a continuous process. A single column with storage tanks can be utilized for minor applications that experience discontinuities. The PSA method is very adaptable for biogas upgrading worldwide, as it is not reliant on the presence of cold or hot sources. Subsequent to this statement, a comprehensive elucidation of the aforementioned procedure will ensue.

3. PRINCIPLESOFPRESSURESWINGADSORPTION(PSA)FORTHEPURIFICATIONOF BIOGAS

A pressure swing adsorption (PSA) unit is used for biogas upgrading, where an adsorbent material is exposed to pressure variations in order to selectively capture and release CO_2. Adsorption is a naturally occurring exothermic process where CO_2 is loaded onto an adsorbent material. The amount of CO_2 that can be loaded onto the adsorbent is determined by the specific qualities of the material used, such as its surface area, composition, and pore size. After the material has been specified, its regeneration should be carried out. It is important to understand that when the material is consistently consumed and regenerated, there is a specific stage in the process called a "cyclic steady state" (CSS). The primary focus of building a PSA process lies in developing a regeneration strategy for the adsorbent that minimizes energy consumption (reducing the energetic penalty) and maximizes efficiency (increasing productivity).

The basic basis of the PSA process is illustrated in Figure 1, which displays two CO_2 isotherms representing distinct materials. Both materials have the ability to adsorb CO_2 up to a certain amount determined by the partial pressure of CO_2 in the feed step (Pfeed), which is qfeed,1 and qfeed,2 for adsorbents 1 and 2, respectively. Once the adsorbent reaches its saturation point, it is regenerated by reducing the pressure to a lower level, known as Preg. This causes the amount of CO2 loaded onto the adsorbent to fall to qreg,1 and qreg,2. Material 1 exhibits a greater capacity for CO_2 absorption than material 2 over the entire range of pressures. However, the disparity in loading between qfeed and qreg (Δq) is greater for material 2, suggesting that the "cyclic capacity" will be superior for this material. The graphic highlights the significance of understanding the isotherm's form for effective design of the PSA process. For PSA applications, it is preferable to have linear or somewhat non-linear isotherms rather than steep isotherms with significant loading. Moreover, if the isotherms have a steep slope, the process of regenerating the adsorbent becomes more challenging due to the increased energy needed to remove CO_2. Figure 1 demonstrates that the choice of regeneration pressure significantly impacts the cyclic operation of the PSA process. In the following sections, we will explain the characteristics of the adsorbents and the processes involved in regenerating the adsorbents in PSA for biogas purification.

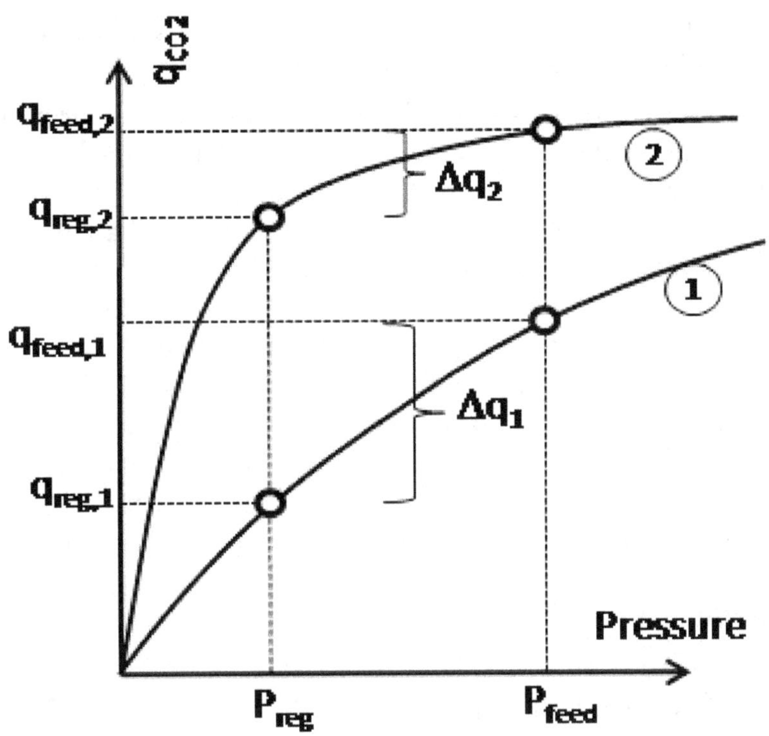

Figure 1. Generic CO_2 isotherms for two different materials (1) and (2) indicating partial pressure of CO_2 in feed step and regeneration (low) pressure.

3.1. Adsorbents

The adsorbent material can be considered as the central component of the PSA unit. The parameters of the cycle, including operating conditions and operating mode, are contingent upon the initial selection of the adsorbent. As previously stated, PSA technique can utilize many materials. The material selected should at least satisfy one of two criteria: have a higher selectivity to CO_2 : this gas should be more "attached" to the surface of the material than CH_4; in most solids CO_2 can create stronger bonds with surface groups than CH_4. This kind of materials will be termed as equilibrium-based adsorbents since its main selectivity is due to differences of interaction forces between CO_2 and $CH_4$with and the surface. The pores of the adsorbent can be adjusted in such a way that CO_2 (kinetic diameter of 3.4 Å) can easily penetrate into their structure while larger CH_4molecules (kinetic diameter of 3.8 Å) have size limitations to diffuse through them. These materials will be termed as kinetic adsorbents since its main selectivity is due to diffusion constrains.

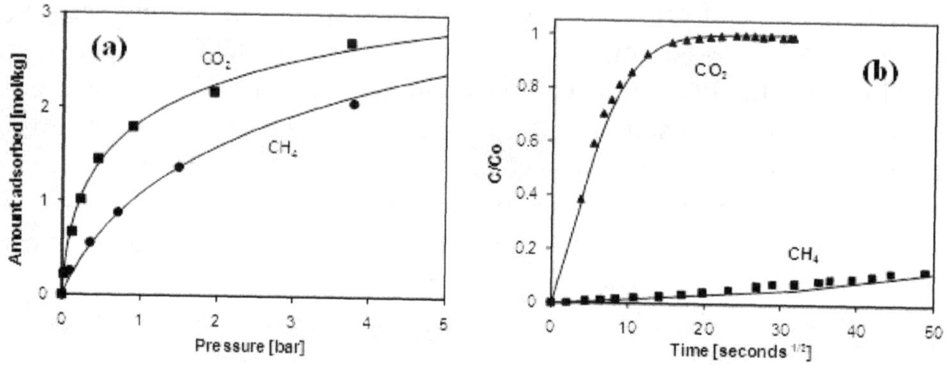

Figure 2. Adsorption of CO_2 and CH_4in carbon molecular sieve 3K (Takeda Corp, Japan) at 298 K: (a) adsorption equilibrium; (b) uptake rate curves (data fromCavenati et al., 2005).

Carbon molecular sieves are widely used for biogas upgrading. The adsorption equilibrium isotherms of carbon dioxide (CO_2) and methane (CH_4) in CMS-3K, a material developed by Takeda Corp. in Japan, are presented in Figure 2 (Cavenati et al., 2005). The material has a distinct preference for CO_2, with its most significant characteristic in CMS-3K being its kinetic selectivity rather than its equilibrium selectivity. CO_2 has a significantly higher rate of adsorption compared to CH_4 in this substance. The adsorption equilibrium of CH_4 is only attained after a duration of two days of solid-gas interaction. The pore mouth of CMS-3K is actually constricted to a size that is very close to the kinetic diameter of CH4. This creates a particular resistance, which refers to the mass transfer that occurs at the micropore mouth. This phenomenon may be observed in the early stages of CH4 uptake, as shown in Figure 2 (b) (Srinivasan et al., 1995). ETS-4 (titanosilicate-4) modified with alkali-earth metals is another material that has a high resistance to the diffusion of CH4. This has been demonstrated in studies by Kuznicki (1990), Marathe et al. (2004), and Cavenati et al. (2009). The pore diameter of this material may be adjusted by varying the heating temperatures, which leads to a phenomenon known as the "molecular gate" effect. This technology has been commercialized by Guild Associates Inc. (USA).

Additional commonly used adsorbents include activated carbons and zeolites. These materials facilitate rapid diffusion of both gases, taking advantage of the disparity in CO_2 and CH_4 loadings.

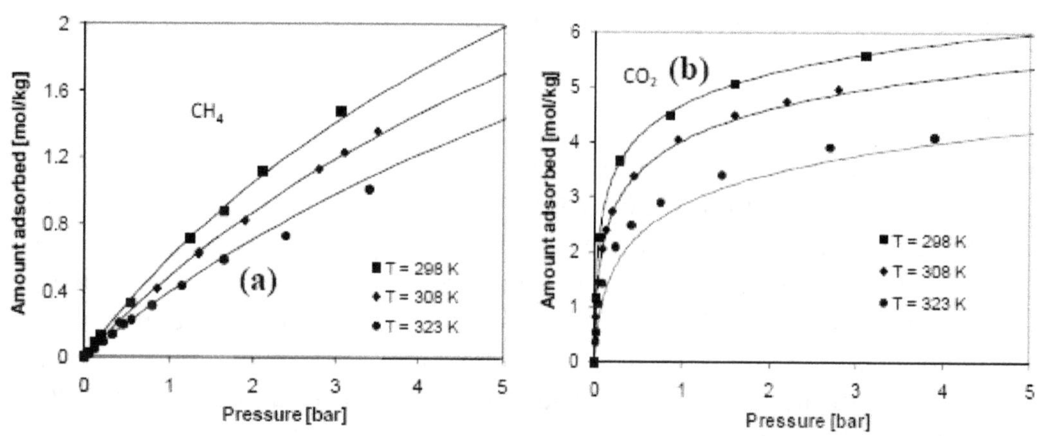

Figure 3. Adsorption equilibrium of CO_2 (a) and CH_4(b) on zeolite 13X at 298, 308 and 323 K (Data fromCavenati et al., 2004).

Figure 3 illustrates an instance of materials that rely on equilibrium, namely the adsorption equilibrium of CO_2 and CH_4 on zeolite 13X (CECA, France), as demonstrated by Cavenati et al. (2004). It should be noted that the amount of CO_2 being loaded is significantly greater than the amount of CH_4 being loaded at the same pressure and temperature conditions. In addition, while analyzing the findings shown in Figure 1, it is evident that the capacity of cyclic CO_2 decreases at lower temperatures compared to higher temperatures. This implies that if zeolite 13X is used at 323 K, it will be more convenient to regenerate than at 298 K.

Contaminant presence is another crucial factor to consider when selecting materials for the PSA process in biogas upgrading. In addition to methane (CH_4) and carbon dioxide (CO_2), biogas also contains hydrogen sulfide (H_2S) and water vapor (H2O). H2S is absorbed irreversibly in nearly all adsorbents, necessitating its removal prior to the PSA process. By using carbonaceous materials, it is feasible to eliminate water (H_2O) and carbon dioxide (CO_2) simultaneously in a single container. Unfortunately, achieving this is not feasible with zeolites due to their high water adsorption rate, which makes desorption extremely challenging.

3.2. Packed-bed performance

Adsorption is an inherently spontaneous process. When a gas comes into contact with the adsorbent, a new equilibrium state is produced, which is determined by the partial pressure of each gas and the overall temperature of the system. Once a state of equilibrium is reached, further adsorption ceases and the adsorbent must be renewed. To ensure the effective absorption of CO_2 in various cycles, it is necessary to periodically regenerate a PSA column. To maintain continuous feed processing, multiple columns are used in parallel. While one column is used for the selective removal of CO_2 from the biogas, the other column(s) are being regenerated. The functioning of a Pressure Swing Adsorption (PSA) process for biogas purification can be elucidated by illustrating the events that occur when a blend of methane (CH_4) and carbon dioxide (CO_2) is introduced into a column containing an adsorbent material. To simplify, we will assume that the column is under the same pressure as the biogas stream and is filled with an inert gas, specifically helium. An instance of such conduct is commonly referred to as "breakthrough experiments".

Figure 4 (Cavenati et al., 2005) displays a breakthrough curve of a combination containing 55% CH_4 and 45% CO_2 in CMS-3K. It is evident that during the early stages, methane molecules move through the column and occupy the gas phase in the spaces between particles, as well as the empty spaces within the particles (macropores), displacing helium. CH_4 adsorption is hindered by the significant resistance to diffuse into the micropores, resulting in rapid breakthrough in the column. Conversely, carbon dioxide (CO_2) requires a significant amount of time to penetrate the column due to its continual adsorption. It is important to observe that before to the breakout of CO_2, there is a specific duration during which only methane is collected at the end of the column product. Figure 4(b) also displays the temperature increase at various places along the column. It should be noted that in this experiment, the rise in temperature is completely caused by the absorption of CO_2. The experiment was conducted in conditions that were not isothermal and nor adiabatic. For larger adsorbers operating under adiabatic conditions, a higher temperature increase will have a more pronounced negative effect on the adsorption of CO_2, resulting in faster breakthrough. Figure 4 also demonstrates the dispersion of the CO_2 curve, which is another significant observation. The feed stream had a sudden spike in CH4 and CO_2 partial pressure, causing a disturbance. The breakthrough result reveals that the output from the column, after being affected by this disturbance, is widely dispersed.

The shape of the adsorption breakthrough curves is associated to diverse factors:

- **Slope of the adsorption isotherms:** comprise the concentration wave if isotherm is favourable (Langmuir Type) and dispersive if the adsorption equilibrium is unfavourable (desorption for Langmuir-type isotherms). No effect if the isotherm is linear,
- **Axial dispersion of the adsorption column:** disperse the concentration wave, Resistance to diffusion within the porous structure of the adsorbent: disperse the concentration wave.

- **Thermal effects:** normally in gas separations the thermal wave travels at the same velocity as the concentration wave (Yang, 1987;Ruthven et al., 1994;Basmadjian, 1997) and its effect is to disperse the concentration wave. Thermal effects can control the shape of the breakthrough curve.

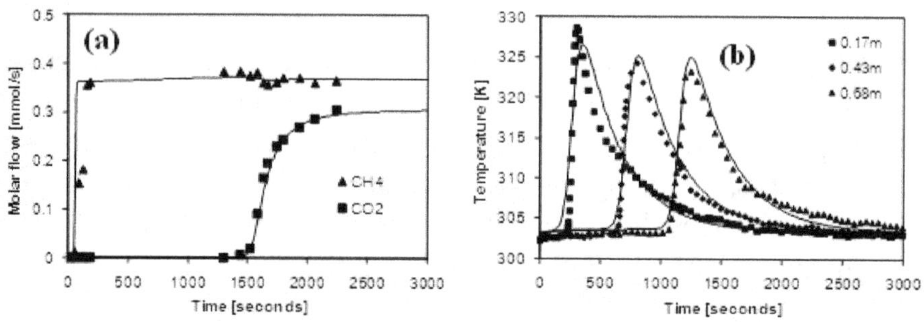

Figure 4. Binary CH_4 (55%) – CO_2 (45%) breakthrough curve experiment in fixed-bed filled with CMS-3K extrudates. Temperature: 303 K; Pressure: 4 bar (data fromCavenati et al., 2004). (a): molar flow of $CH_4$and CO_2 ; (b) temperature evolution in three different points of the column.

Figure 5 shows the temperature effects that occur during the adsorption of CO_2 in zeolite 13X extrudates. This allows for a comparison of the performance of different adsorbents. The breakthrough of CO_2 was conducted as described by Cavenati et al. (2006). The experiment was carried out at a temperature of 299 K and a pressure of 3.2 bar. The CO_2 breakthrough through the bed is characterized by a fast and rapid increase, which can be attributed to the very non-linear CO_2 adsorption isotherm. This non-linearity causes the concentration front to become compressed. Following the first abrupt advancement, the curve becomes significantly scattered as a result of thermal influences. Figure 5(b) illustrates that there is a significant temperature increase at specific spots in the column. This high temperature reduces the amount of CO_2 being loaded and accelerates the breakthrough process beyond what would occur under isothermal conditions. In the process of CO_2 desorption, the temperature in the bed will decrease, causing the adsorption isotherm to become steeper. This will make the desorption process less favorable

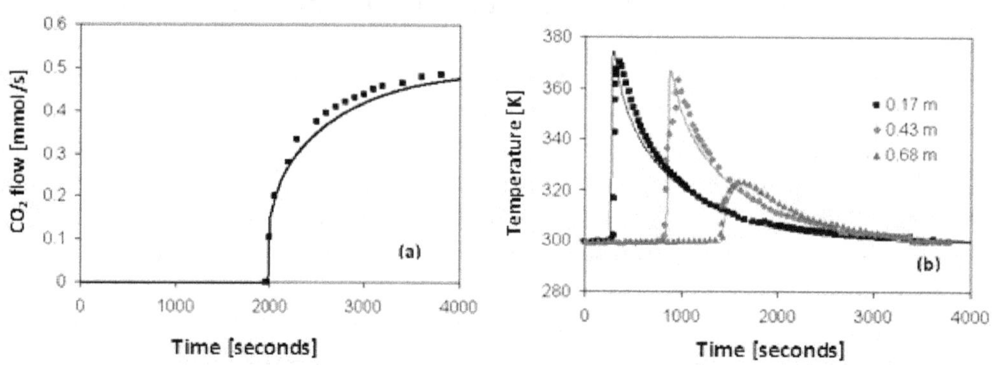

Figure 5. Breakthrough curve of pure CO_2 in fixed-bed filled with zeolite 13X extrudates. Temperature: 299 K; Pressure: 3.2 bar (data fromCavenati et al., 2006). (a): molar flow of CO_2 ; (b) temperature evolution in three different points of the column.

Due to the thermal effects and the steepness of the CO_2 isotherm on zeolite 13X, it was concluded that using a similar PSA cycle, if the temperature of the biogas stream is close to ambient temperature, it is better to use the Carbon Molecular Sieve (CMS-3K) than zeolite 13X (Grande and Rodrigues, 2007).

The solid lines shown in Figures 4 and5, represent the prediction of a mathematical model, based on pure gas adsorption equilibrium and kinetics (Cavenati et al., 2004; Cavenati et al., 2005). The resulting equations for the prediction of the fixed-bed behaviour are (Da Silva, 1999):

- Mass balances in the column, particle and micropores (crystals) of the adsorbent.
- Energy balances in the gas and solid phases and column wall
- Momentum balance (simplified to the Ergun equation)
- Multicomponent adsorption isotherm model.

It is important to understand that the equations governing mass, energy, and momentum balances are partial differential equations that are connected by a usually non-linear equation known as the isotherm model. The mathematical model underwent testing using various adsorbents and operating circumstances to separate CH_4-CO_2 as well as other gas combinations. The mathematical model used is referred to as a "homogeneous model" since it takes into account mass and heat movement in different phases using distinct equations. Both heterogeneous models, such as the single energy balance model, and simplified mass transfer models can accurately predict column behavior (Ruthven, 1984; Yang, 1987; Ruthven et al., 1994).

3.3. Basic cycles for packed-bed regeneration

After selecting the appropriate adsorbent for a specific CH4-CO2 separation, considering the operating conditions of temperature (T), pressure (P), and CO2 concentration (yCO2), there are limited options available to enhance the efficiency of the adsorption process. These options mostly involve optimizing energy transfer. When designing the upgraded PSA, the primary objective is to optimize desorption efficiency. The first publication on Pressure Swing Adsorption technology was authored by Charles W. Skarstrom in 1960 (Skarstrom, 1960). In 1964, Guerin and Domine devised a comparable cycle (Guerin and Domine, 1964). The Skarstrom cycle is commonly used as a benchmark to determine the practicality of using the PSA method to separate a certain mixture.

The Skarstrom cycle is constituted by the following cyclic steps:

- **Feed:** the CH_4-CO_2 mixture is fed to the fixed bed where the adsorbent is placed. Selective adsorption of CO_2 takes place obtaining purified CH_4at the column product end at high pressure.
- **Blowdown:** immediately before CO_2 breaks through, the column should be regenerated. This is done by stopping the feed step and reducing the pressure of the column counter-currently to the feed step. Ideally, this step should be carried out until a new equilibrium state is established as shown inFigure 1. However, the blowdown step is stopped when the flowrate of CO_2 -rich stream exiting the column is small. With the reduction of pressure, CO_2 is partially desorbed from the adsorbent. In this step, the lowest pressure of the system is achieved.
- **Purge:** when the low pressure is achieved, the column will have CO_2 molecules in the adsorbed phase but also in the gas phase. In order to reduce the amount of CO_2 in both phases, a purge step is performed counter-current to feed step. In the purge, some of the purified methane is recycled (light recycle) to displace CO_2 from the CH_4product end.

Pressurization: As the purge is conducted under low pressure, the pressure needs to be raised in order to initiate a new cycle. Pressurization can be performed either in the same direction as the feed stream or in the opposite direction as the purified CH4. Choosing the pressurization approach is a complex decision that can have significant impacts, as demonstrated by Ahn et al. (1999).

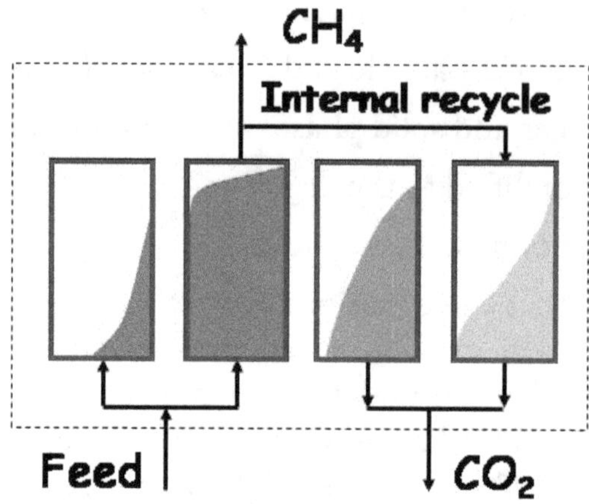

Figure 6. Schematic representation of the different steps in a Skarstrom cycle. The dotted line represents the external boundary used to calculate performance parameters.

Figure 6 illustrates a schematic depiction of the various stages during a single cycle of one column. It should be noted that an exterior boundary was formed in this image. This border is utilized to delineate the performance metrics of the PSA unit, namely CH_4 purity, CH_4 recovery, and unit productivity. The calculations are derived from the following equations:

$$PURITY = \frac{\int_0^{tfeed} C_{CH4}\, u|_{z=L}\, dt}{\left(\int_0^{tfeed} C_{CH4}\, u|_{z=L}\, dt + \int_0^{tfeed} C_{CO2}\, u|_{z=L}\, dt\right)} \tag{1}$$

$$RECOVERY = \frac{\int_0^{tfeed} C_{CH4}\, u|_{z=L}\, dt - \int_0^{tpurge} C_{CH4}\, u|_{z=L}\, dt}{\int_0^{tfeed} C_{CH4}\, u|_{z=0}\, dt + \int_0^{tpress} C_{CH4}\, u|_{z=L}\, dt} \tag{2}$$

$$PRODUCTIVITY = \frac{\left(\int_0^{tfeed} C_{CH4}\, u|_{z=L}\, dt - \int_0^{tpurge} C_{CH4}\, u|_{z=L}\, dt\right).A_{col}}{t_{cycle} w_{ads}} \tag{3}$$

where CCH_4 is the concentration of methane,uis the velocity,tcycleis the total cycle time, Acol is the column area andwadsis the total adsorbent weight. Note that the calculation of CH_4 recovery and unit productivity involves the molar flowrates of the different steps where some $CH_4$is recycled. In the case of changing the cycle configurations, the equations to calculate the process parameters may also be different.

In the cycle developed by Guerin-Domine, a pressure equalization step between different columns take place between feed and blowdown and after the purge and the pressurization. The pressure equalization steps are very advantageous for PSA applications since they help to improve the recovery of the light product, they reduce the amount of gas lost in the blowdown step and as a direct consequence, the purity of the CO_2 -rich stream obtained in the blowdown (and purge) steps increases and also less power is consumed if blowdown is carried out under vacuum. It should be mentioned that in the PSA process for biogas upgrading, it is important to perform some pressure equalization steps to reduce the amount of methane that is lost in the blowdown step. The amount of CH_4lost in the process is termed as CH_4slip and in PSA processes is around 3-12% (Pettersson and Wellinger, 2009). More advanced cycles for other applications also make extensive use of the equalization steps: up to three pressure equalizations between different columns take place in H_2 purification (Schell et al., 2009;Lopes et al., 2011). As an example, inFigure 7, the pressure history over one cycle is shown for the case of a two-column PSA process using a modified Skarstrom cycle with one pressure equalization step (Santos et al., 2011).

Continuing with the example of CMS-3K as selective adsorbent for biogas upgrading, the cyclic performance of a Skarstrom cycle is shown in Figure 8. In this example, the feed was a stream of CH_4(55%) – CO_2 (45%) resembling a landfill gas (T = 306 K), with a feed pressure of 3.2 bar. The blowdown pressure was established in 0.1 bar and pressurization step was carried out co-current with feed stream (Cavenati et al., 2005).Figure 8(a)shows the pressure history over one entire cycle while Figure 8(b) shows the molar flowrate of each gas exiting the column. It can be seen that in the feed step, a purified stream of $CH_4$is obtained. In this experiment, the purity of $CH_4$was 97.1% with a total recovery of 79.4% (Cavenati et al., 2005). An important feature of the CMS-3K adsorbent is related to the very slow adsorption kinetics of CH_4. In Figure 8(c) the simulated amount of CH_4adsorbed is shown. It can be observed that after reaching the cyclic steady state (CSS), the loading of CH_4per cycle is constant: this means that no $CH_4$is adsorbed in the column. This is very important since no CH_4will be adsorbed in the pressurization step, even with a very strong increase in its partial pressure. Unfortunately, the narrow pores also make CO_2 adsorption (and desorption) difficult, reason why only part of the capacity of the bed is employed as shown in Figure 8(d) resulting in small unit productivity.

As can be seen, an important amount of $CH_4$is lost in the blowdown step, since there is no pressure equalization: pressure drops from 3.2 bar to 0.1 bar having at least 55% of CH_4in the gas phase. The main problem of using the Skarstrom cycle for biogas upgrading is that the CH_4slip is quite high. Since the Skarstrom cycle is potentially shorter than more complex cycles, the unit productivity is higher. Keeping this in mind, it may be interesting to employ this cycle in the case of combining the production of fuel (bio-CH_4) and heat or electricity where the gas obtained from the blowdown step can be directly burned or blended with raw biogas.

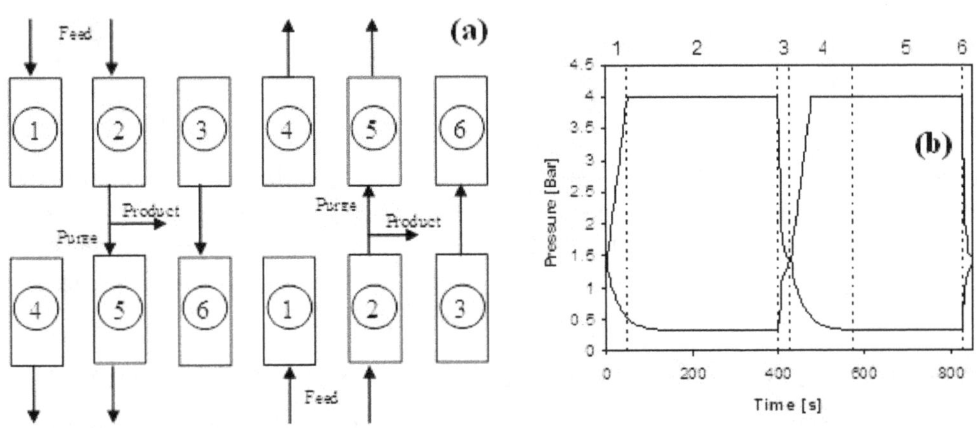

Figure 7. Scheduling of a Skarstrom cycle in a two column PSA unit: (a) step arrangement: 1. Pressurization; 2. Feed; 3. Depressurization; 4. Blowdown; 5. Purge; 6. Equalization. (b) Pressure history of both columns during one cycle.

In order to avoid large CH_4slip, at least, one pressure equalization should be employed to reduce the amount of methane in the gas phase that is lost in the blowdown stream. If such step is performed, it is possible to increase the methane recovery from 79.4% to 86.3% obtaining methane with a similar purity (97.1%). It can be concluded that the increase of number of equalization steps will reduce the methane lost in the blowdown step. Furthermore, if less gas is present in the column when the blowdown step starts, the vacuum pump will consume less power. However, to perform multiple pressure equalizations, the number of columns and the complexity of operation of the unit increase. Furthermore, the time required by the multiple pressure equalization steps will reduce the unit productivity resulting in larger units. A trade-off situation is normally achieved in PSA units with four-columns employing up to two pressure equalization steps before blowdown (Wellinger, 2009).

Another cause of CH_4 slip occurs when the purified CH_4 stream is recycled in the purge stage to remove the leftover CO_2 in the column. Due to its lack of adsorption, CH_4 will quickly penetrate the column. Nevertheless, if the duration of the purge step is insufficient, it adversely affects the efficiency of the PSA cycle. To minimize the emission of CH4 while maintaining an effective purge, one potential approach is to compress and reuse this stream (Dolan and Mitariten, 2003). Moreover, by recycling this stream, the flowrate of the purge can be utilized to regulate the efficiency of the PSA cycle during significant fluctuations in the biogas stream, such as changes in CO_2 content or overall flowrate.

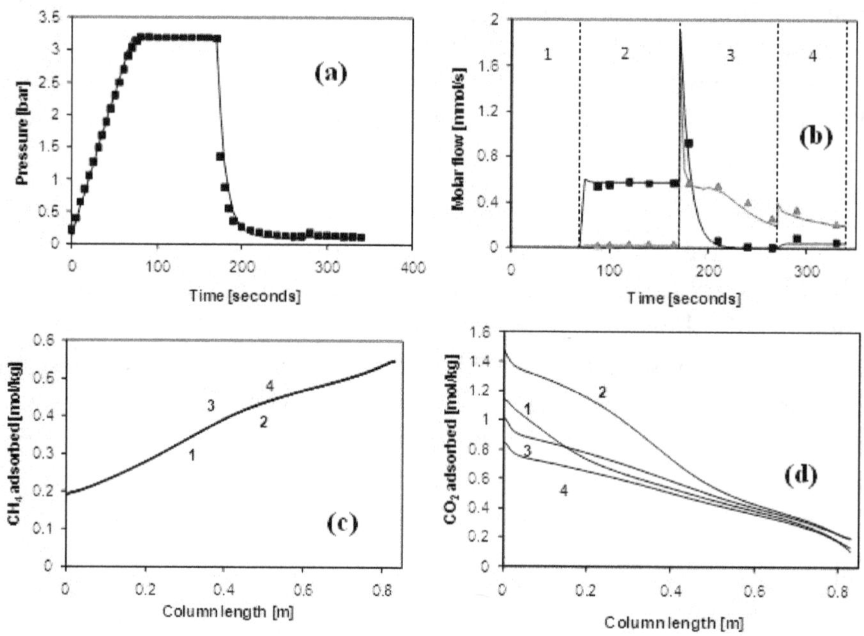

Figure 8. PSA separation of a mixture of CH_4(55%) – CO_2 (45%) using a packed bed filled with CMS-3K operating with a Skarstrom cycle (1. Pressurization; 2. Feed; 3. Blowdown; 4. Purge). Feed pressure: 3.2 bar; blowdown pressure: 0.1 bar. (a) Pressure history over one cycle; (b) molar flowrate exiting the column; (c) loading of CH_4at the end of each step; loading of CO_2 at the end of each step. Data fromCavenati et al., 2005.

4. EMERGING MARKETS AND ADVANCEMENTS IN PSA TECHNOLOGY

As previously stated, the biogas market has significant potential for expansion. Small farms have the potential to significantly drive the rise of PSA development. In such instances, the biogas can be utilized for heating and electricity generation, while a portion of the surplus can be refined into fuel. For such applications, in addition to the standards of process performance, there are six desired features for any technology that is being upgraded:

- Economic for small streams,
- Compact,
- Automated,

- Minimal attendance (by non-expert person most of the time),
- Possible to switch on /off quite fast
- Deliver product specifications even when subjected to strong variations in feed.

The PSA technology has the potential to be used in these applications because it can meet the majority of the defined criteria. For instance, certain plants utilizing Molecular Gate technology can be managed remotely with minimum human presence. These plants are transportable in compact trucks and are designed for processing modest streams of natural gas (Molecular Gate, 2011). Nevertheless, the magnitude of little biogas application is rather diminutive (less than 10 m3/hour). In addition, the literature does not mention any instances of rapidly switching a PSA unit on and off many times. This would undoubtedly necessitate specialized research and PSA design to effectively manage significant fluctuations in feed streams. To effectively address such applications, research should focus on two key areas: the development of novel adsorbents and the application of design engineering principles to the PSA unit.

4.1. New adsorbents

Although there has been a surge in the discovery of new materials with diverse potential, the majority of PSA units currently available in the market continue to rely on the familiar zeolites (4A, 5A, and 13X), activated carbons, carbon molecular sieves, silica gel, and alumina. Given that the choice of adsorbent material is crucial for the design of the PSA unit, it is imperative to utilize more efficient materials in order to meet various market constraints, such as energy consumption and size. An intriguing instance of the potential application of novel materials is the Molecular Gate technology, which involves the use of narrow pore titanosilicates (ETS^{-4}) to achieve successful CH_4 upgrading (Kuznicki, 1990; Dolan and Mitariten, 2003). When alkali-earth metals are partially substituted for ETS^{-4} materials, an interesting phenomenon occurs: the pores constrict as the activation temperature increases (Marathe et al., 2004; Cavenati et al., 2009). This characteristic is crucial as it allows for precise adjustment of the pores to enable complex separations such as CH_4-N_2. Among these inorganic substrates, the aluminophosphates are particularly noteworthy. Although these materials may not have a high CO_2 capacity, they have linear isotherms that make them ideal for use in PSA applications. Additionally, some of these materials exhibit Type V isotherms for water adsorption, indicating that they can tolerate and be regenerated in the presence of small amounts of water (Liu et al., 2011). Recently, a novel class of materials with exceptionally large surface area has been identified (Li et al., 1999; Wang et al., 2002; Millward and Yaghi, 2005; Mueller et al., 2005; Kongshaug et al., 2007). Metal-organic frameworks (MOFs) have a remarkable capacity to adsorb significantly higher quantities of CO2 in comparison to traditional adsorbents. Moreover, it is feasible to modify the arrangement in a manner that reduces the steepness of the isotherm, making the process of regeneration easier. Figure 9 illustrates the impact of high CO2 levels on MOFs, specifically by displaying the isotherms of CO_2 and CH_4 on Cu-BTC at various temperatures (Cavenati et al., 2008). Upon comparing these isotherms with those exhibited by zeolite 13X (Figure 3), it is evident that the steepness of the isotherm is rather gentle, resulting in a significantly higher "cyclic capacity" than zeolite 13X. A number of Metal-Organic Frameworks (MOFs) were examined in order to isolate mixtures of CH_4 and CO_2. The studies conducted by Schubert et al. (2007), Cavenati et al. (2008), Llewellyn et al. (2008), Dietzel et al. (2009), and Boutin et al. (2010) investigated this matter. The majority of them exhibit exceptional properties for adsorbing CO_2, maybe with a slight non-linear behavior in the CO_2 adsorption process. The challenges in commercializing these materials are mostly associated with achieving the appropriate formulation and final shape while minimizing any substantial reduction in their surface area.

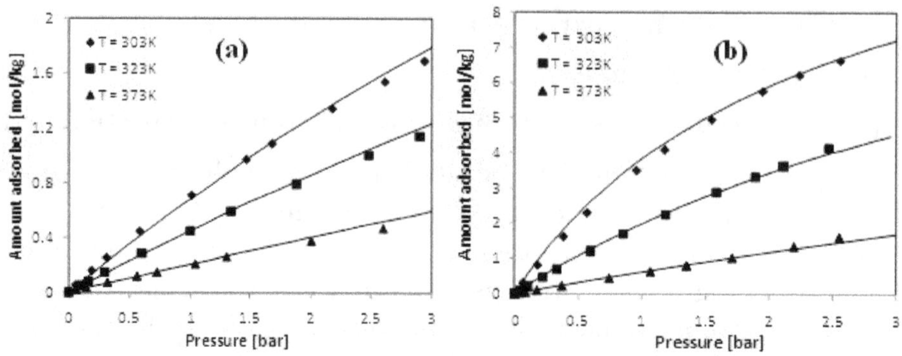

Figure 9. Adsorption equilibrium of CO_2 (a) and CH_4(b) on Cu-BTC MOF at 303, 323 and 373 K (data fromCavenati et al., 2008).

The exceptionally elevated CO2 loading of MOFs suggests that the size of the PSA unit can be substantially diminished by employing this material in lieu of conventional adsorbents. In addition, the rate at which CO2 is adsorbed in many MOFs is highly rapid, allowing for a significant portion of it to be utilized in each cycle. A primary concern about MOFs is the exclusion of water from the system, necessitating its removal at a preceding stage. However, this is not a significant issue as water removal is already a standard procedure.

4.2. Alternative PSA design

One approach to designing a new PSA unit includes choosing the appropriate adsorbent, selecting the PSA cycle, determining the unit size, defining the operating variables for effective adsorbent regeneration, and arranging the multi-column process for continuous operation (Knaebel and Reinhold, 2003). Nevertheless, while creating new applications on a smaller scale, it is possible to take into account other factors, specifically those pertaining to the unit's design. An instance of innovative process design can be seen in the use of a rotary valve by Xebec. This valve has enabled the industrial implementation of rapid-PSA units for biogas upgrading, as described by Toreja et al. (2011). When creating compact structures, the columns can have non-traditional shapes, which can be utilized to optimize the amount of adsorbent required per unit volume. Moreover, in certain instances where there are elevated levels of CO_2, the heat generated during adsorption can cause a rise in the temperature of the adsorbent, leading to a notable drop in its overall capacity. In such instances, the option of achieving efficient thermal transfer with the environment can serve as a viable alternative (Bonnissel et al., 2001), along with the potential to enhance the thermal capacity of the column (Yang, 1987). Another option for enhancing unit productivity when utilizing kinetic adsorbents such as CMS-3K is to incorporate an additional layer of adsorbent with bigger pores, allowing for faster adsorption, and with the capacity to be easily regenerated (Grande et al., 2008). Through the utilization of this stratified configuration, it becomes feasible to effectively confine the CO_2 within the outermost layer for an extended duration, resulting in a twofold increase in the system's overall productivity (while maintaining comparable levels of CH4 purity and recovery). The technique of layering adsorbents can also be utilized to simultaneously eliminate water and CO_2 in the same bed, as demonstrated in other CO_2 applications (Li et al., 2008).

Perhaps the most important engineering challenges of new PSA design are related to the modification of the PSA cycles. Most of the PSA units existing in industry nowadays use the Skarstrom cycle (or small variations of it) with several pressure equalizations to reduce the CH_4slip. The utilization of different cycles can be adjusted for different applications of the biogas stream: production of extremely high CH_4purity, small CH_4slip, combined heat / electricity and/or fuel generation, etc. The possibility of "playing" with the step arrangement in a PSA cycle for a given application is virtually infinite. Extreme variations in

PSA cycles can be achieved with PSA units with three or four columns. An example of such possibilities is given in Figure 10 where a different cycle is presented in order to radically improve the unit productivity of kinetic adsorbents (Santos et al., 2011b). This 4-column PSA cycle was designed keeping in mind that the adsorption should be continuous, that at least one equalization step is necessary to reduce CH_4slip and also to improve the contact time between gas and solid which is particularly important to increase the loading of CO_2 in the adsorbent. To enhance the contact time between the adsorbent and the feed stream, a lead-trim concept is employed (Keller et al., 1987).

In a kinetic adsorbent, the CO_2 breakthrough happens relatively fast and the mass transfer zone is quite large as shown inFigure 8(d). In order to avoid contamination of the CH_4-rich stream, the feed step is normally stopped, but using the lead-trim cycle arrangement, the gas exiting one column is routed to a second column where this residual CO_2 can be adsorbed, giving the first column extra time to adsorb CO_2 . This column arrangement leads to a column with virtually the double of the size (only for some adsorption steps). Also, the column that is ready for regeneration has a higher content of CO_2 , which also result in small CH_4slip. A simulation of the performance of this PSA cycle using CMS-3K is shown in Figure 11.

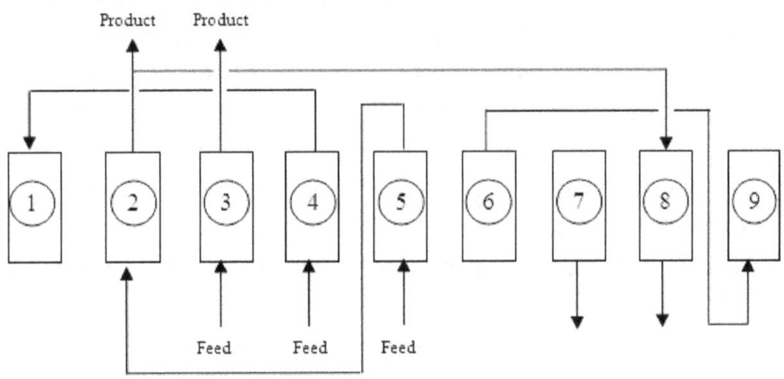

Figure 10. Scheduling of a column for a PSA cycle for biogas upgrading using lead-trim concept. The steps are: 1. Pressurization; 2. Trim feed; 3-4. Feed; 5. Lead adsorption; 6. Depressurization; 7. Blowdown; 8. Purge; 9. Pressure equalization.

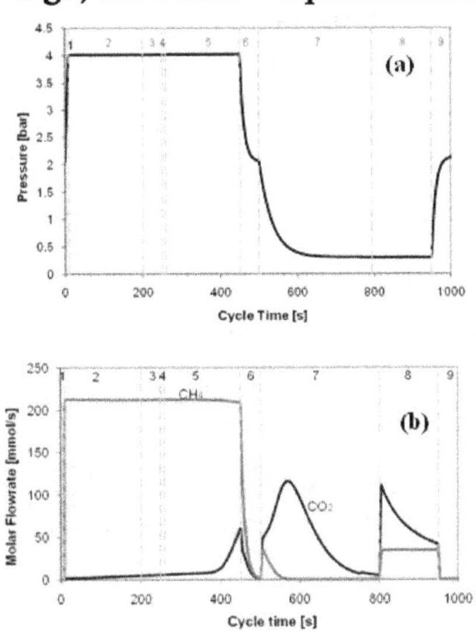

Figure 11. Simulation of a 4-column PSA process using the lead-trim cycle (see Figure 10) with CMS-3K for separation of a mixture of CH_4 (67%) and CO_2 (33%). (a) pressure history of one cycle; (b) molar flow of CH_4 and CO_2 after cyclic steady state was achieved. Feed pressure: 4 bar; Blowdown pressure: 0.3 bar; Temperature: 323 K.Data from Santos et al., 2011b.

Using this column arrangement, CH_4 purity of 98.3% could be obtained with a total recovery of 88.5% and a unit productivity of 5.5 moles of CH_4 per hour per kilogram of adsorbent (Santos et al., 2011b). From the 11.5% of CH_4 lost in blowdown and purge steps, around 7% is lost in the purge step, which means that if this stream is recycled, the CH_4-slip will drop to values lower than 5%. It is important to observe that in Figure 11(b), the CO2 began to penetrate the column at the conclusion of the feeding process. The goal in this scenario was to generate CH4 with a purity exceeding 98%, although this process can be adjusted to achieve even greater levels of purity if necessary. In addition, the cycle is highly efficient and does not necessitate reducing the pressure to 0.1 bar for regeneration. Instead, just 0.3 bar is used, resulting in a substantial reduction in power consumption compared to traditional step arrangements.

5. CONCISE SUMMARY

Pressure Swing Adsorption (PSA) has demonstrated its efficacy as a technique for biogas purification across various operational circumstances. This study provides an overview of the many technologies now available for enhancing the quality of biogas, including water and chemical scrubbing as well as membrane-based methods. It places particular emphasis on the Pressure Swing Adsorption (PSA) technique. This text provides a concise summary of the fundamental principles underlying PSA technology, including information on the types of adsorbents used and the various design options available for PSA units. The concluding section presents many potential enhancements for the design of new applications, specifically tailored to accommodate the lower biogas flow rates typically found in agricultural settings. Undoubtedly, further study is necessary to effectively improve PSA technology for applications with low flowrates. Undoubtedly, a robust correlation between materials science and process engineering can expedite the advancement of this technology. Effectively implementing PSA in this sector will enhance the adoption of biogas as an eco-friendly and sustainable fuel source.

REFERENCES

Ahn, H.; Lee, C-H; Seo, B.; Yang, J.; Baek, K (1999). Backfill Cycle of a Layered Bed H2 PSA Process. Adsorption, Vol. 5, No. 4, (October 1999), pp 419-433, ISSN 0929-5607.

Air Liquide (2011). Biogas Recovery System. 02.04.2011. Available at: http://www.dta.airliquide.com/en/our-offer/decentralized-energies/biogas.html

Basmadjian, D. (1997). The Little Adsorption Book: A Practical Guide for Engineers and Scientists; CRC Press, 1997.

Bonnissel, M.P.; Luo, L.; Tondeur, D (2001). Rapid Thermal Swing Adsorption. Ind. Eng. Chem. Res. Vol. 40, No. 10, (April 2001), pp 2322-2334, ISSN 0888-5885.

Boutin, A.; Coudert, F-X.; Springuel-Huet, M-A.; Neimark, A.V.; Ferey, G.; Fuchs, A.H (2010). The Behavior of Flexible MIL-53 (Al) upon CH_4 and CO_2 Adsorption. J. Phys Chem. C. Vol. 114, No. 50, (December 2010), pp 22237-22244, ISSN 1932-7447.

Cavenati, S.; Grande, C.A.; Rodrigues, A.E (2004). Adsorption Equilibrium of Methane, Carbon Dioxide and Nitrogen on Zeolite 13X at High Pressures. J. Chem. Eng. Data, Vol. 49, No. 4, (June 2004), pp 1095-1101, ISSN 0021-9568.

Cavenati, S.; Grande, C.A.; Rodrigues, A.E (2005). Upgrade of Methane from Landfill Gas by Pressure Swing Adsorption. Energy & Fuels, Vol. 19, No. 6, (August 2005), pp 2545-2555, ISSN 0887-0624.

Cavenati, S.; Grande, C.A.; Rodrigues, A.E (2006). Removal of Carbon Dioxide from Natural Gas by Vacuum Pressure Swing Adsorption. Energy & Fuels, Vol. 20, No. 6, (September 2006), pp 2648-2659, ISSN 0887-0624.

Cavenati, S.; Grande, C.A.; Rodrigues, A.E (2008). Metal Organic Framework Adsorbent for Biogas Upgrading. Ind. Eng. Chem. Res. Vol. 47, No. 16, (July 2008), pp 6333-6335, ISSN 0888-5885.

Cavenati, S.; Grande, C.A.; Lopes, F.V.S.; Rodrigues, A.E (2009). Adsorption of Small Molecules on Alkali-Earth Modified Titanosilicates. Microp. Mesop. Mater, Vol. 121, No. 1-3, (May 2009), pp 114-120, ISSN 1387-1811.

Da Silva, F. A. Cyclic Adsorption Processes: Application to Propane/Propylene Separation. Ph.D. Dissertation, University of Porto, Portugal, 1999.

Dietzel, P.D.C.; Besikiotis, B.; Blom, R (2009). Application of Metal-Organic Frameworks with Coordinatively Unsaturated Metal Sites in Storage and Separation of Methane and Carbon Dioxide. J. Mater. Chem. Vol. 19, (August 2009), pp 7362-7370, ISSN 0959-9428.

Demirbas, M.F.; Balat, M.; Balat, H (2011). Biowastes-to-biofuels. Energy Conv.Management, Vol.52, No. 4, (April 2011), pp 1815-1828, ISSN 0196-8904.

Dolan, W.B.; Mitariten, M.J (2003). CO_2 Rejection from Natural Gas. United States Patent US 2003/0047071, 2003.

Gavala, H.N.; Yenal, U.; Skiadas, I.V.; Westermann, P.; Ahring, B.K. (2003). Mesophilic and Thermophilic Anaerobic Digestion of Primary and Secondary Sludge. Effect of Pretreatment at Elevated Temperature. Water Research. Vol. 37, No. 19, (November 2003), pp 4561-4572, ISSN 0043-1354.

Grande, C.A.; Rodrigues, A.E (2007). Biogas to Fuel by Vacuum Pressure Swing Adsorption I. Behavior of Equilibrium and Kinetic-Based Adsorbents. Ind. Eng. Chem. Res., Vol. 46, No. 13, (May 2007), pp 4595-4605, ISSN 0888-5885.

Grande, C.A.; Cavenati, S.; Rodrigues, A.E (2008). Separation Column and Pressure Swing Adsorption Process for Gas Purification. World Patent ApplicationWO/2008/072215.

Guerin de Montgareuil, P.; Domine, D. Process for Separating a Binary Gaseous Mixture by Adsorption. US Patent 3,155,468, 1964.

Keller, G.E.; Anderson, R.A.; Yon, C.M (1987). Handbook of Separation Process Technology. JohnWiley, New York.

Knaebel, K.S.; Reinhold, H. E (2003). Landfill Gas: From Rubbish to Resource. Adsorption Vol. 9, No. 1, (March 2003), pp 87-94, ISSN 0929-5607.

Kongshaug, K.O.; Heyn, R.H.; Fjellvag, H.; Blom, R (2007). MOF-compounds as Gas Adsorbers. World Patent WO/2007/128994, 2007.

Kuznicki, S.M. (1990). Preparation of Small-Pored Crystalline Titanium Molecular Sieve Zeolites. U.S. Patent 4,938,939. July 1990.

Llewellyn, P.L.; Bourrelly, S.; Serre, C.; Vimont, A.; Daturi, M.; Hamon, L.; De Weireld, G.; Chang, J-S; Hong, D-Y; Hwang, Y.K.; Jhung, S.W.; Ferey, G (2008). High Uptakes of CO_2 and CH_4 in Mesoporous Metal-Organic Frameworks MIL-100 and MIL-101. Langmuir, Vol. 24, No. 14, (July 2008), pp 7245-7250, ISSN 0743-7463.

Li, G.; Xiao, P.; Webley, P.; Zhang, J.; Singh, R.; Marshall, M. (2008). Capture of CO_2 from High Humidity Flue Gas by Vacuum Swing Adsorption with Zeolite 13 X. Adsorption, Vol. 14, No. 2-3, (June 2008), pp 415-422, ISSN 0929-5607.

Li, H.; Eddaoudi, M. O'Keeffe, O. M. Yaghi (1999). Design and Synthesis of an Exceptionally Stable and Highly Porous Metal-Organic Framework. Nature, Vol. 402, (November 1999), pp 276-279, ISSN 0028-0836.

Liu, Q.; Cheung, N.C.O.; Garcia-Bennet, A.E.; Hedin, N (2011). Aluminophosphates for CO_2 Separation. ChemSUSChem, Vol. 4, No. 1, (January 2011), pp 91-97, ISSN 1864-5631.

Lopes, F.V.S.; Grande, C.A.; Rodrigues, A.E. (2011). Activated Carbon for Hydrogen Purification by Pressure Swing Adsorption. Multicomponent Breakthrough Curves and PSA Performance. Chem. Eng. Sci., Vol. 66, No. 3, (February 2011), pp 303-317,ISSN 0009-2509.

Marathe, R.P.; Mantri, K.; Srinivasan, M.P.; Farooq, S. (2004). Effect of Ion Exchange and Dehydration Temperature on the Adsorption and Diffusion of Gases in ETS-4. Ind. Eng. Chem. Res., Vol. 43, No. 17, (July 2004), pp 5281-5290, ISSN 0888-5885.

Millward, A.R.; Yaghi, O.M (2005). Metal-organic Frameworks with Exceptionally High Capacity for Storage of Carbon Dioxide at Room Temperature. J. Am. Chem. Soc. Vol. 127, No. 51, (December 2005), pp 17998-17999, ISSN 0002-7863.

Molecular Gate Adsorption Technology. 02.04.2011. Available at: www.moleculargate.com/molecular-gate-CO_2 -removal-nitrogen-rejection.html.

Monteiro, E.; Mantha, V.; Rouboa, A. (2011). Prospective Application of Farm Cattle Manure for Bioenergy Production in Portugal. Renewable Energy. Vol. 36, No. 2, (February 2011), pp 627-631, ISSN 0960-1481.

Mueller, U.; Lobree, L.; Hesse, M.; Yaghi, O.; Eddaoudi, M (2005). Shaped Bodies Containing Metal-Organic Frameworks. U. S. Patent 6,893,564, 2005.

Pettersson, A.; Wellinger, A (2009). Biogas Upgrading Technologies – Developments and Innovation, 02.04.2011. Available at: http://www.iea-biogas.net/

Ruthven, D.M. (1984) Principles of Adsorption and Adsorption Processes; John Wiley & Sons: New York.

Ruthven, D. M.; Farooq, S.; Knaebel, K. S. (1994) Pressure Swing Adsorption; Wiley-VCH, New York, 1994. ISBN: 0471188182.

Santos, M.P.S.; Grande, C.A.; Rodrigues, A.E (2011). Pressure Swing Adsorption for Biogas Upgrading. Effect of Recycling Streams in PSA Design. Ind. Eng. Chem. Res., Vol. 50, No. 2, (December 2010), pp 974-985, ISSN 0888-5885.

Santos, M.P.S.; Grande, C.A.; Rodrigues, A.E (2011b).New Cycle Configuration to Enhance Performance of Kinetic PSA Processes. Chem. Eng. Sci., Vol. 66, No. 8, (April 2011), pp 1590-1599, ISSN 0009-2509.

Schell, J.; Casas, N; Mazzotti, M (2009). Pre-Combustion CO_2 Capture for IGCC Plants by an Adsorption Process. Energy Procedia, Vol. 1, No. 1, (February 2009), pp 655-660, ISSN 1876-6102.

Schubert M.; Müller, U.; Hesse, M.; Diehlmann, U (2007). Process for Preparing Porous Metal-Organic Framework Materials. World Patent WO/2007/090809.

Skarstrom, C. W. Method and apparatus for fractionating gaseous mixtures by adsorption. U.S. Patent No. 2 944 627; 1960

Srinivasan, R.; Auvil, S.R.; Schork, J.M (1995) mass Transfer in Carbon Molecular Sieves- an Interpretation of Langmuir Kinetics. Chem. Eng. J. Vol. 57, No. 2, (April 1995), pp 137-144, ISSN 0923-0467.

Toreja, J.; VanNostrand, B.; Chan, N.; Dickinson, J.P (2011). Rotary-Valve, Fast-Cycle Pressure-Swing Adsorption Technology Allows West Coast Platform to Meet Tight California Specifications and Recover Stranded Gas. Laurence Reid Gas Conditioning Conference. 02.04.2011.Available at: www.xebecinc.com/pdf/Rotary-Valve-Fast-Cycle-Pressure-Swing-Adsorption-Paper.pdf

Wang, Q.M.; Shen, D.; Bülow, M.; Lau, M.L.; Deng, S.; Fitch, F.R.; Lemcoff, N.O.; Semanscin, J (2002). Metallo-organic Molecular Sieve for Gas Separation and Purification. Microp. Mesop. Mater, Vol. 55, No. 2, (September 2002), pp 217-230, ISSN 1387-1811.

Wellinger, A (2009). Gas Upgrading Issues. European Biomethane Fuel Conference. Göteborg, Sweden, September 2009. Available at: http://www.biogasmax.eu/european-conference-on-biomethane/download/

Yang, R. T (1987) . Gas Separation by Adsorption Processes Butterworths: Boston, 1987. ISBN: 0409900044.

CHAPTER-4

UTILIZATION OF RAPESEED STRAIGHT VEGETABLE OIL AS FUEL MANUFACTURED IN SMALL-SCALE OPERATIONS

1. INTRODUCTION: AN OVERVIEW

The existing reliance on oil in the majority of industrial sectors, particularly in the transportation sector, is not sustainable in either the short or long term. This promotes the exploration of alternative options in many industrial sectors and provides incentives to stimulate the adoption of renewable energy sources. Furthermore, the European Union is actively advocating for and, in some cases, mandating the adoption of renewable energy sources to fulfill the obligations outlined in the Kyoto Protocol. Biodiesel and bioethanol are the most prevalent biofuels used for transportation in Europe. The production of these biofuels primarily comes from industrial-sized facilities and has been found to have significant environmental and social issues, as indicated by various authors (Russi, 2008; Galan et al., 2009). It is imperative to introduce additional biofuels that are now absent from the Spanish market in this particular situation. Straight vegetable oil (SVO) is a type of biofuel that can be manufactured on a small scale using rapeseed grown in dry Mediterranean regions. According to Baquero et al. (2010), small-scale production offers several benefits and is more environmentally friendly compared to large-scale production. This chapter outlines a technique for cultivating rapeseed and extracting rapeseed oil and rapeseed cake meal on a modest scale. Additionally, it demonstrates the utilization of rapeseed oil as a viable source of fuel for self-consumption in diesel engines used in agriculture. The paper also presents a model that examines the production, processing, and utilization of rapeseed oil as a fuel source. It analyzes the environmental and economic aspects of using rapeseed oil as fuel in comparison to other agricultural production options. The results are assessed under arid Mediterranean climate conditions.

2. PRODUCTION OF RAPESEED

Rapeseed, a very oil-rich plant, is found in abundance across the globe. It is capable of thriving and maturing in a temperate climate. Rapeseed is well-suited to a variety of soil types, with rich and well-drained soils being particularly advantageous. However, it has a limited tolerance for flooding. Optimal soil types include loamy soils, which consist of a balanced mixture of clay, silt, and sand. The optimal pH range is between 5.5 and 7, however it may tolerate slight alkalinity up to 8.3. The plant exhibits resistance to drought periods as a result of its extensive taproot and fibrous root system at the surface. Additionally, it demonstrates a strong ability to rebound after experiencing drought conditions (Sattell et al., 1998). Figure 1 displays an image of the rapeseed flower. Rapeseed is a drought-tolerant crop in the area under investigation. Rapeseed, because to its extensive root system, has a greater ability to tap into underground water sources compared to wheat and barley, which are typically cultivated in the examined region. The optimal crop rotation for rapeseed involves planting it every five years, alternating with one year of wheat and three years of barley. Producers may choose to retain rapeseed in the same field for many years, even up to three, if they have high price expectations. However, this practice carries the danger of the crop being susceptible to fungal infections (Provance et al., 2000).

Figure 1. Image of the flower and siliqua of rape (Photo J.F. Marti).

A test was conducted in an experimental and representative field to determine the most suitable rapeseed variety for the Anoia area in Catalonia, Spain. This area was chosen as it is a dry Mediterranean region. A study was conducted during the 2006 harvest to investigate the yield and oil content of 9 different rapeseed cultivars. The experimental area was partitioned into 36 rectangular subdivisions, namely, 4 replicates of each of the 9 rapeseed varieties under study were conducted. This study is now ongoing to calculate the average of the results acquired across multiple years. Table 1 displays the initial findings acquired during the 2006 harvest. The data acquired in 2008 were rendered invalid due to the severe drought experienced throughout the autumn of 2007 and the winter of 2007-2008.

Table 1. Studied varieties of rapeseed. Average oil content and yield.

Variety	Supplier	Average oil content (%)	Rapeseed yield (kg/ha)
Bellini	Aceites Borges Pont	41.6	3636
Pacific	Limagrain Iberica	42.6	4645
Madrigal	Koipesol semillas	39.1	4525
Aviso	Aceites Borges Pont	40.5	4348
Sun	Agrusa	41.5	5251
Potomac	Limagrain Iberica	38.7	5251
Bambin	Agrusa	42.0	-
Royal	Koipesol semillas	39.5	5110
Standing	S.A. Marisa	40.3	4722

The average oil content of the 9 varieties and rapeseed yield are presented in Table 1with an average content of humidity of 9.0% in the harvest of 2006. The analysis was carried out by applying the method described by EUETII-UPC (2006).

It is worth noting that the edge effects observed in tiny rectangular divisions during experiments lead to higher experimental yields compared to those observed in actual arable fields. Based on the data shown in Table 1, it is evident that the Pacific rapeseed variety has a higher oil content. However, the Sun and Potomac types have a greater yield. The Sun rapeseed variety achieved the highest rapeseed oil output in the 2006 harvest survey. The ground was fertilized using an application rate of 450 kg/ha of a fertilizer containing 15% nitrogen, 0% phosphorus, and 15% potassium oxide. In addition, a fertilizer coverage of 260 kg/ha of ammonium nitrosulphate with a nitrogen content of 27% was applied. Prior to planting, a herbicide treatment was sprayed using Trifluralin (48%, 2.5 l/ha), Glyphosate (36%, 1.0 l/ha), and Metazachlor (50%, 3.5 l/ha). The pesticide treatment involved applying Deltamethrin 2.5% at a rate of 0.4 liters per hectare. The cultivation of rapeseed involves the utilization of many inputs such as fertilizers, insecticides, herbicides, fungicides, and rapeseed seeds for planting. Additionally, agricultural activities, primarily tractor work, are carried out to support the cultivation process. When analyzing the specific area, we are considering the arid farming conditions that are suitable for growing rapeseed. The high yields in Table 1 are a result of an experimental investigation, in which the edge effect and other variables contribute to an increase in production value. The average rapeseed yield in this study is 2300 kg/ha. The application of 3 kg/ha of fertilizer and 2 kg/ha of herbicide is being examined. In the field of study, the straw obtained from the gathered seeds is commonly left in the field as a form of fertilizer. Therefore, the straw is regarded as a co-product that serves as fertilizer for the next year.

3. PROCESSING OF RAPESEED

The rapeseed is processed to extract SVO for use as motor fuel by three mechanical steps: seed cleaning, pressing, and purification (refer to Fig. 2). The initial stage involves removing impurities like as stones, metal fragments, and straw from the seeds. It is crucial to minimize the likelihood of causing harm to the press during this procedure.

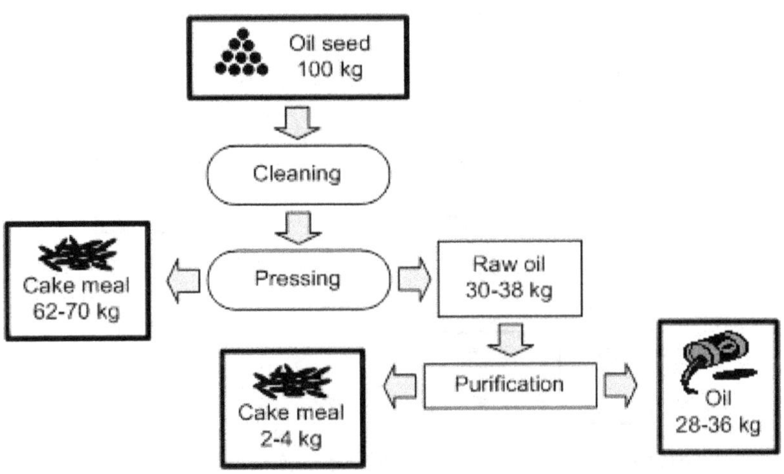

Figure 2. Rape seed oil processing.

The second stage involves using a screw press to cold press the oil seed and extract the oil. It is crucial to perform this step with caution in order to minimize the inclusion of unwanted substances from the solid residue (rapeseed cake). The pressing process affects the levels of phosphorus, calcium, and magnesium, as well as the size and composition of the particles. The fluctuation of the elements is contingent upon the velocity and the applied temperature. A decrease in speed (reduced throughput) leads to an increase in both oil yield and particle content. A high speed (high throughput) has a contrary impact, reducing both the oil production and the particle count. An optimal balance can be achieved by aligning the production requirements with the filtering capacity. According to Ferchau (2000), the oil yield from rapeseed should range from 32% to 36% of the total mass. This is because excessive pressure or doing a second pressing can result in the presence of unwanted particles in the oil. As a final stage, the raw oil obtained from the press must undergo purification. It is advisable to utilize a press filter and do a security filtration subsequent to decantation. Following decantation, it is necessary to perform a standard filtration process to eliminate the dispersed particles from the oil. Typically, a pressure filter is employed, either in the form of a chamber filter or a vertical filter. To complete the process, it is advisable to perform a security filtration using a certain pore size (ranging from 1 to 5 μm) in order to eliminate any remaining small particles present in the oil. It is crucial to successfully complete the quality control outlined in section 4.5 at this step. Once the final stage is completed and the quality control requirements are met, the oil is ready to be used as fuel in a modified diesel engine. Both the cake meal and the filter cake formed during the SVO extraction method have a high protein content and can be effectively used as animal feed. There exists a modified version of this procedure that employs a solvent to extract a greater quantity of oil from the seed. The aforementioned procedure represents the initial stage. Approximately 70% of the oil is obtained from the seed, while the remaining 30% is left in the cake meal. The subsequent phase involves the utilization of hexane as a solvent for the extraction process. It has a maximum extraction rate of 95% for the seed oil. At this phase, a hexane solvent is combined with rapeseed cake. The solvent effectively dissolves the residual oil present in the rapeseed cake. Once the solvent has evaporated, it is retrieved for reuse. Figure 3 illustrates the schematic of the process. When hexane extraction is used, the resulting cake meal contains a lower protein content compared to simply pressing the seed. Although there are no issues with using it as animal feed.

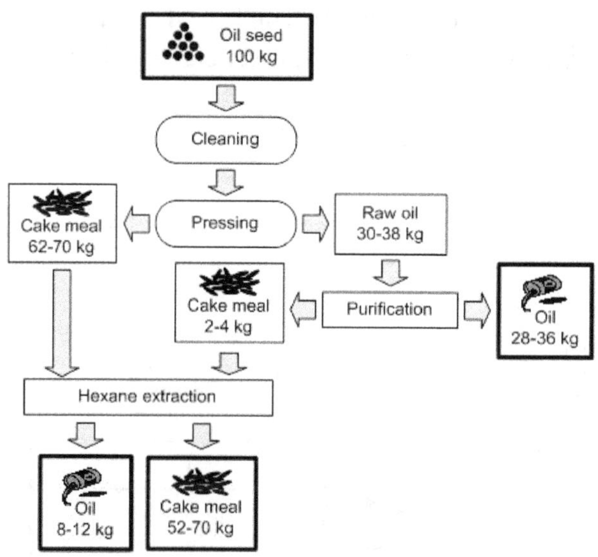

Figure 3. Rapeseed oil hexane extraction process.

4. USE OF RAPESEED OIL AS FUEL

4.1. Use of rapeseed SVO in diesel engines

Rapeseed oil is a viable alternative fuel for diesel engines. Additional vegetable oils can also serve as straight vegetable oil (SVO) to power diesel engines due to their comparable characteristics. Table

2 displays the characteristics of various oils. The disparities in the oil characteristics are minimal. Nevertheless, in order to substitute diesel fuel, certain alterations must be made to adapt the physical characteristics of the oil that will be fed into the engine and crushed in diesel common injectors.

Table 2. Physical and chemical specifications of some vegetable oil fuels.

Fuel type		Diesel fuel	Rapeseed oil	Corn oil	Soybean oil
LHVa(MJ/kg)b		43.35	37.62	37.83	39.62
Density 20ºC (kg/m^3)c		828	915	920	920
Energy content (MJ/l)b,c		35.81	34.42	34.80	36.45
Viscosity (mm2/s)c	20°C	4.64	75.27	70.8	64.37
	80°C	1.64	12.27	11.65	11.29
Cetane numberb		47	37.6	37.6	37.9
Flame point (°C)b		58	275-290	270-295	230
Chemical formulab		$C_{16}H_{34}$	$C_{57}H_{105}O_6$	$C_{56}H_{103}O_6$	$C_{56}H_{102}O_6$
aLHV: Lower Calorific Value;b(Altin et al., 2001);c(Riba et al., 2010)					

The adjustments are intended to elevate the temperature of the rapeseed oil in order to decrease its viscosity and density. During the initial phase of operation, the vehicle utilizes diesel fuel in order to prevent the engine from operating at low temperatures using pure vegetable oil. After the engine reaches its optimal operating temperature, it will have the capability to generate heat and effectively utilize SVO (Straight Vegetable Oil). It is important to avoid prolonged engine shutdown when using SVO (Straight Vegetable Oil) as it can make it difficult to restart the engine when it is cold.

The components that need to be installed in the fuel supply system:

- An additional deposit for the start-up diesel
- A water-oil heat exchanger
- A temperature sensor
- Two solenoid valves to select the fuel to be used
- Filters for oil and diesel fuels

The utilization of vegetable oil as a source of fuel commenced in the distant past. In 1900, Rudolf Diesel utilized peanut oil as fuel to operate a diesel engine during the World Exhibition in Paris (Baquero et al., 2010). In addition, he proposed that vegetable oils had the potential to serve as a future alternative fuel for diesel engines. However, diesel fuel derived from oil has replaced vegetable oil as a result of its greater availability and lower cost.

The use of SVO in diesel engines carries also some difficulties, namely:

- Difficulties in operating the motor itself because of the different ignition temperatures of the two fuels. These difficulties can be solved just by preheating the vegetable oil.
- Problems of engine durability due to deposit formation in the combustion chamber and mix of the vegetable oil with the engine lubricating oil. The first problem is solved by increasing the vegetable oil temperature, so it decreases its viscosity and density, which allows a correct injection and burning of the vegetable oil. The second problem is solved by reducing the life of the engine lubricant, (Agarwal et al., 2008; Vaitilingom et al., 2008).

Although facing these challenges, it is worth mentioning that both rapeseed SVO and diesel fuel have almost same energy content, with rapeseed SVO containing 34.42 MJ/l and diesel fuel containing 35.81 MJ/l. This results in the engine's performance and fuel consumption being nearly identical for both types of fuel. Comparing the performance of different fuels in the same engine, empirical data indicates that diesel fuel performs best at low loads, whereas vegetable oil performs best at high loads.

4.2. Quality control of oil used as fuel

To utilize rapeseed oil as a fuel source, several criteria regarding the oil's physical and chemical qualities must be satisfied. It is important to consider the description of these qualities and their impact on the diesel engine. Therefore, it is necessary to adhere to the German standard DIN 51605. This regulation sets the upper and lower limits for the chosen factors that determine whether a rapeseed vegetable oil may be considered a suitable biofuel for replacing diesel in adapted engines. The characteristics consist of certain inherent features of rapeseed oil, as well as some that are subject to variation and serve as indicators of adequate processing of the oil. The parameters that indicate the degradation of vegetable oil are acid value, iodine index, and oxidation time.

4.3. Use of SVO as fuel

The authors experience in the use of a car with a modified diesel engine is described in this section. The car which engine was adapted to run with SVO is a VW Caddy 2.0 SDI using the parts described in section 4.1.

Table 3 presents the results of a test performed by the authors of this paper with the modified VW Caddy 2.0 SDI after 45000 km of trial. The consumption of this vehicle using diesel is nearly the same as with SVO, as the calorific value of both fuels are almost the same.

Table 3. SVO consumption as fuel.

Own average consumption (l/100km)	7.54
Total distance (km)	45000
Fuel consumption (l)	3393
Average rate oil/total (%)	91.36%
SVO consumption (l)	3100

From the technical data available from Volkswagen, the urban consumption for this vehicle is 7.5 l/100km, the extra-urban is 5.3 l/100km and the combined consumption is 6.1 l/100km. The test carried out with the above-mentioned 70 HP vehicle shows that maintaining an average speed of 70-80 km/h leads to an average consumption of about 6 l/100km. Driving faster, maintaining 120 km/h during long periods of the ride, leads to a consumption of about 9 l/100km.

5. USE OF RAPESEED CAKE FOR ANIMAL FEEDING

Given its substantial protein content, the utilization of rapeseed cake for animal nutrition is worth considering. Many studies have examined the use of cake meal in animal feed, consistently finding that cake meal is a good supplement for animal feed. The use of rapeseed cake into animal feed has been extensively researched. Several studies have been conducted, and the findings indicate that incorporating small amounts of rapeseed cake (up to 10-15%) in animal feed does not cause any notable alterations in parameters such as nitrogen, lipid, and mineral metabolism, as well as the overall health of the animals (Gopfert et al., 2006). The study conducted by Simek et al. (2000) reported no notable variations in the fat, protein, casein, solids, and non-solids fat content of cow milk when cows were fed with fodder containing

15% rape cake. Additional research on rapeseed, conducted by Brzoska (2008), Kracht et al. (2004), and Rinne et al. (1999), has demonstrated that it does not have any detrimental impact on animals or the quality of the meat and milk they produce. Currently, rapeseed is utilized as an ingredient in the feed of several animals. In Spain, there is no legal requirement for determining the limit proportion. However, the Spanish Animal Nutritional Foundation (FEDNA, 2003) has provided some suggestions for different species and ages. The mean chemical composition of rapeseed meal is displayed in Table 4, as reported by Moss and Givens in 1994.

Table 4. Mean chemical composition of rapeseed meal.

Crude protein (g/kg DMa)	397
Crude fiber (g/kg DMa)	106
Cellulose (g/kg DMa)	177
Starch (g/kg DMa)	45
Water-soluble carbohydrates (g/kg DMa)	115
Gross Energy (MJ/kg DMa)	19.7
aDM: Dry matter	

The predominant groups of livestock in the surveyed region consist of cattle, pigs, and poultry (IDESCAT, 2008). The prospective fodder demand is computed by considering the total number of animals and the specific intake of each species. Table 5 displays the fodder consumption values for sample groups in the Anoia region. The cake meal proportion in feed was determined using the guidelines provided by FEDNA (2003). The cake meal yield (1500 kgcake/ha) is determined by multiplying the yield of rapeseed in the regions (2300 kg/ha) by the oil extraction rate by pressing (35% from rapeseed w/w), as explained in section 2 and section 3, respectively.

Table 5. Rapeseed land requirement.

Animal group	Bovines	Pigs	Poultries	Total
Number of livestock per year	6779	89439	607491	-
Fodder (t/year)	16321.8	62093.2	22958.4	-
Maximum cake meal in fodder (%)	17%	7%	5%	-
Maximal cake meal consumption (t/year)	2774.7	4346.5	1147.9	8269.2
Cake meal yield (kg/ha)	1500	1500	1500	-
Rapeseed land (ha)	1849.8	2897.7	765.3	5512.8

If around 3000 hectares of arable land were allocated to rapeseed production, the fodder need in the region under consideration could be fully met by the amount of rapeseed cake meal produced. According to the data presented in Table 5, the land area needed for growing rapeseed to meet the maximum demand for cake meal in the examined region is approximately 5500 hectares.

6. PROPOSED MODEL FOR CROP CULTIVATION AND AGRICULTURAL UTILIZATION

The preceding sections illustrate the cultivation of rapeseed, the extraction of oil by rapeseed processing, and the utilization of the residual cake flour derived from the seed processing. This data can be utilized to construct a cropping model that involves including rapeseed into the existing agricultural rotation, which is mostly focused on wheat and barley (referred to as WBBB, where W represents wheat and B represents barley). The suggested rotation plan involves maintaining a cycle of three years of barley after one year of wheat in each section of the field, with the addition of one year of rapeseed before planting wheat (RWBBB). Rapeseed cultivation leads to a 10% increase in wheat output and a 3% rise in barley yield for typical weather conditions in the two subsequent crops. In addition to incorporating rapeseed into the crop rotation, processing the seeds into oil and cake meal would enable the use of the oil as a direct fuel source for the farm tractor.

The graphical representation of the proposed paradigm for small-scale biofuel self consumption exploitations is depicted in Figure. 4, where the fundamental model, the processing of rapeseed, and the outcome of the many products obtained are depicted. Prior to designing this model, it is necessary to formulate some hypotheses. Initially, the focus is placed on small-scale producers. The farmer is expected to cultivate approximately 100 hectares of fertile land. The concept entails allocating roughly 10% of the cultivable land for self-sufficiency purposes. In the surveyed region, which is characterized by a dry Mediterranean climate, the presence of irrigated areas is almost non-existent. The predominant crops cultivated in this area are traditionally wheat and barley. It is suggested to grow rapeseed as a rainfed crop to alleviate strain on water resources. Furthermore, the use of crop rotation in conjunction with direct sowing will be employed. Therefore, rapeseed can be sown in the same plot of land once every five years. Only the seeds are harvested, while the remainder of the plant is crushed during the seed collection process and left in the fields to decompose. By doing this, the soil replenishes a portion of the nutrients found in the plant's straw, hence reducing the need for a certain quantity of fertilizer. Ultimately, the farmers transport the harvested crops to the nearby farmer's cooperative, where an industrial press is used to extract the oil from the rapeseed yield.

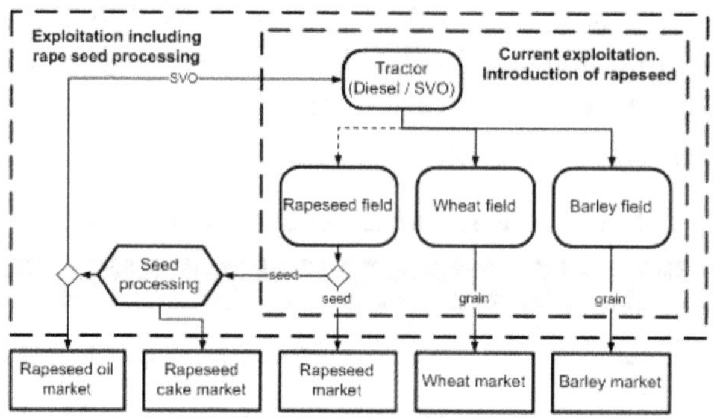

Figure 4. Exploitation model and products fate.

The Food and Agriculture Organization (FAO) of the United Nations is a significant agency that promotes and advocates for the use of effective agricultural practices to reduce the adverse effects, specifically on carbon emissions, soil quality, and water resources. These methods include no tillage and direct seeding, maintaining soil cover, growing numerous crops, selecting appropriate crops, and implementing crop rotations. In the examined area, three primary harvest strategies coexist: traditional seeding, minimum cultivation, and direct seeding. Among these, direct seeding is preferred due to its lower environmental impact, improved carbon retention in the soil, and reduced fuel usage. This model is based on general assumptions. For instance, it is generally accepted that the press is able to extract an average of 80-85% of the total oil content from the seeds. After pressing, the seeds are transformed into 35% oil and 65% meal cake. Furthermore, based on a survey conducted by farmers in the Anoia region (EUETII-UPC, 2010), the average rapeseed production in this area is at least 2300 kg of rapeseed per hectare. If a direct harvesting style of cultivation is used, the fuel usage would amount to around 7000 liters per 100 hectares. The production of rapeseed SVO is expected to reach 875 liters per hectare, as previously stated. Hence, allocating 10% of the cultivable land for rapeseed cultivation is sufficient to meet our own fuel needs.

Additionally, there is a slight surplus of SVO that could be marketed for alternative purposes. According to Hancsok et al. (2008), vegetable oils can also be utilized to create additives that have practical applications in many industries. Section 5 calculates that the 15000 kg of rapeseed cake per 10 ha will be utilized to provide animal feed in this specific location.

7. ENVIRONMENTAL AND ECONOMIC ANALYSES

Life cycle assessment (LCA) is a commonly employed tool for assessing the environmental impact of various processes and the production of different goods (Hsu et al., 2010; Huo et al., 2009; Lardon., 2009; Schmidt, 2010). The literature also utilizes economic assessment using the LCA technique, as demonstrated by Lee et al. (2009), Huppes et al. (2010), Ouyang et al. (2009), and Nassen et al. (2008).

7.1. Environmental analysis

According to the FAO (FAO, 2008), several countries have a policy objective of reducing climate change by promoting the use of bioenergy. On the other hand, life-cycle assessments, which assess emissions during the whole bioenergy production process, indicate significant variations in carbon balances depending on the technologies employed, locations, and production methods. Therefore, further investigation should be conducted in this domain. According to FAO, significant contributors to emissions include land conversion, automation, and fertilizer use during feedstock production. Additionally, the use of non-renewable energy in processing and transport also plays a role. In order to assess the environmental consequences of the proposed model in this study, a comprehensive examination of several aspects can be conducted, including energy and water demands, biodegradability, comparable CO2 emissions (contributing to global warming), tailpipe engine emissions, and deforestation. In addition, the LCA technique (Schmidt, 2010) is employed to assess environmental impacts in a comparable manner. The proposed technology significantly reduces the effects associated with fuel processing and transportation, hence lowering energy needs. In contrast, fossil fuels are imported from distant nations together with the necessary resources for producing large-scale first-generation biofuels. In addition, both fossil fuels and first-generation biofuels necessitate intricate processing that demands substantial energy consumption. Hence, the suggested model effectively minimizes energy usage. In addition, the production of rapeseed contributes to crop rotation and direct seeding. This is strongly suggested as it decreases the number of land working steps, hence lowering power requirements. This leads to a decrease in fuel consumption for each crop, which is a favorable method for reducing emissions. Regarding water needs, according to FAO (2008), numerous feedstocks demand a significant amount of water, which implies that their growth is expected to intensify the competition for this scarce resource, depending on the location and production techniques. The proposed strategy aims to transition to a dry land usage by replacing grains with rapeseed, which can thrive in similar conditions. However, SVO manufacturing does not require any water, while according to (Pate et al., 2007), the current technique for bioethanol production requires around 4 liters of water each liter of bioethanol produced. The consumptive water use in petroleum refining is approximately 1.5 liters per liter, while biodiesel refining necessitates around 1 liter of fresh water each liter of biodiesel produced. Furthermore, in terms of biodegradability, regularly utilized SVOs, such as rapeseed oil, has the ability to decompose naturally and are not harmful to living organisms. This characteristic renders them valuable for transportation purposes in extremely delicate settings, such as marine ecosystems and mining enclosures, as exemplified by West et al. (2008). This indicates a reduced level of danger when storing the fuel and a decreased impact on biodiversity in the event of inadvertent contamination. In order to compare the CO_2 emissions of two models, it is necessary to take into account their disparities. The primary difference between the two systems is in the utilization of SVO instead of petrodiesel, assuming that the machinery, fertilizer, and herbicide requirements remain relatively consistent. The emissions resulting from the transportation, manufacture, and combustion of one liter of petrodiesel amount to 3.16 kg of carbon dioxide per liter (Flessa et al., 2002). Around 10% of these emissions stem from the extraction, manufacture, and transportation of diesel fuel, while the remaining 90% are a result of its combustion. According to local farmers, the fuel usage for direct sowing and conventional seeding is 70 and 140 liters of fuel per hectare, respectively. Therefore, the amount of CO2 released as a result of tractor diesel use is twice as high when utilizing traditional seeding compared to the direct seeding approach. Conversely, the carbon dioxide (CO2) released during the combustion of SVO in a diesel engine was offset by the absorption of CO2 by the crop during its growth, making it carbon neutral. Therefore, the emissions are offset by the absorption of photosynthesis. The manufacture of SVO is characterized by its simplicity and low energy demand, as previously observed. Therefore, the carbon dioxide emissions connected with this stage are far lower than those from petrodiesel. Based on these findings, the suggested solution effectively prevents the release of over 200 kg of CO2 per hectare. A future study will conduct a life-cycle assessment of this model to comprehensively consider all emissions within the examined area. Life-cycle assessments quantify the emissions generated at every stage of the bioenergy production chain. Regarding tailpipe engine emissions, conflicting findings have been reported by Krahl et al. (2007) and Thuneke & Emberger (2007). Regarding CO, CO2, and Particulate Matter (PM) emissions, the SVO (Straight Vegetable Oil) is undeniably superior to petrodiesel. However, when examining the levels of NOx and HC, it remains uncertain whether the utilization of SVO leads to a decrease or an increase in emissions. Therefore, further investigation is required to examine this field with more thoroughness. The large-scale production of biofuels, driven by the high demand for fertile soil, will result in deforestation, particularly in tropical forests (Russi, 2008). Conversely, the provided small-scale manufacturing plant utilizes only a fraction of the available area to manufacture biofuel, hence preventing the aforementioned impact. This study adheres to the overall framework for conducting a Life Cycle Assessment (LCA) as outlined in ISO 14040 (2006) and ISO 14044 (2006) to ensure accurate and representative results. The LCA effect assessment was conducted using the Gabi 4 program (PE International 2010), based on the cropping model described in section 6. The consideration of fuel consumption and the associated fuel emissions includes the utilization of diesel or straight vegetable oil (SVO) as fuel for the tractor. The selection of crop kinds is determined based on the desired crop rotation for each scenario. Crop work data, fertilization requirements, and yield information were collected from the Anoia region, a dry Mediterranean location in northern Spain. Various cropping systems are being examined while keeping the functional unit constant at 109kcal of energy output. This amount of energy is equivalent to what is received from around 100 hectares of land. The direct cropping technique is presumed. An energy functional unit is the optimal choice for assessing a system that yields various outputs, specifically barley grain, wheat grain, rapeseed seed, cake, and oil. The system border encompasses an agricultural enterprise that takes into account various sorts of crops. The destiny of the acquired things is not taken into account, solely the energy that each acquired product signifies. The boundaries encompass three main components: (i) the materials inputs, which consist of fertilizers, herbicides, insecticides, fungicides, diesel fuel, and planting seeds, (ii) the cropping stages, which involve fertilizing, herbicide, insecticide, and fungicide treatments, as well as sowing, harvesting, and transportation of seeds/grains to cooperative installations, and (iii) the rapeseed processing stage, which includes transportation, pressing, filtering, and degumming processes. This environmental study considers three scenarios, each based on grouping three crop types: barley, wheat, and rapeseed. The barley, wheat, and rapeseed models focus on the cultivation and harvesting of the respective grains and seeds. In addition, the rapeseed model includes the process of extracting rapeseed oil from the seeds, which can then be used as a biofuel (SVO) in the operation. One scenario also considers the utilization of SVO as fuel. Therefore, the first scenario refers to the present manner of exploitation (current scenario). The second scenario involves the integration of rapeseed into crop rotation, but it only relies on diesel fuel. The third component also encompasses the processing of rapeseed and the utilization of SVO fuel in the SVO scenario. The emissions of the model under consideration are grouped together into impact categories using an internationally recognized manner throughout the impact assessment process. The CML technique, developed by the Environmental Sciences Institute at Leiden University, was used for this study because to its ability to generate greater worldwide consensus and minimize subjectivity (Guinée et al., 2001; Alvaro-Fuentes et al., 2009). The cause-effect method described by Dreyer et al. (2003) is used to reduce uncertainty in groups by categorizing impacts. It computes the escalation of harm and measures its impacts (Garraín, 2009). Figure 5 displays the results of the environmental impact categories utilizing 6 CML non-toxicological impact categories, energy consumption, and land use for each scenario, with the current scenario serving as the reference point. The incorporation of rapeseed into the traditional crop rotation and its utilization for the production of straight vegetable oil (SVO) for self-consumption of fuel somewhat mitigates certain environmental consequences that are taken into account. The crop energy ratio indicator demonstrates a clear preference for scenarios fueled by SVO (Straight Vegetable Oil), with the ratio being 21.6% higher for the SVO scenario compared to both the current scenario and the diesel seed scenario. The SVO scenario has adverse environmental implications, namely in terms of Ozone Depletion Potential (ODP) and Photochemical Ozone Creation Potential (POCP), which are only 8.5% and 9.8% more severe compared to the reference scenario. A marginal increase in land requirements is observed in both diesel and SVO scenarios, although it is not significant, amounting to less than 1.7%. The SVO scenario has resulted in a significant reduction in both the ADP and GWP environmental impacts, with a decrease of 21.2% and 6.2% respectively compared to the reference scenario. AP and EP exhibit a decrease of 2.7% and 1.9% respectively compared to the reference scenario, although these reductions are not highly indicative. However, the impact of the diesel scenario relative to the reference scenario varies by less than 4%, making them nearly identical. The higher diesel scenario has a greater influence on global warming potential (GWP), increasing it by 3.7%, while in the SVO scenario, it decreases by 6.2%. The sensitivity analysis conducted indicates minimal difference in trend. However, the influence of electrical energy consumption in the SVO scenario can be mitigated by utilizing renewable electrical energy.

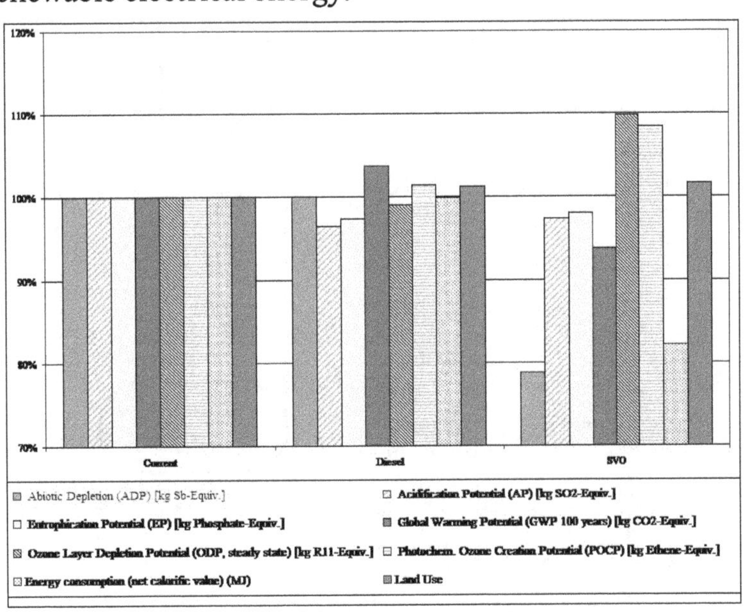

Figure 5. Environmental impact category results

7.2. Analysis of economic factors

Following the same model as in the environmental analysis, an economic assessment is done to evaluate The many cropping options being evaluated for the region. Therefore, it is possible to attain a comparative economic outcome. The life cycle costing (LCC) technique allows for the collection and assessment of the resources, outcomes, and possible financial advantages of a product system over its entire life cycle (Lee et al., 2009). Various life cycle costing (LCC) methodologies exist, which vary based on their specific objectives, associated expenses, and the overall context of the LCC. LCC is widely applied in various domains, including construction methodologies and restoration (Nassen et al., 2008; Ouyang et al., 2009), as well as military hardware (Huppes et al., 2004). However, the adoption of a product-oriented technique in agricultural processes is rare. The costs encompass investments that are evenly divided over the years of utilization, as this study aims to provide an average economic advantage for a one-year period of operation (a full season). The benefit is demonstrated as a representative indicator of the feasibility of each scenario. Figure 6 depicts a straightforward illustration of the rapeseed processing and the utilization of its by-products. This diagram also illustrates the production of rapeseed, including the many inputs involved and the subsequent processing. Both wheat and barley are subjected to the same input method. Utilizing the LCA-based LCC methodology, this analysis considers all the steps of the process. The benefit calculation is formulated by simulating each crop variety and the subsequent rapeseed processing phase. Every crop variety necessitates specific fertilization and crop protection chemicals. The rapeseed processing stage is only relevant when there is a need to alter the seed, as in the case of the straight vegetable oil (SVO) situation. When considering the consumption and accompanying fuel emissions, it is important to take into account the usage of diesel or SVO in the tractor. Each scenario is derived from the amalgamation of various crop partitions, each possessing its own distinct characteristics, as previously elucidated. The divisions consist of Barley, Barley-2, Fallow, Rapeseed, Wheat, and Wheat-1. Barley-2 represents barley that is grown two years after rapeseed, while Wheat-1 represents wheat that is grown one year after rapeseed. The advantages acquired in each situation are illustrated in Figure 7.

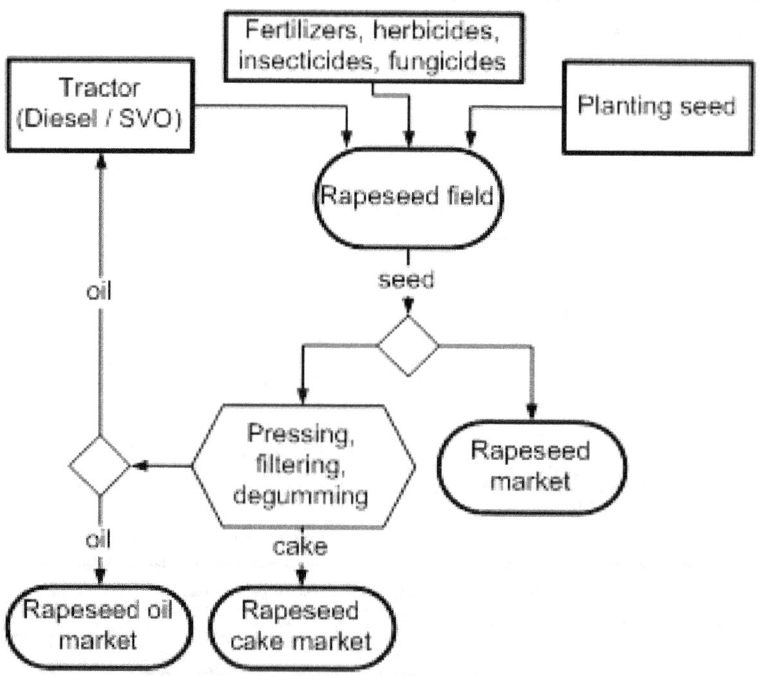

Figure 6. Rapeseed production scheme

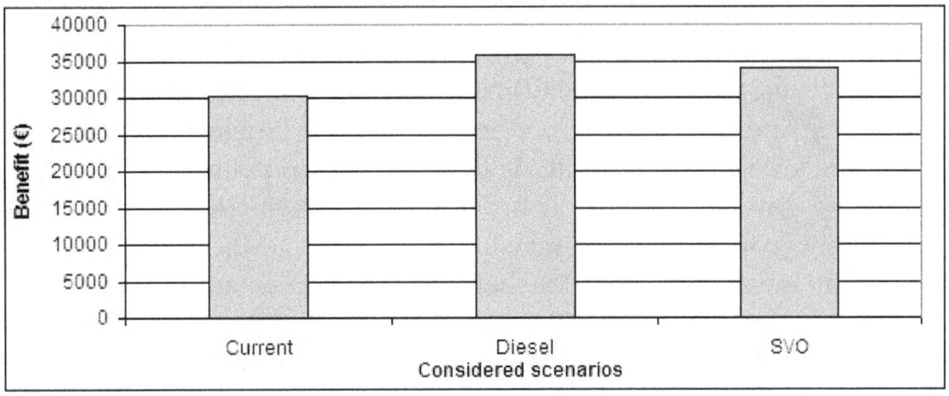

Figure 7. Economic benefit results for each scenario

The contributions to the benefit vary depending on the proportion of each crop split. Unprocessed rapeseed yields a greater benefit per hectare compared to other types of crops. The results indicate that the economicviabilityofsellingrapeseedoilthroughsmall-scaleprocessingisinferiortodirectlysellingrapeseed. An analysis has been conducted to evaluate the use of diesel or SVO as fuel options, specifically focusing on the sensitivity of diesel prices and taxes. The results indicate that the provision of diesel for agriculture in Spain is hindering the adoption of SVO (Straight Vegetable Oil) as a fuel for agricultural operations. Existing rules do not particularly promote the use of self-supply fuels for agriculture, which makes it difficult for them to compete with the already established use of diesel fuels. Under the present economic circumstances, it is evident that utilizing crop rotation RWBBB with diesel as fuel (diesel seed scenario) is presently the most optimal choice. A similar option to this is the SVO seed scenario, which follows the same rotation strategy but allocates a portion of the harvested seed to provide fuel for the agricultural machines. Consequently, it decreases the quantity of diesel utilized in the operation. The benefits of the Diesel and SVO scenarios are respectively 15% and 11% larger than the reference scenario (current situation).

8. CONCISE SUMMARY

This chapter provides an overview of the production and utilization of rapeseed oil as a self-produced agricultural biofuel. It also examines the application of rapeseed oil in the specific research region. The given approach is also assessed from an environmental and economic perspective. The initial three sections present the methodology for acquiring rapeseed oil using small-scale production technologies and conducting an analysis to determine the most suitable rapeseed variety for a particular region. Further research is required to analyze the optimal variety for each specific region being studied. The fourth section provides a concise overview of the required alterations in diesel engines to enable their operation on straight vegetable oil. Additionally, it presents actual fuel consumption statistics obtained from a vehicle that has been modified accordingly. Additionally, it demonstrates the utilization of rapeseed cake as animal feed. Section 6 introduces an exploitation model that incorporates rapeseed into the traditional cropcycleofwheatandbarley.Thismodelalsoincludestheintegrationofseedprocessingandoilutilization. Section 7 and 8 present the environmental and economic outcomes of the proposed model in comparison to both the traditional rotation and the exclusive reliance on diesel in the proposed rotation. In the suggested exploitation model, all byproducts derived from the processing of the rapeseed plant (including straw, rapeseed cake, oil, and seeds) have specific use (such as being used in the field, as animal feed, for biofuel production, or for seed sales) and are assigned a predetermined market value. Consequently, there are no byproducts produced. Moreover, the process of getting SVO (Straight Vegetable Oil) is more environmentallyfriendlycomparedtobiodieselmanufacturingduetoitsreducedenergyconsumptionand the absence of chemical usage, such as methanol. The results of using SVO and diesel fuel in the proposed rotation with rapeseed are being compared to the current rotation. The life cycle assessment presents the environmental impact category results by utilizing six CML non-toxicological impact categories. The environmental assessment indicates a clear preference for SVO (Subject-Verb-Object) in the majority of categories, while there are some areas that yield unfavorable outcomes. The implementation of this exploitation model should consider the latter in order to reduce them. However, the economic viability is uncertain given the current economic circumstances. However, it could be viable in future situations where the availability of fossil fuels was restricted. Furthermore, the implementation of small-scale production and consumption of straight vegetable oil (SVO) has the potential to rejuvenate rural economies and reduce their reliance on diesel fuel. Moreover, this technique might also prove beneficial in underdeveloped nations, where the availability or accessibility of diesel fuel may be limited. Furthermore, the promotion of research in various domains like crop sustainability, crop emissions, new plant varieties, modifications to diesel engines, and the development of new types of lubricants is encouraged. Further investigation is particularly necessary to evaluate the sustainability of the suggested model across its whole life cycle. Therefore, the utilization of SVO (Straight Vegetable Oil) in diesel engines is a tangible option that should be considered when contemplating small-scale biofuel manufacturing.

REFERENCES

Agarwal D. Agarwal A.K. 2007 Performance and emissions characteristics of Jatropha oil (preheated and blends) in a direct injection compression ignition engine. Appl. Therm. Eng., 2713 September 2007) 231423231359-4311

Altin R. Cetinkaya S. Yucesu H. S. 2001 The potential of using vegetable oil fuels as fuel for diesel engines. Energy Conversion and Management,425March 2001) 5295380196-8904.

Alvaro-FuentesJ. Lopez M.V. ArrueJ. L. Moret D. Paustian K. 2009 Tillage and cropping effects on soil organic carbon in Mediterranean semiarid agroecosystems: Testing the Century model. Agriculture Ecosystems & Environment, 1343-42112170167-8809.

Baquero G. Esteban B. Rius A. RibaJ. R. Puig R. 2010 Small-scale production of straight vegetable oil from rapeseed and its use as biofuel in the Spanish territory. Energy Polic, 3811891960301-4215

Brzoska F.2008 Milk production and composition as influenced by soybean meal, rapeseed meal or rapeseed cake in concentrates for dairy cows. Annals of Animal Science, 82 March 2008) 1331431642-3402.

Dreyer L.C. Niemann A.L. Haus child M. Z. 2003 Comparison of three different LCIA methods: EDIP97, CML 2001 and Eco-indicator 99-Does it matter which one you choose? International Journal of Life Cycle Assessment,841912000948-3349

Esteban B. Baquero G. Puig R. Riba-RJ. Rius A. 2011 Is it environmentally advantageous to use vegetable oil directly as biofuel instead of converting it to biodiesel? Biomass & Bioenergy, 353 March 2011), 131713280961-9534.

EUETII-UPC. PNT GRA 105, Normalised working procedure: Determination of the total amount of oil extractable from rape seed.2006 Escola d' Enginyeria d' Igualada. Igualada, Spain.

EUETII-UPC. Survey conducted between 2009 2010 in the Anoia region, to the farmers of Unio de Pagesos and Coperal agriculture cooperatives. (2010 Escola d' Enginyeria d' Igualada. Igualada, Spain.

FAO. Reports on international commodity prices. Food and Agricultures Organization of the United Nations 2.03.2011 Available from: Available from: www.fao.org/es/esc/prices/WeeklyPricesServlet.jsp.

FEDNA.2003 Fundacition española para el desarrollo de la nutricition animal, Tables of composition for animal food and quality control, In Fundación Española el Desarrollo de la Nutricition Animal. 2.03.2011, Available from:http://www1.etsia.upm.es/fedna/introtabla.htm.

Ferchau E. 2000 Equipment for decentralised cold pressing of oil seeds, In: Folkecenter for Renewable Energy, 2.03.2011, Available from: http://www.folkecenter.net/mediafiles/folkecenter/pdf/dk/efdcpos_ef.pdf.

Flessa H. Ruser R. Dorsch P. Kamp T. Jimenez M.A. MunchJ. C. Beese F. 2002 Integrated evaluation of greenhouse gas emissions (CO_2, CH_4, N2O) from two farming systems in southern Germany. Agriculture Ecosystems & Environment, 911-31751890167-8809

GalanM. I. Bonet J. Sire R. Reneaume J. M. Plesu A. E. 2009 From residual to useful oil: Revalorization of glycerine from the biodiesel synthesis. Bioresour. Technol, 10015 August 2009), 37753778ISS 0960-8524

Garration D. 2009 Desarrollo y aplicacition de las categorías de impacto ambiental de ruido y de uso de suelo en la metodologtia de análisis de ciclo de vida. In Universitat Jaume I. 2.03.2011, Available from: http://www.tdr.cesca.es/TESIS_UJI/AVAILABLETDX-063010901132garrain.pdf

Gopfert E. Trckova M. Dvorak R. 2006 The use of treated rape cake in a calf starter diet, Czech J. of Anim Sci, 5111 August 2006), 4915011212-1819

GuinéeJ. B. Gorrée M. Heijungs R. Huppes G.K leijn R. Koning A.d. Oers L. v. Sleeswijk A.W. Suh S. Haes H.A.U. d. Bruijn H. d. Duin R. v. Huijbregts M.A.J. 2001 Life cycle assessment: an operational guide to the ISO standards, Center of Enviromental Science- Leiden University 2.03.2011, Available from:http://media.leidenuniv.nl/ legacy/ new-dutch-lca-guide-part-1.pdf.

Hancsok J. Bubalik M. Beck A. Baladincz J. 2008 Development of multifunctional additives based on vegetable oils for high quality diesel and biodiesel. Chemical Engineering Research & Design, 867A September 2007) 7937990263-8762.

Hsu D. Inman D. Heath G.A. Wolfrum E. J. Mann M. K. Aden A. 2010 Life Cycle Environmental Impacts of Selected US Ethanol Production and Use Pathways in 2022. Environmental Science & Technology, 4413 June 2010) 528952970001-3936X.

Huo H.W ang M.Bloyd C. Putsche V. 2008 Life-Cycle Assessment of Energy Use and Greenhouse Gas Emissions of Soybean-Derived Biodiesel and Renewable Fuels. Environmental Science &Technology, 433February 2009) 7507560001-3936X.

Huppes G.Rooijen M.v. Kleijn R. Heijungs R. Koning A. d.Oers L. 2004 Life Cycle Costing and the Environment. Report of a project commissioned by the Ministry of VROM-DGM for the RIVM Expertise Centre LCA. 2.03.2011, Available from: http://www.rivm.nl/milieuportaal/images/Report%20LCC%20April%20%202004%20final.pdf.

ISO 14040: Environmental management-- Life cycle assessment- Principles and framework. 2006 International Organization for Standardization Geneve, Switzerland ISO 14044: Environmental Manangement-Life Cycle Assessment-Requirements and guidelines. 2006 International Organization for Standardization Geneve, Switzerland IDESCAT database, Agricultural census,1999Institut d'Estadística de Catalunya, 2.03.2011, Available from:http://www.idescat.cat/KrachtW. Danicke S. Kluge H. KellerK.MatzkeW.HennigU.SchumannW.2004Effect of dehulling of rapeseed on feed value and nutrient digestibility of rape products in pigs. Archives of Animal Nutrition,585August 2006)3894040000-3942

Krahl J. Munack A. Bünger J. 2007 Emissions of Biodiesel and Vegetable Oil, In: UFOP, 2.03.2011, Available from: http://www.ufop.de/downloads/report.pdf

Lardon L. Helias A. Sialve B. Stayer J.P. Bernard O. 2009 Life-Cycle Assessment of Biodiesel Production from Microalgae. Environmental Science & Technology, 4317647564810001-3936X

LeeJ. Y.YooM.ChaK.LimT. W.HurT.2009Life cycle cost analysis to examine the economical feasibility of hydrogen as an alternative fuel. International Journal of Hydrogen Energy, 3410424342550360-3199

MossA. R. Givens D. I.1994 The Chemical-Composition, Digestibility, Metabolizable Energy Content and Nitrogen Degradability of Some Protein-Concentrates. Animal Feed Science and Technology, 473-4June 1994) 3353510377-8401

Nassen J. Sprei F. HolmbergJ. 2008 Stagnating energy efficiency in the Swedish building sector-Economic and organisational explanations. Energy Policy, 3610381438220301-4215

Ouyang J. L. Ge J. Hokao K. 2009 Economic analysis of energy-saving renovation measures for urban existing residential buildings in China based on thermal simulation and site investigation. Energy Policy, 3711401490301-4215

Pate R. Hightower M. Cameron C. Einfeld W. 2007 Overview of Energy-Water Interdependencies and the Emerging Energy Demands on Water Resources, Report SAND20071349C, 2.03.2011, Available from:http://www.sandia.gov/energy-water/ PE_International2010In Institut für Kunststoffprüfung und Kunststoffkunde; GMBH, P. E., Ed. Stuttgart, Echterdingen, 2010. 2000 World Rapeseed Production in 2000/01 Likely to Fall Below Record 1999/ 2000 Levels, **In:** Feature commodotity articles, 2.03.2011, Available from: http://ffas.usda.gov/wap/circular/2000/0003/wap2.htm

Riba J. Esteban B. Baquero G. Puig R. Rius A. 2010 Characterization of physical properties of vegetable oils to be used as fuel in diesel engines. A finidad 67546100106

Rinne M. Jaakkola S. Varvikko T. Huhtanen P. 1999 Effects of type and amount of rapeseed feed on milk production. Acta Agriculturae Scandinavica Section a-Animal Science, 493September 1999), 1371480906-4702

Russi D.2008 An integrated assessment of a large-scale biodiesel production in Italy: Killing several birds with one stone? Energy Policy, 363 March 2008), 11691180ISS 0301-4215 1998 Rapeseed (Brassica campestris/ Brassica napus), In: 2.03.2011, Available from: http://extension.oregonstate.edu/catalog/html/em/em8700

Schmidt J. H.2010Comparative life cycle assessment of rapeseed oil and palm oil. International Journal of Life Cycle Assessment,1521831970948-3349

SimekM.SustalaM.VrzalovaD.TrinactyJ.2000The effect of rape cakes in feed mixtures on the content parameters of cow milk. Czech Journal of Animal Science,454April 2000)1611671212-1819

Thuneke K. Emberger P. 2007 Exhaust gas emission characteristics of rapeseed oil fuelled tractors-Investigations at a test stand. Conference: Agricultural Engineering 2007,200147520083-5560

Vaitilingom G. Perilhon C. Liennard A. Gandon M. 1998 Development of rape seed oil burners for drying and heating. Industrial Crops and Products,72-3January 1998), 273279ISS 0926-6690

WestA. H. Posarac D. Ellis N. 2008 Assessment of four biodiesel production processes using HYSYS. Plant. Bioresour. Technol.,9914658766010960-8524

CHAPTER-5

NANOTECH BIOFUELS AND FUEL ADDITIVES

1. INTRODUCTION: AN OVERVIEW

The author of this chapter was inspired to write it after giving a presentation at the Chemindix conference in Bahrain in October 2010. The 8th International Conference & Exhibition on Chemistry in Industry was organized by the Saudi Chapter of the American Chemical Society and Aramco. This text examines the application of nanotechnology in the manufacturing of biofuels and the use of gasoline additives, including those made from renewable materials. The text provides an introduction to the topic by presenting the wider context of petroleum fuels and biofuels. An intelligent approach to the future of oil refining would involve using profits to fund a shift away from oil and towards alternative sources of transportation, including biofuels. Future prospects for liquid biofuels entail the widespread adoption of advanced technology beyond the initial generation, as well as the ongoing enhancement of existing first-generation methods. However, nanotechnologies are being considered as potential technologies for the future development of biofuels. The field of nanotechnology is extensive and its applications are limitless. Subsequently, there is a concise examination of advancements in nanotechnology, particularly in relation to liquid particles, in addition to the more prevalent applications involving solid particles. The paper presents and discusses the growth, harvesting, and conversion of algae, highlighting their considerable potential in contributing to an energy future where biofuels play a crucial role. The majority of the present endeavor in the advancement of second generation liquid biofuels focuses on the conversion of biomass cellulosics into ethanol and biodiesel. The chapter will examine and evaluate the ongoing pursuit of nano processes. Currently, the methods employed to transform oils and animal fat into biodiesel involve trans-esterification using methanol or ethanol. This process unavoidably produces glycerol, which must be either sold or disposed of correctly. The utilization of nanoscale techniques could prove beneficial in tackling this problem. There are speculations regarding how liquid nanoparticles of fuel additives can improve the efficiency of biofuel/ fuel mixes with additives. These speculations are based on the impacts of surface and combustion. Public apprehensions over the effects of nanotechnologies on security, health, and the environment are also acknowledged and deliberated. However, a prudent and hopeful perspective is offered regarding the significant advantages of a meticulous integration of nanotechnologies in the field of biofuels and fuel additives, as well as in several other applications, particularly those related to human health.

2. THE FUTURE OF OIL REFINING [1] AND THE SHIFT OF THE OIL INDUSTRY TO OTHER ENERGY SOURCES [2]

The oil business is facing significant challenges as the demand for energy services such as transportation, lighting, and heating continues to grow. This increased demand is leading to a greater need for liquid fossil fuels, given the current state of technology. Oil producers and refiners encounter challenges in sourcing ample quantities of high-quality crude oil at reasonable prices to meet the increasing demand in the long term. Meanwhile, users and the general public are advocating for immediate environmental enhancements, such as improved air quality. Furthermore, apprehensions regarding the consequences of climate change resulting from the amplified release of greenhouse gases during the production and utilization of liquid hydrocarbons may ultimately necessitate a shift towards energy systems that are environmentally sustainable. This presents biofuels with an opportunity to enter the market as we shift into an uncertain future in terms of energy. Given this context, the oil refining industry in the United States and the European Union has experienced a lack of growth or progress. The construction of new refineries has been hindered by environmental barriers presented by an assertive public, compounded by a prevailing "not in my backyard" mentality. Furthermore, the decrease in profits for processed goods may have caused prominent companies to prioritize activities in the first stages of production. Conversely, the process of refining is growing in several regions around the globe, including India, China, Brazil, and the Middle East. This expansion is driven by the development of these countries and the efforts of oil producers to enhance the worth of their resources. One indication of the transition of refining operations to developing economies is the presence of the world's largest refinery in India, which is owned by Reliance. However, globally, the rising long-term marginal cost of oil, along with environmental constraints and more stringent government laws and mandates, are expected to result in a decrease in the importance of oil in the global energy mix, with natural gas becoming more favored. The transition of dominance from wood and coal to oil has occurred over the course of the last two centuries. Oil businesses are progressively rebranding themselves as energy companies. Some companies will utilize their existing oil production profits to gradually shift towards alternate energy sources. An economically advantageous shift to alternative sources in the oil industry necessitates a steady diversion of oil revenues towards environmentally friendly investments and the extension of the lifespan of existing oil reserves.

3. CHALLENGES RELATED TO LIQUID BIOFUELS [3]

The worldwide daily use of liquid fuels is around 80 million barrels per day, which is equivalent to 12.7 million cubic meters per day. To create an equivalent volume of ethanol, approximately 700 million hectares of sugar cane land is needed, assuming a yield of 6.5 m3/ha/year of ethanol. This region is equal to 100 times the size of the sugar cane cultivation area in Brazil, which is the second largest producer of bio-ethanol globally. Biofuels undeniably encounter a challenge in terms of scale. In 2010, the combined use of fuel ethanol and biodiesel only accounted for a small 3% reduction in global oil consumption. Figure 1[4] below demonstrates the scale problem by depicting the amount of land required for the USA to produce its own fuel.

Algae seem to require the smallest amount of space to fulfill the substantial need for liquid fuels in the USA, while soybeans require an area larger than the 48 continental states of the USA. The land area needed for growing corn is significant. These findings indicate that the existing biofuels production capacity in the USA would be insufficient to satisfy the demand, necessitating imports to fulfill the enormous energy needs of America.

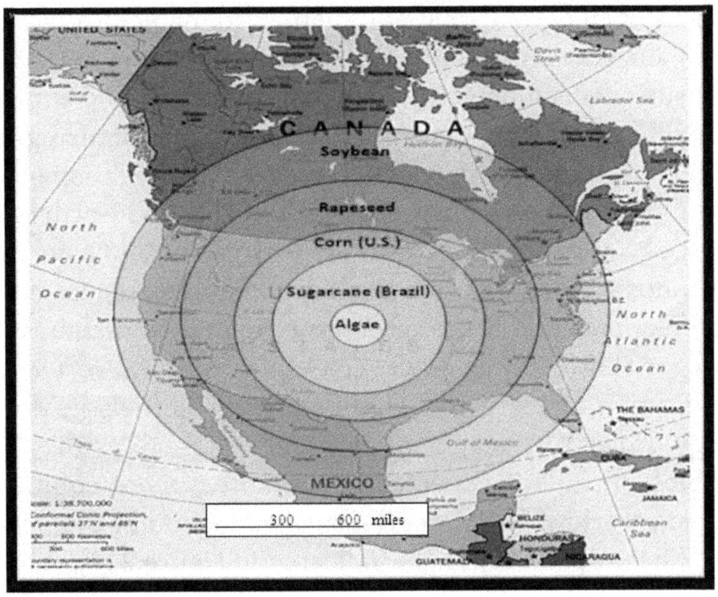

Figure 1. How much land would it take for the USA to grow its own fuel?

The primary obstacle faced by biofuels is the issue of scalability, which encompasses the labor, management, land, water, and sunlight necessary for biomass production and the subsequent processing involved. These resources are limited in quantity and are necessary for the production of food, animal feed, and materials such as fiber to fulfill diverse human needs. These resources have a trade-off in terms of the potential benefits they could have in other markets. Developing biofuels on a commercial scale necessitates substantial funding, which might also be utilized for other purposes. One of the challenges that biofuels need to overcome in order to succeed is the mobilization of private money in the face of market and other uncertainty.

Local environmental concerns arising from the development of biofuels require attention. For example, in the context of sugar cane ethanol, the liquid residue of distillation, known as stillage, possesses a significant chemical and biological oxygen demand. Consequently, it necessitates suitable treatment prior to its ultimate disposal. When planned and managed well, a biofuels production system would contribute just a small amount to greenhouse gas emissions, considering the viewpoint of global climate change. However, in reality, numerous biofuel production systems worldwide are causing a net increase in greenhouse gas (GHG) emissions. An ongoing issue in the growth of the biofuels sector is the current conflict over the limited resources used as raw materials by both the food and fuel industries. All edible vegetable oils often used in food preparation can be converted into biodiesel. The same scenario applies to fuel ethanol, particularly when it comes to the feedstocks derived from starch, such as corn and wheat. However, the increase in food prices that occurred worldwide in 2007/8 and is still occurring in 2010/1 is primarily a result of other factors such as droughts and other climate-related events, elevated oil costs, and market speculation. The success of the biofuels economy relies heavily on the availability of feedstocks at the right quantities, timing, and prices, as feedstock cost is the main factor influencing the overall cost of biofuels. The industry is moving towards non-food feedstocks and the use of second generation technologies to convert cellulosic biomass into liquid biofuels due to the food versus fuel dilemma and the requirement for reliable and cost-effective access to feedstocks. Consuming markets of biofuels are expressing concern about the sustainability of the systems used to produce them. This concern is causing suppliers to feel pressured to comply with sustainability norms that are subject to certification. The sustainability of biofuels is contingent upon the promotion of unrestricted global trade, as this would gradually replace unsustainably produced biofuels with those from regions that adhere to sustainable production standards. In 2008, the Rockefeller Foundation organized a significant discussion at its Bellagio Centre, which resulted in the establishment of a consensus on sustainable biofuels. The goal was to comprehend the various factors that influence the sustainable commerce, consumption, and manufacturing of biofuels, as well as the advantages of supplying regions in conjunction with the need and technology of consuming regions [5]. Nevertheless, there are still numerous tasks that need to be accomplished in order to attain unrestricted global commerce in biofuels. The Doha rounds of the World Trade Organization have come to a deadlock. At now, biodiesel is classified as an industrial product, but gasoline ethanol is categorized as an agricultural product, which grants it greater protectionism. A necessary requirement is the establishment of a comprehensive approach to biofuels, categorizing them as environmental goods and services. However, despite these disadvantages, a clear indication of a growing significance of biofuels in the future is the emergence of new biofuels technology projects by major oil corporations, such as BP, Chevron, Exxon, and Shell. The expansion of global biofuel trade is expected to promote a more equitable distribution of biofuel production and consumption worldwide. Currently, the production of biofuels is mostly focused in the United States, Brazil, and the European Union, as indicated in Figure 2 [6].

4. THE VASTNESS OF NANOTECHNOLOGY

Nanotechnology is the field of constructing machines or devices at the molecular scale, typically a few nanometers (10-9m) in size, which is significantly smaller than a cell. Table 1 presented below illustrates several practical uses of nanotechnologies, highlighting the extensive scope of their area [7]: Nanotechnologies give rise to nano-structured polymers in the field of polymers, which have practical significance. These polymers find applications in several areas such as support structures, industrial processes, diagnostics and treatments, medicines, medical and dental prostheses, and thin films for surface treatment. The primary chemical constituents of nano-structured polymers include poly-oxides, poly-acrylates, poly-vinylics, poly-saccharides, and poly-ethylenes. The primary constituents integrated into polymer nano-matrices are silicon, chromium, and carbon.The user's text is "[8]".

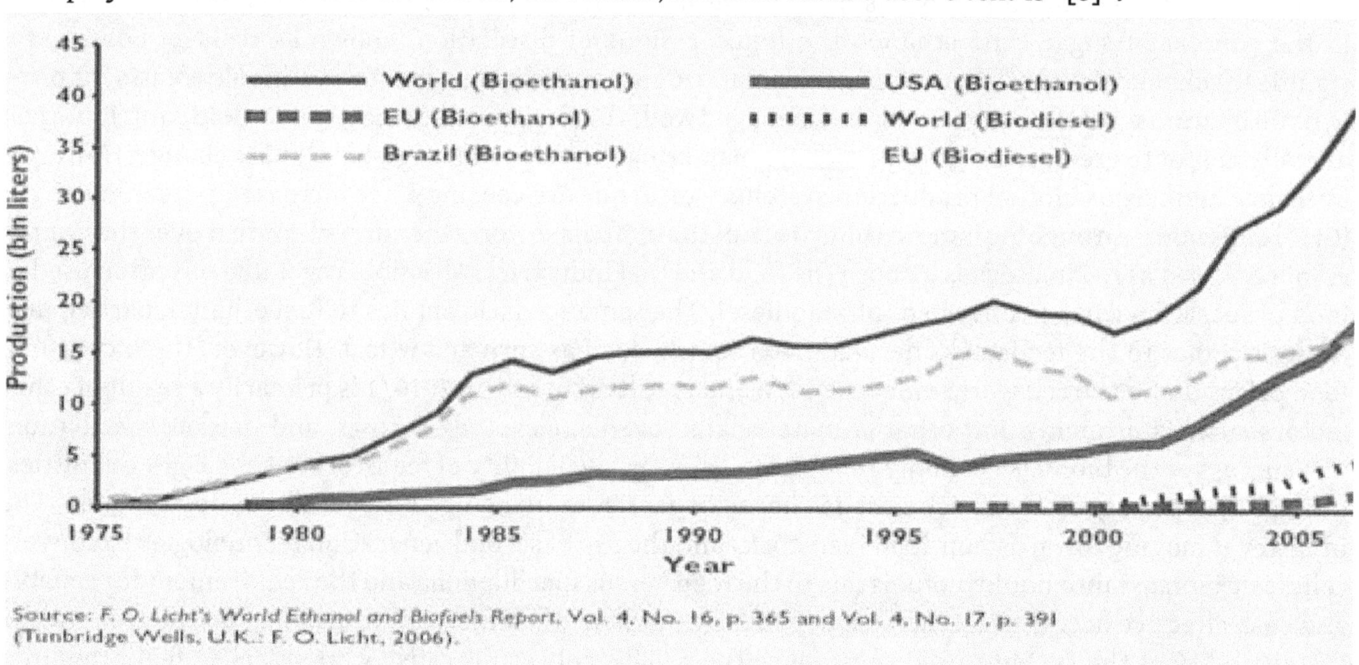

Figure 2. World's biofuels production is concentrated in the US, Brazil and the EU

Table 1. Some applications of nanotechnology

Nanocrystalline	DrugsOxide	Nanoparticles
Nanofilm	CoatingsCarbon	Nanotubes
CosmeticsSilicon		Nanomaterials
NanocatalystsMetal		nanoparticles
CeramicsPolymer		Nanocomposites
RubbersInorganic Nanocrystals		

5. CONVERTING ALGAE INTO BIOFUELS

According to Figure 1, biofuels produced from algae have significant promise due to their ability to achieve high yields and require less land space. Furthermore, algae can contribute to carbon mitigation by utilizing carbon-dioxide (CO_2) as a nutrient source, along with water and sunlight, throughout their growth process. Algae can also be nourished with alternative substances. In order to grow efficiently using carbon dioxide, concentrated sources of the gas are required, such as those present in the emissions from power plants that burn fossil fuels Oil can constitute up to 50% of the algal mass, which is significantly higher compared to oil palm trees, the most oil-rich plants, where less than 20% of the biomass is composed of oil. Algae polysaccharides can be converted into ethanol or transformed into bio-gas by gasification, resulting in methane or hydrogen [9]. However, the progress of algae in becoming viable sources of commercial biofuels is hindered by several obstacles that need to be addressed. As algae require water for growth, the increase in algal production may lead to a conflict between using water for fuel production or for other purposes, similar to the preceding discussion on the conflict between using food for fuel. Another obstacle arises from the relatively low concentration of natural carbon dioxide in the atmosphere. Therefore, it is necessary to explore alternative sources of carbon to support the growth of algae in a commercial biofuels system. One possible solution to these difficulties could involve the application of nanotechnology to convert algae into biofuels. For instance, in 2009, QuantumSphere was awarded a grant by the California Energy Commission to advance the development of nano-catalyzed algae biogasification. In California, the Salton Sea is subject to significant volumes of agricultural runoff, which may result in the formation of extensive algae blooms. These algae and related biomass have been converted into methane, hydrogen, and other gasses through experimental means [10]. An example of nanotechnology used to algal development involves the utilization of nanoparticles to extract biofuel oils from algae without causing any damage. This concept is seen in Figure 3 [11]. The nanoparticles are depicted on the left side of the shot prior to the introduction of oil-laden algae. The image on the right depicts the interaction between the algae and the nano particles, leading to the extraction of oil without causing any harm to the algae. Ensuring the survival of the algae can significantly decrease production expenses and the time it takes to generate new batches.

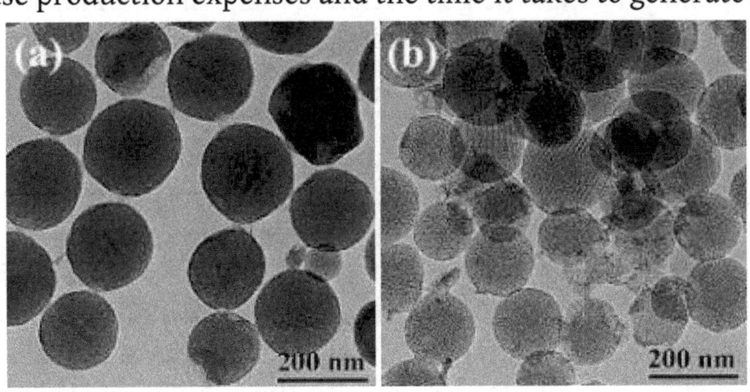

Figure 3. Nano-particles harvesting oil from algae without harming the organism

One potential drawback of the nano-harvesters is the possibility of their unintentional release into the environment, despite the fact that the spherical nano-particles are composed of calcium compounds and sand [12]. The spheres possess holes that are coated with chemicals, enabling the extraction of algal oil without causing damage to the cell membrane. However, before nano-harvesting can be widely used in the commercial market, it is necessary to conduct thorough investigations to assure the safety of these technologies.

6. THE APPLICATION OF NANOTECHNOLOGY TO LANDFILL FACILITIES[13]

The organic matter in landfills tend to undergo anaerobic fermentation yielding methane and CO_2 [14], which if naturally vented into the atmosphere would add to the greenhouse emissions that warm the climate. And the climate change impact of methane is 25 times larger than that of carbon dioxide for a time horizon of 100 years [15]. Thus, there is a need to sequester the carbon present in landfill methane. Nano-catalysts can crack methane into elemental carbon and hydrogen. The carbon can be produced in high-purity nano-graphite for use in aerospace, automobile, batteries, etc. This approach to handling methane can considerably improve the economics of landfills as well as of anaerobic digester plants that generate electricity from biogas fueled electricity.

7.THEUTILIZATIONOFNANOTECHNOLOGYFORTHECONVERSIONOFBIOMASSINTO BIOFUELS.

Decoupling the production of biofuels from food crops is an essential prerequisite for increasing the worldwide market share of biofuels. One of the difficulties this technique encounters is the natural resistance of cellulosic feedstocks to being converted into simpler sugars that can be fermented into ethanol. The potential lies in utilizing nano-particles as immobilizing beds for costly enzymes, enabling them repeated usage in breaking down complex cellulose polymers into simpler fermentable sugars [16]. Louisiana Tech University, led by Dr. James Palmer, is actively involved in this global initiative. They are partnering with esteemed scholars Dr. Yuri Lvov, Dr. Dale Snow, and Dr. Hisham Hegab [17]. The emphasis is placed on non-consumable cellulosic biomass, such as wood, grass, stalks, and similar materials, with the purpose of transforming them into ethanol. This ethanol production method can decrease greenhouse gas (GHG) emissions by approximately 86% compared to fossil fuels. The subject of nanotechnology research focused on converting biomass into biofuels is expanding rapidly. In 2007, the oil firm BP provided a research fund of $500 million to the University of California, Berkeley, and the University of Illinois. The purpose of the fund was to investigate the process of converting grain, plant material, algae, and switch grass into fuel [18]. Previously, Berkeley employed nanotechnology in their efforts to develop affordable solar panels [19]. However, the newly established Energy Biosciences Institute (EBI) at Berkeley will prioritize the generation of fuel while minimizing environmental impacts and carbon emissions. An integrated strategy is being implemented, which encompasses advancements in crop production methods, enhanced processing of feedstocks, and the innovation of novel biofuels. The objective of using this strategy is to enhance the quality of raw materials used for feeding, by decomposing plant matter into sugars and subsequently converting them into ethanol. EBI's success in this field is anticipated to prompt further exploration of nanotechnology for the development of alternative fuels, including butanol and renewable hydrocarbon fuels. Another significant use of nanotechnology involves the utilization of nano-catalysts to facilitate the trans-esterification process of converting fatty esters derived from vegetable oils or animal fats into biodiesel and glycerol [20]. The nano-catalyst spheres serve as a substitute for the often employed sodium methoxide. The spheres contain acidic catalysts to react with the free fatty acids and basic catalysts to react with the oils. This method removes multiple stages of the traditional process, such as acid neutralization, water rinses, and separations. The aforementioned stages result in the dissolution of the sodium methoxide catalyst, rendering it non-reusable. On the other hand, the catalytic nanospheres have the ability to be retrieved and reused. The final outcome is a more cost-effective, uncomplicated, and streamlined procedure. Summarizing, the process purports to be cost-effective, environmentally friendly, operating at moderate temperatures and pressures, capable of handling both low and high levels of free fatty acids (FFA) in the raw materials, yielding more environmentally friendly biodiesel and glycerol, significantly reducing water usage and environmental pollutants, and compatible with current facilities.

8. NANOTECH LIQUID ADDITIVES

The preceding presentation and discussion focused on the utilization of solid nano-particles as catalysts for the production of biofuels from algae, landfill methane, and biomass. The subsequent sections will analyze the tangible prospects available for liquid nano-particles or droplets. The user's text is "[21]." Examine the use of versatile liquid additives that improve lubrication by creating a single layer across the surfaces that come into contact with the treated fuel. The user's text is "[22]". The pace at which lubricity is affected is determined by the concentration at which adsorption reaches its maximum capacity. Hypothesize that the enhanced ability to clean and dissolve water is achieved by the creation of nano emulsions. Furthermore, it can be hypothesized that the enhanced combustion and subsequent improvement in fuel economy can be attributed to the behavior of nanoscale droplets. The formation of these nano droplets is caused by the surfactant properties of the additive in the fuel mixture, as well as the presence of a small amount of water that is typically found in all commercial fuel systems, often owing to condensation that occurs in the evening.

According to a study conducted by Wulff and colleagues [23], nano emulsions, referred to as micro emulsions by the authors, consist of fuel (including biofuel), water, and surfactant.

- Thermodynamically stable and
- Microscopically isotropic, and
- Nano-structured (thus, nano emulsions).

Their research concluded that:

- The use of these nano structures with fuel, water and surfactant is able to break the usual trade off between reduction of soot and NOxemissions, by achieving them simultaneously, and
- For the same fuel consumption, higher efficiency is obtained.
- Strey and collaborators filed patent applications for what they call micro-emulsions used as fuel [24]. The interpretation offered for the behaviour of stable diesel (and most likely biodiesel)-water-surfactant nano emulsions is as follows:
- The surfactant components– oleic acid and nitrogen containing compounds (amines)–dissolve readily in diesel (and possibly in biodiesel) fuel and bind water to it without stirring;
- The water droplets are as small as a nanometer across, helping stabilize the emulsion.
- The result is a "liquid sponge", can be stored indefinitely, like ordinary diesel fuel, without risk of phase separation.
- This fuel formulation, when burned, results in the near-complete elimination of soot, and a reduction of up to 80% in nitrogen-oxide emissions.
- The surfactant in the formulation also burns without creating emissions beyond water, carbon dioxide and nitrogen.

9. PUBLIC APPREHENSIONS OVER NANOTECHNOLOGY ENCOMPASS ISSUES OF SECURITY, HEALTH, AND THE ENVIRONMENT

Like any emerging technologies, there may be legitimate reasons to be concerned about the potential consequences on security, health, and the environment. The implementation of nanotechnologies has been extensively assessed to anticipate potential effects for both humans and the environment. The Woodrow Wilson Center conducted a Nanotechnology project [25] starting in 2005. The project administrators stated that changing materials at the atomic level can have significant and far-reaching consequences, both beneficial and detrimental. The lack of precise knowledge on the potential effects of nanotechnology was the driving force behind the establishment of the Project on Emerging Nanotechnology at the Woodrow Wilson Center.

The International Risk Governance Council (IRGC) Nanotechnology project [26] introduced another initiative. Two specialized workshops were conducted. The initial discussion in May 2005 centered around the process of conceptualizing nanotechnology, as well as evaluating its potential hazards and advantages. The nanotechnologies were categorized into two frames: Frame One, which refers to passive or classical technology assessment, and Frame Two, which refers to active or the social desirability of innovation. The second event, held in January 2006, focused on identifying deficiencies in the management of risks associated with nanotechnology and formulating suggestions for enhancing risk governance. A symposium regarding the topic occurred in Zurich during July of 2006. The policy implications of Frame One, as seen in Figure 4, were examined in a presentation by Ortwin Renn[27]. Most individuals lack distinct associations about nanotechnology. Their expectations revolve around economic gains rather than revolutionary advancements in technology. Risks are frequently left unmentioned, although there is a worry over unforeseeable adverse consequences. There is a growing apprehension that industry, science, and politics are forming an alliance that may be detrimental to the public's interests. Furthermore, a single adverse event has the potential to significantly influence public sentiments in a negative manner.

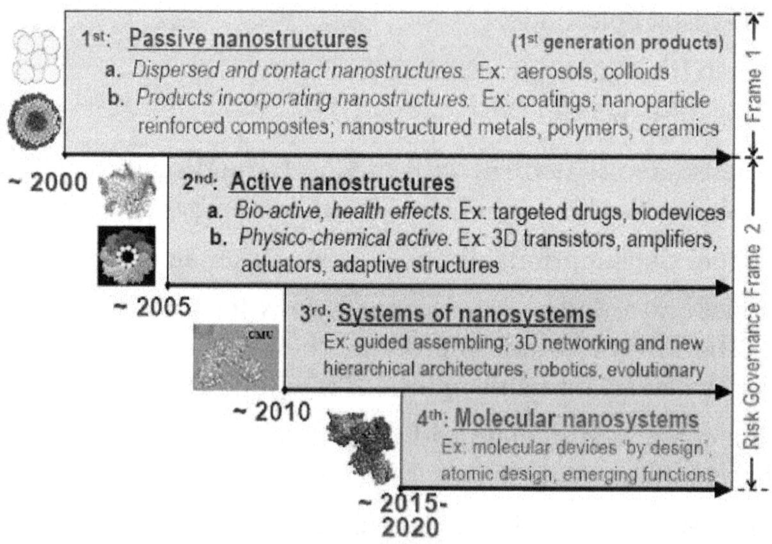

Figure 4. Frames of reference of nanotechnology generations

The Nanotechnology project conducted by the IRGC has determined that while communicating about the benefits and risks of nanotechnology, it is important to differentiate between passive and active nano-materials and products. It emphasizes that separate strategies for risk management are necessary for each category. It is important to be cautious in addressing any social concerns over the potential effects of active nano-materials in Frame Two. However, we should avoid unduly generating anxiety about

Frame One goods that just use passive nanostructures. Renn [29] provides a more detailed explanation of this concept. Frame One passive nanostructures can be observed in various applications such as easy-to-clean surfaces, paints, and cosmetics. Frame Two pertains to dynamic nanostructures and molecular systems that possess the capability to actively interact or might be perceived as biosystems that undergo autonomous evolutionary changes in their attributes.

Indeed, nanotechnologies are currently encountering obstacles. The UK Soil Association has prohibited the use of artificially produced nano-materials in all of its certified organic products. The 2008 annual report of the Soil Association of the UK includes the statement [30]: The Soil Association has released the world's inaugural set of regulations prohibiting the use of nanotechnology. The hazards associated with nanotechnology remain mostly uncertain, unverified, and indeterminate. Preliminary scientific research indicates adverse impacts on living species, and three years ago scientists cautioned the Government against the discharge of nanoparticles, advising that it should be minimized to the greatest extent feasible. There are numerous similarities between the development of nanotechnology and GM, especially in terms of how commercial opportunities have outpaced scientific comprehension and governmental oversight. In addition, although nano-substances are being quickly introduced to the market, there is now no formal evaluation procedure or labeling system for these items, which is a more severe issue compared to genetically modified organisms (GM).

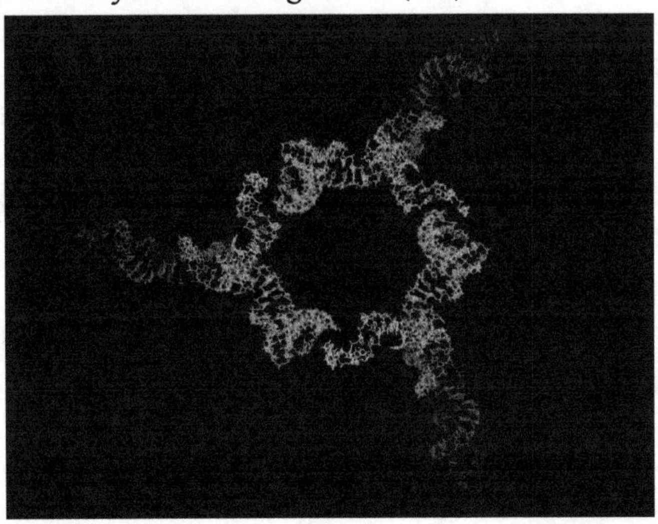

Figure 5. RNA nano-particle created by Peixuan Guo, Purdue University [31]

The potential toxicity of health and beauty products containing nanoparticles is a worry when they penetrate the skin. There are comparable concerns involving food and textiles. Undoubtedly, further research on the health and environmental effects is necessary in order to address public apprehensions. However, the potential of nanotechnologies to bring about positive outcomes is significant. Therefore, it is necessary to do Frame One and Two assessments when new applications emerge. This includes exploring the possibility of using nanotechnologies to enhance medicine delivery for combating diseases in both humans and animals. Figure 5 depicts an RNA nano-particle developed by Peixuan Guo from Purdue University, serving to exemplify the concept. Strands are joined together by combining two types of RNA - a framework and a seeker that detects cancer cells. This nanostructure has demonstrated efficacy in inhibiting cancer proliferation in live mice, as well as in vitro human nasopharyngeal carcinoma and breast cancer cells.

10. CONCISE SUMMARY

In the coming decades, the rising need for energy services will necessitate a greater availability of liquid fuels, even with efforts to enhance energy efficiency and diversify energy systems, such as the increasing use of electricity in transportation. Biofuels are crucial in this context. Nevertheless, it is imperative that the future production of biofuels be sufficiently large to necessitate the extensive use of non-food feedstocks and novel technologies. Nanotechnologies are highly likely to have a significant impact on the future of energy. Their role will be to facilitate the introduction of liquid biofuels, such as renewable hydrocarbons derived from algae, carbohydrates, fatty esters, and biogas, into the market. Nanotechnologies will enhance the efficiency of utilizing present and future liquid fuels, particularly biofuels, by improving the burning of nanodroplets. Although any new technology carries inherent hazards, the world is now more capable of evaluating and responding to these risks. Therefore, it appears feasible to progress in the development of nanotechnologies for biofuels without compromising security, public health, or the environment. However, nanotechnologies have a wide scope that extends far beyond biofuels and have promise in other areas, most notably human health.

REFERENCES

1. Trindade, Sergio C. (2010). Refining will definitely survive. Pipeline Magazine, 29 August 2 010
2. Trindade, Sergio C. (2010). Renewable Energy Perspective – a profitable pathway from oil, Exploration and Processing, Fall 2010 [8-9], Sep.
3. Trindade, Sergio C. (2010). International Biofuels Trade: Issues and Options. International Biofuels Conference_São Paulo, 26-28 May.
4. Santana, G. and S. Quirk (2009). Growing Green: An In-Depth Look at the Emerging Algae Industry, Greener Dawn Research, 22 July, 16p.
5. A Sustainable Biofuels Consensus (2008). Statement from a conference hosted by the Rockefeller Foundation Bellagio Study and Conference Center, Bellagio, Italy, 24-28 March 2008
6. www.defra.gsi.gov.uk (2007), In: F.O. Lichts's World Ethanol and Biofuels Report, Vol. 4, No. 16, p.365 and Vol. 4, No. 17, p.391, Turnbridge Wells, U.K.: F.O. Licht, 2006.
7. Carvalho da Silva, Flávio; Paulo Roberto da Costa Brum and Taís Neno dos Santos (2005). Nanotechnology/ Nanoscience Knowledge Managament emphasizing nanostructured polymers. Presentation, School of Chemistry, UFRJ, Brazil.
8. Borschiver, Suzana; Maria José O. C. Guimarães, Taís N. dos Santos, Flávio C. da Silva, Paulo Roberto C. Brum (2005). Patenteamento em Nanotecnologia: Estudo do Setor de Materiais Poliméricos Nanoestruturados., Polímeros: Ciência e Tecnologia, vol. 15, n° 4, p. 245-248.
9. http://ecolocalizer.com/2009/04/23/nanotechnology-to-aid-the-commercial-viabilityof- algal-bio-fuel-production, April 23, 2009
10. http://www.qsinano.com/news/releases/2009_02_24.php
11. http://www.ameslab.gov/news/news-releases/nanofarming-technology-extractsbiofuel-oil-without-harming-algae
12. http://www.ameslab.gov/news/news-releases/nanofarming-technology-extractsbiofuel-oil-without - harming-algae

13. http://biomassmagazine.com/articles/2354/dudek-catalyx-nanotech-to-build-landfillfacilities
14. http://journalstar.com/news/local/article_6d5b6a34-e86f-11df-ae58-001cc4c002e0.html Nov. 4, 2010
15. http://en.wikipedia.org/wiki/Global_warming_potential
16. http://www.sciencedaily.com/releases/2009/10/091008131858.htm
17. Nanotechnology Used In Biofuel Process to Save Money, Environment Science Daily (Oct.10, 2009)
18. http://berkeley.edu/news/media/releases/2007/02/01_ebi.shtml
19. (U.S. Department of Energy. Berkeley Lab Helios Project. (n.d.) Helios Solar Energy Research Center. Goals and challenges. Retrieved December 10, 2009 from http://www.lbl.gov/LBL-Programs/heliosserc/html/goals.html
20. http://www.public.iastate.edu/~nscentral/news/2007/jun/catilin.shtml
21. Cleaner diesel engines – pouring water on troubled oils, The Economist, June 3rd, 2010, p.86 http://www.economist.com/node/16271415
22. http://www.internationalfuel.com
23. Wulff, Pascal; Lada Bemert, Sandra Engelskirchen and Reinhard Strey (2008). Waterbiofuel microemulsions. Institute for Physical Chemistry, University of Cologne. http://strey.unikoeln.de/fileadmin/user_upload/Download/WATER___BIOFUE L_MICROEMULSIONS.pdf http://strey.uni-koeln.de/333.html?&L=1
24. Strey, R. et al (2007). Microemulsions and use thereof as a fuel. US Patent Application 2007/028507 , Feb. 8. http://www.rexresearch.com/strey/strey.htm
25. http://www.loe.org/shows/segments.htm?programID=05-P13-00050&segmentID=3
26. http://www.irgc.org/-Nanotechnology-.html
27. http://www.yasni.ch/ext.php?url= http %3A% 2F%2 Fwww.irgc.org% 2FIMG% 2Fpdf% 2F Ortwin_Renn_ Nanotechnology_ Frame_ 1_ Policy_Implications_.pdf&name=Ortwin+Renn&cat=document&showads=1
28. http://www.irgc.org/Policy-Recommendations,188.html
29. http://ec.europa.eu/health/ph_risk/documents/ev_20081002_rep_en.pdf, p.14
30. http://www.soilassociation.org/LinkClick.aspx? fileticket= Moyw3 Q7H%2 Fp4%3D & taid=303, p.22
31. http://www.eng.uc.edu/nanomedicine/Papers/1NCI.pdf

CHAPTER-7

UTILIZING BIORESOURCES FOR ADVANCED (THIRD-GENERATION) BIOFUELS

1. INTRODUCTION: AN OVERVIEW

The well-being of modern societies is heavily dependent on fossil fuels. The existing energy paradigm, which heavily relies on the widespread use of fossil fuels, is confronted with both economic and environmental challenges. According to the 2009 report from the United States Department of Energy, it is projected that worldwide energy consumption would increase by 100% in the next 20 years (Conti, 2009). However, according to the 2009 assessment of the European Commission, the handling of climate change issues in Europe has been generally unfavorable on a global scale since 2000. However, there are some encouraging indications, such as a 1.4% decrease in CO2 emissions in 2007 compared to the data collected from 2000 to 2004 in the European Union of Fifteen (E-15). However, when examining the 27 European nations (E-27) and their use and production of renewable energy and biofuels, it becomes evident that the decrease in emissions has not met the objectives set by the European Union. The unfavorable rating can be attributed to several factors, including a decline in enterprises' productivity, an increase in transport and industry emissions, and a drop in research and development areas (Radermacher, 2009). First- and second-generation biofuels have the potential to alleviate or resolve the issues related to the depletion of fossil fuels. However, there have been concerns raised about their recent implementation. The primary issues related to biofuels include the contentious debate between food and fuel production, the allocation of agricultural and forestry land, and the overall sustainability of biofuel production. Possible solution could be found in third-generation biofuels, which involve the microbial processing of agricultural, urban, and industrial leftovers. Nevertheless, there are other technological challenges that need to be addressed in order to ensure the economic feasibility and accessibility of these technologies for the industry (Robles-Medina et al., 2009).

2. FIRST-GENERATION BIOFUELS

The simultaneous increase in energy requirements due to diminishing oil sources and the escalating levels of greenhouse gas emissions poses a significant threat to biodiversity, as well as to the food security and overall well-being of humanity. Therefore, there is a requirement for a novel energy framework that relies on environmentally friendly and renewable energy sources, as well as cleaner and more sustainable fuel technology (Fortman et al., 2008; Jegannathan et al., 2009).

2.1. Biogas, syngas, mixes of vegetable oils, and Fischer Tropsch liquids

The first response of heavy industry to the current energy and environmental problems includes some old systems, such as syngas and Fischer Tropsch liquids. Current advances in technology and engineering could bring new opportunities to these classical chemistry and biochemistry solutions, associated with fuel shortage situations such as the Arab oil embargo of the 1970s, or the Second World War. Some of these will be detailed below.

2.1.1. Biogas

Biogas is an attractive source of energy primarily because it is renewable and enables the recycling of organic waste. The production of biogas from manure can help to manage the problems associated with this residue, contributing to the reduction of the greenhouse gas methane. Besides, biomethanation is not only useful for energy production, but also for cleaning up solid waste in urban areas. Compared with bioethanol from wheat and biodiesel from rapeseed, biogas production based on energy crops could generate about twice the net energy yield per hectare per year. Furthermore, biogas could be produced from the by-products generated by the current bioethanol and biodiesel industries (Jegannathan et al., 2009).

Biogas production is based on bacterial methanogenesis in the absence of air of organic matter in a water solution. The process occurs in three steps. The first, hydrolysis, is carried out by strict anaerobes such asBacteroidesorClostridia, and facultative anaerobes such asStreptococci. It involves the enzymatic transformation of insoluble organic material and higher molecular mass compounds such as lipids, polysaccharides, proteins, nucleic acids etc. into soluble organic materials — energy and cell carbon sources such as monosaccharides and amino acids, among others. In the second step, acidogenesis, other types of microorganisms ferment the mentioned products to acetic acid, hydrogen, carbon dioxide and other lower weight simple volatile organic acids, such as propionic and butyric acid, which are converted to acetic acid. Finally, organic acids, hydrogen and carbon dioxide are converted into a mixture of methane and carbon dioxide by the methanogenic bacteria such asMethanosarcina spp. orMethanothrix spp. (consuming acetate), as well as microorganisms such asMethanobacterium sp. andMethanococcus sp., or others that consume hydrogen and formate to yield methane (Jegannathan et al., 2009).

In spite of its attractions, biogas has only been used in rural areas of developing countries and has received investment from governmental and non-profit organizations. The absence of private investment is due to some technical limitations that hamper its economic viability. The process is relatively slow and unstable, and requires large volumes of digester. The decrease in gas generation during the winter season is a serious problem, and can lead to the clogging of the reactor. Other causes for the reduction in gas production are pH and temperature variations, so the loading rate and solid concentration have to be continuously maintained (Jegannathan et al., 2009).

2.1.2. Syngas, biosyngas and Fischer-Tropsch derivatives

Synthetic gas, known as syngas, is a mixture of H2, CO and CO_2 in different proportions. Traditionally, syngas was produced through gasification of coal at high temperatures, but it can also be produced by methane reformation (submitting the methane to a high temperature water steam stream, or hydrocracking) or by gasification of biomass. In the latter case, the obtained gas is called biosyngas. Syngas and biosyngas can be used directly as fuel, but they also can serve as precursors for other fuels, such as hydrogen, obtained by the compression of carbon monoxide and dioxide. Also, by Fischer-Tropsch synthesis (FTS), short and long chain hydrocarbons can be obtained from the aforementioned H2, CO and CO_2 mixture (Srinivas et al., 2007). Fischer-Tropsch synthesis was discovered in the first half of the twentieth century and developed for large-scale production during the Second World War. It is based on the polymerization, through successive stages, of H2with CO and CO_2 , yielding linear hydrocarbons. Iron, cobalt or ruthenium can be used as catalysts (Huber et al., 2006). FTS can be developed at high or low temperature. The high temperature FTS is performed at 330–350ºC yielding mostly short-chain hydrocarbons (gasolines) and light olefins in a fluidized-bed reactor. On the other hand, low temperature FTS develops at 220–250ºC in a slurry bubble column reactor, and waxes and long-chain hydrocarbons are obtained (Bludowsky & Agar, 2009). As FTS is an extremely exothermic reaction, it can be coupled with biomass gasification. However, FTS has some drawbacks, such as the fact that complex mixtures of different chain lengths are always obtained. Thus, FTS products have to be separated prior to subsequent processes (Huber et al., 2006).

2.1.3. Vegetable oil blends

It is feasible to use crude or filtered vegetable oils directly as fuel for diesel engines by combining them with traditional diesel fuels in an appropriate proportion. These mixtures are readily attainable and can be maintained in a stable condition for temporary usage. However, vegetable oils exhibit elevated viscosity, acid contamination, and the presence of free fatty acids, which might result in the development of gum due to oxidation, polymerization, and carbon deposition (Ranganathan, 2008). Consequently, the extended use of vegetable oils as fuel results in the obstruction of filters, blockage of nozzles, and the accumulation of deposits in the combustion chamber (Sidibé et al., 2010). In addition to the persistent issues with injection systems, filters, and combustion chambers, concerns regarding the viability of utilizing crude vegetable oil as a fuel must also be taken into account. Utilizing vegetable oils directly in engines or as feedstock for petro-diesel alternatives would face similar economic and environmental challenges as the conventional biodiesel and bioethanol sectors, as stated by UNCTAD in 2010. Additionally, vegetable oils are costly.

A more interesting solution is the usage of waste cooking oil (WCO; also called waste frying oil, WFO). Waste cooking oil is widely produced, inedible, and could serve as a low-cost and almost ready-to-use substitute for fossil origin diesel. As crude vegetable oil, waste cooking oil has a high viscosity. Besides, it is enriched with free fatty acids and, hence, can generate clogging problems in unmodified diesel vehicles, especially in temperate climates and during the ignition of the engine. Viscosity problems are usually bypassed by blending WCO with petrol diesel or by using transesterification to produce biodiesel (Pugazhvadivu et al., 2005;Al-Zuhair et al., 2009;Chen et al., 2009). Al-Zuhair et al. studied the production of biodiesel with lipases fromCandida antarcticaandBurkholderia cepacia, both free and immobilized in ceramic beads, with or without solvents. They found that clay micro-environments protected immobilizedB. cepacialipase from methanol damage (Al-Zuhair et al., 2009). Also, Pugazhvadivu et al. proposed solving the injection and filter-clogging problems by preheating the waste cooking oil (Pugazhvadivu et al., 2005), by comparing the performance of a diesel engine when using conventional diesel and waste frying oil, preheated at different temperatures, as fuel. They found that preheating the waste frying oil to 135ºC improved the overall yield of the engine. In particular, the brake specific energy consumption and brake thermal efficiency were improved, and the engine exhaust CO and smoke density were reduced considerably compared to WFO preheated at 75ºC. They concluded that WFO could be used as a diesel fuel by preheating it to 135ºC.

2.2. Bioethanol and biodiesel

Bioethanol and biodiesel are commonly touted as the most viable substitutes for fossil fuels. Renewable fuels may be produced in large quantities, and both the distribution of fossil fuels and engines can be readily modified to accommodate mixtures of ethanol and gasoline, diesel and biodiesel, or even pure ethanol and pure biodiesel (Da Costa et al., 2010). However, in order to have a substantial impact on replacing fossil fuels, these companies that produce renewable fuels must address technological limits in the efficiency of their production processes and resolve concerns linked to the materials used as feedstock (UNCTAD, 2010). Decisions regarding the selection of feedstock, catalytic technique, and energy gain during the production process are crucial for the successful manufacture of biodiesel and bioethanol.

2.2.1. Bioethanol and biodiesel production

Bioethanol is derived from raw materials that are abundant in simple sugars or starch. These sugars are easily extracted and processed, and then fermented to make ethanol. Bioethanol synthesis involves three distinct stages. Initially, the intricate sugars undergo hydrolysis to liberate glucose. Afterwards, the glucose undergoes a subsequent fermentation process conducted by yeasts such Saccharomyces cerevisiae, resulting in the production of ethanol and carbon dioxide. The third phase involves a thermochemical process that focuses on distilling the diluted ethanol to obtain ethanol with a high concentration. In order to utilize lignocellulosic raw materials like agricultural leftovers (such as maize stover, straw, and sugar cane bagasse), forestry waste, wastepaper, and other cellulosic wastes, it is necessary to undergo a chemical or enzymatic hydrolysis pretreatment to break down the lignin. This supplementary measure diminishes the effectiveness of the procedure. Cellulases from the Trichoderma genus fungi have been engineered to achieve improvements. Additionally, microorganisms that can express both cellulase and the enzymes required for ethanol fermentation, such as pyruvate decarboxylases and alcohol dehydrogenases, have been utilized. These advancements have been documented by Fukuda et al. (2006) for cellulases and by Lu et al. (2006), van Zyl et al. (2007), Jegannathan et al. (2009), Rahman et al. (2009), and van Dam et al. (2009) for the utilization of microorganisms. Nevertheless, these enhancements have yet to provide a streamlined and financially viable procedure.

With regard to biodiesel, it consists of a mixture of fatty acid alkyl esters (FAAE) obtained by the transesterification of fatty acids and straight chain alcohols (generally ethanol or methanol), mainly from vegetable oils. When methanol is the alcohol of choice, the term used to refer to the biodiesel is fatty acid methyl esters (FAME), while the ethanol-derived biodiesel is known as fatty acid ethyl esters (FAEE). The properties of the biodiesel obtained from ethanol or methanol are very similar, but methanol is the preferred alcohol in spite of its toxicity and fossil fuel origin because of its low cost and wide availability (Ranganathan et al., 2008;Fjerbaek et al., 2009).

The commercially delivered biodiesel is mainly obtained by the chemical transesterification of the triglycerides contained in sunflower, rapeseed or palm oil. This process can be carried out by acid and alkaline liquid catalysts (Kawahara & Ono, 1979;Jeromin et al., 1987;Aksoy et al., 1988; Fukuda et al., 2001), or heterogeneous solid catalysts such as supported metals, basic oxides or zeolites (Cao et al., 2008). The preferred catalysts are the liquid ones, particularly the alkaline ones, because these catalysts are cheap, very versatile and yield less corrosive fuel than the acid catalysts. Also, liquid catalysts are preferred because the reusable solid catalysts are still withdrawn with mass transfer and reactant diffusion problems. However, the alkaline catalysis has several limitations, especially the futile consumption of the catalyst, problems of viscosity, mass transfer and recovery of biodiesel and by-products owing to the saponification of the catalyst and free fatty acids in the presence of water (Freedman et al., 1984;Zhang et al., 2003; Jaruwat et al., 2010). These problems are bypassed by high temperature reaction conditions, addition of organic solvents to manage the water presence or enhance the interface formation, or increase of the alcohol: catalysts ratio (Kawahara & Ono, 1979; Fukuda et al., 2001). Therefore, the procedure necessitates substantial energy inputs to sustain elevated temperatures that are favorable for achieving effective transesterification rates, as well as to separate methanol. In addition, the procedure produces alkaline wastewater that must be treated before it can be disposed of (Jaruwat et al., 2010). Collectively, these adverse circumstances cast doubt on the long-term viability and ecological advantages of the biodiesel sector.

2.2.2. Advantages and disadvantages of bioethanol and biodiesel

The widespread use of bioethanol and biodiesel has resulted in a variety of economic, social, and environmental problems (Guerrero-Compeán, 2008). It is important to highlight the significant reliance of the biofuels industries on crops that are also used for human consumption and animal feed (UNCTAD, 2010). Despite the numerous patents that have been suggested to address various technical issues, the recent surge in biofuel demand has shown significant technological constraints in the existing production methods. Consequently, there is a rising disagreement about the true sustainability and environmental friendliness of the current biofuels business (Fortman et al., 2008; Abdullah et al., 2009; Demirba, 2009; Yee et al., 2009; Jaruwat et al., 2010). Furthermore, the impact of biofuel production on farming practices and food markets, as well as the actual reduction of greenhouse gas emissions throughout the biofuel life cycle, is a significant concern that is often not adequately addressed. It is important to consider parameters such as the specific type of biofuel being studied, the feedstock used, and the energy inputs required to sustain the transformation process. In addition, it is necessary to incorporate the potential for cogeneration of power or energy exchange between the biofuel synthesis and feedstock transformation processes into the model. Consequently, significant discrepancies in the net energy acquisition and resource use can arise due to the diverse assumptions employed in calculating the overall advantages and disadvantages. Timilsina and his colleagues provide a comprehensive overview of this subject based on the OECD estimates. These authors argue that the most effective method for producing biofuel is through the use of sugarcane-based bioethanol in Brazil. This method results in a 90% reduction in greenhouse gas emissions compared to gasoline. The high efficiency of this process is mostly due to the high crop output and the utilization of sugarcane as both an energy source for production plants and for the cogeneration of electricity. Cellulosic feedstock-derived second-generation biofuels offer a significant decrease of 70-90% in greenhouse gas (GHG) emissions compared to gasoline or diesel. When combined with power cogeneration, this particular type of biofuel has the potential to further enhance the reduction of greenhouse gas emissions. However, it is important to note that this biofuel is still in the developmental stage. The GHG reduction of ethanol derived from sugar beet ranges from 40 to 60%, whereas wheat-based ethanol exhibits a GHG decrease of 30-50%. The manufacture of bioethanol from corn has the lowest greenhouse gas (GHG) reduction compared to other biofuels. It has a low efficiency in lowering GHG emissions, ranging from 0 (or even negative in some situations) to 50% when compared to gasoline (OECD, 2008; Timilsina & Shrestha, 2010).

3. SECOND-GENERATION BIOFUELS

In theory, the implementation of biofuels in transportation and industry should address, or at least enhance, the ecological and economic issues arising from the unsustainable nature of the fossil fuel-dependent energy model. Nevertheless, new observations in the field suggest a somewhat more intricate situation. The market economy and imbalanced relationships between various sectors of the economy and national markets create unpredictable fluctuations in the cost of fuel raw materials. In this scenario, the emergence of new commercial and industrial prospects has modified the already volatile dynamics of the global agriculture markets. The surge in demand for grain, driven by its utilization as a primary ingredient for ethanol production, has resulted in a quick and significant rise in corn prices (Fischer et al., 2009). The demand for palm oil has led to the expansion of palm oil tree and soy cultivation in various regions, particularly in South-East Asia. This expansion has had significant consequences, including threats to biodiversity and negative social impacts. The unregulated and intensive agricultural practices associated with the biofuels industry have resulted in the switch from using palm oil for traditional human nutrition, housekeeping, and livestock feed. These issues have been highlighted in several studies (Abdullah et al., 2009; Jaruwat et al., 2010; Fortman et al., 2008; Guerrero-Compeán, 2008; Demirba, 2009; UNCTAD, 2010; Yee et al., 2009).

3.1. Feedstock costs and biofuel competition

Biodiesel typically has a price exceeding 0.5 US$/l, while petroleum-based diesel is priced at around 0.35 US$/l (Demirbaş et al., 2009). The primary factor contributing to the expensive nature of biodiesel is the substantial expense of unused vegetable oil (Krawczyk, 1996; Connemann & Fischer, 1998). As an illustration, the current price of soybean oil is $1.27 per liter, whereas the price of palm oil is $1.18 per liter (World-Bank, 2011). now, biodiesel made from animal fat is the most cost-effective option, priced at 0.4-0.5 US$/l. In comparison, the traditional process of converting vegetable oil into biodiesel through transesterification is now priced at roughly 0.6-0.8 US$/l (Bender, 1999). Zhang et al. (2007) asserted that there is no worldwide demand for ethanol. Several factors contribute to this, including the types of crops grown, agricultural methods used, labor expenses for land work, scale of production facilities, processing technology, and government laws. The current total cost of ethanol production in a dry mill facility is 0.44 US$/l. Corn accounts for 66% of the running costs, while the energy required to power the production plant, including electricity and natural gas, accounts for around 20% of the operational costs. However, ethanol derived from sugar cane, primarily manufactured in poor nations with tropical climates, is typically more cost-effective to generate compared to ethanol derived from grain or sugar beet (Bender, 1999). Due to this factor, nations such as Brazil and India are witnessing a growing trend towards the use of sugar cane-derived ethanol as a more economically viable substitute for petroleum-based fuels. Conversely, the production of ethanol from cellulosic feedstock through enzymatic hydrolysis necessitates more extensive processing compared to feedstock generated from starch or sugar. However, the cost of employing grasses and trees as feedstock is often cheaper than that of grain and sugar crops. If specific reductions in conversion costs are successfully attained, the overall cost of manufacturing cellulosic ethanol in EOCD nations may decrease to a level lower than that of grain ethanol.

Estimations indicate that ethanol in the European Union (EU) becomes economically viable when the price of oil reaches 70 US dollars per barrel. In contrast, in the United States, ethanol becomes economically viable at a range of 50 to 60 US dollars per barrel. For countries like Brazil, Pakistan, Swaziland, and Zimbabwe that efficiently produce sugar, the cost of ethanol is significantly lower, ranging from 25-30 US$/barrel. Nevertheless, anhydrous ethanol, which may be mixed with gasoline, remains costlier, despite the fact that costs in India have decreased and are getting closer to the price of gasoline. While feedstock prices make up the largest portion of the overall cost of biofuels, the scale of the production plant can decrease the end cost of the fuel. Therefore, the larger conversion plants in the USA generally produce biofuels, specifically ethanol, at a lower cost compared to those in Europe. Countries with warm climates, like Brazil, have significantly cheaper production costs compared to Europe, with expenses less than half of those in Europe. However, despite the lower manufacturing costs, ethanol from Brazil remains competitive with gasoline due to the significant output of sugar cane and the simultaneous creation of energy (Demirbaş et al., 2009).

3.2. Comparison of Brazilian and USA approaches for implementing the bioethanol industry

Since the Arab oil embargo of the 1970s, Brazil has made an incomparable effort in the reduction of its energy dependency by intensifying and extending sugar cane-based bioethanol production. Although the alternative periods of scarcity and abundance of oil have marked fluctuations in the strength of the Brazilian Alcohol National Programme (Proalcool), the global trend has been an ascending progression in the total production of alcohol, as well as in the yield per hectare of sugar cane, and the implantation of this alcohol as transportation fuel. Today, Brazil is the second largest worldwide ethanol producer. In this way, Brazil has reduced its energy dependency, and has become the first ethanol exporter. According to Brazilian Government data, this milestone has been achieved on the basis of rural employment and welfare improvement. The key aspects of the Proalcool programme are a combination of technological advances, social planification and projection of the bioethanol industry. According to the Brazilian Government (Da Costa et al., 2010), owing to the high productivity of sugar cane, Brazil has expanded ethanol production and use without a significant increment in the fertile land surface used to cultivate sugar cane, or a food vs. fuel competition. However, there are several authors who are not so enthusiastic with the success of the Brazilian model, and point to the sugar cane industry as one of the reasons for the losses in biodiversity and the expansion of agricultural land over doubtfully catalogued marginal land, which is more relevant and dangerous than the Brazilian Government data indicates (Coelho et al., 2008;Gauder et al., 2011). On the other hand, the American bioethanol industry choice of corn grain as its raw material has been followed by a dramatic rise in the prices of corn derivatives. Although the USA production of bioethanol supersedes the Brazilian one, the production:consumption ratio of the former (1:3) is much smaller than the latter (8:3). Despite its commercial orientation, the global efficiency of the USA model is low compared with the Brazil system and relies on the high importation taxes that protect the American industry from foreign ethanol inputs (Da Costa et al., 2010). Finally, the narrow margin of the USA production:consumption ratio suggests that the model has reached a production glass ceiling that blockades the medium-term implantation of biofuels in American society and hampers their exportation (UNCTAD, 2010;Da Costa et al., 2010).

3.3. Chemically accelerated biodiesel production in Europe and Asia.

The European and Asian strategy to improve climate change and fossil fuel depletion problems is based mainly on the chemically catalyzed biodiesel obtained from vegetable oils. There is a variety of feedstocks for the production of this biofuel, from inedible oils, (mainly rapeseed oil in Europe or jatropha oil in Asia), to edible oils (principally sunflower oil in Europe and palm oil or soybean oil in Asia, although corn, peanut, cotton seed or canola oil can also be cited) (Ranganathan et al., 2008;Abdullah et al., 2009). As the elected method for industrial biodiesel production is chemical catalysis, these vegetable oils are preferred to other heterogeneous lipids sources. These other lipids need pretreatment prior to their use (Peterson, 1986;Fortman et al., 2008), and include waste frying oils, waste-activated bleaching earth from the oil refinery industry, and even animal origin lipids such as beef tallow, lard, yellow grease and poultry grease or fat from fat traps, septic tanks, or waste water sludges. The need for economically viable vegetable oils for biodiesel production implies the cultivation of greater areas with oil-producing crops such as sunflowers or palm oil trees. Thus, the previously mentioned rising corn prices, owing to the derivation of huge amounts of grain for the industrial production of bioethanol, is neither an isolated case in developing biofuel industries nor the only aspect of the biofuel industry issue. Like the bioethanol industry, the European and Asian biodiesel industries have the energy and chemical problems associated with the current biofuels model. These limitations can be summarized according to nearly obsolete technology, being strongly dependent on chemical catalysis, non-renewable materials and promotion of non-sustainable market and farming practices (Guerrero-Compeán, 2008;Demirbaş, 2009; UNCTAD, 2010).

3.4. Technical factors involved in the manufacture of biodiesel

The industrial production of biodiesel needs to solve several technical problems in order to obtain this kind of biofuel in an efficient and sustainable way. The physical factors to consider can be summarized by pH, temperature, hydric activity, solvents and supports. Depending on the catalyst used to drive the transesterification reaction, some of the cited factors have different impacts on the global efficiency and feasibility of the process. A non-optimal configuration of the system can reduce significantly the biodiesel yield and compromise the viability of the production plant, especially if the upstream by-products, excess catalyst or auxiliary devices for solvent recovery hinder an easy, clean and rapid downstream processing of the biofuel.

3.4.1. pH, temperature and hydric activity

As mentioned above, the chemical catalysis of the transesterification reaction requires high temperatures to achieve an acceptable reaction rate. In the case of the alkaline catalysis, the minimal temperature to produce conventional biodiesel is 60ºC, while in the acid catalysis the temperature ranges from 50 to 80ºC (Robles-Medina et al., 2009). Acid catalysis is slower than the alkaline one and generates a more corrosive fuel, so alkaline catalysts are preferred by the industry. It incurs a great energy cost in order to initiate and maintain the reaction (Kawahara & Ono, 1979;Aksoy et al., 1988;Cao et al., 2008). However, the utilization of sodium hydroxide as a catalyst has a serious limitation in the form of saponification of free fatty acids if water is present. This drives the increased consumption of the catalyst and downstream processing problems, such as the separation of glycerol and unreacted precursors. The solutions to manage this problem include using only virgin oils, often edible vegetable oils, instead of oils with high free fatty acids and water content, such as waste cooking oils or animal origin fats, as well as other residual fats. Higher temperatures, up to 120ºC, and the addition of organic solvents, or additional steps for free fatty acids esterification with sulphuric acid before performing the alkali-catalyzed transesterification are quite common as well (Jeromin et al., 1987).

When lipases are used as catalysts, it is possible to get over the saponification problems owing to their ability to transesterificate alcohols with both triacylglycerols and free fatty acids. Besides, lipases work as well in the presence of water. In fact, they need a certain hydric activity to maintain their tridimensional structure, so the presence of water is not a problem with this kind of catalyst — although excessive hydric activity affects the transesterification reaction because the substrates are water insoluble (Jaeger & Eggert, 2002;Shah et al., 2004;Gilham & Lehner, 2005;Fjerbaek et al., 2009). Lipases can operate at low or relatively low temperatures in the range of 20 to 70ºC, and at even lower temperatures if the enzyme has been obtained from psycrophilic microorganisms (Dabkowska & Szewczyk, 2009). Depending on the chosen lipase and preparation (free, immobilized orwhole cell catalyst), lower temperatures (below 65ºC) can be applied to avoid the thermal denaturation of the enzyme, thus saving in production costs (Fukuda et al., 2008). Within the thermostable lipases, we can citeBurkholderia cepacialipase (Amano PS lipase, from Amano Pharmaceutical Co., Japan), that reaches its highest activity at 60ºC (Dabkowska & Szewczyk, 2009), and the lipases obtained fromThermoanaerobacter thermohydrosulfuricusSOL1 andCaldanaerobacter subterraneoussubsp. tengcongensis, which show their activity maximum at 75ºC and tolerate temperatures as high as 95ºC (Royter et al., 2009).

On the other side of the spectrum, the lipase from Bacillus sphaericus MTCC 7526 presents its optimal temperature at 15ºC, keeping stable until 30ºC, and the Microbacterium phyllosphaeraelipase presents the optimal temperature at 20ºC and deactivates when the temperature exceeds 35ºC, with the pH value fixed at 8 for both psycrophilic enzymes (Joseph et al., 2006;Srinivas et al., 2009). Therefore, pH plays an important role in the enzymatic production of biodiesel because it influences both the reaction rate and the thermal stability or solvents' susceptibility of the lipases. An adequate pH can facilitate the optimization of the operation temperature and improve the activity of the enzyme. Gutarra and collaborators reported a high stability of thePenicillium simplicissimumlipase in the pH range 4.0–6.0, that showed the maximal activity at 50ºC and remained stable and active (although with a lower activity) even at 70ºC (Gutarra et al., 2009).

3.4.2. Heterogeneous catalysts and immobilized enzymes are two types of substances used in chemical reactions.

An alternative to the chemical transesterification of low quality oils with a relatively high concentration of water or free fatty acids consists of heterogeneous catalysis using acidic cation-exchange resins, supported metals (Zabeti et al., 2009), basic oxides or zeolites (Knezevic et al., 1998;Suppes et al., 2004). Even low cost alternatives such as waste eggshell have been proposed as well (Wei et al., 2009). These kind of catalysts are considered as an intermediate and relatively low-cost solution between the traditional homogeneous catalysts and the lipases. However, the cited heterogeneous catalysts are affected by the slow diffusion of the triglycerydes through their pores and require a higher alcohol:oil ratio to accelerate the reaction, in order to increase the production (Zabeti et al., 2009). Nevertheless, the heterogeneous catalysts can serve to improve the reusability and efficiency of immobilized enzymes andwhole cell catalysts. Immobilization of enzymes on inert materials such as porous ceramic beads (Iso et al., 2001) or polymeric resins (Dizge et al., 2008; Dizge et al., 2009) can improve their performance. This improvement is owing to the protection that the pores' microenvironment brings to the enzyme, avoiding the inhibition or damage of the enzyme caused by methanol or solvents. Another attractive approach to the immobilization is the so-called protein-coated microcrystals technology (PCMC). PCMC is based on the use of crystalized proteins as a support to the lipases, or in the direct use of crystalized lipases as solid catalysts (Raita et al., 2010). However, the real increase in reaction rate and enzyme stability with these immobilization techniques is usually lower than the theoretically expected. One of the reasons for these lower rate issues responds to the blockage of the pores of the used material as support because of the precipitation of glycerol or the insufficient circulation of substrates around the enzyme (Zabeti et al., 2009). The knowledge generated by the intense research on production and use of supports, resins and porous metallic alloys can be useful for enzymatic production of biodiesel. With this comparative approach, the optimization of the immobilized enzyme technology could be a reality in the short rather than medium term.

3.4.3. Ratio of alcohol to oil and types of solvents

Depending on the kind of catalyst used and the selected operation conditions in the biodiesel production plant, the alcohol to oil molar ratio will present a wide variation. Adding excess alcohol is a common practice, and could serve as reference. However, excess alcohol use implies higher reactant associated costs, especially when the alcohol of choice is ethanol, which is more expensive than methanol. Thus, a more detailed approach to the system optimization in terms of minimal alcohol consumption is needed. Besides, a fine adjustment of the alcohol to oil ratio allows the maximal biodiesel production in the shortest possible time span and with the lowest energy input (Shieh et al., 2003).

The optimization is a relatively simple task when homogeneous catalysts such as sulphuric acid or sodium hydroxide are used to perform the conventional transesterification of vegetable oils with methanol. High yields are achieved with a methanol to oil ratio of 1:1 with an alkaline catalyst (although to improve the yield this proportion rises to 6:1) and a 30:1 ratio when an acid catalyst is used (Zhang et al., 2003).

However, in the case of lipase-catalyzed biodiesel, the situation is more complex and the molar ratio of alcohol to oil varies depending on the type of lipase, the use of an immobilized or free enzyme, and the alcohol used. Similar to the chemical catalysts, an increase of the molar alcohol:oil ratio elevates the efficiency of the reaction, but an excessive alcohol content inhibits and even damages the enzyme, especially when using methanol and free enzymes. Although the lipase-based solvent-free systems are under intensive research, owing to advantages such as the direct saving in solvents and the indirect cost reductions in downstream processes, the utilization of lipases does not necessary mean abandoning the use of a certain amount of solvents. The addition of solvents liket-butanol, diesel oil, hexane or dioxane to the precursors of biodiesel usually allows a better mixing of the reactants. Thus, solvents relieve the problems associated with the different water solubility of lipids and alcohols. In addition, solvents provide a more durable interaction between the enzyme and its substrates, and can favour the circulation of reactants through resins and support pores in immobilized enzyme systems. This improved circulation confers some protection to the lipases against inhibition by substrates and damages by excessive alcohols. However, solvents' addition has to be carefully studied, since an excess of solvent or an inadequate amount of solvent can affect the enzyme activity and stability. For example, Shieh et al. studied the optimal operation conditions to transesterificate soybean oil with methanol byRhizomucor mieheilipase immobilized on macroporous weak anionic resin beads. They found that the best transesterification rate was obtained when the methanol:oil molar proportion was 3.4:1 at 36.5ºC (Shieh et al., 2003). Raita et al. studied the transesterification of palm oil with ethanol by Thermomyces lanuginosalipase-coated microcrystals in the presence oft-butanol. In this case, the optimal conditions were ethanol to fatty acids 4:1 molar ratio andt-butanol:tryacylclycerides 1:1 molar ratio, at 45ºC (Raita et al., 2010. However, Tongboriboon et al. worked on the solvent-free transesterification of used palm oil withThermomyces lanuginosaandCandida antarcticalipases immobilized in porous polypropylene powder, reporting that the best yield was achieved at an ethanol to oil ratio of 3:1, and the yield decreased when the molar ratio was increased to 4:1 at 45ºC (Tongboriboon et al., 2010). These authors pointed to the inhibition of the enzymes by an excessive amount of ethanol, although it is worth emphasizing that they worked on a solvent-free system, so the enzyme was relatively vulnerable to alcohol-driven damage. On the other hand, Shah et al used 4:1 ethanol to oil molar ratio as standard reaction settings in their study about the transesterification of jatropha oil with ethanol at 40ºC. The experimental design consisted of a solvent-free system and three different lipases (free and immobilized on Celite), namelyChromobacterium viscosum,Candida rugosaandPorcine pancreaslipases, although they did not try different alcohol to oil molar ratios (Shah et al., 2004).

4. THIRD-GENERATION BIOFUELS

As a response to the problems associated with the recent worldwide implantation of second-generation biofuels, some authors propose focusing on the processes involved in the production of such biofuels. This new approach consists of the utilization of microbial enzymes to achieve the current chemical pretreatment steps of cellulosic or starchy raw materials (Carere et al., 2008). Microorganisms deal with the degradation of lignocellulose, hemicellulose or lipid-rich materials by means of enzyme catalyzed processes at near to room temperature. Therefore, microbial enzymes could be used to make the current biofuels industry cleaner and greener. Furthermore, the production of biofuels would be coupled with the management of woody and oily wastes, converting these residues into suitable and cheap raw materials (Steen et al., 2010).

4.1. Production of biodiesel using microalgae

Another promising lipids source, still not implemented but currently being studied worldwide, is represented by microalgae. Microalgae have a high potential as biodiesel precursors because many of them are very rich in oils, sometimes with oil contents over 80% of their dry weight, although not all species are suitable as biodiesel production oils (Chisti, 2008;Manzanera, 2011). Besides, these microorganisms are able to double their biomass in less than 24 hours, achieving a reduction between 49 and 132 fold in the medium culture time required by a rapeseed or soybean field. Furthermore, microalgae cultures require low maintenance and can grow in wastewaters, non-potable water or water unsuitable for agriculture, as well as in seawater (Mata et al., 2010). The production of microalgae biodiesel could be combined with the CO_2 removal from power generation facilities (Benemann, 1997), the treatment of waste water from which microalgae would remove NH4+, NO3-and PO43-(Aslan & Kapdan, 2006), or the synthesis of several valuable products, from bioethanol or biohydrogen to organic chemicals and food supplements (Banerjee et al., 2002;Chisti, 2007;Rupprecht, 2009;Harun et al., 2010). However, microalgae biomass-based biofuels have several problems ranging from the optimization of high density and large surface units of production to the location of the microalgae production unit. Anyway, the main decisions to take are the adoption of open or closed systems, and the election of batch or continuous operation mode. As will be discussed below, depending on the system and mode of operation choice, there will be different advantages and drawbacks.

4.1.1. Open vs. closed systems

Microalgae can be grown in open-culture systems, such as lakes or ponds, or in closed-culture systems known as photobioreactors (PBRs). Open-culture systems often offer cost savings in construction and operation, increased durability, and higher production capacity compared to PBRs. Nevertheless, open systems incur higher energy costs due to challenges in distributing nutrients caused by mass transfer issues. Additionally, their depth is restricted to 15 cm to guarantee sufficient light exposure for the growth of microalgae. In addition, ponds are highly susceptible to fluctuations in weather conditions, and it is not possible to effectively regulate temperature, evaporation, and light intensity. In addition, these open systems necessitate a larger land area compared to PBRs and are more prone to contamination, both from bacteria and microalgae found in the vicinity of the culture installations (Manzanera, 2011).

On the other hand, Photobioreactors (PBRs) are highly adaptable and require a lot of acreage. They may be adjusted to meet the specific physical and chemical needs of the chosen algae, making it possible to grow species that are not suitable for open ponds. Photobioreactors can achieve nutrient homogeneity, even light distribution, pH regulation, temperature control, and precise management of CO2 and O2 levels. Therefore, closed systems offer more stability and optimal growth conditions, resulting in larger cell densities and reduced contamination. However, PBRs face many technological issues that prevent them from being competitive in applications that can be accomplished in raceway ponds. The issues associated with this system include overheating, bio-fouling, shearing stress, oxygen accumulation, scaling-up challenges, and the substantial expenses involved in construction, operation, and maintenance (Chen et al., 2011).

These issues are particularly noteworthy: capital building investment and high operation expenditures. The biomass production costs of PBRs may be ten times higher than those of open systems. PBRs can be competitive if the added value of biomass is high. Alternatively, open ponds will be the favored choice. Nevertheless, assessing the effectiveness of open and closed systems is intricate and contingent upon various aspects, such as the specific algae species or the method used to calculate productivity. Productivity in microalgae culture facilities is generally evaluated using three metrics. Firstly, volumetric productivity (VP) refers to the productivity of a reactor per unit of volume, measured in grams per liter per day (g/l d). The second metric is area productivity (AP), which is defined as the productivity per unit of ground area occupied by the reactor, measured in grams per square meter per day (g/m2 d). The third factor is illuminated surface productivity (ISP), which refers to the productivity per unit of reactor illuminated surface area (g/m2 d). However, the decision to choose closed or open systems for an election depends on factors other than only productivity, as will be explained further (Richmond, 2010).

4.1.2. Continuous vs. batch operation mode

PBRs can be operated either in batch mode or in continuous mode. There are numerous benefits associated with utilizing them in continuous mode. First and foremost, continuous culture offers a greater level of control compared to batch mode. Additionally, growth rates can be controlled and maintained at a constant level for extended durations, while the concentration of biomass can be adjusted through the control of dilution rate. Furthermore, continuous reactors maintain a stable condition, resulting in more reliable and reproducible outcomes. Additionally, this approach leads to higher quality manufacturing (Molina et al., 2001).

Nevertheless, there exist constraints that may render the continuous process inappropriate for certain situations. One of the limits is the challenge in regulating the manufacturing of certain non-growth-related products. For example, the system frequently necessitates feed-batch culturing and ongoing nutrient input, which can result in wash-out. Due to the viscosity and variety of the culture media, it can be challenging to cultivate filamentous organisms in continuous PBRs. An additional issue arises when the original strain is displaced by a more rapidly proliferating contaminant, resulting in its loss. The risk of contamination and decrease in reliability of the bioreactor becomes more significant when longer incubation durations are required. Consequently, the initial investment in higher quality equipment may increase, potentially hindering the economic feasibility of the production unit (Mata et al., 2010).

The potential collaboration in producing valuable compounds could potentially resolve the aforementioned issues, but it necessitates the careful evaluation of various factors and alternatives. The microalgae production units will undergo significant modifications, both in terms of operations (such as temperature, sunlight exposure, wind, contamination by microalgae, bacteria, or fungi) and in terms of commercial aspects (fluctuations in the value of by-products, advancements in centrifugation or extraction techniques, or the development of non-algal biofuels, etc). Considering all the characteristics described above, it can be determined that every microalgae-based biodiesel project is distinct. Therefore, it is imperative to approach these initiatives with a mindset that emphasizes the need for a versatile, and potentially multifunctional and adjustable, setup (Richmond, 2010).

4.2. Production of biodiesel from oily biomass

Microalgae are not the sole alternative for the production of biofuels from oily feedstock. Both prokaryotes and eukaryotes have the ability to amass significant quantities of lipids. However, similar to what happened with microalgae, not all species are appropriate for the generation of biodiesel due to variations in the kind of storage lipids. According to Waltermann & Steinbüchel (2010), numerous prokaryotes produce polymeric substances like poly(3-hydroxybutyrate) (PHB) or other polyhydroxyalkanoates (PHAs). However, only a small number of genera store triacylglycerols (TAGs) and wax esters (WEs) as lipid bodies within their cells. Conversely, eukaryotes commonly possess storage TAGs, although PHAs are not present, and WE accumulation has only been shown in jojoba (Simmondsia chinensis). These lipids serve as energy and carbon storage molecules, ensuring the viability of metabolism during periods of hunger. Just like the production of PHAs, TAGs, and WE, synthesis is stimulated by cellular stress and uneven growth. For example, when there is a lack of nitrogen but an excess of carbon source (Kalscheuer et al., 2004).

The prokaryote genera that are most notable for their accumulation of TAGs include nocardioforms, including Mycobacterium sp., Nocardia sp., Rhodococcus sp., Micromonospora sp., Dietzia sp., and Gordonia sp. Additionally, streptomycetes are known to accumulate TAGs in both their cells and mycelia. The storage of TAGs is also commonly observed in members of the gram-negative genus Acinetobacter. However, in this particular case, we are the predominant components of the inclusion bodies. (Waltermann & Steinbüchel, 2010). Among eukaryotes, algae apart, the most noteworthy yeasts for producing biodiesel feedstocks are Candida (non albicans), Saccharomyces, and Rhodotorula.

Steinbüchel and colleagues conducted research on the expression of the non-specific acyl transferase WS/DGAT from Acinetobacter calcoaceticus ADP1 in Saccharomyces cerevisiae H1246, a mutant strain that cannot accumulate TAGs (Kalscheuer et al., 2004). The researchers discovered that the yeast regained the capacity to store TAGs, as well as fatty acid ethyl esters and fatty isoamyl esters. This discovery demonstrated that the Acinetobacter calcoaceticus transferase possesses significant promise for the biotechnological synthesis of diverse lipids, in both prokaryotic and eukaryotic organisms. In Section 4.3, the researchers focused on Escherichia coli TOP 10 (Invitrogen) and successfully developed a modified strain that can directly manufacture fatty acid ethyl esters (biodiesel) from oleic acid and glucose (Kalscheuer et al., 2006).

Cheirsilp et al. (2011) has suggested a potential approach of mixing microalgae and yeast biomass. The authors conducted a study on a combination of oleaginous yeast Rhodotorula glutinis and microalga Chlorella vulgaris in industrial waste. The effluents used consist of seafood processing wastewater and molasses from a sugar cane mill. A synergistic effect was discovered in the hybrid culture. R. glutinis exhibited enhanced growth and lipid accumulation in the presence of C. vulgaris, which served as an oxygen source for the yeast. Conversely, the microalgae got excess CO2 from the yeast. The most favorable conditions for lipid formation were a microalga to yeast ratio of 1:1, an initial pH of 5.0, a concentration of molasses at 1%, shaking at 200 rpm, and a light intensity of 5.0 klux under a 16:8 hours light and dark cycle (Cheirsilp et al., 2011).

4.3. Whole cell catalysts

Enzymes derived from microorganisms, either in their pure form or immobilized, have the potential to decrease the energy expenditures associated with the manufacturing of ethanol and biodiesel in industrial settings. However, the cellulases utilized for the treatment of (ligno)cellulosic materials, such as forestry residues, waste paper, or straw, pose challenges in terms of purification, similar to the lipases employed for the transesterification of lipids to produce biodiesel. Therefore, the cost of these items remains too expensive, making their practical use economically impractical (Shieh et al., 2003; Ranganathan et al., 2008). An additional constraint on the utilization of enzymes is the deactivation and suppression caused by reactants and substrates. The limitations mentioned are now being addressed by protein engineering (Ebrahimpour et al., 2008) to enhance the stability and activity of enzymes, enabling their reutilization. The research focus is on immobilization using various supports or utilizing genetically engineered microorganisms, known as whole cell catalysts, that possess the required enzymes. This approach prevents the enzymes from being exposed to inhibitory substrates and allows them to function as microrefineries (Kalscheuer et al., 2006). The authors of this study, which focuses on the microbiological production of biodiesel, refer to this type of biofuel as 'Microdiesel'. The synthesis of biodiesel through microbial means necessitates the development of genetically engineered microbes capable of catalyzing the transesterification of ethanol with lipids. Ideally, these bacteria should also possess the ability to autonomously synthesize ethanol in order to enhance the overall efficiency of the process. Steinbüchel and his colleagues have laid down the principles for the advancement of the microdiesel business since their research in 2006, which involved producing microdiesel on a small scale in a laboratory setting using a modified strain of Escherichia coli. Their method involved expressing the genes from Zymomonas mobilis, which encode for pyruvate decarboxylase (pdc) and alcohol dehydrogenase (adhB), in E. coli. They also expressed the non-specific acyl transferase ADP1 (atfA) from Acinetobacter baylyi. The strain developed was capable of performing aerobic ethanol fermentation from sugars, as well as enzymatic transesterification of this alcohol with fatty acids produced from lipid metabolism. This process resulted in the production of FAEE, which the scientists referred to as 'microdiesel' (Kalscheuer et al., 2006). Elbahloul and Steinbüchel have recently conducted a pilot plant study employing the microdiesel generating E. coli stated earlier. They employed glycerol as a carbon source and sodium oleate as a source of fatty acids. The findings of the study were promising (Elbahloul & Steinbüchel, 2010). However, the findings from both research suggest that there is still significant progress needed before the practical implementation of their results in industry. The technique must be adjusted to ensure that the engineered strains can effectively adapt to various sources of lipids and lignocellulosic raw materials. These improvements would enable the utilization of forestry and agricultural residues, so enhancing the versatility of the biodiesel manufacturing process to be on par with chemical transesterification.

4.4. Production of microdiesel from waste materials

Vegetable oils are costly and necessitate extensive farmland for cultivation, making their direct utilization for biodiesel manufacture both expensive and unsustainable. Nevertheless, there exist numerous untapped alternate sources of fatty acids. Similarly, the manufacturing of bioethanol, whether for direct use as a biofuel or as a precursor for biodiesel, necessitates substantial quantities of maize grain or sugar cane. However, industrial leftovers, such as waste from vegetable oil refineries, along with waste from agricultural, forestry, livestock, and residential sources, are extensive and significant sources of lipids and carbon (Chen et al., 2009; Dizge et al., 2009). Wang et al. suggested utilizing the soybean oil deodorizer distillate (SODD), a by-product generated during soybean oil refining, which accounts for 0.3–0.5% of the processed soybean oil, for the production of biodiesel. Based on their analysis, the authors found that approximately 80% of the SODD can be converted into biodiesel through a transesterification process using methanol and the lipases from Thermomyces lanuginosa and Candida antarctica. This transformation occurs in the presence of tertbutanol and 3Å molecular sieve. The SODD consists of 45-55% triglycerides and 25-35% free fatty acids (Wang et al., 2006). Park et al. employed waste-activated bleaching earth (ABE), a byproduct of the rapeseed or palm oil refinery sector that contains 35–40% oil and has the potential to produce various large-scale chemicals, such as biodiesel. In the Wang example, the authors used methanol as the alcohol, fuel oil and kerosene as the solvent, Candida cylindracea lipase as the catalyst, and extracted the produced FAME using a filter press (Park et al., 2008). Al-Zuhair and his colleagues conducted a study on the synthesis of biodiesel using simulated waste cooking oil (SWCO). They used Candida antarctica and Burkholderia cepacia lipases, which were either free or immobilized on ceramic beads. The lipases were tested with or without the presence of an organic solvent. The researchers achieved the highest yield when they utilized B. cepacia without an organic solvent. They also noted that the system performed more effectively when the enzymes were immobilized. This was likely due to the clay's structural microenvironments providing protection to the lipases against denaturation caused by methanol (Al-Zuhair et al., 2009). In a recent study, Steen et al. and other researchers have suggested a method of directly fermenting cellulosic biomass to create biodiesel, fatty alcohols, waxes, and other useful chemical compounds (Steen et al., 2010). Their methodology integrates waste management practices and the principles outlined by Steinbüchel et al. with the latest advancements in synthetic biology and established bioprocessing techniques. This interdisciplinary method introduces a versatile and adaptable toolkit consisting of genetically engineered FAEE synthetic strains. These strains possess the necessary enzymes to convert raw (hemi)cellulosic materials into ethanol, or to combine it with fatty acids through transesterification. Additionally, they are capable of directly synthesizing both fatty acids and ethanol from cellulose (Steen et al., 2010).

4.5. Microdiesel derived from wastewater sludges

The microdiesel idea, introduced by Steinbüchel et al., can be integrated with the management and reuse of waste waters through the use of microbial lipases to convert the lipids found in dairy industry or urban wastewater sludges into biodiesel through transesterification. The lipidic fraction of sludges derived from urban wastewater treatment accounts for approximately 17 to 30% of the total weight after removing moisture. The lipidic fraction is created through the direct absorption of fats found in the water by the sludge particles. It is also formed by the release of phospholipids from the cell membranes of microorganisms, as well as from metabolites and by-products of cell lysis (Boocock et al., 1992; Shen & Zhang, 2003; Jardé et al., 2005).

Pretreatment is necessary for lipid-rich wastewaters to decrease lipid content and facilitate subsequent conventional treatment. Typically, pretreatment involves employing physical methods, with the most prevalent ones being fat traps, tilted plate separators (TPS), and dissolved air flotation (DAF) units. Furthermore, centrifuges and electroflotation systems are employed on an intermittent basis (Willey, 2001). Fat traps are vessels, either rectangular or circular in shape, that facilitate the passage of wastewater under laminar-flow conditions. This flow rate enables the lipids to ascend to the surface in close proximity to the outlet end of the trap. The separation principle is founded on Stoke's law, which establishes a correlation between the upward velocity of a particle and its diameter. Consequently, the theoretical effectiveness of separation is contingent upon the depth. Typically, fat traps have a depth of 1.5 m. However, if there is an anticipated buildup of sludge at the bottom, an extra 0.5 m is added to the overall liquid depth. When supplying the trap, it is preferable to use gravity flow instead of pumping in order to decrease the emulsification of wastewater. Fat traps are utilized in the food business and restaurant settings (Willey, 2001).

Subsequently, tilting plate separators were devised in the petrochemical sector, operating on the principle that the surface area, not the depth, is the determining factor for oil separation. By incorporating inclined plates into a container, numerous gravity separators are created in tandem, resulting in a shallow tank with a large surface area relative to its capacity. Generally, TPS systems require less than 10% of the space required for installing a traditional fat trap, albeit having a few drawbacks. If the effluent contains solid or semi-solid fat, the plates may become fouled and require cleaning. In such cases, a crane is necessary to remove the plate pack for cleaning. In addition, it is crucial to meticulously choose and regulate the pumping systems in order to prevent surges and changes in the depth of the liquid (Zeevalkink & Brunsmann, 1983; Willey, 2001). Dissolved air flotation units utilize microbubble clouds (with a diameter of 60-70 μm) generated by injecting air at a pressure of 6 bar through nozzles into water to facilitate the flotation of lipids. Microbubbles adhere to the surface of the fat and oil particles, enhancing their ascent speed. These systems are utilized in both the food business (Willey, 2001) and in the treatment of wastewater in mining (Tessele, 1998). After the pre-treatment process is complete, wastewater has the option to undergo additional treatment before being disposed of or undergoing biological treatment. Chemical treatment can be employed to decrease the overall amount of fatty substances in wastewater. This treatment employs aluminium sulfate, ferric chloride, or more commonly, lime, to disrupt the emulsion and cause the fat particles to form clumps. Afterwards, the fats can be separated through the process of flotation or sedimentation. To enhance the rate of sedimentation, a second-stage flocculation can be employed. This process involves adding small amounts of polyelectrolyte (0.5-5.0 mg/l) to the wastewater after coagulation has occurred (Willey, 2001).

The utilization of sludges for the production of biofuels is not a novel concept. However, the existing literature primarily emphasizes the generation of methane through anaerobic fermentation, a process currently employed in most waste water treatment plants (WWTP) to supply energy to these facilities. Another area of focus is the production of fermentative biohydrogen, although this method is not yet commercially accessible (van Groenestijn et al., 2002; Wang et al., 2003). Multiple research groups have investigated the in situ transesterification of WWTP sludges, with a specific emphasis on the chemical catalysis of the transesterification using methanol (Haas & Foglia, 2003; Mondala et al., 2009). Nevertheless, this approach is still subject to the same constraining variables that impact the chemical transesterification of edible vegetable oils (Freedman et al., 1984). Despite their chemical methodology, these studies offer valuable insights into several parts of the biodiesel manufacturing process, particularly during the first stages. One prevalent issue in chemical and enzymatic biodiesel synthesis is the requirement to preprocess the feedstock in order to enhance the accessibility of its lipids to the catalyst. For wastewater treatment sludge, the pretreatment step often involves the application of organic or non-polar solvents to extract the lipids from the organic debris (Antczak et al., 2009; Siddiquee & Rohani, 2011).

The predominant techniques utilize chloroform:methanol mixtures, such as the Folch's method (Folch, 1957), where a 2:1 chloroform:methanol mixture is combined with the sample, with water serving as a third component to create an emulsion. After reaching equilibrium with a fourth volume of saline solution, the emulsion divides into two distinct phases. The lower phase consists of chloroform:methanol:water in the ratios 86:14:1, and it contains the lipids. The upper phase consists of the same solvents in the ratios 3:48:47, and it carries the non-lipidic components of the sample. The Bligh and Dyer method is a simpler version of the previous procedure, however it necessitates the further extraction of the remaining sample with chloroform (Bligh & Dyer, 1959). However, there are alternative methods to Folch's reagent that produce similar results but use less toxic substances. These methods involve using pure hexane or combinations of hexane with other solvents, such as the hexane-isopropanol blend (3:2) suggested by Hara and Radin (1978), or the ethyl acetate-ethanol mixture (2:1) employed by Lin et al. (2004). To obtain a thorough review of the extraction techniques using solvents, refer to the works of Kuksis (1994), Murphy (1994), and Kates (1996).

Although the cited solvents are marginally less toxic than chloroform, they nonetheless pose significant hazards and entail substantial risks in terms of handling and control to warrant exploring alternative extraction methods. Several authors propose using solvent-free methods that rely on ionic liquids, boiling the sludge, or exposing it to supercritical gases such as t-butanol, propane, syngas, CO2, or subjecting it to extreme pressures and temperatures (cracking). Each of these options is prohibitively expensive and not viable given the current state of technology (Siddiquee et al., 2011). An alternative that is both practical and effective is the utilization of hot ethanol for the extraction process. This method allows for the extraction of lipids without the need for additional solvents. An exemplification of this extraction method can be demonstrated by the research conducted by Holser and Akin (2008) or Nielsen and Shukla (2004), among other scholars. While the authors have primarily studied the use of ethanol-based extraction to obtain valuable lipids from flax processing wastewater and egg yolk powder, their results can also be extrapolated and utilized for biodiesel generation from wastewater sludges. Nielsen and Shukla discovered that the application of ethanol at normal room temperature resulted in the extraction of almost all phospholipids, along with cholesterol and a small portion of triacylglycerols, without the need for specialized extraction and filtration equipment. Alternatively, Holser and Akin conducted a sequential extraction of the lipids found in flax wastewater using three distinct temperature settings (50, 80, 90, and 100ºC). The researchers discovered that the optimal extraction occurred when the mixture of sample and ethanol was heated to a temperature of 90ºC and left to react for a duration of 15 minutes (Holser & Akin, 2008).

Based on the aforementioned findings, and considering that the enzymatic production of biodiesel typically necessitates high ratios of alcohol to oil, in order to enhance the dissolution of lipids and the creation of interfaces between water-oil, enzyme-oil, and enzyme-alcohol-oil, we suggest a feasible approach for producing biodiesel from WWTP sludge. This approach involves employing a pressurized pretreatment tank, in which the sludge is immersed in ethanol, maintained at a temperature of 90ºC while being stirred, and subjected to three extraction cycles through refluxing. After extraction, the lipids and a portion of the ethanol from the previous step are combined with the enzyme in a reaction tank. The mixture is then incubated at the ideal temperature and pH to maintain enzyme stability and achieve a satisfactory rate of microdiesel production. Heat exchangers can be used between the two tanks to conserve energy by utilizing the heat emitted prior to entering the second tank to preheat the sludge before it enters the first tank.

If the microorganism used to produce the lipase enzyme is capable of simultaneously producing ethanol, or if a third reactor is connected to the mentioned tanks where bioethanol is produced from sugars in the non-lipidic products obtained in the pretreatment tank using ethanol-producing yeast or bacteria strains, the system could potentially operate autonomously in terms of ethanol requirements. When faced with economic limitations, one way to reduce costs in the short term is to replace the pressurized tank used for pretreatment with a non-pressurized tank. Additionally, keeping the temperature of the extraction mix below 79ºC can also help cut costs. However, it is important to note that this approach may result in medium-term economic losses. This is because the lower extraction efficiency would require more extraction cycles at a higher reflux rate and a larger volume of ethanol in the pretreatment tank. Alternatively, it may be necessary to use at least two consecutive pretreatment tanks.

5. CONCISE SUMMARY

In reaction to the immediate effects of greenhouse gas emissions and the unsustainable nature of the fossil fuel energy system, the industry has created readily available alternatives to traditional fossil fuels. These substitutes are commonly referred to as 'bio' products in commercial settings, although the phrase is often used vaguely. However, the initial and subsequent iterations of what are known as biofuels are not entirely derived from biological sources nor are they derived from renewable and environmentally sustainable raw materials. Furthermore, the production procedures heavily depend on substantial amounts of energy, both during the cultivation of feedstock (such as rapeseed, soybean, or palm oil) and in the synthesis of biofuels (acid-catalyzed biodiesel or corn bioethanol serve as clear examples of the energy gain challenges). In addition to existing issues, novel and intricate difficulties have arisen. There are two main factors contributing to the rise in prices of grain and vegetable oils. Firstly, the prices have increased due to their use in both biofuel production and human and livestock consumption. Secondly, there has been an expansion of agricultural land to meet the high demand for sugar cane and vegetable oils, driven by the significant increase in biofuel production. Therefore, it is necessary to develop cleaner and more sustainable biofuels in order to completely replace traditional fossil fuels with third-generation biofuels. These biofuels should be produced without using non-renewable resources or inefficient industrial methods that directly and indirectly harm the environment, as well as pose a threat to biodiversity and food security (UNCTAD, 2010).

A wide range of domestic, agricultural, and industrial residues, including lignocellulosic forestry and agriculture waste, as well as fatty acid-rich waste waters from the dairy, poultry, and vegetable oil refinery industries, can serve as precursors for biofuels. Additionally, sludges from urban waste waters can also be utilized for this purpose. The utilization of these residues might be integrated with the manufacture of third-generation biofuels by enzymatic catalysis, since the economical expense of enzymes could be offset by the inexpensive nature of the residues (or perhaps by the existence of incentives for residue reduction and management). However, the widespread implementation of these ideas necessitates a range of technical and biotechnological advancements. These include enhancing the extraction of lipids and sugars, refining the processes for preparing feedstock, designing biofuel production facilities, developing effective catalysts and techniques for immobilizing enzymes, improving the performance and reusability of lipases, alcohol dehydrogenases, and hydrolases through protein engineering, and genetically modifying microbes to streamline precursor pretreatment, biofuel synthesis, and purification.

REFERENCES

Abdullah, A.Z., Salamatinia, B., Mootabadi, H. & Bathia, S. (2009). Current status and policies on biodiesel industry in Malaysia as the world's leading producer of palm oil. Energy Policy 37(12): 5440-5448

Aksoy, H.A., Kahraman, I., Karaosmanoglu, F., & Civelekoglu, H. (1988). Evaluation of Turkish sulphur olive oil as an alternative diesel fuel. Journal of the American Oil Chemists Society 65: 936-938.

Alauddin, Z.A.B.Z., Lahijani, P., Mohammadi, M., & Mohamed, A.R. (2010). Gasification of lignocellulosic biomass in fluidized beds for renewable energy development: A review. Renewable and Sustainable Energy Reviews 14: 2852-2862.

Al-Zuhair, S., Dowaidar, A. & Kamal, H. (2009). Dynamic modelling of biodiesel production from simulated waste cooking oil using immobilized lipase. Biochemical Engineering Journa.l 44: 256-262.

Amaretti, A., Raimondi, S. & Rossi, M. (2010). Production of Single Cell Oils from Glycerol By Oleaginous Yeasts. Journal of Biotechnology 150: 389.

Antczak, M.S., Kubiak, A., Antczak, T., & Bielecki, S. (2009). Enzymatic biodiesel synthesis: Key factors affecting efficiency of the process. Renewable Energy. 34: 1185-1194.

Aslan, S., & Kapdan, I.K. (2006) Batch kinetics of nitrogen and phosphorus removal from synthetic waste water by algae. Ecological Engineering. 28: 64-70.

Banerjee, A., Sharma, R., Chisti, Y., & Banerjee, U.C. (2002). Botryococcus braunii: A renewable source of hydrocarbons and other chemicals. Critical reviews in Biotechnology. 22: 245-279.

Bender, M. (1999). Economic feasibility review for community-scale farmer cooperatives for biodiesel. Bioresource Technology. 70: 81.

Benemann, J.R. (1997). CO2 mitigation with microalgae systems. Energy Conversion and Management. 38: S475-S479.

Bligh, E.G. & Dyer, W.J. (1959). A rapid method of total lipid extraction and purification. Canadian Journal of Biochemistry and Physiology. 37: 911-917, 1959.

Bludowsky, T. & Agar, D.W. (2009). Thermally integrated bio-syngas-production for biorefineries. Chemical Engineering Research and Design. 87: 1328.

Boocock, D.G.B., Konar, S.K., Leung, A. & Ly, L.D. (1992). Fuels and chemicals from sewage sludge. The solvent extraction and composition of a lipid from a raw sewage sludge. Fuel. 71: 1283.

Cao, F., Chen, Y., Zhai, F., Li, J., Wang, J., Wang, X., Wang, S., & Zhu, W. (2008). Biodiesel production from high acid value waste frying oil catalyzed by superacid heteropolyacid. Biotechnology and Bioengineering. 101: 93- 100.

Carere, C.R., Sparling, R., Cicek, N. & Levin, D.B. (2008). Third-generation biofuels via direct cellulose fermentation. International Journal of Molecular Sciences. 9: 1342-1360.

Cescut, J., Severac, E., Molina-Jouve, C. & Uribelarrea, J.L. (2011). Optimizing pressurized liquid extraction of microbial lipids using the response surface method. Journal of Chromatography A. 1218: 373.

Cheirsilp, B., Suwannarat, W. & Niyomdecha, R. (2011). Mixed culture of oleaginous yeast Rhodotorula glutinis and microalga Chlorella vulgaris for lipid production from industrial wastes and its use as biodiesel feedstock. New Biotechnology In Press, Corrected Proof.

Chen, Y., Xiao, B., Chang, J., Fu, Y., Ly, P. & Wang, X. (2009). Synthesis of biodiesel from waste cooking oil using immobilized lipase in fixed bed reactor. Energy Conversion and Management. 50: 668-673.

Chen, C-Y., Yeh, K-L., Aisyah, R., Lee, D-J. & Chang, J-S. (2011). Cultivation, photobioreactor design and harvesting of microalgae for biodiesel production: A critical review. Bioresource Technology. 102: 71.

Chisti, Y. (2008). Biodiesel from microalgae beats bioethanol. Trends in Biotechnology. 26: 126- 131.

Chisti, Y. (2007). Biodiesel from microalgae. Biotechnology Advance., 25: 294-306.

Chisti, Y. (1992). Assure bioreactor sterility. Chemical Engineering Progress. 88: 80-85.

Choi, K.H., Chisti, Y. & Moo-Young, M. (1995). Split-channel rectangular airlift reactors: Enhancement of performance by geometric modifications. Chemical Engineering Communications. 138: 171-181.

Coelho, C.H., Francisco, J.G., Nogueira, R.F.P. & Campos, M.L.A.M. (2008). Dissolved organic carbon in rainwater from areas heavily impacted by sugar cane burning. Atmospheric Environment, 42: 7115.

Connemann, J., & Fischer, J. (1998). Biodiesel in Europe 1998: Biodiesel Processing Technologies. Paper resented at the International liquid biofuels congress. July 19-22, 1998, Curitiba-Parana, Brazil

Conti, J.J. (2009). Annual Energy Outlook 2009 With Projections to 2030. USDE, EIA. Energy Information Administration Office of Integrated Analysis and Forecasting US. Department of Energy Washington, DC 20585.

Da Costa, A.C.A., Junior, N.P. & Aranda, D.A.G. (2010). The situation of biofuels in Brazil: New generation technologies, Renewable and Sustainable Energy Reviews. 14: 3041-3049.

Dabkowska, K. & Szewczyk, K.W. (2009). Influence of temperature on the activity and enantioselectivity of Burkholderia cepacia lipase in the kinetic resolution of mandelic acid enantiomers. Biochemical Engineering Journal 46: 147-153.

Demirbaş, A. (2009). Political, economic and environmental impacts of biofuels: A review. Applied Energy, 86: S108-S117.

Dizge, N., Aydiner C., Imer D.Y., Bayramoglu, M., Tanriseven, A. & Keskinler, B. (2009). Biodiesel production from sunflower, soybean, and waste cooking oils by transesterification using immobilized lipase. Bioresource Technology, 100: 1983-1991.

Dizge, N. & Keskinler, B. (2008). Enzymatic production of biodiesel from canola oil using immobilized lipase. Biomass and Bioenergy, 32: 1274-1278.

Domínguez, A., Fernández, Y., Fidalgo, B., Pis, J.J. & Menéndez, J.A. (2008). Bio-syngas production with low concentrations of CO2 and CH4 from microwave-induced pyrolysis of wet and dried sewage sludge. Chemosphere, 70: 397.

Ebrahimpour, A., Rahman, R.N.Z.R.A., Ch'ng, D.H.E., Basri, M. & Salleh, A.B. (2008). A modelling study by response surface methodology and artificial neural network on culture parameters optimization for thermostable lipase production from a newly isolated thermophilic Geobacillus sp. strain ARM. BMC Biotechnology. 8: 96.

Elbahloul, Y. & Steinbüchel, A. (2010). Pilot-scale production of fatty acid ethyl esters by an engineered Escherichia coli strain harboring the p(Microdiesel) plasmid. Applied and environmental microbiology. 76: 4560-4565.

Fischer, G., Hizsnyik, E., Prieler, S., Shah, M. & van Velthuizen, H. (2009). Biofuels and food security. Prepared by the International Institute for Applied Systems Analysis (IIASA) for OPEC Fund for International Development (OFID). OFID Pamphlet Series.

Fjerbaek, L., Christensen, K.V. & Norddahl, B. (2009). A review of the current state of biodiesel production using enzymatic transesterification. Biotechnology and Bioengineering 102: 1298-1315.

Folch, J., Lees, M. & Sloane-Stanley, G.H. (1957). A simple method for the isolation and purification of total lipides from animal tissues. The Journal of Biological Chemistry, 226: 497-509.

Fortman, J.L., Chhabra, S., Mukhopadhyay, A., Chou, H., Lee, T.S., Steen, E. & Keasling, J.D. (2008). Biofuel alternatives to ethanol: Pumping the microbial well. Trends in Biotechnology. 26: 375-381.

Freedman, B., Pryde, E.H. & Mounts, T.L. (1984). Variables affecting the yields of fatty esters from transesterified vegetable oils. Journal of the American Oil Chemists Society. 61: 1638-1643.

Fukuda, H., Kondo, A. & Noda, H. (2001). Biodiesel fuel production by transesterification of oils. Journal of Bioscience and Bioengineering. 92: 405-416.

Fukuda, H., Hama, S., Tamalampudi, S. & Noda, H. (2008). Whole-cell biocatalysts for biodiesel fuel production. Trends in Biotechnology. 26: 668-673.

Fukuda, T., Ishikawa, T., Ogawa, M., Shiraga, S., Kato, M., Suye, S-I. & Ueda, M. (2006) Enhancement of cellulase activity by clones selected from the combinatorial library of the cellulose-binding domain by cell surface engineering. Biotechnology progress 22: 933-938.

Gauder, M., Graeff-Hönninger, S. & Claupein, W. (2011). The impact of a growing bioethanol industry on food production in Brazil. Applied Energy. 88: 672.

Gilham, D. & Lehner, R. (2005). Techniques to measure lipase and esterase activity in vitro. Methods. 36: 139-147.

Grothe, E. & Chisti, Y. (2000). Poly(ß-hydroxybutyric acid) thermoplastic production by Alcaligenes latus: Behavior of fed-batch cultures. Bioprocess Engineering. 22: 441-449.

Guerrero-Compeán, R. (2008). Regional, economic, and environmental effects of traditional and biotechnologically enhanced ethanol production processes in Brazil. Thesis (M.C.P.) -- Massachusetts Institute of Technology, Dept. of Urban Studies and Planning. Fulltext available at http://hdl.handle.net/1721.1/45426

Gutarra, M.L.E., Godoy, M.G., Maugeri, F., Rodrigues, M.I., Freire, D.M.G. & Castilho, L.R. (2009). Production of an acidic and thermostable lipase of the mesophilic fungus Penicillium simplicissimum by solid-state. ioresource Technology. 100: 5249-5254.

Ha, S.H., Lan, M.N., Lee, S.H., Hwang, S.M. & Koo, Y-M. (2007). Lipase-catalyzed biodiesel production from soybean oil in ionic liquids. Enzyme and Microbial Technology. 41(4): 480-483.

Haas, M.J. & Foglia, T.A. (2003). In situ production of fatty acid alkyl esters. United States patent and trademark office pre-grant publication, Dec 2003 patno: US20030229237

Hara, A. & Radin, N.S. (1978). Lipid extraction of tissues with a low-toxicity solvent. Analytical Biochemistry. 90: 420.

Harun, R., Singh, M., Forde, G.M. & Danquah, M.K. (2010). Bioprocess engineering of microalgae to produce a variety of consumer products. Renewable and Sustainable Energy Reviews. 14: 1037-1047.

Helwani, Z., Othman, M.R., Aziz, N., Fernando, W.J.N. & Kim, J. (2009). Technologies for production of biodiesel focusing on green catalytic techniques: A review. Fuel Processing Technology. 90: 1502.

Holser R.A. & Akin D.E. (2008). Extraction of lipids from flax processing waste using hot ethanol. Industrial Crops and Products 27: 236.

Huber, G.W., Iborra, S. & Corma, A. (2006). Synthesis of transportation fuels from biomass: chemistry, catalysts, and engineering. Chemical reviews 106 (9): 4044-4098.

Hung, T-C., Giridhar, R., Chiou, S-H. & Wu, W-T. (2003). Binary immobilization of Candida rugosa lipase on chitosan. Journal of Molecular Catalysis B: Enzymatic. 26: 69-78.

Iso, M., Chen, B.X., Eguchi, M., Kudo, T. & Shrestha, S. (2001). Production of biodiesel fuel from triglycerides and alcohol using immobilized lipase. Journal of Molecular Catalysis B-Enzymatic. 16: 53-58.

Jardé, E., Mansuy, L. & Faure, P. (2005). Organic markers in the lipidic fraction of sewage sludges. Water research. 39: 1215.

Jaeger, K.E. & Eggert, T. (2002). Lipases for biotechnology. Current Opinion in Biotechnology. 13:390-397.

Jaruwat, P., Kongjao, S. & Hunsom, M. (2010). Management of biodiesel waste water by the combined processes of chemical recovery and electrochemical treatment. Energy Conversion and Management. 51: 531-537.

Jegannathan, K.R., Chan, E-S. & Ravindra, P. (2009). Harnessing biofuels: A global Renaissance in energy production? Renewable and Sustainable Energy Reviews. 13: 2163-2168.

Jeromin, L., Peukert, E. & Wollmann, G. (1987). Process for the pre-esterification of free fatty acids in fats and oils US Patent-4698186. 6 October, 1987. 327. Henkel Kommanditgesellschaft auf Aktien.

Joseph, B., Ramteke, P.W. & Kumar, P.A. (2006). Studies on the enhanced production of extracellular lipase by Staphylococcus epidermidis. Journal of General and Applied Microbiology. 52: 315-320.

Kalscheuer, R., Luftmann, H. & Steinbüchel, A. (2004). Synthesis of novel lipids in Saccharomyces cerevisiae by heterologous expression of an unspecific bacterial acyltransferase. Applied and Environmental Microbiology 70 (12): 7119-7125.

Kalscheuer, R., Stolting, T. & Steinbuchel, A. (2006). Microdiesel: Escherichia coli engineered for fuel production. Microbiology-Sgm. 152: 2529-2536.

Kates, M. (1996). Review of "Advances in Lipid Methodology-Three" (Ed. Christie, WW) pp. 1 177, 1996, The Oily Press, Dundee, Scotland. Chemistry and Physics of Lipids 83: 87- 89. Book review.

Kawahara, Y. & Ono, T. (1979). Process for producing lower alcohol esters of fatty acids. United States Patent and Trademark Office granted patent, Aug 1979 patno:US4164506.

Knezevic, Z., Mojovic, L. & Adnadjevic, B. (1998). Palm oil hydrolysis by lipase from Candida cylindracea immobilized on zeolite type Y. Enzyme and Microbial Technology. 22: 275-280.

Krawczyk, T. (1996). Biodiesel: Alternative fuel makes inroads but hurdles remain. INFORM. 7: 801-829.

Kuksis, A. (1994). Advances in Lipid Methodology-Two Edited by W.W. Christie, The Scottish Crop Research Institute, Innergowrie, Dundee DD22 5DA, Scotland Published by The Oily Press, Dundee, 1993. Chemistry and Physics of Lipids 71: 115-116. Book review.

Lin, J-H., Liu, L-Y., Yang, M-H. & Lee, M-H. (2004). Ethyl acetate/ethyl alcohol mixtures as an alternative to Folch reagent for extracting animal lipids. Journal of Agriculture and Food Chemistry. 52: 4984-4986.

Lu, Y.P., Zhang, YHP. & Lynd, L.R. (2006). Enzyme-microbe synergy during cellulose hydrolysis by Clostridium thermocellum. Proceedings of the National Academy of Sciences of the United States of America 103: 16165-16169.

Manzanera, M. (2011). Biofuels from Oily Biomass, In: Carbon-neutral fuels and energy carriers.

M. Muradov & N. Veziroghu (Ed.), 635-663, CRC, Taylor & Francis Group, ISBN 978-143-9818-57-2 Orlando,USA.

Mata, T.M., Martins, A.A. & Caetano, N.S. (2010). Microalgae for biodiesel production and other applications: A review. Renewable and Sustainable Energy Reviews. 14: 217-232.

Matsumura, M. & Murakami, S. (2002). Method and equipment of refining plant oil and waste vegetable oil into diesel engine fuel. Japan Patent JP2000219886 (A). Also published as EP1026224 (A1) and US6364917 (B1).

Molina-Grima, E., Acién-Fernández, F.G. & Chisti, Y. (2001). Tubular photobioreactor design for algal cultures. Journal of Biotechnology. 92: 113-131.

Molina-Grima, E., Acién-Fernández, F.G., García-Camacho, F., Camacho-Rubio, F. & Chisti, Y. (2000) Scale-up of tubular photobioreactors. Journal of Applied Phycology. 12: 355-368.

Mondala, A., Liang, K., Toghiani, H., Hernandez, R. & French, T. (2009). Biodiesel production by in situ transesterification of municipal primary and secondary sludges. Bioresource Technology. 100: 1203-1210.

Murphy, E.J. (1994). Advances in Lipid Methodology One. Edited by W.W. Christie, The Oily Press, Ayr, Scotland ISBN 0 9514171 1 8. Chemistry and Physics of Lipids 69: 179-180. Book review.

Nielsen, H. & Shukla, VKS. (2004). In situ solid phase extraction of lipids from spray-dried egg yolk by ethanol with subsequent removal of triacylglycerols by cold temperature crystallization. LWT - Food Science and Technology. 37: 613.

OECD (2008). Biofuel Support Policies: An Economic Assessment. OECD Publishing. Also available in French under the title Politiques de soutien des biocarburants: Une évaluation économique. Online publication.

http: //dx.doi.org /10.1787/ 9789264050112-en Park, E.Y., Sato, M. & Kojima, S. (2008). Lipase-catalyzed biodiesel production from waste activated bleaching earth as raw material in a pilot plant. Bioresource Technology. 99: 3130-3135.

Peterson, C.L. (1986). Vegetable oil as a diesel fuel: Status and research priorities. Transactions of the Asae. 29: 1413-1422.

Pugazhvadivu, M. & Jeyachandran, K. (2005). Investigations on the performance and exhaust emissions of a diesel engine using preheated waste frying oil as fuel. Renewable Energy. 30: 2189.

Radermacher, W. & Director-General E. European Commission (2009). Measuring progress towards a more sustainable Europe statistical books 2009 monitoring reportof the EU sustainable development strategy. Office for Official Publications of the European Communities. ISBN 978-92-79-12695-6

Rahman, Z., Shida, Y., Furukawa, T., Suzuki, Y., Okada, H., Ogasawara, W. & Morikawa, Y. (2009). Application of Trichoderma reesei cellulase and xylanase promoters through homologous recombination for enhanced production of extracellular betaglucosidase I. Bioscience, biotechnology, and biochemistry. 73: 1083-1089.

Raita, M., Champreda, V. & Laosiripojana, N. (2010). Biocatalytic ethanolysis of palm oil for biodiesel production using microcrystalline lipase in tert-butanol system. Process Biochemistry. 45: 829.

Rakopoulos, C.D., Antonopoulos, K.A. & Rakopoulos, D.C. (2006). Multi-zone modelling of Diesel engine fuel spray development with vegetable oil, biodiesel or diesel fuels. Energy Conversion and Management, 47: 1550.

Ranganathan, S.V., Narasimhan, S.L. & Muthukumar, K. (2008). An overview of enzymaticproduction of biodiesel. Bioresource Technology. 99: 3975-3981.

Ribeiro, A.M., Santos, J.C. & Rodrigues, A.E. (2010). PSA design for stoichiometric adjustment of bio-syngas for methanol production and co-capture of carbon dioxide. Chemical Engineering Journal. 163: 355.

Richmond, A. (Ed.) (2010). Handbook of Microalgal Culture, Blackwell Science. Iowa State Press, a Blackwell Publishing Company, 2121 State Avenue, Ames, Iowa 50014-8300, USA. ISBN 0-632-05953-2 .

Robles-Medina, A., Gonzalez-Moreno, P.A., Esteban-Cerdan, L. & Molina-Grima, E. (2009). Biocatalysis: Towards ever greener biodiesel production. Biotechnology Advances. 27:398-408.

Rosa, C.D., Morandim, M.B., Ninow, J.L., Oliveira, D., Treichel, H. & Oliveira, J.V. (2008). Lipase-catalyzed production of fatty acid ethyl esters from soybean oil in compressed propane. Journal of Supercritical Fluids. 47: 49-53.

Royon, D., Daz, M., Ellenrieder, G. & Locatellu, S. (2007). Enzymatic production of biodiesel from cotton seed oil using t-butanol as a solvent. Bioresource Technology. 98: 648–653.

Royter, M., Schmidt, M., Elend, C., Höbenreich, H., Schäfer, T., Bornscheuer, U.T. & Antranikian, G. (2009). Thermostable lipases from the extreme thermophilic anaerobic bacteria Thermoanaerobacter thermohydrosulfuricus SOL1 and Caldanaerobacter subterraneus subsp. tengcongensis. Extremophiles. 13: 769-783.

Rupprecht, J. (2009). From systems biology to fuel-Chlamydomonas reinhardtii as a model for a systems biology approach to improve biohydrogen. Journal of Biotechnology. 142: 10- 20.

Saka, S. & Kusdiana, D. (2001). Biodiesel fuel from rapeseed oil as prepared in supercritical methanol. Fuel. 80: 225-231.

Shah, S., Sharma, S. & Gupta, M.N. (2004). Biodiesel preparation by lipase-catalyzed transesterification of Jatropha oil. Energy & Fuels. 18: 154-159.

Shen, L. & Zhang, D-K. (2003). An experimental study of oil recovery from sewage sludge by low-temperature pyrolysis in a fluidised-bed. Fue.l 82: 465.

Shieh, C.J., Liao, H.F. & Lee, C.C. (2003). Optimization of lipase-catalyzed biodiesel by response surface methodology. Bioresource Technology. 88: 103-106.

Sidibé, S.S., Blin, J., Vaitilingom, G. & Azoumah, Y. (2010). Use of crude filtered vegetable oil as a fuel in diesel engines state of the art: Literature review. Renewable and Sustainable Energy Reviews 14: 2748-2759.

Siddiquee, M.N. & Rohani, S. (2011). Lipid extraction and biodiesel production from municipal sewage sludges: A review. Renewable and Sustainable Energy Reviews 15:1067–1072.

Silvis, T.W. (1977). Waste vegetable oil as a diesel fuel. Thesis. The Ohio State University. Department of Mechanical Engineering.

Srinivas, S., Malik, R.K. & Mahajani, S.M. (2007). Fischer-Tropsch synthesis using bio-syngas and CO2. Energy for Sustainable Development. 11: 66.

Srinivas, T.N.R., Nageswara Rao, S.S.S., Vishnu-Vardhan-Reddy, P., Pratibha, M.S., Sailaja, B., Kavya, B., Hara-Kishore, K., Begum, Z., Singh, S.M. & Shivaji, S. (2009). Bacterial diversity and bioprospecting for cold-active lipases, amylases and proteases, from culturable bacteria of Kongsfjorden and Ny-Alesund, Svalbard, Arctic. Current microbiology. 59: 537-547.

Steen, E.J., Kang, Y., Bokinsky, G., Hu, Z., Schirmer, A., McClure, A., B.-del-Cardayre, S. & Keasling, J.D. (2010). Microbial production of fatty-acid-derived fuels and chemicals from plant biomass. Nature. 463: 559-563.

Suppes, G.J., Dasari, M.A., Doskocil, E.J., Mankidy, P.J. & Goff, M.J. (2004). Transesterification of soybean oil with zeolite and metal catalysts. Applied Catalysis A: General. 257: 213-223.

Tessele, F., Misra, M. & Rubio, J. (1998). Removal of Hg, As and Se ions from gold cyanide leach solutions by dissolved air flotation. Minerals Engineering. 11: 535.

Timilsina, G.R. & Shrestha, A. (2010). How much hope should we have for biofuels? Energy. Corrected Proof. doi:10.1016/j.energy.2010.08.023

Tirado-Acevedo, O., Chinn, M.S. & Grunden, A.M. (2010). Production of biofuels from synthesis gas using microbial catalysts. Advances in Applied Microbiology. 70: 57.

Tongboriboon, K., Cheirsilp, B. & H-Kittikun, A. (2010). Mixed lipases for efficient enzymatic synthesis of biodiesel from used palm oil and ethanol in a solvent-free system. Journal of Molecular Catalysis B: Enzymatic. 67: 52.

UNCTAD, (2010). United Nations Conference on Trade and Development - Harvard – Belfer Center for Science and International Affairs. United Nations. New York and Geneva.

van Dam, J., Faaij, APC., Hilbert, J., Petruzzi, H. & Turkenburg, W.C. (2009). Large-scale bioenergy production from soybeans and switchgrass in Argentina: Part A: Potential and economic feasibility for national and international markets. Renewable and Sustainable Energy Reviews. 13: 1710-1733.

van Groenestijn, J.W., Hazewinkel, J.H.O., Nienoord, M. & Bussmann, P.J.T. (2002). Energy aspects of biological hydrogen production in high rate bioreactors operated in the thermophilic temperature. International Journal of Hydrogen Energy. 27: 1141-1147.

van Zyl, W.H., Lynd, L.R., den Haan, R. & McBride, J.E. (2007). Consolidated bioprocessing for bioethanol production using Saccharomyces cereviside. Biofuels. 108: 205-235.

Waltermann, M. & Steinbüchel, A. (2010). Neutral lipid bodies in prokaryotes: Recent insights into structure, formation, and relationship to eukaryotic lipid depots. Journal of Bacteriology 187 (11): 3607-3619.

Wang, C.C., Chang, C.W., Chu, C.P., Lee, D.J., Chang, B.V. & Liao, C.S. (2003). Producing hydrogen from waste-water sludge by Clostridium bifermentans. Journal of Biotechnology. 102: 83-92.

Wang, L., Du, W., Liu, D.H., Li, L.L. & Dai, N.M. (2006). Lipase-catalyzed biodiesel production from soybean oil deodorizer distillate with absorbent present in tertbutanol system. Journal of Molecular Catalysis B-Enzymatic. 43: 29-32.

Wei, Z., Xu, C. & Li, B. (2009). Application of waste eggshell as low-cost solid catalyst for biodiesel production. Bioresource Technology. 100(11): 2883-2885.

Willey, R. (2001). Fats, oils, and greases: the minimization and treatment of wastewaters generated from oil refining and margarine production. Ecotoxicology and Environmental Safety 50: 127-133.

World-Bank (2011). World Bank Commodity Price Data (Pink Datasheet). 1818 H Street, N.W. , MSN MC 2-200 Washington, D.C. 20433 U.S.A.

Yee, K.F., Tan, K.T., Abdullah, A.Z. & Lee, K.T. (2009). Life cycle assessment of palm biodiesel: Revealing facts and benefits for sustainability. Applied Energy. 86: S189-S196.

Zabeti, M., Wan Daud, WMA. & Aroua, M.K. (2009). Activity of solid catalysts for biodiesel production: A review. Fuel Processing Technology. 90: 770-777.

Zeevalkink, J.A. & Brunsmann, J.J. (1983). Oil removal from water in parallel plate gravitytype separators. Water research. 17: 365.

Zhang, Y., Dube, M.A., McLean, D.D. & Kates, M. (2003). Biodiesel production from waste cooking oil: 1. Process design and technological assessment. Bioresource Technology. 89: 1-16.

Zhang, Y., Habibi, S. & MacLean, H.L. (2007). Environmental and economic evaluation of bioenergy in Ontario, Canada. Journal of the Air & Waste Management Association 57: 919-933.

Zhao, Z.K., Zhao, X., Hu, C., Shen, H. & Wu, S. (2009). Towards industrial lipid production using oleaginous yeasts. New Biotechnology. 25: S233.

CHAPTER-7

CORN-BASED FUEL ETHANOL COPRODUCTS: PRODUCTION AND USE

1. INTRODUCTION: AN OVERVIEW

Contemporary cultures encounter numerous obstacles, such as expanding populations and heightened requirements for sustenance, attire, shelter, consumer products, and the necessary resources to manufacture them. Furthermore, there is an increasing demand for energy, which can be conveniently fulfilled through the utilization of fossil fuels such as coal, natural gas, and petroleum. In 2008, the total energy consumption in the United States was 99.3 x 1015 Btu (1.05 x 1014 MJ), with fossil sources supplying 84% of this need. During this period, transportation fuels constituted 28% of the total energy consumed, with fossil sources contributing to about 97% of this amount. The United States produced 4.96 million barrels of crude oil per day domestically, while importing 9.76 million barrels per day, which accounted for approximately two-thirds of the entire U.S. consumption (U.S. EIA, 2011). There is a widespread belief that this situation cannot be maintained over a long period of time, and that different solutions are necessary. Biofuels, being sustainable energy sources, can assist in fulfilling a portion of these growing demands. They can be produced from a range of materials that contain carbohydrates or lipids, including cereal grains (such as corn, barley, and wheat), oilseeds (such as soybean, canola, and flax), legumes (such as alfalfa), perennial grasses (such as switchgrass, miscanthus, prairie cord grass, and others), agricultural residues (such as corn stover and wheat stems), algae, food processing wastes, and other biological materials. The lignocellulosic ethanol industry is expected to require significant amounts of biomass in the future, as indicated by several studies (Agrawal et al., 2007; Alexander and Hurt, 2007; Cassman, 2007; Cassman et al., 2006; Cassman and Liska, 2007; Dale, 2007; De La Torre Ugarte et al., 2000; Dewulf et al., 2005; Lynd and Wang, 2004). Currently, the primary feedstock utilized for biofuel production in the United States is corn grain. Large-scale manufacture of alcohol from maize starch is easily achieved, and at a cheaper expense (often ranging from \$1/gallon to \$1.4/gallon), compared to other biomass sources in the United States. The dry grind process is the most often employed method for producing gasoline ethanol from corn. The principal byproduct of this process is distillers dried grains with solubles (DDGS) (Figure 1), which will be further addressed later. The utilization of corn-based ethanol as a liquid fuel for transportation has a history of over 150 years, although the industry has remained relatively limited until recent times. The contemporary ethanol business, which relies on corn as its primary source, has grown to a size that can increase the country's stock of transportation fuels. For instance, in 2008, ethanol replaced around 321 million barrels of oil (Urbanchuk, 2009), or almost 5% of total oil imports. It is only in recent times that this industry has been genuinely apparent to the ordinary citizen. This can be attributed, to some extent, to the increasing need for transportation fuels, rising prices at fuel stations, beneficial economic impacts in rural America, and the presence of inquiries and disputes about the manufacturing and utilization of maize ethanol.

Figure 1. Corn-based distillers dried grains with solubles (DDGS), which is currently available from most U.S. fuel ethanol plants.

To help meet the increasing demand for transportation fuels, the number of ethanol plants has been rapidly increasing in recent years, as has the quantity of fuel ethanol produced (Figure 2).

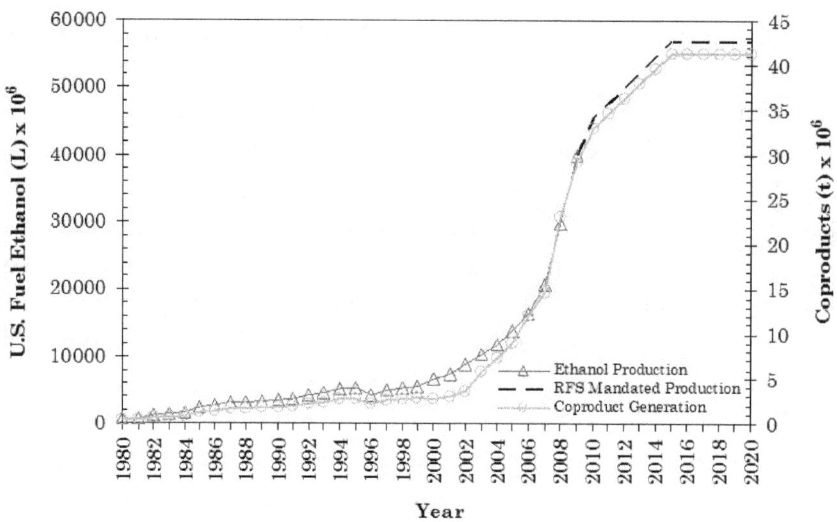

Figure 2. U.S. fuel ethanol (L) and DDGS (t) production over time; RFS denotes levels mandated by the Renewable Fuel Standard. Inset shows number of U.S. ethanol plants over time (adapted fromRFA, 2009a,2009b,2011)

In 2005, 87 manufacturing plants in the U.S. had an aggregate production capacity of 13.46 billion L/y (3.56 billion gal/y). At the beginning of 2011, however, that number had risen to 204 plants with a production capacity of nearly 51.1 billion L/y (13.5 billion gal/y), which is an increase of nearly 380% in six years (RFA, 2011). Most new ethanol plants have been dry-grind facilities (Figure 3), which will be discussed subsequently. And, over the next several years, the Renewable Fuel Standard (RFS) requires the utilization of 15 billion gallons per year (56.8 billion liters per year) of renewable biofuels, mainly derived from corn-based ethanol. However, the RFS also includes provisions for the increased use of advanced and cellulosic biofuels. Due to the dynamic and evolving nature of the industry, the current output quantities mentioned will inevitably become outdated before the publication of this book.

Figure 3. U.S. dry grind corn-to-ethanol manufacturing plants. A. 450 x 106L/y plant. B. 80 x 106L/y plant.

As the production volume increases, the processing residues, commonly referred to as "distillers grains," will also increase in proportion, as depicted in Figure 2. The U.S. gasoline ethanol business is expected to produce more than 40 million metric tonnes (t) of distillers grains (both wet and dry) once output achieves a state of balance, thanks to the RFS. Indeed, it is a fact that the consumption of corn has increased in parallel with the growth of the industry (Figure 4). Since 2008, more than 30% of the corn yield in the United States has been utilized for ethanol production. When analyzing these figures, it is crucial to consider a few significant factors: exports have remained relatively stable over time, there has been a slight decrease in the amount of corn used for animal feed, and the total quantity of corn produced by American farmers has significantly increased over time. Therefore, it seems that the corn used for ethanol production primarily comes from the expanding corn supply. It is worth mentioning that the corn that is diverted from animal feed is really being substituted by DDGS and other ethanol byproducts in these animal diets. Therefore, coproducts, particularly DDGDS, play a crucial role in ensuring the long-term viability of both the ethanol and livestock sectors. Put simply, the need for fuel, feed, and food may all be satisfied at the same time.

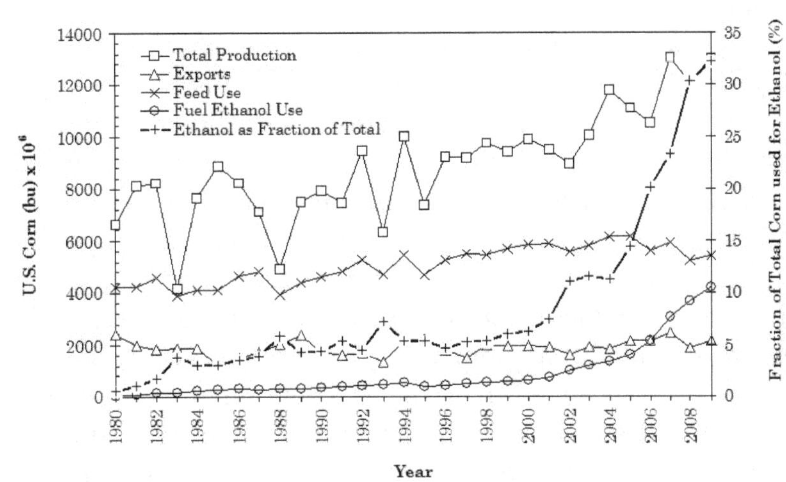

Figure 4. Historic U.S. corn production (bu) and major categories of use (adapted fromERS, 2011).

2. THE STUDY AIMS TO ACHIEVE SPECIFIC GOALS AND OBJECTIVES

The objectives of this chapter are threefold: The objectives of this discussion are: 1) to provide a concise overview of the manufacturing processes involved in U.S. fuel ethanol and coproduct production; 2) to emphasize the significance of coproducts in both the fuel ethanol and livestock sectors; and 3) to illustrate the ongoing improvement in coproduct quality and the expanding range of potential applications as the ethanol industry progresses.

3. METHODS OF PRODUCTION

There are three commercial procedures for converting corn into fuel ethanol: wet milling, dry milling, and dry grind processing. In the past ten years, there has been a significant increase in the construction of fuel ethanol facilities (Figure 2). The industry has also witnessed notable advancements in production methods, final products, and the utilization of raw materials, water, and energy. A multitude of these advancements have emerged with the introduction of dry grind processing. Most modern ethanol plants are dry grind facilities instead of the older style mills due to several benefits, such as reduced capital and operating expenses, including energy use. For instance, in 2002, half of the ethanol plants in the United States were dry grind. By 2004, this percentage had increased to 67%. In 2006, dry grind plants accounted for 79% of all facilities, and by 2009, this proportion had surpassed 80% (RFA, 2009a). The dry grind process (Figure 5) involves multiple essential phases, such as grain reception, distribution, storage, cleaning, grinding, cooking, liquefaction, saccharification, fermentation, distillation, ethanol storage and loadout, centrifugation, coproduct drying, and coproduct storage and loadout. Other essential systems include energy/heat recovery, waste management, grain aeration, CO2 scrubbing and extraction, dust control, facility sanitation, instrumentation and controls, and sampling and inspection. Figure 5 illustrates the integration of all these components in a commercial facility. The processes of grinding, heating, and liquefying serve to liberate and convert corn starch into glucose, which is then absorbed by yeast (Sacchharomyces cerevisiae) during the fermentation process. Following the process of fermentation, the ethanol is extracted from the water and nonfermentable residues (comprising of proteins, fibers, oils, and minerals derived from maize kernels) through distillation. Subsequent to that, downstream processes such as dewatering, separation, evaporation, mixing, and drying are employed to exclude water from the solid residues and generate several streams of byproducts (referred to as distillers grains as a whole): either wet or dry, with or without the inclusion of condensed solubles (CDS). Distillers dried grains with solubles (DDGS) is commonly dried to around 10% moisture content, or even lower at certain facilities, to enhance its shelf life and flowability. It is then either sold to nearby livestock producers or transported by truck or rail to different locations across the country. DDGS is experiencing a growing trend of being sold to international markets as well. Distillers wet grains (DWG) have been increasingly popular among livestock producers located near ethanol plants in recent years. In fact, it has been estimated that over 25% of distillers grains sales countrywide are currently comprised of DWG. However, due to the high moisture content, typically exceeding 50 to 60%, the longevity of these products is significantly restricted, particularly during the summer season, and transporting substantial volumes of water incurs high costs. DDGS remains the dominant form of distillers grain in the market. The process of dry grind ethanol manufacturing yields three basic products: ethanol, the main final product; residual nonfermentable corn kernel components, which are commercially marketed as distillers grains; and carbon dioxide. It is generally accepted that when 1 kilogram of maize is processed, approximately 1/3 kg of each of the constituent streams will be generated. According to another general principle, every bushel of maize, which weighs around 56 pounds (25.4 kilograms), can produce up to 2.9 gallons (11.0 liters) of ethanol, around 18 pounds (8.2 kilograms) of distillers grains, and nearly 18 pounds (8.2

kilograms) of carbon dioxide. Naturally, these factors will inevitably differ to a certain extent over time as a result of production methods, equipment configurations, duration of storage, levels of substances, maintenance plans, equipment conditions, environmental circumstances, the composition and quality of the original corn, the geographical origin of the corn, and the specific growing season that yielded the corn. During fermentation, carbon dioxide is produced by the metabolic conversion of carbohydrates into ethanol by the yeast. The residual stream can be collected and marketed to compressed gas industries, such as manufacturers of beverages or dry ice. Frequently, nevertheless, it is emitted into the sky due to the impracticality of selling and marketing this gas, caused by factors such as location and logistics. In the future, the output of carbon dioxide could potentially be influenced by restrictions and laws on greenhouse gas emissions.

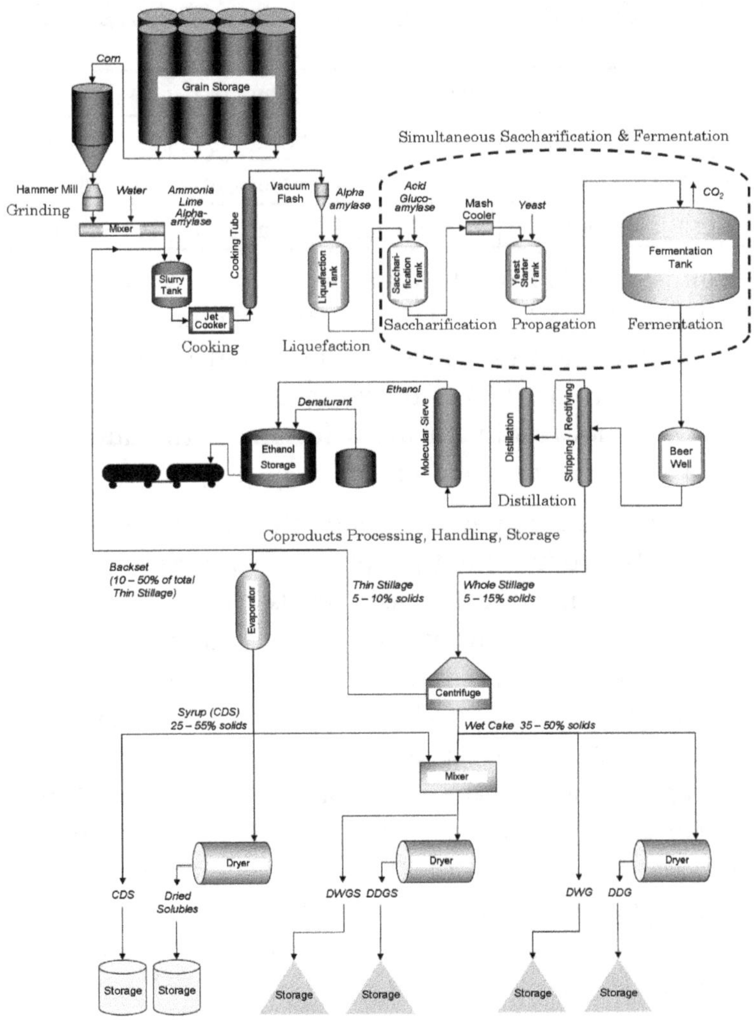

Figure 5. Flow chart of typical corn dry grind fuel ethanol and coproducts processing.

For more in-depth information on the processing phases of ethanol and DDGS, refer to the following sources: Tibelius (1996), Weigel et al. (1997), Dien et al. (2003), Jaques et al. (2003), Bothast and Schlicher (2005), Rausch and Belyea (2006), and Ingledew et al. (2009).

4. SIGNIFICANCE OF COPRODUCTS

DDGS derived from advanced U.S. fuel ethanol facilities generally consists of approximately 30% protein, 10% fat, a minimum of 40% neutral detergent fiber, and a maximum of 12% starch (Rosentrater &

Muthukumarappan, 2006). The composition of plants can differ both across different plants and within the same plant over time, owing to various influences. Table 1 provides a concise overview of the composition of DDGS samples obtained from five ethanol plants located in South Dakota. The crude protein levels on a dry basis ranged from 28.3 to 31.8%, while the crude fat content varied between 9.4 and 11.0%. The ash content ranged from 4.1 to 13.3%. Regarding the variability within the plant, the levels of crude protein, crude lipid, and starch showed relatively little variation, while neutral detergent fiber (NDF), acid detergent fiber (ADF), and ash shown significantly larger variability.

Table 1. Composition (% db) of DDGS from five ethanol plants in South Dakota (± 1 standard deviation in parentheses). Statistically significant differences among plants for a given nutrient are denoted by differing letters, α=0.05, LSD (adapted fromBhadra et al., 2009).

Plant	Protein	Lipid	NDF	ADF	Starch	Ash
1	28.33b(1.25)	10.76a(1.00)	31.84b(4.02)	15.56a(2.29)	11.82a(1.20)	13.27a(3.10)
2	30.65a(1.20)	9.75a(1.05)	39.90a(3.95)	15.21a(3.95)	9.81a(1.52)	12.84a(2.56)
3	28.70a(1.32)	10.98a(0.95)	38.46a(4.01)	17.89a(4.01)	11.59a(1.42)	11.52a(3.05)
4	30.65a(1.23)	9.40b(0.16)	36.73a(1.07)	15.28a(0.49)	9.05b(0.33)	4.13b(0.21)
5	31.78a(0.63)	9.50b(0.41)	38.88a(0.86)	17.24a(1.12)	10.05a(0.65)	4.48b(0.22)

Furthermore, DDGS from 49 plants from 12 states were analyzed for proximate composition (Table 2)and amino acid profiles (Table 3)(UMN, 2011). Dry matter content varied from 86.2% to 92.4%, while protein varied from 27.3% to 33%. Crude fat content displayed even higher variability, and ranged from 3.5% to 13.5%; crude fiber ranged from 5.37% to 10.58%; and ash content varied from 2.97% to 9.84%. On average, geographic trends were not readily apparent for any of the nutrient components. In terms of amino acids, lysine ranged from 0.61% to 1.19%, but again, no geographic trends were apparent.

Some plants are beginning to implement various fractionation processes (either pre-fermentation or post-fermentation) in order to produce multiple product streams (RFA, 2009a). These new processes can lead to additional differences in DDGS nutrient levels. For example, various techniques for dry fractionation and wet fractionation have been developed to concentrate protein, fiber, and oil components from the endosperm (which contains the starch). This allows a highly-concentrated starch substrate to be introduced to the fermentation process, and it allows the other components to be used for human food applications.Singh and Johnston (2009) have provided an extensive discussion regarding various pre-fermentation fractionation approaches. On the other hand, post-fermentation fractionation techniques have also been examined. For example,Srinivasan et al. (2005) used a combination of (air classification and sieving to separate fiber particles from DDGS. Processes have also been developed to remove corn oil from thin stillage and CDS; although the resulting corn oil fractions cannot be used as food-grade oil, they can readily be converted into biodiesel. All of these approaches, if implemented commercially, will alter the composition of the resulting DDGS.

Table 2. Composition (% db) of DDGS samples from 49 ethanol plants from 12 states (adapted fromUMN, 2011).

State	Plants Sampled	Dry Matter (%)	Crude Protein (%)	Crude Fat (%)	Crude Fiber (%)	Ash (%)
Minnesota	12	89.03	30.70	11.73	6.96	6.63
Illinois	6	89.72	29.98	11.48	7.26	5.60
Indiana	2	90.55	29.40	12.80	8.07	5.86
Iowa	7	88.92	31.23	10.27	7.57	5.76

State	Plants Sampled	Dry Matter (%)	Crude Protein (%)	Crude Fat (%)	Crude Fiber (%)	Ash (%)
Kentu ck y	3	90.57	29.43	9.77	9.28	4.47
Michigan	1	89.60	32.60	11.00	7.37	6.06
Missouri	2	87.90	30.45	10.25	7.17	5.39
Nebraska	4	89.02	30.40	11.35	8.13	4.23
New York	1	88.21	30.00	9.60	7.87	4.55
North Dakota	4	89.21	31.75	11.70	6.89	6.32
South Dakota	4	88.61	31.80	11.53	6.65	4.78
Wisconsin	3	89.68	31.70	11.63	7.59	5.77
Overall Average	49 (Total)	89.25	30.79	11.09	7.57	5.45

Historically, the principal market for distillers grains in the U.S. ethanol business has been as a commodity animal feed. Typically, this has primarily been in the form of DDGS, and to a lesser extent in the form of DWG. The quantities of other coproducts supplied are significantly lower than those of DDGS or DWG, and some of them are not consistently produced. Utilizing ethanol coproducts as animal feed is a realistic approach due to their high nutrient content and digestibility by most livestock, albeit to different extents. Furthermore, including DDGS in animal feeds as a substitute for maize grain serves to counterbalance the corn that has been diverted towards ethanol production. More than 80% of all distillers grains are used by cattle and dairy animals in their diets.

Table 3. Amino acid profiles (% db) of DDGS samples from 49 ethanol plants from 12 states (adapted fromUMN, 2011).

State	Plants Sampled	Agrinine (%)	Histidine (%)	Isoleucine (%)	Leucine (%)	Lysine (%)	Methionine (%)
Minnesota	12	1.39	0.84	1.20	3.63	0.99	0.61
Illinois	6	1.37	0.82	1.15	3.45	0.94	0.63
Indiana	2	1.19	0.79	1.08	3.28	0.85	0.60
Iowa	7	1.34	0.86	1.20	3.63	0.95	0.61
Kentu ck y	3	1.35	0.79	1.09	3.33	0.89	0.66
Michigan	1	1.28	0.86	1.18	3.67	0.87	0.71
Missouri	2	1.35	0.83	1.18	3.68	0.89	0.73
Nebraska	4	1.46	0.88	1.18	3.61	1.05	0.65
New Yor k	1	1.46	0.85	1.21	3.64	1.04	0.61
North Dakota	4	1.37	0.88	1.24	3.76	0.97	0.65
South Dakota	4	1.47	0.87	1.22	3.70	1.08	0.62
Wisconsin	3	1.45	0.86	1.24	3.75	1.07	0.59
Overall Average	49	1.37	0.84	1.18	3.59	0.96	0.64

State	Plants Sampled	Phenylalanine (%)	Threonine (%)	Tryptophan (%)	Valine (%)	Tyrosine (%)
Minnesota	12	1.59	1.17	0.24	1.62	1.20
Illinois	6	1.51	1.11	0.22	1.52	1.22
Indiana	2	1.45	1.04	0.21	1.44	-
Iowa	7	1.57	1.14	0.25	1.60	-
Kentu ck y	3	1.48	1.09	0.26	1.43	-
Michigan	1	1.52	1.15	0.25	1.57	-
Missouri	2	1.53	1.15	0.24	1.58	-
Nebraska	4	1.58	1.15	0.26	1.58	1.14
New Yor k	1	1.63	1.11	0.20	1.59	1.19
North Dakota	4	1.62	1.19	0.25	1.67	-
South Dakota	4	1.67	1.19	0.23	1.63	1.35
Wisconsin	3	1.65	1.14	0.22	1.64	1.25
Overall Average	49	1.56	1.13	0.24	1.57	1.22

Ruminant animals, which can efficiently digest high amounts of fiber, have become the primary users of DDGS. However, with the growing understanding of livestock production and animal nutrition among producers and experts, the demand for swine and poultry products in the market is also on the rise (UMN, 2011). Over the years, multiple research studies have investigated the utilization of coproducts in cattle diets, including both ruminant and monogastric feeds. Table 4 provides a compilation of selected research findings. Various studies have demonstrated that all types of cattle may flourish when their diet includes 10% DDGS. Furthermore, most animals can handle doses of up to 20% or even higher.

Table 4. Summary of livestock research on fuel ethanol coproducts.

Species Beef	Citation	Species Dairy	Citation
	Loy et al., 2007		Kleinschmit et al., 2007
	MacDonald et al., 2007		Anderson et al., 2006
	Martin et al., 2007		Kleinschmit et al., 2006
	Roeber et al., 2005		Leonardi et al., 2005
	Al-Suwaiegh et al., 2002		Birkelo et al., 2004
	Peter et al., 2000		McKendrick et al., 2003
	Lodge et al., 1997a		Al-Suwaiegh et al., 2002
	Lodge et al., 1997b		Liu et al., 2000
	Fron et al., 1996		Huang et al., 1999
	Klopfenstein, 1996		Schingoethe et al., 1999
	Ham et al., 1994		Batajoo and Shaver, 1998
	Larson et al., 1993		Nichols et al., 1998
	Donaldson et al., 1991		Clark and Armentano, 1997
	McCann et al., 1991		DePeters et al., 1997
			O'Mara et al., 1997
			Zhu et al., 1997

Species Beef	Citation	Species Dairy	Citation
			Arosemena et al., 1995
			Murphy et al., 1995
			Powers et al., 1995
			Ham et al., 1994
			Clark and Armentano, 1993
Swine		**Poultry**	
	Stein and Shurson, 2009		Waldroup et al., 2007
	Pedersen et al., 2007		Wang et al., 2007a
	Widmer et al., 2007		Wang et al., 2007b
	Fastinger et al., 2007		Wang et al., 2007c
	Stein et al., 2006		Batal and Dale, 2006
	Whitney et al., 2006a		Fastinger et al., 2006
	Whitney et al., 2006b		Martinez-Amezcua et al., 2006
	Whitney et al., 2006c		Noll, 2006
	Whitney et al., 2006d		Lumpkins and Batal, 2005
	Nyachoti et al., 2005		Lumpkins et al., 2005
	Whitney and Shurson, 2004		Roberson et al., 2005
	Gralapp et al., 2002		Biggs et al., 2004
	Spiehs et al., 2002		Lumpkins et al., 2004
	Nicolai et al., 1999		Martinez Amezcua et al., 2004
	Cromwell et al., 1993		Batal and Dale, 2003
			Roberson, 2003
			Cromwell et al., 1993

The utilization of DDGS in livestock diets has steadily risen throughout the years. Projections for the maximum amount of DDGS that could be utilized in the U.S. beef, dairy, swine, and poultry markets range from 40 to 60 million tons per year. These estimates rely on the specific inclusion rates for each species and are based on studies conducted by Staff (2005), Cooper (2006), and the U.S. Grains Council (2007). On a global scale, there is an increasing demand for animal diets that are rich in protein. In 2008, a total of 23 million tons of DDGS were generated, according to the RFA (2009b). Out of this, 4.5 million tons were exported to international markets, as reported by the FAS (2009). This export volume represented approximately 20% of the total DDGS production in the United States that year, as shown in Figure 6. The U.S. Grains Council (2007) predicts that the possibility of worldwide exports will continue to grow in the next years.

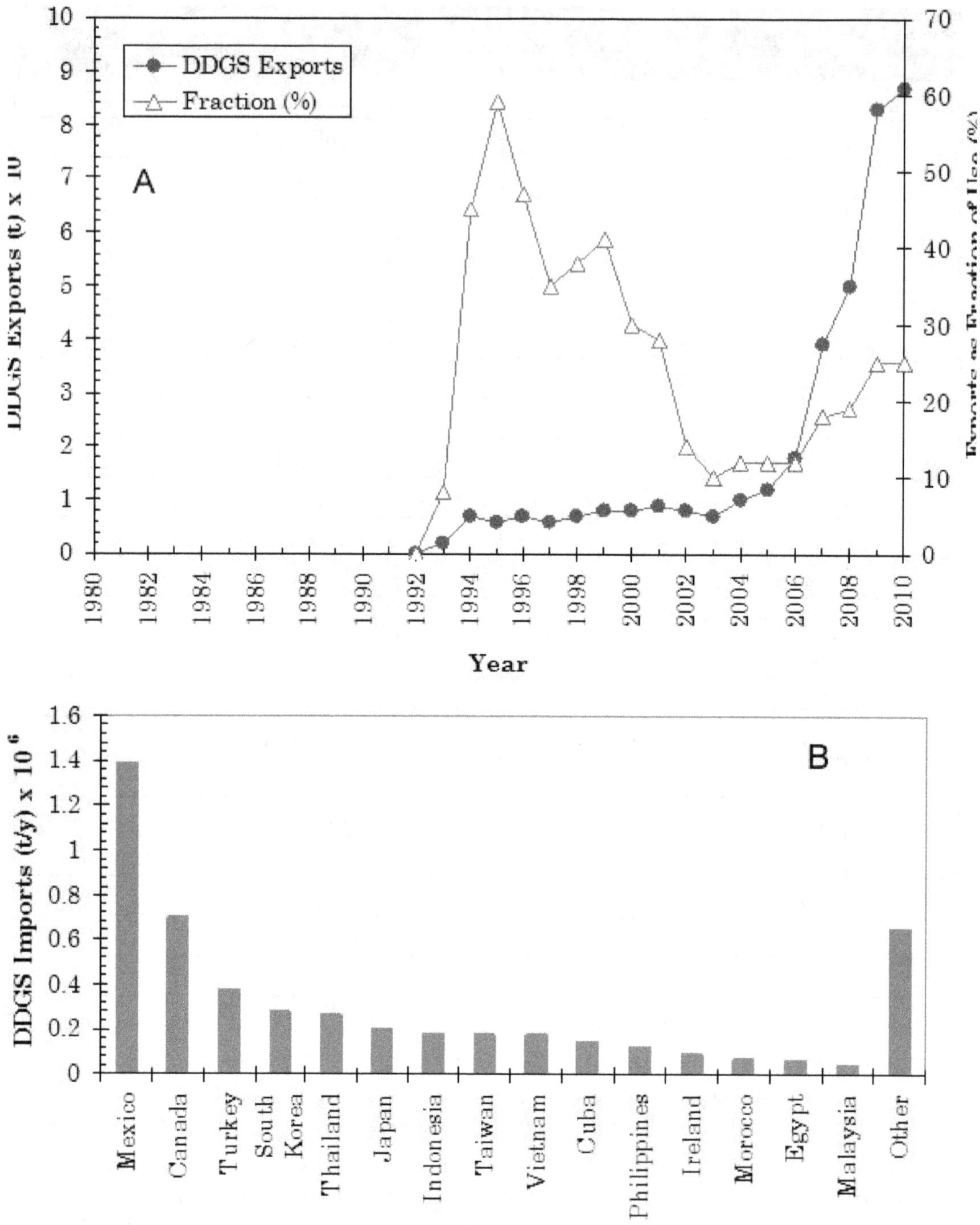

Figure 6. A. U.S. DDGS exports in 2008. B. Countries who imported DDGS in 2008 (adapted fromHoffman and Baker, 2010).

Coproducts play a crucial role in both the livestock business, where they are used as feed additives, and in ensuring the long-term viability of the fuel ethanol industry. The sale of distillers grains, both dry and wet, significantly contributes to the economic sustainability of ethanol plants. Typically, these sales account for 10 to 20% of a plant's total revenue stream, although in certain circumstances it can reach as high as 40%. The extent of this contribution depends on the market conditions for corn, ethanol, and distillers grains. These process wastes are called "coproducts" rather than "byproducts" or "waste products" since they are considered separate products in addition to the fuel.

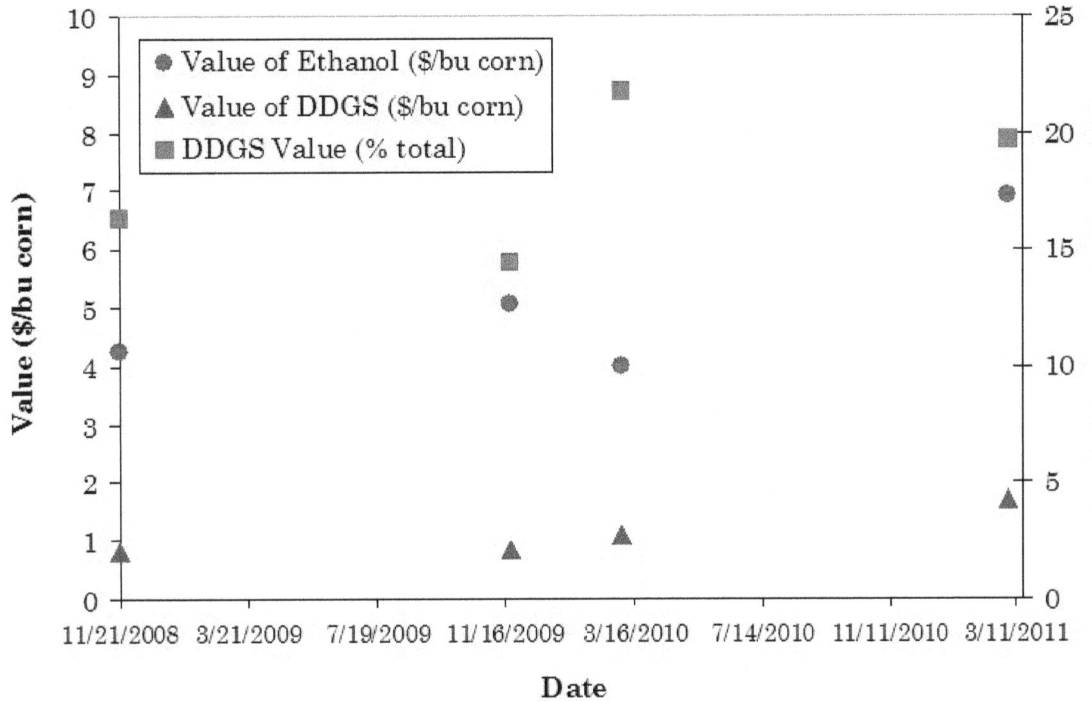

Figure 7. Some relative comparisons of the value of DDGS and fuel ethanol to ethanol plant profits (adapted fromDTN, 2011).

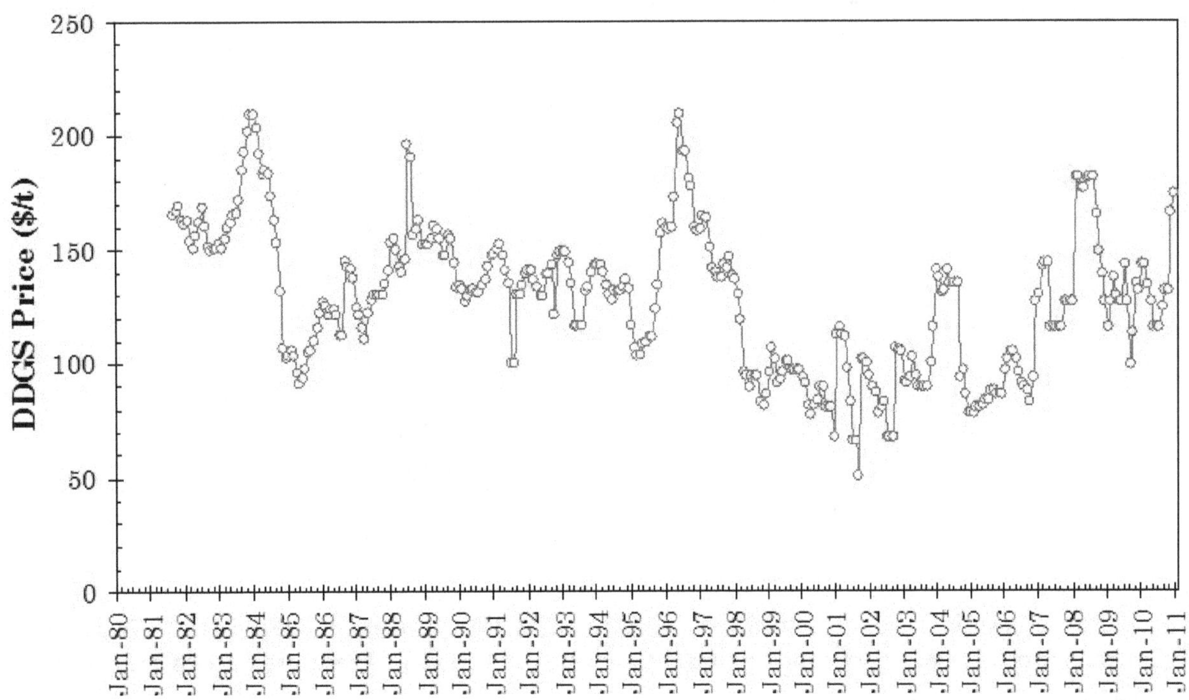

Figure 8. DDGS sales price over time (monthly averages) (adapted fromERS, 2011).

So the sales price of DDGS is important to ethanol manufacturers and livestock producers alike. Over the last three decades, the price for DDGS has ranged from approximately $50.71/t up to $209.44/t (Figure 8). DDGS and corn prices have historically paralleled each other very closely (Figure 9). This relationship has been quite strong over the last several

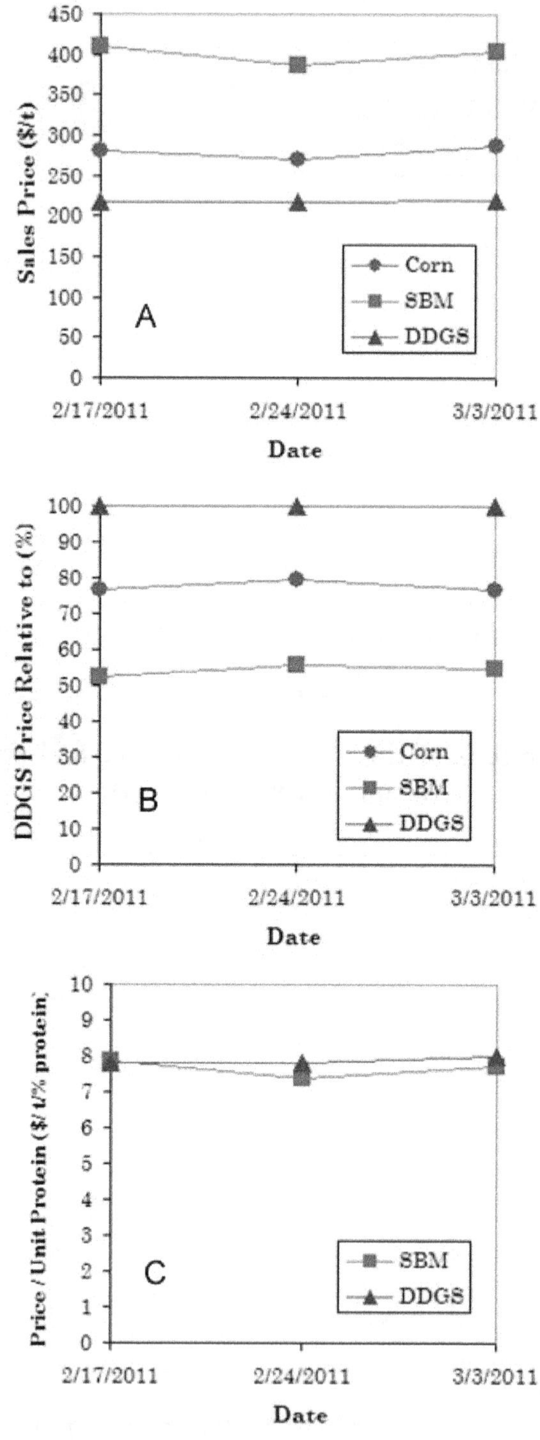

Figure 9. A. Some comparisons of DDGS, soybean meal (SBM), and corn sales prices. B. Relative price comparisons. C. Cost comparisons on a per unit protein basis (adapted fromDTN, 2011).

years. This is unsurprising, given that DDGS is commonly utilized as a substitute for maize in cattle diet formulations. DDGS is being utilized more and more as a substitute for soybean meal, mainly because it is a good source of protein. However, DDGS has consistently been sold at a lower price compared to both maize and soybean meal. This has been consistently seen both in terms of volume per unit and in terms of protein per unit (Figure 9).

5. EVOLUTION OF COPRODUCTS

The ethanol sector is constantly changing and adapting to address the issues related to the processing and usage of both fuel and coproducts (Rosentrater, 2007). A contemporary dry grind ethanol plant differs significantly from the inefficient and resource-intensive Gasohol facilities of the 1970s. Recent advancements and technological breakthroughs encompass various improvements such as enhanced enzymes, increased conversion of starch, superior fermentation processes, cold cooking technologies, upgraded drying systems, reduced energy consumption across the facility, improved water efficiency and recycling, and decreased emissions. Energy and mass balances are improving progressively. The design and operation of contemporary ethanol plants are responsible for many of these advancements. A significant portion can also be attributed to computerized instrumentation and control systems. Extensive research, both official and informal, has focused on modifying current procedures to enhance and maximize the quality of the coproducts that are generated. Ethanol firms have acknowledged the necessity of manufacturing more uniform and superior grade DDGS to better meet the requirements of livestock producers. The sale of Distillers Dried Grains with Solubles (DDGS) and other byproducts has been a crucial factor in the industry's current performance and will remain vital for the industry's long-term viability. While beef and dairy cattle now consume most of the DDGS, its usage in monogastric diets, particularly in pigs and poultry, is steadily growing.

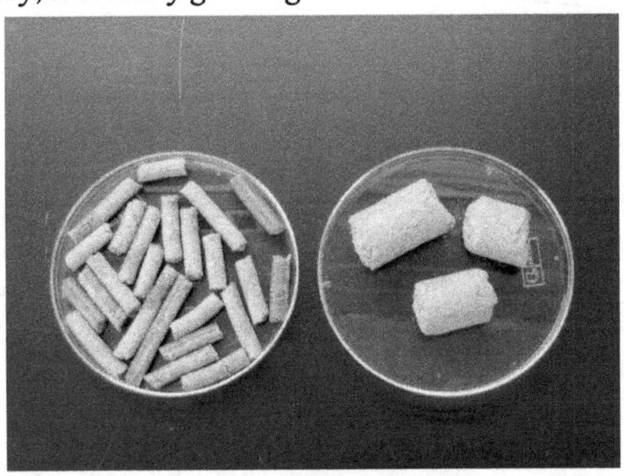

Figure 10. Pelleting is a unit operation that can improve the utility of DDGS, because it improves storage and handling characteristics, and allows more effective use in dairy cattle feeding and range land settings for beef cattle.

The utilization of non-traditional species, including fish, horses, and pets, has also been on the rise. Furthermore, there has been significant interest in creating enhanced methods for delivering and feeding DDGS to animals, specifically through pelleting or densification (as shown in Figure 10). This processing operation has the potential to greatly improve the storage and handling properties of the DDGS. Additionally, it would dramatically reduce the cost of rail transportation and logistics by increasing the bulk density and improving flowability. (See Figure 11) Pelleting could also enhance the utilization of DDGS both domestically (e.g., by improving its suitability for feeding rangeland beef cattle and dairy cattle) and worldwide (e.g., by increasing its bulk density, which would lead to significant cost savings in terms of freight for bulk boats and containers). Furthermore, there are numerous ongoing advancements in the development of coproducts. These processes will ultimately lead to the creation of additional value streams from the maize kernel through upstream fractionation, as well as from the resultant distillers grains through downstream fractionation (Figure 12). Efficient fractionation can lead to the segregation of components with high, medium, and low values. A number of plants have started incorporating the ability to concentrate nutrient streams, such as oil, protein, and fiber, into distinct fractions. These fractions can then be utilized for specific markets and purposes. The use of these novel procedures is leading to the emergence of novel varieties of distillers grains (Figure 13).

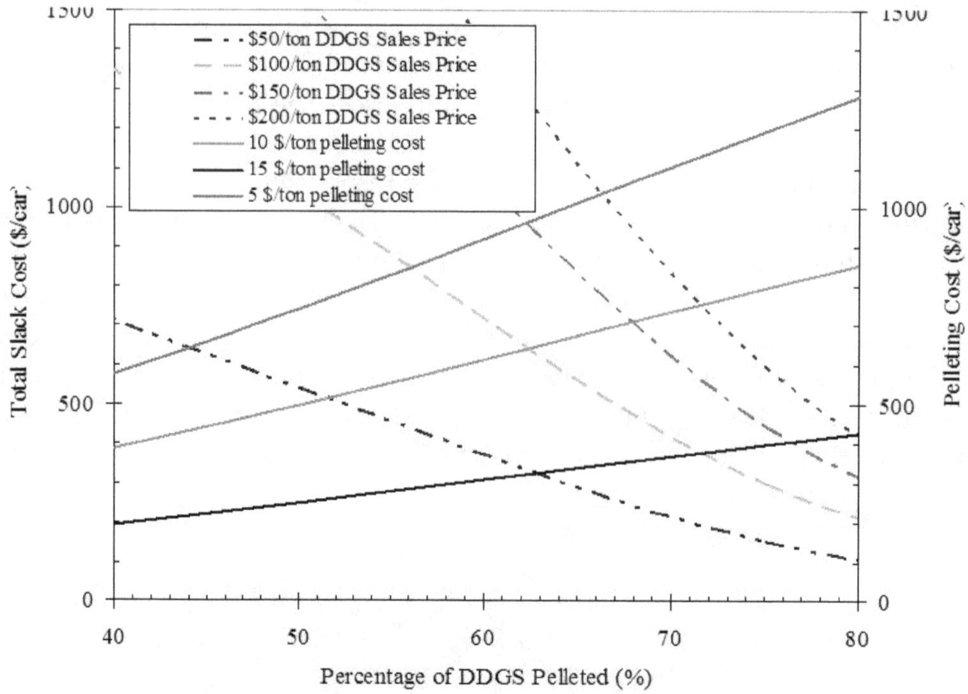

Figure 11. By pelleting, empty space in rail cars is minimized during shipping. Techno-economic analysis of the resulting slack (i.e., wasted space) costs and costs of pelleting for each rail car due to differing DDGS sales prices and pelleting costs indicates the proportion of DDGS which needs to be pelleted in order to achieve breakeven for this process (adapted fromRosentrater and Kongar, 2009).

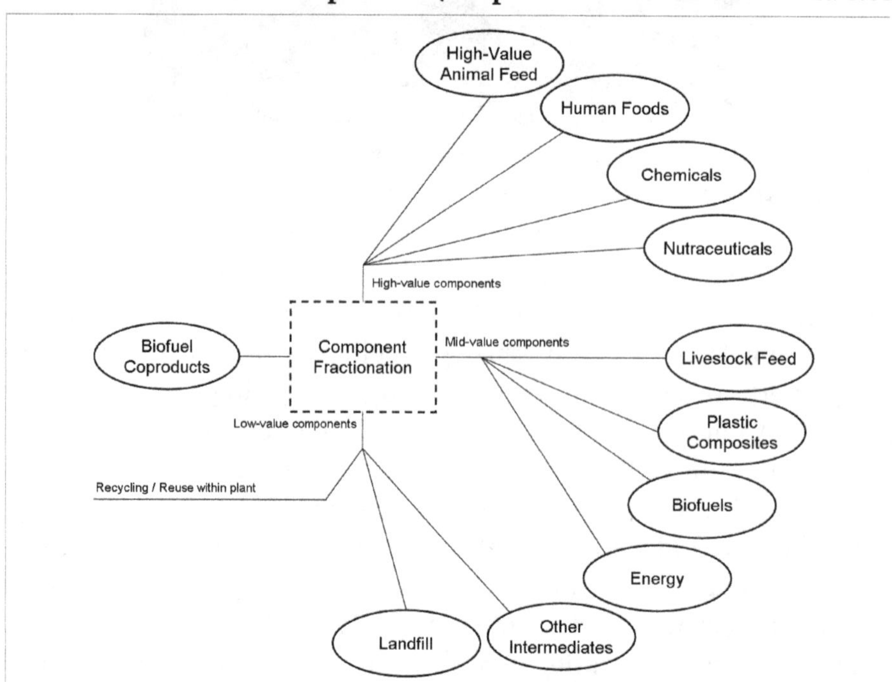

Figure 12. Fractionation of DDGS into high-, mid-, and low-value components offers the opportunity for new value streams.

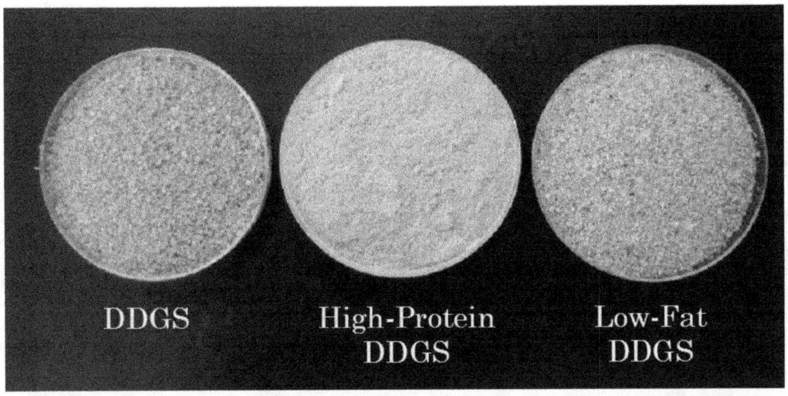

Figure 13. Examples of traditional, unmodified DDGS and some fractionated products (e.g., high-protein and low-fat DDGS) which are becoming commercially available in the marketplace.

For instance, when the lipids are extracted from the DDGS (as shown in Figure 14), they can be easily transformed into biodiesel. However, due to their significant structural degradation, they are unsuitable for producing food-grade maize oil. Another illustration is the utilization of concentrated proteins in the production of animal feeds with elevated value, such as those used in aquaculture or pet foods, as well as other feed applications that necessitate high protein content. Furthermore, DDGS proteins have the potential to be utilized in human food products (Figure 15). In addition, other constituents such as amino acids, organic acids, and nutraceutical substances (such as phytosterols and tocopherols) can be extracted and utilized in valuable applications. Intermediate-value elements, like fiber, can serve as biofillers in plastic composites (Figure 16). They can also be used as raw materials for generating bioenergy, such as heat and electricity, through thermochemical conversion at ethanol plants (Figure 17). Additionally, after undergoing pretreatment to dismantle the lignocellulosic structures, they can be utilized as substrates for producing ethanol or other biofuels. Regarding the prospective applications of the low-value components, it is desirable that methods are devised to modify their structures and make them valuable, thereby avoiding the need to dispose of them in landfills. Fertilizers are essential for the sustainable maintenance of corn grain supply to the ethanol plant, making land application a suitable option for disposing of the low-value components. As the process adjustments are created, verified, and put into use, there will be enhancements in the byproducts produced and the creation of distinctive materials. Naturally, a thorough examination will be necessary to ascertain the most effective methods of utilizing these novel items and to accurately measure their market value.

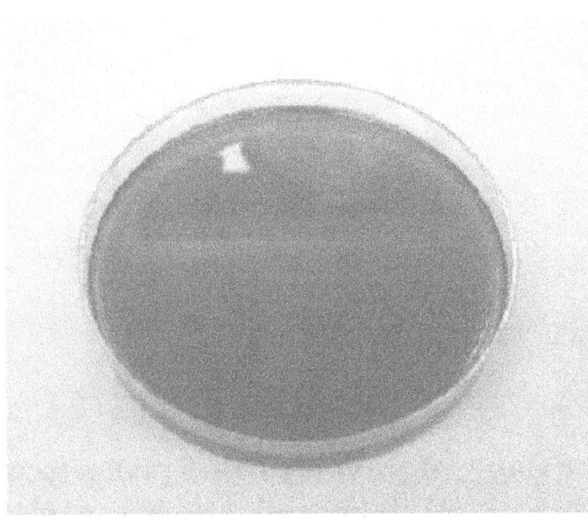

Figure 14. Corn oil which has been extracted from DDGS can be used to manufacture biodiesel.

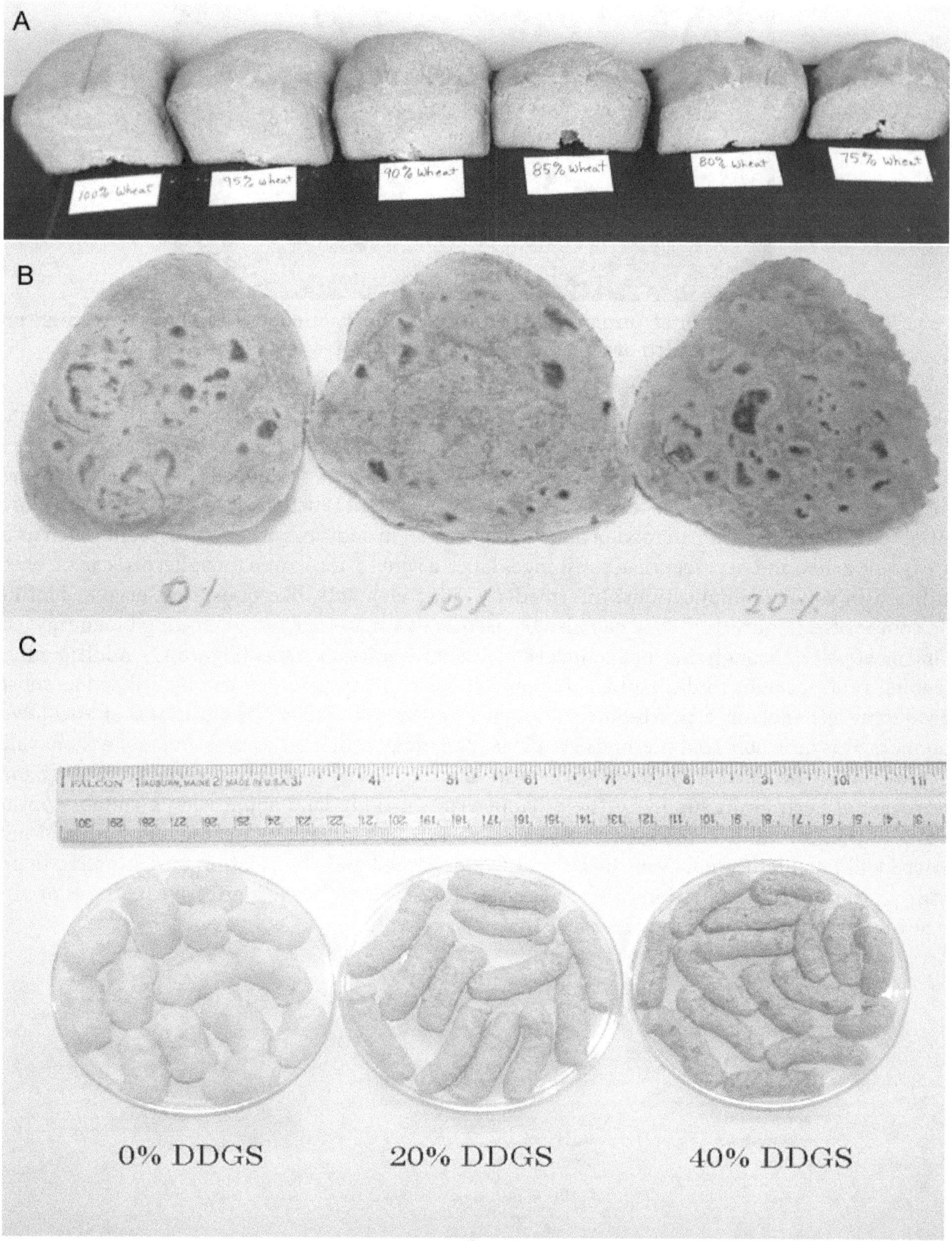

Figure 15. As a partial substitute for flour, high-value DDGS protein can be used to improve the nutrition of various baked foods such as (A) bread, (B) flat bread, and (C) snack foods, by increasing protein levels and decreasing starch content.

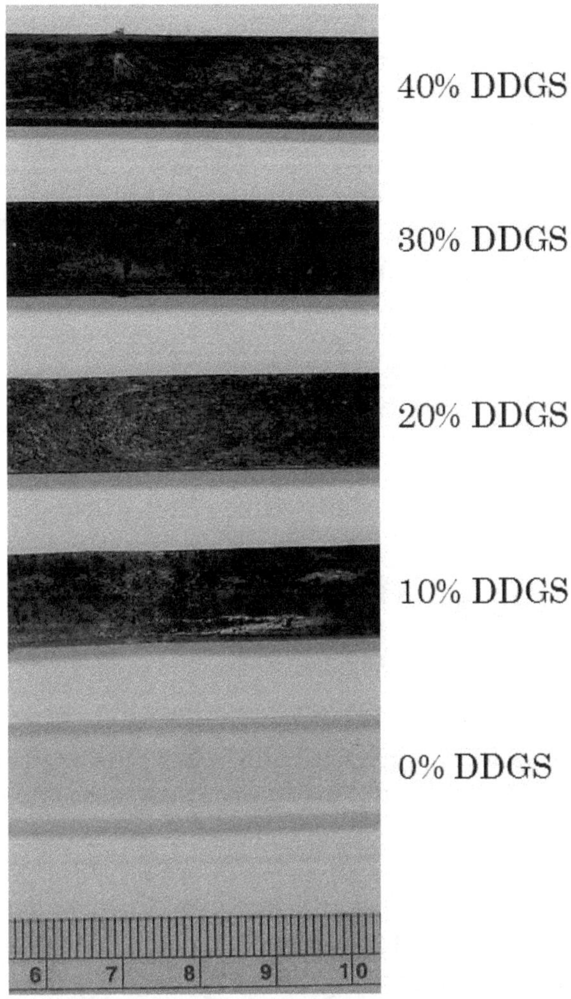

Figure 16. Mid-value or low-value fractions from DDGS (such as fiber) have been shown to be an effective filler in plastics, replacing petroleum additives and increasing biodegradability. Scale bar indicates mm.

Figure 17. Mid-value or low-value fractions from DDGS (such as fiber) can be thermochemically converted into biochar, which can subsequently be used to produce energy, fertilizer, or as a precursor to other bio-based materials.

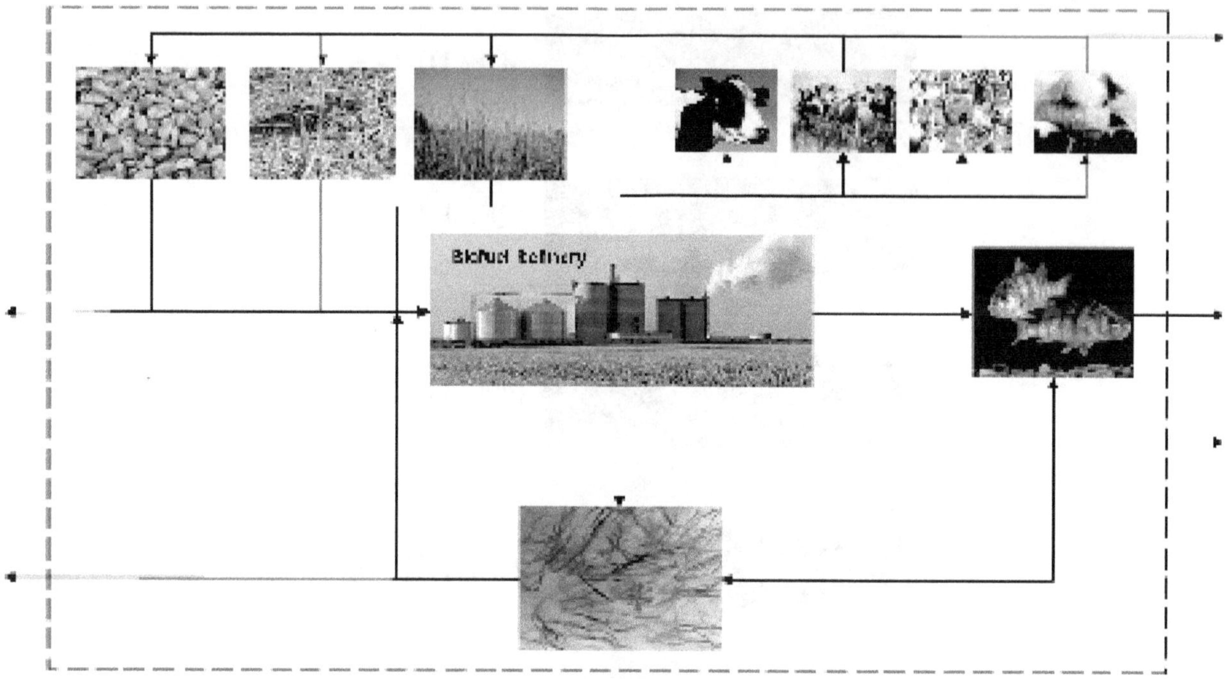

Figure 18. Coproducts such as DDGS will continue to play a key role as the biofuel industry evolves and becomes more fully integrated. This figure illustrates one such concept.

6. CONCISE SUMMARY

The fuel ethanol sector has experienced significant growth in recent years, primarily driven by government regulations and a rising demand for alternative fuels. This has become particularly evident as the cost of gasoline has risen and changed significantly, leading consumers to view fuel prices as troublesome. Corn-based ethanol alone is insufficient to meet our transportation fuel requirements. However, it is undeniably a crucial element in achieving the ultimate objective of attaining energy self-sufficiency. Corn ethanol will remain a prominent player in the developing bioeconomy, as it has demonstrated the efficacy of large-scale biotechnology and bioprocessing in fuel generation. Furthermore, it has established the foundation for cutting-edge biorefineries and manufacturing methods that will generate the subsequent iterations of sophisticated biofuels. As the biofuel industry progresses, coproduct materials, derived from various biomass substrates, will continue to play a crucial role in ensuring resource and economic sustainability. An effective approach to achieve sustainability involves the use of integrated systems (Figure 18), where material and energy streams are continuously cycled and recycled. This means that the outputs from one part of the system become inputs for another part. These integrated systems can be used in various components such as biorefineries, animal feeding operations, energy production systems (including heat, electricity, steam, etc.), feedstock production systems, and other systems. By incorporating these different elements, a diversified portfolio will provide not only fuel, but also fertilizer, feed, food, industrial goods, energy, and most notably, it will be self-sufficient.

REFERENCES

Agrawal, R., N. R. Singh, F. H. Ribeiro, and W. N. Delgass. (2007). Sustainable fuel for the transportation sector. Proceedings of the National Academy of Sciences 104(12): 4828-4833.

Alexander, C. and C. Hurt. (2007). Biofuels and their impact on food prices. Bioenergy ID- 346-W. Department of Agricultural Economics, Purdue University: West Lafayette, IN.

Al-Suwaiegh, S., K. C. Fanning, R. J. Grant, C. T. Milton, and T. J. Klopfenstein. (2002). Utilization of distillers grains from the fermentation of sorghum or corn in diets for finishing beef and lactating dairy cattle. J. Anim. Sci. 80: 1105-1111.

Anderson, J. L., D. J. Schingoethe, K. F. Kalscheur, and A. R. Hippen. (2006). Evaluation of dried and wet distillers grains included at two concentrations in the diets of lactating dairy cows. J. Dairy Sci. 89: 3133–3142.

Arosemena, A., E. J. DePeters, and J. G. Fadel. (1995). Extent of variability in nutrient composition within selected by-product feedstuffs. Animal Feed Sci. and Technology 54: 103-120.

Batajoo, K. K. and R. D. Shaver. (1998). In situ dry matter, crude protein, and starch degradabilities of selected grains and by-product feeds. Animal Feed Science Technology 71: 165-176.

Batal, A. and N. M. Dale. (2003). Mineral composition of distillers dried grains with solubles. J. Appl. Poult. Res. 12: 400-403.

Batal, A. B. and N. M. Dale. (2006). True metabolizable energy and amino acid digestibility of distillers dried grains with solubles. J. Appl. Poult. Res. 15: 89-93.

Bhadra, R., K. A. Rosentrater, and K. Muthukumarappan. (2009). Cross-sectional staining and the surface properties of DDGS and their influence on flowability. Cereal Chemistry 86(4): 410-420.

Biggs, P. E., M. E. Persia, K. W. Koelkebeck, and C. M. Parsons. (2004). Further evaluation of nonfeed removal methods for molting programs. Poultry Sci. 83(5) 745-752.

Birkelo, C. P., M. J. Brouk, and D. J. Schingoethe. (2004). The energy content of wet corn distillers grins for lactating dairy cows. J. Dairy Sci. 87: 1815-1819.

Bothast, R. and M. Schlicher. (2005). Biotechnological processes for conversion of corn into ethanol. Applied Microbiology and Biotechnology 67(1): 19-25.

Cassman, K. G. (2007). Climate change, biofuels, and global food security. Environmental Research Letters 2(011002): DOI # 10.1088/1748-9326/2/1/011002.

Cassman, K. G. and A. J. Liska. (2007). Food and fuel for all: realistic or foolish? Biofuels, Bioproducts and Biorefining 1(1): 18-23.

Cassman, K. G., V. Eidman, and E. Simpson. (2006). Convergence of agriculture and energy: implications for research policy, QTA2006-3. Council for Agricultural Science and Technology: Ames, IA.

Clark, P. W. and L. E. Armentano. (1993). Effectiveness of neutral detergent fiber in whole cottonseed and dried distillers grains compared with alfalfa haylage. J. Dairy Sci. 76: 2644-2650.

Clark, P. W. and L. E. Armentano. (1997). Replacement of alfalfa neutral detergent fiber with a combination of nonforage fiber sources. J. Dairy Sci. 80: 675-680.

Cooper, G. (2006). A brief, encouraging look at 'theoretical' distillers grains markets. Distillers Grains Quarterly 1(1): 14-17.

Cromwell, G. L., K. L. Herkelman, and T. S. Stahly. (1993). Physical, chemical, and nutritional characteristics of distillers dried grains with solubles for chicks and pigs. J. Anim. Sci. 71: 679-686.

Dale, B. E. (2007). Thinking clearly about biofuels: ending the irrelevant 'net energy' debate and developing better performance metrics for alternative fuels. Biofuels, Bioproducts, and Biorefining 1: 14-17.

De La Torre Ugarte, D. G., M. E. Walsh, H. Shapouri, and S. P. Slinsky. (2000). The economic impacts of bioenergy crop production on U.S. agriculture, Agricultural Economic Report 816. USDA Office of the Chief Economist, U.S. Department of Agriculture: Washington, D.C.

DePeters, E .J., J. G. Fadel, and A. Arosemena. (1997). Digestion kinetics of neutral detergent fiber and chemical composition within some selected by-product feedstuffs. Animal Feed Science Technology 67: 127-140.

Dewulf, J., H. Van Langenhove, and B. Van De Velde. (2005). Energy-based efficiency and renewability assessment of biofuel production. Environmental Science and Technology 39(10): 3878-3882.

Dien, B. S., R. J. Bothast, N. N. Nichols, and M. A. Cotta. (2003). The U.S. corn ethanol industry: an overview of current technology and future prospects. In: The Third International Starch Technology Conference – Coproducts Program Proceedings, eds. M. Tumbleson, V. Singh, and K. Rausch, 2-4 June, 2003, University of Illinois, Urbana, IL, pp. 10-21.

Donaldson, R. S., M. A. McCann, H. E. Amos, and C. S. Hoveland. (1991). Protein and fiber digestion by steers grazing winter annuals and supplemented with ruminal escape protein. J. Anim. Sci. 69: 3067-3071.

DTN. (2011). DTN Weekly Distillers Grains Update. Available online: www.dtnprogressivefarmer.com.

ERS. (2011). Feed Grains Database: Yearbook Tables. Economic Research Service, U.S.

Department of Agriculture: Washington, D.C. Available online: www.ers.usda.gov/data/feedgrains/. FAS. (2009). Foreign Agricultural Service, U. S. Department of Agriculture: Washington,

D.C. Available online: www.fas.usda.gov/. Fastinger, N. D. and D. C. Mahan. (2006).

Determination of the ileal amino acid and energy digestibilities of corn distillers dried grains with solubles using grower-finisher pigs. J. Anim. 84: 1722–1728.

Fastinger, N. D., J. D. Latshaw, and D. C. Mahan. (2006). Amino acid availability and true metabolizable energy content of corn distillers dried grains with solubles in adult cecectomized roosters. Poultry Sci. 85: 1212-1216.

Fron, M., H. Madeira, C. Richards, and M. Morrison. (1996). The impact of feeding condensed distillers byproducts on rumen microbiology and metabolism. Animal Feed Sci. Technology 61: 235-245.

Gralapp, A. K., W. J. Powers, M. A. Faust, and D. S. Bundy. (2002). Effects of dietary ingredients on manure characteristics and odorous emissions from swine. J. Anim. Sci. 80: 1512-1519.

Ham, G. A., R. A. Stock, T. J. Klopfenstein, E. M. Larson, D. H. Shain, and R. P. Huffman. (1994). Wet corn distillers byproducts compared with dried corn distillers grains with solubles as a source of protein and energy for ruminants. J. Anim. Sci. 72: 3246-3257.

Hoffman, L. and A. Baker. (2010). Market issues and prospects for U.S. distillers' grains: supply, use, and price relationships. Report FDS-10k-01. United States Department of Agriculture, Economic Research Service: Washington, D.C. Available online: www.ers.usda.gov.

Huang, H. J., P. Wen-Shyg Chiou, C. R. Chen, J. K. Chiang, and B. Yu. (1999). Effects of dried rice distillers' and grain supplementation on the performance of lactating cows. Animal Feed Science and Technology 77: 303-315.

Ingledew, W. M., D. R. Kelsall, G. D. Austin, and C. Kluhspies. (2009). The Alcohol Textbook, 5th Edition. W. M. Ingledew, D. R. Kelsall, G. D. Austin, and C. Kluhspies, ed. Nottingham University Press: Nottingham, UK.

Jaques, K. A., T. P. Lyons, and D. R. Kelsall. (2003). The Alcohol Textbook, 3rd ed. Nottingham University Press: Nottingham, UK.

Kleinschmit, D. H., D. J. Schingoethe, A. R. Hippen, and K. F. Kalscheur. (2007). Dried distillers grains plus solubles with corn silage or alfalfa hay as the primary forage source in dairy cow diets. J. Dairy Sci. 90: 5587-5599.

Kleinschmit, D. H., D. J. Schingoethe, K. F. Kalscheur, and A. R. Hippen. (2006). Evaluation of various sources of corn dried distillers grains plus solubles for lactating dairy cattle. J. Dairy Sci. 89: 4784-4794.

Klopfenstein, T. (1996). Distillers grains as an energy source and effect of drying on protein availability. Animal Feed Science Technology 60: 201-207.

Larson, E. M., R. A. Stock, T. J. Klopfenstein, M. H. Sindt, and R. P. Huffman. (1993). Feeding value of wet distillers byproducts for finishing ruminants. J. Anim. Sci. 71: 2228-2236.

Leonardi, C., S. Bertics, and L. E. Armentano. (2005). Effect of increasing oil from distillers grains or corn oil on lactation performance. J. Dairy Sci. 88: 2820-2827.

Liu, C., D. J. Schingoethe, and G. A. Stegeman. (2000). Corn distillers grains versus a blend of protein supplements with or without ruminally protected amino acids for lactating cows. J. Dairy Sci. 83: 2075–2084.

Lodge, S. L., R. A. Stock, T. J. Klopfenstein, D. H. Shain, and D. W. Herold. (1997a). Evaluation of corn and sorghum distillers byproducts. J. Anim. Sci. 75: 37–43.

Lodge, S. L., R. A. Stock, T. J. Klopfenstein, D. H. Shain, and D. W. Herold. (1997b). Evaluation of wet distillers composite for finishing ruminants. J. Anim. Sci. 75: 44- 50.

Loy, T. W., J. C. MacDonald, T. J. Klopfenstein, and G. E. Erickson. (2007). Effect of distillers grains or corn supplementation frequency on forage intake and digestibility. J. Anim. Sci. 85: 2625–2630.

Lumpkins, B. S. and A. B. Batal. (2005). The bioavailability of lysine and phosphorus in distillers dried grains with solubles. Poultry Science 84: 581-586.

Lumpkins, B. S., A. B. Batal, and N. M. Dale. (2004). Evaluation of distillers dried grains with solubles as a feed ingredient for broilers. Poultry Science 83: 1891-1896.

Lumpkins, B., A. Batal, and N. Dale. (2005). Use of distillers dried grains plus solubles in laying hen diets. J. Appl. Poult. Res. 14: 25-31.

Lynd, L. R. and M. Q. Wang. (2004). A product-nonspecific framework for evaluating the potential of biomass-based products to displace fossil fuels. Journal of Industrial Ecology 7(3-4): 17-32.

MacDonald, J. C., T. J. Klopfenstein, G. E. Erickson, and W. A. Griffin. (2007). Effects of dried distillers grains and equivalent undegradable intake protein or ether extract on performance and forage intake of heifers grazing smooth bromegrass pastures. J. Anim. Sci. 85: 2614–2624.

Martin, J. L., A. S. Cupp, R. J. Rasby, Z. C. Hall, and R. N. Funston. (2007). Utilization of dried distillers grains for developing beef heifers. J. Anim. Sci. 85: 2298–2303.

Martinez Amezcua, C., C. M. Parsons, and S. L. Noll. (2004). Content and relative bioavailability of phosphorus in distillers dried grains with solubles in chicks. Poultry Sci. 83: 971-976.

Martinez-Amezcua, C., C. M. Parsons, and D. H. Baker. (2006). Effect of microbial phytase and citric acid on phosphorus bioavailability, apparent metabolizable energy, and amino acid digestibility in distillers dried grains with solubles in chicks. Poultry Sci. 85: 470–475.

McCann, M. A., R. S. Donaldson, H. E. Amos, and C. S. Hoveland. (1991). Ruminal escape protein supplementation and zeranol implantation effects on performance of steers grazing winter annuals. J. Anim. Sci. 69: 3112-3117.

McKendrick, E. J., D. J. Roberts, and N. W. Offer. (2003). The value of malt distillers' grains ensiled with molassed sugar beet pellets as a feed for dairy cows. Grass and Forage Sci. 58: 287-294.

Murphy, J. J., J. F. Connolly, and G. P. McNeill. (1995). Effects on milk fat composition and cow performance of feeding concentrates containing full fat rapeseed and maize distillers grains on grass-silage based diets. Production Science 44: 1-11.

Nichols, J. R., D. J. Schingoethe, H. A. Maiga, M. J. Brouk, and M. S. Piepenbrink. (1998). Evaluation of corn distillers grains and ruminally protected lysine and methionine for lactating dairy cows. J. Dairy Sci. 81: 482-491.

Nicolai, R. E., M. J. Spiehs, G. C. Shurson, and M. H. Whitney. (1999). Deep pit simulator modeling protocol for individual metabolism crates during diet studies. ASAE Paper No. 994135. ASAE, St. Joseph, MI.

Noll, S. (2006). Maximizing alternative protein ingredient use in market turkey diets. Midwest Poultry Research Program Paper. Available online: www.ddgs.umn.edu.

Nyachoti, C. M., J. D. House, B. A. Slominski, and I. R. Seddon. (2005). Energy and nutrient digestibilities in wheat dried distillers' grains with solubles fed to growing pigs. J. Sci. Food Agric. 85: 2581-2586.

O'Mara, F. P., J. J. Murphy, and M. Rath. (1997). The amino acid composition of protein feedstuffs before and after ruminal incubation and after subsequent passage through the intestines of dairy cows. J. Anim. Sci. 75: 1941-1949.

Pedersen, C., M. G. Boersma, and H. H. Stein. (2007). Digestibility of energy and phosphorus in ten samples of distillers dried grains with solubles fed to growing pigs. J. Anim. Sci. 85: 1168–1176.

Peter, C. M., D. B. Faulkner, N. R. Merchen, D. F. Parrett, T. G. Nash, and J. M. Dahlquist. (2000). The effects of corn milling coproducts on growth performance and diet digestibility by beef cattle. J. Anim. Sci. 78: 1–6.

Powers, W. J., H. H. Van Horn, B. Harris, Jr., and C. J. Wilcox. (1995). Effects of variable sources of distillers dried grains plus solubles on milk yield and composition. J. Dairy Sci. 78: 388-396.

Rausch, K.D. and R.L. Belyea. (2006). The future of coproducts from corn processing. Applied Biochemistry and Biotechnology 128: 47-86.

RFA. (2009a). Growing Innovation. 2009 Ethanol Industry Outlook. Renewable Fuels Association. Washington, D.C. Available at: www.ethanolrfa.org.

RFA. (2009b). Industry resources: co-products. Renewable Fuels Association: Washington, D.C. Available online: www.ethanolrfa.org.

RFA. (2011). Biorefinery locations. Renewable Fuels Association: Washington, D.C. Available online: www.ethanolrfa.org.

Roberson, K. D. (2003). Use of dried distillers' grains with solubles in growing-finishing diets of turkey hens. International Journal of Poultry Sci. 2 (6): 389-393.

Roberson, K. D., J. L. Kalbfleisch, W. Pan, and R. A. Charbeneau. (2005). Effect of corn distiller's dried grains with solubles at various levels on performance of laying hens and egg yolk color. International Journal of Poultry Sci. 4 (2): 44-51.

Roeber, D. L., R. K. Gill, and A. DiCostanzo. (2005). Meat quality responses to feeding distiller's grains to finishing Holstein steers. J. Anim. Sci. 83: 2455-2460.

Rosentrater, K. A. (2007). Ethanol processing coproducts – a review of some current constraints and potential directions. International Sugar Journal 109(1307): 1-12.

Rosentrater, K. A. and E. Kongar. (2009). Modeling the effects of pelleting on the logistics of distillers grains shipping. Bioresource Technology 100: 6550-6558.

Rosentrater, K. A. and K. Muthukumarappan. (2006). Corn ethanol coproducts: generation, properties, and future prospects. International Sugar Journal 108(1295): 648-657.

Schingoethe, D. J., M. J. Brouk, and C. P. Birkelo. (1999). Milk production and composition from cows fed wet corn distillers grains. J. Dairy Sci. 82: 574-580.

Singh, V. and D. B. Johnston. (2009). Fractionation technologies for dry-grind corn processing. Pages 193-207 in: The Alcohol Textbook, 5th Ed. M. W. Ingledew, D. R. Kelsall, G. D. Austin, and C. Kluhspies, ed. Nottingham University Press: Nottingham, UK.

Spiehs, M. J., M. H. Whitney, and G. C. Shurson. (2002). Nutrient database for distiller's dried grains with solubles produced from new ethanol plants in Minnesota and South Dakota. J. Anim. Sci. 80: 2639-2645.

Srinivasan, R., R. A. Moreau, K. D. Rausch, R. L. Belyea, M. D. Tumbleson, and V. Singh. (2005). Separation of fiber from distillers dried grains with solubles (DDGS) using sieving and elutriation. Cereal Chemistry 82: 528-533.

Staff, C. H. (2005). Question and answer. Biofuels Journal 3(4): 26-27.

Stein, H. H. and G. C. Shurson. (2009). The use and application of distillers dried grains with solubles in swine diets. J. Anim. Sci. 87: 1292-1303. doi:10.2527/jas.2008-1290.

Stein, H. H., M. L. Gibson, C. Pedersen, and M. G. Boersma. (2006). Amino acid and energy digestibility in ten samples of distillers dried grain with solubles fed to growing pigs. J. Anim. Sci. 84: 853-860.

Tibelius, C. (1996). Coproducts and Near Coproducts of Fuel Ethanol Fermentation from Grain. Agriculture and Agri-Food Canada – Canadian Green Plan Ethanol Program: Starchy Waste Streams Evaluation Project. Available online: http://res2.agr.ca/publications/cfar/index_e.htm.

U.S. EIA. (2011). Annual Energy Review. Energy Information Administration, U.S. Department of Energy: Washington, D.C. Available online: www.eia.doe.gov/emeu/aer/.

U.S. Grains Council. (2007). An Independent Review of US Grains Council Efforts to Promote DDGS Exports. U.S. Grains Council: Washington, D.C. Available online: www.grains.org/ddgs-information.

UMN. (2011). The value and use of distillers grains by-products in livestock and poultry feeds. University of Minnesota: Minneapolis, MN. Available online: www.ddgs.umn.edu/.

Urbanchuk, J. M. (2009). Contribution of the Ethanol Industry to the Economy of the United States. LECG: Wayne, PA.

Waldroup, P. W., Z. Wang, C. Coto, S. Cerrate, and F. Yan. (2007). Development of a standardized nutrient matrix for corn distillers dried grains with solubles. Internatl. Journal of Poultry Sci. 6 (7): 478-483.

Wang, Z., S. Cerrate, C. Coto, F. Yan, and P. W. Waldroup. (2007a). Effect of rapid and multiple changes in level of distillers dried grain with solubles (DDGS) in broiler diets on performance and carcass characteristics. International Journal of Poultry Sci. 6 (10): 725-731.

Wang, Z., S. Cerrate, C. Coto, F. Yan, and P. W. Waldroup. (2007b). Use of constant or increasing levels of distillers dried grains with solubles (DDGS) in broiler diets. International Journal of Poultry Sci. 6 (7): 501-507.

Wang, Z., S. Cerrate, C. Coto, F. Yan, and P. W. Waldroup. (2007c). Utilization of distillers dried grains with solubles (DDGS) in broiler diets using a standardized nutrient matrix. International Journal of Poultry Sci. 6 (7): 470-477.

Weigel, J. C., D. Loy, and L. Kilmer. (1997). Feed Co-Products of the Dry Corn Milling Process. Iowa State University, Iowa Corn Promotion Board, Iowa Department of Agriculture, Renewable Fuels Association, National Corn Growers Association. Available online: www.iowacorn.org/ethanol/ethanol_17.html.

Whitney, M. H. and G. C. Shurson. (2004). Growth performance of nursery pigs fed diets containing increasing levels of corn distiller's dried grains with solubles originating from a modern Midwestern ethanol plant. J. Anim. Sci. 82: 122-128.

Whitney, M. H., G. C. Shurson, and R. C. Guedes. (2006a). Effect of dietary inclusion of distillers dried grains with solubles on the ability of growing pigs to resist a Lawsonia intracellularis challenge. J. Anim. Sci. 84: 1860–1869.

Whitney, M. H., G. C. Shurson, and R. C. Guedes. (2006b). Effect of including distillers dried grains with solubles in the diet, with or without antimicrobial regimen, on the ability of growing pigs to resist a Lawsonia intracellularis challenge. J. Anim. Sci. 84: 1870–1879.

Whitney, M. H., G. C. Shurson, and R. C. Guedes. (2006c). Effect of dietary inclusion of distillers dried grains with solubles, soybean hulls, or a polyclonal antibody product on the ability of growing pigs to resist a Lawsonia intracellularis challenge. J. Anim. Sci. 84: 1880-1889.

Whitney, M. H., G. C. Shurson, L. J. Johnston, D. M. Wulf, and B. C. Shanks. (2006d). Growth performance and carcass characteristics of grower-finisher pigs fed high-quality corn distillers dried grain with solubles originating from a modern Midwestern ethanol plant. J. Anim. Sci. 84: 3356-3363.

Widmer, M. R., L. M. McGinnis, and H. H. Stein. (2007). Energy, phosphorus, and amino acid digestibility of high-protein distillers dried grains and corn germ fed to growing pigs. J. Anim. Sci. 85: 2994-3003.

Zhu, J. S., S. R. Stokes, and M. R. Murphy. (1997). Substitution of neutral detergent fiber from forage with neutral detergent fiber from by-products in the diets of lactating cows. J. Dairy Sci. 80: 2901-2906.

CHAPTER-8

BIOREFINERY PROCESSES FOR THE CONVERSION OF BIOMASS INTO LIQUID FUEL

1. INTRODUCTION: AN OVERVIEW

The utilization of biomass-derived products is becoming a significant driving factor for global economic development. The increase in oil costs and the uncertainty surrounding the security of current fossil fuel supplies, along with the growing apprehension about global climate change, have generated a demand for alternative transportation fuels and the production of bioproducts as substitutes for materials derived from fossil fuels. Currently, the United States has an annual use of over 140 billion gallons of transportation fuels. The transformation of cellulosic biomass into biofuels presents significant economic, environmental, and strategic advantages. According to the predictions made by DOE and USDA, the biomass resources in the U.S. have the potential to supply around 1.3 billion dry tons of feedstock for biofuels. This amount would be sufficient to meet around 40% of the annual fuel demand for transportation in the U.S. (Perlack et al., 2005). In January 2010, U.S. President Barack Obama made a formal plea to Congress during his State of the Union speech, urging them to maintain their financial support for biofuels and renewable energy technology. In light of this situation, biofuels have emerged as a highly significant and sustainable fuel option due to their ability to enhance supply security, decrease vehicle emissions, and offer a stable source of revenue for farmers. Various biorefinery techniques have been created to manufacture biofuels and chemicals from the original biomass source material. Liquid fuels are highly convenient for storage and transportation, making them one of the most favorable kinds of energy. Additionally, they are well-suited to the current gasoline distribution infrastructure. This chapter provides a thorough examination of the current state, utilization, and limitations of biorefinery methods employed for the production of liquid fuels, notably bioethanol and bio-oil. Additionally, it highlights the obstacles that may hinder the future success of biofuels.

2. CONCEPT OF BIOREFINERY

2.1 Definition of Biorefinery

A biorefinery is a facility that integrates biomass conversion processes and equipment to produce fuels, power, heat, and value-added chemicals from biomass. The biorefinery concept is analogous to today's petroleum refinery, which produces multiple fuels and products from petroleum (Smith & Consultancy, 2007).

The IEA Bioenergy Task 42 on Biorefineries has defined biorefining as the "sustainable processing of biomass into a spectrum of bio-based products (food, feed, chemicals, materials) and bioenergy (biofuels, power and/or heat)." The biorefinery is not a single or fixed technology. It is collection of processes that utilize renewable biological or bio-based sources, or feedstocks, to produce an end product, or products, in a manner that is a zero- waste producing, and whereby each component from the process is converted or utilized in a manner to add value, and hence sustainability to the plant. Several different routes from feedstocks to products are being developed and demonstrated, and it is likely that multiple biorefinery designs will emerge in the future.

By producing multiple products, a biorefinery takes advantage of the various components in biomass and their intermediates, thereby maximizing the value derived from the biomass feedstock. A biorefinery could, for example, produce one or several low-volume, but high- value chemical or nutraceutical products and a low-value, but high-volume liquid transportation fuel such as biodiesel or bioethanol (see also alcohol fuel), while also generating electricity and process heat through combined heat and power (CHP) technology for its own use, and perhaps enough for sale of electricity to the local utility. In this scenario, the high-value products increase profitability, the high-volume fuel helps meet energy needs, and the power production helps to lower energy costs and reduce greenhouse gas emissions, as compared to traditional power plant facilities. Although some facilities exist that can be called biorefineries, the technology is not commonplace. Future biorefineries may play a major role in producing chemicals and materials that are traditionally produced from petroleum.

2.2 Two biorefinery platforms

Biomass can be transformed into various types of valuable energy through multiple methods. Figure 1 illustrates the existence of two main biorefinery platforms: sugar and thermochemical. Both platforms have the capability to generate various chemicals and fuels, such as methanol, ethanol, and polymers. The "sugar platform" is established by converting biomass into aqueous sugars by chemical and biological methods. The fermentable carbohydrates can undergo fermentation, dehydration, and aqueous-phase processing to produce ethanol, aromatic hydrocarbons, or liquid alkanes, respectively. The remaining substances, primarily lignin, can be utilized for generating electricity by co-firing, or can be refined to create alternative products such as etherified gasoline. In the thermochemical platform, biomass undergoes gasification to turn it into synthesis gas, or pyrolysis and hydrothermal conversion (HTC) to produce bio-oils. Bio-oils have the potential to undergo additional processing in order to be transformed into liquid fuels including methanol, gasoline, diesel fuel, and various other compounds.

3. PRODUCTION OF BIOETHANOL FROM LIGNOCELLULOSIC BIOMASS

Ethanol is widely regarded as the most promising fuel for future transportation, and it is already being manufactured in large amounts from corn and sugar cane through a process called fermentation. Using lignocellulosic biomass as a raw material is considered the next crucial move in greatly increasing the capacity for ethanol production. Nevertheless, there are technological obstacles that must be overcome in order to convert lignocellulosic biomass into bioethanol efficiently. These barriers include pretreatment, enzyme hydrolysis, saccharification of the cellulose and hemicellulose matrices, and simultaneous fermentation of hexoses and pentoses. Despite the need for significant technical advancements to make lignocellulose-to-ethanol economically feasible, the production of ethanol through fermentation has the intrinsic disadvantage of requiring distillation from a mixture containing 82% to 94% water. This section will examine the latest advancements in addressing these technological obstacles.

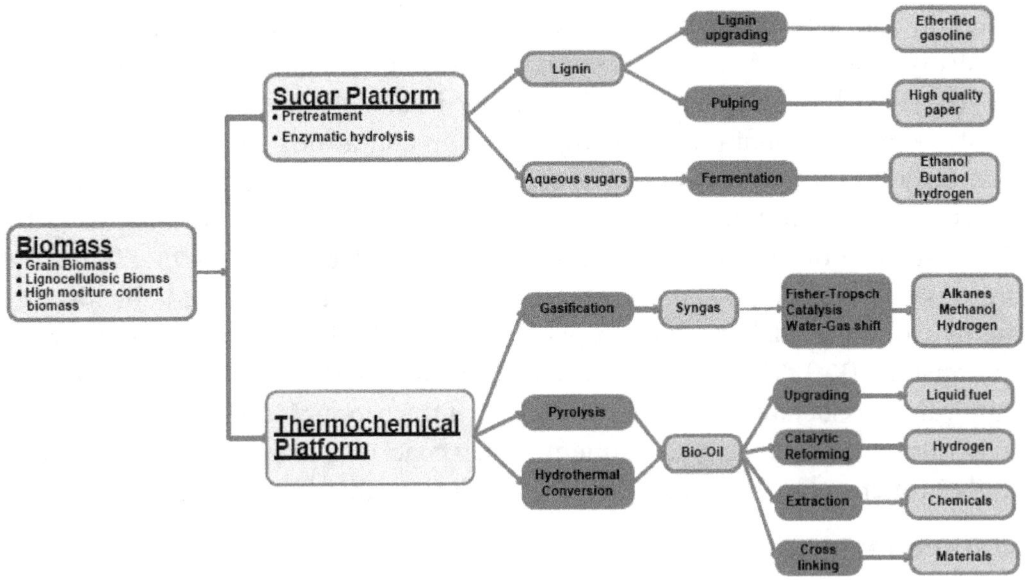

Fig. 1. Primary routes for biofuels conversion

3.1 Pretreatment

Pretreatment is necessary to disrupt the crystalline structure of cellulosic biomass in order to enhance its accessibility to enzymes. These enzymes can then bind to the cellulose and break down the carbohydrate polymers into sugars that can be fermented. The objective of pretreatment is to extract hemicellulose prior to breaking the lignin seal and releasing the cellulose from the matrix of the plant cell wall. Pretreatment is recognized as a costly stage in cellulosic ethanol processes, but it has significant promise for improvement and cost reduction through research and development (Lynd et al., 1996; Lee et al., 1994; Mosier et al., 2005). Several pretreatment technologies have been created and assessed for different types of biomass resources. Nevertheless, every pretreatment technique possesses its unique merits and drawbacks, and there is no universal pretreatment method suitable for all types of biomass feedstocks. Below, we shall cover three commonly utilized pretreatment technologies.

3.1.1 Alkaline pretreatment

The utilization of alkaline chemicals, such as weak sodium hydroxide, aqueous ammonia, and lime, has been well recognized as an effective method for enhancing the digestibility of cellulose by removing lignin (Li et al., 2004). Sodium hydroxide (NaOH) is commonly utilized for pretreatment due to its superior alkalinity compared to other alkaline reagents. However, it is also costly and the recovery process is intricate. The subsequent investigations on several feedstocks exemplify this: Cattails that have not been treated have a cellulose content of 32.0%, hemicellulose content of 18.9%, and lignin content of 20.7%. In a study conducted by Zhang et al. (2010a), it was shown that a 4% NaOH solution was able to remove 54.8% of cattail lignin and 43.7% of the hemicellulose. The glucose production obtained from cattails treated with 4% NaOH was approximately 80% of the total cellulose content. Supplementing NaOH with additional chemicals has the potential to enhance the effectiveness of pretreatment. The application of a NaOH and H2O2 solution aids in further elimination of lignin by oxidizing it. The highest total sugar production achieved from high lignin hybrid poplar was 80%, using a mixture of 5% sodium hydroxide and 5% hydrogen peroxide at a temperature of 80°C (Gupta, 2008). In their study, Zhao et al. (2008) found that a pretreatment using NaOH-urea can partially remove lignin, hemicelluloses, and cellulose from lignocellulosic materials. This process also disrupts the bonds between hemicelluloses, cellulose, and lignin, and modifies the structure of the treated biomass, making cellulose more easily accessible to hydrolysis enzymes. The enzymatic hydrolysis efficiency of spruce can be significantly improved by treating it with a solution of NaOH or NaOH/urea. A glucose yield of 70% was achieved during the cold temperature pretreatment (-15°C) using a solution containing 7% NaOH and 12% urea. However, glucose yields of only 20% and 24% were produced at temperatures of 238°C and 608°C, respectively. The enzyme kinetics of sodium hydroxide pretreated wheat straw were studied using two theoretical approaches. The influence of enzyme concentrations ranging from 6.25 to 75 g/L on the synthesis of reducing sugars was also described. The initial method employed a modified Michaelis-Menten equation to ascertain the hydrolysis model and kinetic parameters, including the maximal velocity (Vemax) and half-saturation constant (Ke). The second approach, known as the Chrastil approach, was utilized to examine the time values associated with the rate at which the product is formed. This approach considers the fact that in a system with varying components, such reactions are limited by diffusion and the time curves are heavily influenced by the structures that limit the rate of the enzyme system.

3.1.2 Hot-water pretreatment

Autohydrolysis is a commonly used term for the process of hot water pretreatment. This process has several significant benefits, including reduced costs, decreased equipment corrosion, and less xylose degradation. As a result, there are fewer byproducts containing inhibitory chemicals in the extracts (Huang et al., 2008). Pressurized hot water has the ability to infiltrate the cellular composition of biomass, saturate cellulose, and eliminate hemicellulose. Hot water pretreatment significantly enhances the enzymatic digestibility of cellulose in biomass. Under ideal circumstances, 90% of the cellulose derived from maize stover that has undergone hot-water pretreatment can be converted into glucose through hydrolysis (Mosier et al., 2003). After subjecting cattails to a pretreatment temperature of 463K for a duration of 15 minutes, all of the hemicellulose was successfully eliminated, while 21.5% of the cellulose was dissolved in the aqueous phase. The efficiency of the procedure could be enhanced with further optimization (Zhang et al., 2010b). An investigation was conducted to examine the pretreatment of bagasse at temperatures ranging from 170 to 203°C, and for durations ranging from 1 to 46 minutes. An 80% conversion yield was attained, and the presence of hydrolysis inhibitors was identified (Laser et al., 2002). The utilization of hot water pretreatment has been documented to enhance the enzymatic digestibility of switchgrass, leading to a glucose yield of 80% (Kim et al., 2008). The most effective hot-water pretreatment conditions for hybrid poplar with a solids concentration of 15% (wt/vol) were found to be 200°C for 10 minutes. This resulted in the highest yield of fermentable sugars, ranging from 54% to 67% (Kim et al., 2008).

3.1.3 Dilute-acid pretreatment

The use of acid hydrolysis for the conversion of cellulose to glucose is a process that has been studied for the last 100 years. Dilute acid (0.5-1.0% sulfuric acid) pretreatment at moderate temperatures (140-190°C) can effectively remove and recover most of the hemicellulose as dissolved sugars. Furthermore lignin is disrupted and partially dissolved, increasing cellulose susceptibility to enzymes (Yang and Wyman, 2004). Under this method, glucose yields from cellulose increase with hemicellulose removal to almost 100% (Knappert et al., 1981). Dilute acid hydrolysis consists of two chemical reactions. One reaction converts cellulosic materials to sugar and the other converts sugars into other chemicals, many of which inhibit the growth of downstream fermentation microbes. The same conditions that cause the first reaction to occur simultaneously cause over-degradation of sugars and lignin, creating inhibitory compounds such as organic acids, furans, and phenols.

Partial cellulose may be degraded as oligomers or monomers during the acid pretreatment process. Sugar (glucose and xylose) yields were often reported for the pretreatment and enzyme hydrolysis stage separately, and as the total for both stages. Lloyd and Wyman (2005) reported that up to 92% of the total sugars originally available in corn stover could be recovered via coupled dilute acid pretreatment and enzymatic hydrolysis. Conditions achieving maximum individual sugar yields were often not the same as those that maximized the total sugar yields, demonstrating the importance of clearly defining pretreatment goals when optimizing the process.

Dilute-sulfuric acid pretreatment of cattails was studied using a Dionex accelerated solvent extractor (ASE) at varying acid concentrations of 0.1 to 1%, treatment temperatures of 140 to 180 °C, and residence times of 5 to 10 min. The yield of extractable products obtained from the pretreatment process increased as the final temperature, treatment time, or acid concentration increased. The highest glucose yield from the pretreatment was 55.4% of the cellulose at 180°C for 15 min with 1% sulfuric acid. The highest glucose yield from the enzyme hydrolysis stage (82.2% of the cellulose) and the highest total glucose yield for both the pretreatment and enzyme hydrolysis stages (97.1% of the cellulose) were reached at a temperature of 180°C, a sulfuric acid concentration of 0.5%, and a time of 5 min. When switchgrass was pretreated for 60 min with 1.5% acid, the highest glucan conversion yield of 91.8% was obtained (Yang et al. 2009).

3.2 Enzyme hydrolysis

After pretreatment, hydrolysis converts the carbohydrate polymers into monomeric sugars. Although a variety of process configurations have been studied for conversion of cellulosic biomass into ethanol, enzymatic hydrolysis of cellulose provides opportunities to improve the technology so that biomass ethanol is competitive with other liquid fuels(Wyman, 1999). Novozymes (www.novozymes.com) and Genencor (www.genencor.com) are two companies leading research & development for advanced cellulosic ethanol enzymes. In early 2010, Novozymes said its new Cellic® CTec2 enzymes enable the biofuel industry to produce cellulosic ethanol at a price below US$ 2.00 per gallon for the initial commercial- scale plants that are scheduled to be in operation in 2011. This cost is on par with gasoline and conventional ethanol at current US market prices. According to Novozymes, the new enzyme can be used on different types of feedstock including corn cobs and stalks, wheat straw, sugarcane bagasse, and woodchips. The enzyme is designed to break down cellulose in biomass into sugars that can be fermented into ethanol. Genencor, a division of Danisco also introduced its enzyme Accellerase®, which is designed to do the same thing.

The choice of enzymes must align with the pretreatment technology, feedstock, and process. In the case of employing a diluted acid pretreatment, the majority of the hemicellulose is broken down, rendering hemicellulases unnecessary. Nevertheless, in the event that an alkaline or hot-water pretreatment is employed, the hemicellulose must still undergo hydrolysis, necessitating the usage of hemicellulases. Another challenge lies in the cellulose component of the biomass. To effectively decompose it, a combination of many enzymes with diverse activity is necessary. This blend comprises three fundamental categories of enzymes.

- Endoglucanases break bonds between adjacent sugar molecules in a cellulose chain, fragmenting the chain into shorter lengths. Endoglucanases act randomly along the cellulose chain, although they prefer amorphous regions where the chains are less crystalline.
- Cellobiohydrolases attack cellulose chains from the ends of the chain. This exo- or processive action releases mainly cellobiose (glucose dimer). Because endoglucanases create new ends for cellobiohydrolases to act upon, the two classes of enzymes interact synergistically.

- β-glucosidases break down short glucose chains, such as cellobiose, to release glucose. β-glucosidases are important as they act on cellobiose, which inhibits the action of the other cellulases as it builds up the hydrolysis reactor.

3.3 Utilizing fermentation for the manufacture of bioethanol

Saccharomyces cerevisiae, commonly known as baker's yeast, has been employed in the industrial synthesis of ethanol from hexoses, which are sugars with six carbon atoms, for a millennium. Nevertheless, a substantial quantity of pentoses (C5 sugars) obtained from the hemicellulose component of the lignocellulosic biomass is found in the hydrolysate resulting from the pretreatment procedure. Modern biotechnologies allow fermenting microorganisms to utilize both C5 and C6 sugars present in the hydrolysate. This enhances the economic competitiveness of ethanol production and other bio-products derived from cellulosic biomass. In recent times, metabolic engineering techniques have been employed to genetically modify microorganisms like Saccharomyces cerevisiae, Zymomonas mobilis, and Escherichia coli for the purpose of producing cellulosic ethanol. In their study, Lau et al. (2010) conducted a comparison of the fermentation performance of three microorganisms, namely Escherichia coli KO11, Saccharomyces cerevisiae 424A(LNH-ST), and Zymomonas mobilis AX101, for the manufacture of cellulosic ethanol. Three microorganisms exhibited a metabolic yield, final concentration, and rate that exceeded 0.42 g/g of eaten sugars, 40 g/L, and 0.7 g/L/h (0-48 h), respectively. The researchers determined that Saccharomyces cerevisiae 424A(LNH-ST) is the most favorable strain for commercial production due to its capacity to ferment both glucose and xylose. In their study, Vasan et al (2011) successfully incorporated a cellulase gene from Enterobacter cloacae into the Zymomonas mobilis strain. As a result, the recombinant bacterium exhibited a cellulase activity of 0.134 filter paper activity unit (FPU)/ml units. When carboxymethyl cellulose and 4% NaOH prepared bagasse were used as substrates, the recombinant strain yielded 5.5% and 4% (V/V) ethanol, respectively. This amount was three times greater than what was achieved with the original E. cloacae isolate.

In 2010, Purde University scientists improved a strain of yeast that can produce more biofuel from cellulosic plant material by fermenting all five types of the plant's sugars: galactose, manose, glucose, xylose and arabinose. Arabinose makes up about 10 percent of the sugars contained in cellulosic biomass (Casey et al., 2010).

3.4 Closing remarks

Ethanol serves as the initial prototype for the commercialization of biofuels. Nevertheless, in order to ensure the economic feasibility of the cellulosic ethanol process, both government subsidies and scientific research and development (R&D) are necessary. It is widely acknowledged that ethanol alone is unable to supply society's long-term energy requirements (Hill et al., 2006). It has a relatively low energy density, cannot be carried through pipelines, and is quite expensive to extract from fermentation broths. This is facilitating the exploration of several alternative molecules to ethanol, hence enabling the discovery of novel fuel compounds through the use of microbial biotechnology.

4. PRODUCTION OF BIO-OIL FROM LIGNOCELLULOSIC BIOMASS AND BIOMASS WITH HIGH MOISTURE CONTENT

Bioethanol is only one of the products that may be extracted from lignocellulosic feedstocks. Other forms of energy and a full range of value-added bioproducts may be produced from biomass by thermochemical means. Thermochemical conversion processes include pyrolysis, hydrothermal conversion and gasification. The major product of pyrolysis and hydrothermal conversion, known as "bio-oil" or "biocrude", can be used as a boiler fuel or as fuel in combustion engines. Alternatively, the bio-oil can serve as a raw material for the production of chemicals and biomaterials. One of the major technical obstacles to large scale thermochemical conversion of biomass into bio-oil is its poor oil quality and low biofuel production rate. This section intensively reviewed current technologies used to produce bio-oil and technologies development towards improving the bio-oil yield and quality.

4.1 Presently, the methods used to convert biomass into bio-oils

Two main types of processes for production of bio-oils from biomass are flash pyrolysis and hydrothermal conversion, as shown in Fig.1. Both of the processes belong to the thermochemical platform in which feedstock organic compounds are converted into liquid products. An advantage of the thermochemical process is that it is relatively simple, usually requiring only one reactor, thus having a low capital cost. However, this process is non- selective, producing a wide range of products including a large amount of char (Huber & Dumesic, 2006). The characteristic and technique feasibility of the two thermochemical processes for bio-oil production are compared in table 1. Flash pyrolysis is characterized by a short gas residence time (~1s), atmospheric pressure, a relatively high temperature (450-500 °C). Furthermore, feedstock drying is necessary. Hydrothermal processing (also referred to in the literature as liquefaction, hydrothermal pyrolysis, depolymerisation, solvolysis and direct liquefaction), is usually performed at lower temperatures (300-400 °C), longer residence times (0.2-1.0 hr.), and relatively high operating pressure (5-20 Mpa). Contrary to flash pyrolysis and gasification processes, drying the feedstock is not needed in the hydrothermal process, which makes it especially suitable for naturally wet biomass. However, a reducing gas and/or a catalyst is often included in the process in order to increase the oil yield and quality.

Table 1. Comparison of two typical thermochemical processes for bio-oil production

Methods	Treatment condition/ requirement	Reaction mechanism / process description	Technique Feasibility	
			Pros.	Cons.
Flash/Fast Pyrolysis	Relatively high temperature (450- 500 °C); a short residence time (~1s); atmosphere pressure; drying necessary	The light small molecules are converted to oily products through homogeneous reactions in the gas phase	High oil yield up to 80% on dry feed; lower capital cost; Commercialized already	Poor quality of fuels obtained
Hydrothermal Processing (HTU)/ liquefaction /hydrotherma l pyrolysis	Lower temperature (300-400 °C); longer residence time (0.2-1.0 hr.); High pressure (5-20 Mpa); drying unnecessary	Occurs in aqueous medium which involves complex sequences of reactions	Better quality of fuels obtained (High PTU, low moisture content)	Relatively low oil yield (20-60%); Need high pressure equip, thus higher capital cost

The reaction mechanisms of the two processes are different, which have been studied by many investigators (Demirbaş, 2000a; Minowa et al., 1998). The hydrothermal process occurred in aqueous medium which involves complex sequences of reactions including solvolysis, dehydration, decarboxylation, and hydrogenation of functional groups, etc. (Chornet and Overend, 1985). The decomposition of cellulose was studied by Minowa et al. (1998). The effects of adding a sodium carbonate catalyst, a reduced nickel catalyst, and no catalyst addition in the decomposition of cellulose in hot-compressed water were investigated. They found that hydrolysis can play an important role in forming glucose/oligomer, which can quickly decompose into non-glucose aqueous products, oil, char and gases (Fig. 2). Without a catalyst, char and gases were produced through oil as intermediates. However, in the presence of an alkali catalyst, char production was inhibited because the oil intermediates were stabilized, resulting in oil production. Reduced nickel was found to catalyze the steam reforming reaction of aqueous products as intermediates and the machination reaction. Typical yields of liquid products for hydrothermal conversion processes were in the range of 20-60%, depending on many factors including sustrate type, temperature, pressure, residence time, type of solvents, and catalysts employed (Xu and Etcheverry, 2008).

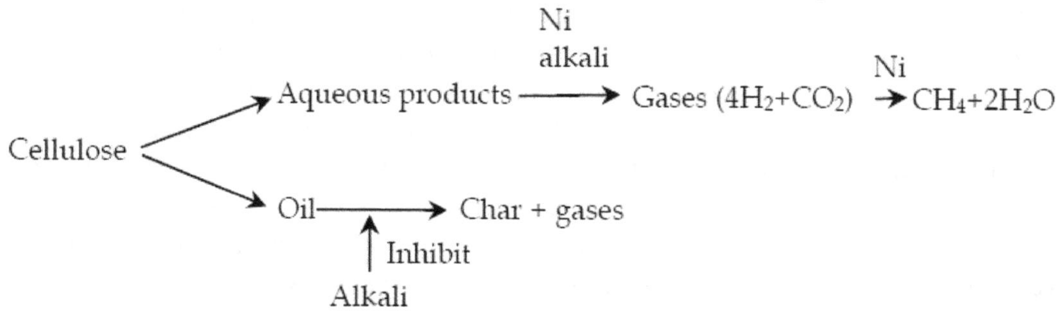

Fig. 2. Reaction pathway for the hydrothermal processing of cellulose

With flash pyrolysis, the light small molecules are converted to oily products through homogeneous reactions in the gas phase. The principle of the biomass flash pyrolysis process is shown in Fig.3. Biomass is rapidly heated in the absence of air, vaporizes, and quickly condenses to bio-oil. The main product, bio-oil, is obtained in yields of up to 80% wt on dry feed, together with the by-product char and gas (Bridgewater and Peacocke, 2000).

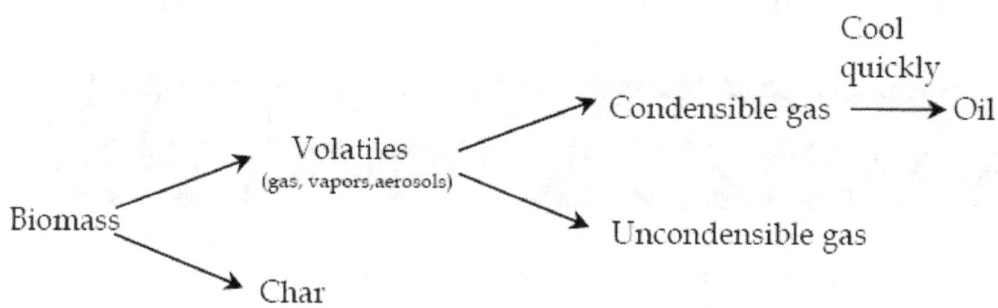

Fig. 3. Reaction pathway for the biomass flash pyrolysis process

4.2 The research focuses on the development of flash pyrolysis and hydrothermal processes.

Flash pyrolysis for the production of liquids has developed considerably since the first experiments in the late 1970s. Several pyrolysis reactors and processes have been investigated and developed to the point where fast pyrolysis is now an accepted, feasible and viable route to renewable liquid fuels, chemicals and derived products. Since the 1990s, several research organizations have successfully established large-scale fast pyrolysis plants. Bridgwater and Peacocke (2000) have intensively reviewed the key features of fast pyrolysis and the resultant liquid product, and described the major reaction systems and processes that have been developed over the last 20 years.

Unlike flash pyrolysis, technological developments in the area of hydrothermal conversion present new ways to turn wastes to fuel. Hydrothermal processing was initially developed for turning coal into liquid fuels, but recently, the technique has been applied to a number of feedstocks, including woody biomass, agricultural residues, and organic wastes (e.g., animal wastes, municipal solid wastes (MSW), and sewage sludge). Table 2 summarizes representative literature data of hydrothermal processing of common types of biomass and the most influential operating parameters. As can be seen from Table 2, organic waste materials are more favourable than woody biomass and agricultural residues for hydrothermal processing, owing to their higher oil yield and the higher heating value of their bio-oil products. This earlier work was very promising, showing that hydrothermal technology can be used as an efficient method to treat different types of biomass and produce a liquid biofuel. In particular, hydrothermal conversion processes present a unique approach to mitigate the environmental and economic problems related to disposing of large volumes of organic wastes. It not only reduces the pollutants, but also produces useful energy in the form of liquid fuel. Compared with flash pyrolysis, hydrothermal conversion is at an early developmental stage, and the reaction mechanisms and kinetics are not yet fully understood.

Table 2. Overview of literature on hydrothermal processing of common types of biomass

Raw Materials	Reactor Capacity	Temp. (°C)	Pressure (Mpa)	Time (min)	Oil Yield (%)	Heating Value(MJ/kg)	Reference
a) Woods							
Beech	--	277-377	--	25	13.8-28.4	27.6-31.3	Demirbaş, et al., 2005a
Spruce	--	277-377	--	25	13.8-25.8	28.3-33.9	Demirbaş, et al., 2005b
Sawdust	0.2 L	280	N/A		7.2	--	Karagöz et al., 2005
b) Agricultural residues							
Corn stalk	0.3 L	300	10 Mpa	30	28.3 on organic basis	29.7	Minowa et al., 1998
Rice husk	0.3 L	300	10 Mpa	30	28.8 on organic basis	30.8	Minowa et al., 1998
Rice straw	1.0 L	260-350	6-18 Mpa	3-5	13.0-38.35	27.6-35.8	Yuan et al., 2007
c) organic wastes							
Swine manure	1-L autoclave	260-340		0-90	14.9-24.2	36.1	Xiu et al., 2010a
Swine manure	Continuous mode	285-305	9-12	40-80	2.8-53.3	25.2-33.1	Ocfemia et al., 2006
Dairy manure	Batch/ continuous mode	250-380	10-34	--	50	--	Appell, et al., 1980
Sewage sludge	5 t/d	300	10	--	48	37-39	Itoh, et al., 1994
Garbage	0.3 L autoclave	250-340	6-18	6-120	27.6	36	Minowa, et al., 1995
Sewage sludge	0.3 L autoclave	150-300	--	0-180	44.5	35.7	Suzuki, et al., 1990
Sewage sludge	4.2L microwave	250-350	8-20	--	30.7	36.4	Bohlmann, et al., 1999
MSW	autoclave	260-340	13-34	--	32	46	Gharieb, et al., 1995
MSW	autoclave	295-450	--	20-90	35-63.3	--	Kranich et al., 1984
Sewage sludge	20 kg/hr.	300-360	10-18	5-20	--	30-35	Goudriaan et al., 2000

4.3 Characteristics of bio-oils

Variations in processing conditions lead to substantial disparities in both the quantity and chemical makeup of bio-oils. Lu et al. (2009) conducted a thorough examination of the fuel qualities of fast pyrolysis oils and discussed the impact of these properties on the use of bio-oils. Bio-oils typically consist of intricate combinations of volatile hydrocarbons, alcohols, organic acids, aldehydes, ketones, ethers, furans, phenols, and other chemicals that are not easily evaporated. The volatile components in bio-oil have the ability to reorganize themselves by undergoing condensation, cyclization, and polymerization processes, resulting in the formation of different compounds, including aromatics. Table 3 provides an overview of the key characteristics of bio-oils derived from the hydrothermal liquefaction of swine manure and the pyrolysis of wood. Table 3 also included the parameters of heavy petroleum fuel oil for the sake of comparison.

Table 3. Comparison of selected properties of bio-oils produced by hydrothermal liquefaction of swine manure and pyrolysis of wood and heavy fuel oil

Properties		Liquefied bio-oil from swine manure(xiu et al., 2010a)	Pyrolysis bio-oil from wood pyrolysis (Zhang et al., 2007)	Heavy petroleum fuel oil (Oasmaa et al., 1999)
Moisture content (wt%)		2.37	15-30	0.1
PH		--	2.5	-
Specific gravity		1	1.2	0.94
Elemental composition (wt%)	C	72.58	54-58	85
	H	9.76	5.5-7.0	11
	O	13.19	35-40	1.0
	N	4.47	0-0.2	0.3
Ash		0.78	0-0.2	0.1
HHV(MJ/kg)		36.05	16-19	40
Viscosity(at 50 0C)(cP)		843	40-100	180
Solids (wt%)		--	0.2-1	1
Distillation residue (wt%)		63	Up to 50	1

Table 3 demonstrates that liquefied oils exhibit significantly lower levels of oxygen and moisture, resulting in a considerably higher energy value as compared to oils derived from quick pyrolysis. The higher heating value (HHV) of liquefied oil derived from swine dung is 36.05 MJ/kg, which is almost 90% of the HHV of heavy fuel oil (40 MJ/kg). The characteristics of bio-oil derived from both procedures exhibit notable distinctions when compared to those of heavy petroleum fuel oil. The bio-oils possess several unfavorable characteristics for fuel applications when compared to heavy petroleum fuel oil, including elevated viscosity, increased water and ash concentrations, elevated oxygen content, and reduced heating value. Pyrolysis oil has an acidic character, is polar, and does not mix with ordinary crude oil. Furthermore, the oil exhibits instability due to the (re)polymerization of organic materials, resulting in a gradual rise in viscosity. In general, bio-oils are unsuitable for use as transportation fuels because they have a high viscosity, high water and ash content, low heating value, instability, and high corrosiveness. Hence, bio-oil must undergo the process of upgrading in order to be suitable for use as a fuel for engines.

4.4 Common bio-oil upgrading technologies and their constraints.

Considering the above discussion on the properties, it is obvious that the fuel quality of bio- oils is inferior to that of petroleum-based fuels. There have been intensive studies on bio-oil upgrading research and various technologies have been developed for bio-oil upgrading. Table 4 summarizes current techniques in bio-oil upgrading. The characteristics, as well as recent progress, advantages, and disadvantages of each technique are also described as follows:

Table 4. Brief description, treatment condition, and technical feasibility of the current techniques used for upgrading bio-oil

Upgrading methods	Treatment condition/ requirement	Reaction mechanism /process description	Technique Feasibility	
			Pros.	Cons.
Hydrotreating /hydrofining	Mild conditions, (~500°C /low pressure) chemical needed: H_2/CO, catalyst (e.g., CoMo, HDS, NiMo, HZSM-5)	Hydrogenation without simultaneous cracking (eliminating N, O and S as NH_3, H_2O and H_2S)	Cheaper route, Commercialized already	high coking (8-25%) and poor quality of fuels obtained
Hydro-cracking /hydrogenolysis /catalytic cracking	Severe conditions, (>350 °C, 100~2000 Psi), chemical needed: H2/ CO or H2donor solvents, catalyst (e.g., Ni/Al2O3-TiO_2)	Hydrogenation with simultaneous cracking Destructive(resulting in low molecular product)	Makes large quantities of light products	Need complicated equipment, excess cost, catalyst deactivation, reactor clogging
Supercritical fluid	Mild conditions, organic solvents needed such as alcohol, acetone, ethyl acetate, glycerol	Promotes the reaction by its unique transport properties: gas-like diffusivity and liquid-like density, thus dissolved materials not soluble in either liquid or gaseous phase of solvent	Higher oil yield, better fuel quality (lower oxygen content, lower viscosity)	Solvent is expensive
Solvent addition (direct add solvent or esterification of thethe oil with alcohol and acid catalysts	Mild conditions, polar solvents needed such as water, methanol, ethanol, and furfural	Reduces oil viscosity by three mechanisms: (1) physical dilution (2) molecular dilution or by changing the oil microstructure; (3) chemical reactions like esterification and acetalization	The most practical approach (simplicity, the low cost of some solvents and their beneficial effects on the oil properties)	Mechanisms involved in adding solvent are not quite understand yet
Emulsification /Emulsions	Mild conditions, need surfactant (e.g. CANMET)	Combines with diesel directly. Bio-oil is miscible with diesel fuels with the aid of surfactants	Simple, less corrosive	Requires high energy for production
Steam Reforming	High temperature (800-900 °C), need catalyst (e.g. Ni)	Catalytic steam reforming + water-gas shift	Produces H2as a clean energy resource	Complicated, requires steady, dependable, fully developed reactors
Chemical extracted from the bio-oils	Mild conditions	Solvent extraction, distillation, or chemical modification	Extract valuable chemicals	Low cost separation and refining techniques still needed

4.4.1 Hydrotreating

The quality of petroleum products, particularly fuel products, is generally acknowledged to improve as the hydrogen level increases. This understanding has prompted the implementation of a hydrogenation process in the refinery, which involves adding hydrogen. Hydrotreating is currently the predominant method for converting petroleum and petroleum products through hydrogenation. Hydrotreating (HDT) is a hydrogenation procedure that enhances product quality without significantly changing the boiling range. It has emerged as the prevailing method in contemporary petroleum refineries. Bio-crude can also be refined using a traditional refinery and potentially mixed with petroleum crude. Hydrotreating can be used to eliminate the oxygen present in bio-oils. The catalysts typically employed for hydrotreating are CoMo/Al2O3 and NiMo/Al2O3 systems in sulfide form (Nava et al., 2009). Hydrotreating necessitates gentle operating conditions, but the production output of bio-oil is quite modest. The process additionally generates a substantial quantity of char, coke, and tar, leading to catalyst deactivation and reactor obstruction.

4.4.2 Hydro-cracking

Hydro-cracking is less widely utilized than hydrotreating in the petroleum sector. Hydro-cracking is a high-temperature process (>350 °C, >660°F) in which cracking is accompanied with hydrogenation. The process involves the use of relatively high pressure, ranging from 100 to 2000 Psi. As a result, there is typically a modification in the nature or excellence of the final products (Ancheyta and Speight, 2007). Dual-purpose catalysts are used to carry out this process. Silica-alumina (or zeolite) catalysts are responsible for the cracking function, while platinum and tungsten oxide catalyze the reactions. Alternatively, nickel can be used to give the hydrogenation function. Alumina is the most commonly utilized support material. Hydro-cracking is a highly efficient method for producing a significant quantity of light products. However, it necessitates more rigorous conditions, like elevated temperatures and hydrogen pressure, to handle acids. These circumstances are not cost-effective or energy-efficient.

4.4.3 Supercritical fluids (SCFs)

A fluid is classified as supercritical when its temperature and pressure exceed its critical point. Superconducting films (SCFs) exhibit distinctive transport characteristics. They have the ability to permeate solids in a manner similar to gases and can dissolve substances in a manner similar to liquids. Specifically, supercritical fluids (SCFs) possess the capacity to dissolve substances that are typically insoluble in either the liquid or gaseous state of the solvent. As a result, they facilitate the gasification or liquefaction reactions (Xu and Etcheverry, 2008).

Recently, SCFs have been utilized to enhance the quantity and quality of oil, showing significant promise in the production of bio-oil or bio-crude with higher energy content and reduced thickness. Water, although the most affordable and often employed supercritical fluid in hydrothermal processing, has certain disadvantages when utilized as a solvent for biomass liquefaction: 1) The water-insoluble oil product has reduced yields. 2) The resulting bio-oil is highly viscous and contains a significant amount of oxygen. In order to increase both the quantity and quality of oil, researchers have utilized various organic solvents such as ethanol, ethyl acetate, acetone, 2-propanol, 1,4-dioxane, methanol, and butanol. Each of these solvents has had a notable impact on both the quantity and quality of bio-oil produced. According to Minami and Ska (2003, 2005), 90% of beech wood was effectively destroyed in supercritical methanol. The aforementioned supercritical organic solvent fluids are mostly utilized in hydrothermal treatment processing to enhance the yield and quality of bio-oil. Additionally, it was employed for the enhancement of pyrolysis bio-oil. Tang et al. (2009) found that using supercritical ethanol (with a temperature of 243.1°C and pressure of 6.37 MPa) can improve the quality of lignin-derived oligomers in pyrolysis oil, resulting in a decrease in the production of tar or coke. Despite the fact that supercritical fluids (SCFs) may be generated at comparatively lower temperatures and the production method is environmentally sustainable, the high cost of these organic solvents renders large-scale economic viability unattainable. In recent times, scientists have been endeavoring to evaluate more cost-effective organic solvents as alternatives to supercritical fluids (SCFs).

Crude glycerol, which is a low-value by-product of biodiesel production, has demonstrated highly encouraging outcomes when used as a supercritical fluid (SCF) solvent. Glycerol has been utilized as an organic solvent in various applications such as biomass delignification, bio-oil separation, and enhancing the efficiency of biomass conversion into bio-oil. Xiu et al. discovered that the interaction between swine manure and crude glycerol had a major impact on the hydrothermal process. They also observed that the utilization of crude glycerol led to a substantial increase in bio-oil production, raising the yield from 23.9% to 70.92% (Xiu et al., 2010c, Fig.4). Furthermore, it was found that the unbound fatty acid in the impure glycerol is the crucial factor responsible for increasing the amount of oil produced (Fig.5). Furthermore, the quality of the oil was enhanced, exhibiting a reduced density and viscosity.

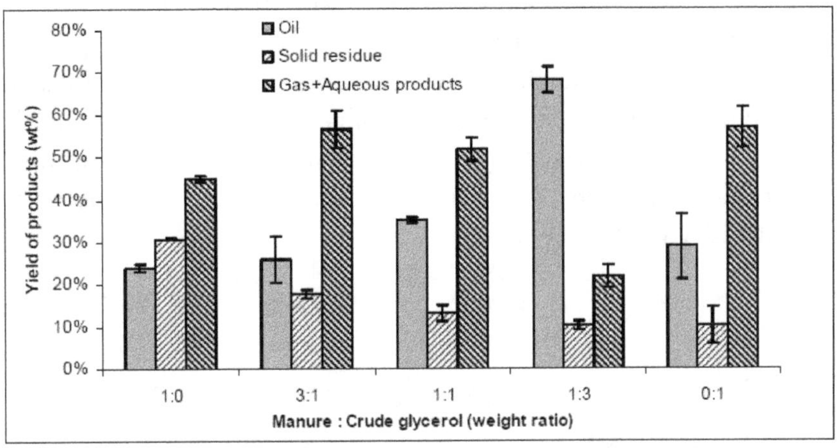

Fig. 4. Effect of swine manure to crude glycerol ratio (weight ratio) on the products yield in hydrothermal pyrolysis of swine manure. (Xiu et al., 2010c)

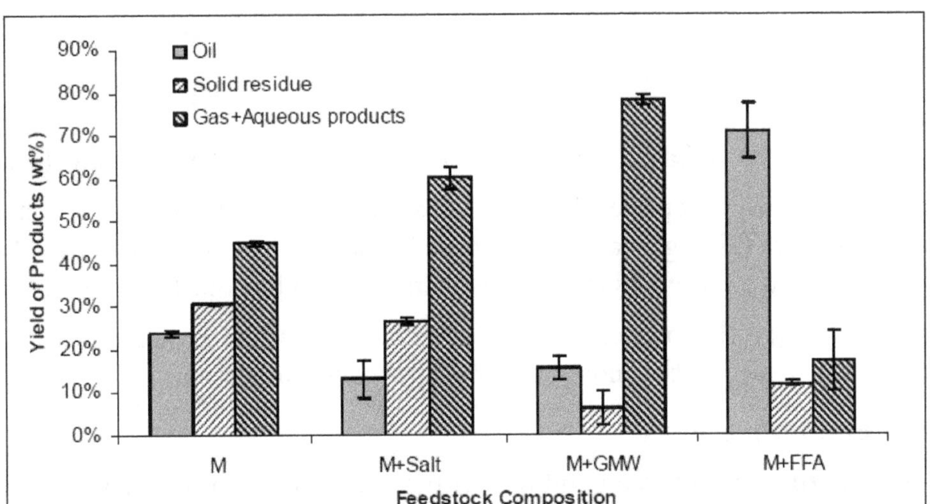

Fig. 5. Effect of crude glycerol fractionated components on the products yield in hydrothermal pyrolysis of swine manure.(Xiu et al., 2010c). M-manure; GMW- glycerol,methnol and water; FFA-free fatty acids.

4.4.4 The process of adding a solvent or carrying out an etherification reaction.

Polar solvents, including methanol, ethanol, and furfural, have long been employed to homogenize and decrease the viscosity of biomass oils (Radlein et al., 1996; Diebold and Czernik, 1997; Oasmma, 2004; Boucher et al., 2000). Adding these polar solvents has an immediate impact of reducing viscosity and increasing heating value. The rise in heating value of bio-oils when mixed with solvents is due to the fact that the solvent possesses a higher heating value compared to the majority of bio-oils. The addition of solvent decreases the viscosity of the oil through three mechanisms: (1) physical dilution without impacting the rates of chemical reactions; (2) reduction of the reaction rate through molecular dilution or alteration of the oil's microstructure; (3) chemical reactions between the solvent and the oil components that hinder further chain growth (Oasmaa and Czernik, 1999). Many research have employed the direct addition of solvents following pyrolysis, which effectively reduces viscosity and enhances stability and heating value. Recent studies have demonstrated that when oil is combined with alcohol (such as ethanol) and acid catalysts (such as acetic acid) under mild conditions using reactive distillation, the quality of the resulting bio-oil is improved (Mahfud et al., 2007; Xu et al., 2008; Tang et al., 2008; Oasmma, et al., 2004; Xu and Etcheverry, 2008). The literature refers to this procedure as catalytic etherification or etherification treatment (Xiong et al., 2009; Wang et al., 2010; Hilten et al., 2010; Yu et al., 2009). Esterification and acetalization are the chemical reactions that can take place between bio-oil and methanol or ethanol (Fig.6). Under these circumstances, the reactive compounds found in bio-oil, such as organic acids and aldehydes, undergo reactions with alcohols to form esters and acetals, respectively. Therefore, apart from the reduction in viscosity and the slowing down of the aging process, these factors also result in other favorable alterations, including decreased acidity, enhanced volatility and heating value, and greater compatibility with diesel fuels.

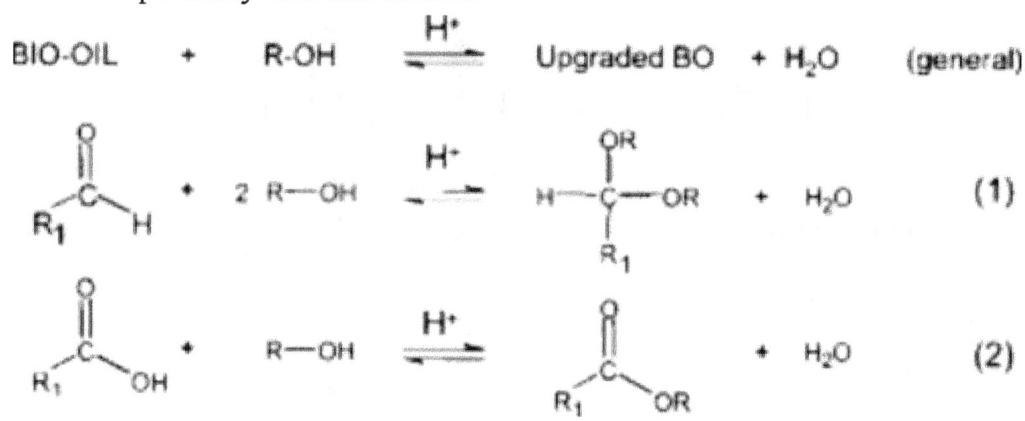

Fig. 6. Reactions involved in bio-oil alcoholysis: (1) acetalization, (2) esterification. (Mahfud et al., 2007)

Most environmental catalysts applied in bio-oil upgrading are heterogeneous catalysts. Solid acid catalysts, solid base catalysts (Zhang et al., 2006), ionic liquid catalysts (Xiong et al., 2009), HZSM-5, and aluminum silicate catalysts were investigated for esterification of bio-oils (Peng et al., 2008, 2009).

Considering the simplicity, the low cost of some solvents such as methanol and their beneficial effects on the oil, this method seems to be the most practical approach for bio-oil quality upgrading.

4.4.5 Process of Emulsification

A technique for utilizing bio-oil as a combustion fuel in transportation or boilers involves creating an emulsion by combining it with other fuel sources. Pyrolysis oils do not mix well with hydrocarbon fuels, although they can be combined with diesel oil by using surfactants to create an emulsion. Several researchers (Chiaramonti et al., 2003a, b; Ikura et al., 2003; Jiang & Ellis, 2010; Garcia-Perez et al., 2010) have studied the process of enhancing bio-oil by combining it with diesel oil by emulsification. The CANMET Energy Technology Centre has devised a method for creating stable microemulsions containing 5-30% bio-oil in diesel fuel (Oasmaa & Czernik, 1999; Ikura, et al., 1998). These emulsions exhibit reduced corrosiveness and demonstrate encouraging igniting properties. Jiang and Ellis (2010) examined the process of emulsifying bio-oil with biodiesel, while excluding the pyrolytic lignin phase. An emulsion of bio-oil and biodiesel was successfully created by utilizing octanol as an emulsifying agent, resulting in a stable mixture. Additionally, the impact of several process variables on the stability of the combination was investigated. The researchers discovered that the most favorable conditions for achieving a consistent blend of bio-oil and biodiesel are as follows: a 4% volume dosage of octanol surfactant, an initial bio-oil/biodiesel ratio of 4:6 by volume, a stirring intensity of 1200 rpm, a mixing time of 15 minutes, and an emulsifying temperature of 30 °C. The emulsion exhibits superior values in acid number, viscosity, and water content in comparison to the original bio-oil. Ikura et al (1998) also reported a decrease in viscosity and corrosiveness of the emulsion.

Chiaramonti et al. (2003b) conducted experiments to evaluate the performance of emulsions made from biomass pyrolysis liquid and diesel fuel when used in engines. Their findings indicate that the primary issue is corrosion, which is expedited by the rapid and turbulent flow in the spray channels. A nozzle made of stainless steel has been constructed and effectively tested. Nevertheless, long-term validation remains necessary. In a more recent study, He et al. (2010) employed a unique high-pressure homogenization (HPH) process to enhance the physicochemical characteristics and storage durability of switchgrass bio-oil. In contrast to the standard emulsification approach involving the mixing of bio-oil with diesel oil, the HPH process enhances the inherent characteristics of bio-oil by reducing its viscosity and enhancing its stability during storage. Nevertheless, there were no alterations observed in the heating value, water content, density, pH value, or ash content. In general, the process of enhancing bio-oil by combining it with diesel oil through emulsification is relatively uncomplicated. It offers a temporary solution for utilizing bio-oil in diesel engines. The emulsions exhibited favorable ignition characteristics, but, they did not meet the desired specifications for fuel parameters such as heating value, cetane number, and corrosivity. Furthermore, this technique necessitated a significant amount of energy for its manufacture. The design, fabrication, and testing of injectors and fuel pumps fabricated from stainless steel or alternative materials are necessary.

4.4.6 Steam reforming

The term 'reforming' initially referred to the thermal conversion of petroleum fractions into more volatile compounds with higher octane values. It encompassed various simultaneous events, including cracking, dehydrogenation, and isomerization (Yaman, 2004). Reforming is the process of converting hydrocarbon gases and vaporized organic molecules into gases that contain hydrogen, such as synthesis gas. Synthesis gas is a combination of carbon monoxide and hydrogen. Synthesis gas can be generated from natural gas using techniques such as steam reforming, where natural gas is reformed in the presence of steam (Klass, 1998). An efficient method for improving the quality of biomass pyrolysis oils involves rapidly heating the biomass, then subjecting it to catalytic steam reforming and shift conversion processes to produce hydrogen from the resulting bio-oil. The National Renewable Energy Laboratory (NREL) conducted thorough investigations into the production of hydrogen from bio-oil reforming. This included studying the reactions in both a fixed bed and a fluidized bed, as documented by Wang et al. (1997, 1998) and Czernik et al. (2007). The study conducted by Ekaterini and Lemonidou in 2008 found that commercial nickel catalysts were highly effective in converting biomass generated liquids.

4.4.7 Chemicals derived from the bio-oils

Various compounds can be derived from bio-oil, including phenols utilized in the resin industry, volatile organic acids, nitrogen heterocycles, and n-alkanes (Ross et al., 2010; Gallivan and Matschei, 1980). In their latest study, Cao et al. (2010) successfully isolated triacetonamine (TAA) from a bio-oil obtained from the fast pyrolysis of sewage sludge. They achieved a high yield of 27.9% and a high purity of 80.4% by utilizing acetone as the absorption solvent. Hydrothermal bio-oil, with its asphalt content reaching up to 50 wt %, is highly suitable for utilization in the asphalt sector. In a recent study, Fini and her colleagues separated and chemically altered the bio-oil to create a highly efficient alphalt bio-binder. This bio-binder shows great promise in being able to substitute or enhance the use of petroleum-based road asphalt (Fini et al, 2010).

The sole existing commercially significant utilization of bio-oil compounds is in the form of wood flavor or liquid smoke (Mohan et al., 2006).Successful commercialization of specialized chemicals derived from bio-oils necessitates a greater commitment to the development of dependable, cost-effective methods for separation and refining.

5 CONCLUSION

Flash pyrolysis processes are so far the only commercially practiced technology for production of bio-oil or bio-crude from biomass. However, pyrolysis oils consist of high oxygen/water contents and hence only about half the caloric value of petroleum (20-25 MJ/kg. In addition, they are strongly acidic and corrosive. Hydrothermal liquefaction with a suitable solvent (water or organics) is superior to pyrolysis. It can potentially produce liquid oils with much higher caloric values. In particular, liquefaction to produce bio-oil from organic wastes is a promising way to not only create value, but also reduce pollutants associated with sludge. There are intensive studies on bio-oil upgrading and several techniques have been developed.

Solvent addition (esterification) appears to be the most practical approach due to simplicity, the low cost of some solvents and their beneficial effects on the oil properties. However, none of these bio-oil upgrading techniques has been commercialized due to low biofuel efficiency and their limitations. Therefore, novel refinery processes are needed to systematically upgrade bio-oils into transportation fuels that have desirable qualities, while producing other value-added co-products to make the economics work.

REFERENCES

Ancheyta, J. & Speight, J.G. (2007). Hyroprocessing of Heavy Oils and Residua. CRC press, ISBN -13: 978-0-8493-7419-7, FL, U.S.A

Appell, H.R., Fu Y.C., Friedman S., Yavorsky P.M., and Wender I. 1980. Converting of organic wastes to oil: a Replenishable Energy Source. U.S. Bureau of Mines. Report of Investigations 7560. Washington, DC.

Bao,G.R. ; Saka, S. & Wang, H. (2008). Cellulose Decomposition Behavior in Hot- Compressed Arotic Solvents. Sci. China Ser B-Chem. Vol.51, No.5, pp.479-486.

Bridgwater, A.V.; Peacocke, G.V.C. (2000). Fast pyrolysis processes for biomass. Renewable and Sustainable Energy Reviews, Vol.4, No.1, pp.1-73.

Bohlmann, B.J.T., Lorth C.M., Drews A., and Buchholz, R. 1999. Microwave high pressure thermochemical conversion of sewage sludge as an alternative to incineration. Chem.Eng.Technol. 21(5): 404-409.

Boucher, M.E.; Chaala, A. & Roy, C. (2000). Bio-oils obtained by vacuum pyrolysis of softwood bark as a liquid fuel for gas turbines. Part I: Properties of bio-oil and its blends with methanol and a pyrolytic aqueous phase. Biomass & Bioenergy, Vol.19, No.5, pp. 337–350.

Cao, J.P.; Zhao, X.; Morishita, K.; Li, L.; Xiao, X.; Obara,R.; Wei, X.& Takarada, T. (2010). Triacetonamine formation in a bio-oil from fast pyrolysis of sewage sludge using acetone as the absorption solvent. Bioresoure Technology, Vol. 101, pp. 4242-4245.

Carrillo, F.; Lis, M.J.; Colom, X.; López-Mesas, M. & Valldeperas, J. (2005). Effect of alkali pretreatment on cellulase hydrolysis of wheat straw: kinetic study. Process Biochem. 40:3360–3364.

Casey, E.; Sedlak, M.; Ho, W.Y. N. & Mosier, S. N. (2010). Effect of acetic acid and pH on the cofermentation of glucose and xylose to ethanol by a genetically engineered strain of Saccharomyces cerevisiae. FEMS Yeast Research.; 10 (4): 385

Cemek, M. & Kucuk, M.M. (2001). Liquid products from verbascum stalk by supercritical fluid extraction. Energy Conver. Manage., Vol.42, pp.125-130.

Chiaramonti, D.; Bonini, M.; Fratini, E.; Tondi, G.; Gartner, K.; Bridgwater, A. V.; Grimm, H P.; Soldaini, I.; Webster, A.; Baglioni, P. (2003a). Development of emulsions from biomass pyrolysis liquid and diesel and their use in engines—Part 1 : emulsion production. Biomass Bioenergy, Vol. 1, pp. 85– 99

Chiaramonti, D.; Bonini, M.; Fratini, E.; Tondi, G.; Gartner, K.; Bridgwater, A. V.; Grimm, H. P.; Soldaini, I.; Webster, A.; Baglioni, P. (2003b). Development of emulsions from biomass pyrolysis liquid and diesel and their use in engines—Part 2: tests in diesel engines. Biomass Bioenergy, Vol. 1, pp.101– 111.

Chornet, E., and R. P. Overend. (1985). Biomass liquefaction: an overview. In Fundamentals of Thermochemical Biomass Conversion, eds. R. P. Overend, T. A. Milne, and L. K. Mudge, 967-1002. New York: Elsevier Applied Science.

Czernik, S.; Evansa, R. & Frencha, R. (2007). Hydrogen from biomass-production by steam reforming of biomass pyrolysis oil. Catalysis Today, Vol.129, No. 3-4, pp. 265-268

Demirbaş, A. (2000a). Mechanisms of liquefaction and pyrolysis reactions of biomass. Energy Conversion & Management, Vol.41, pp.633-646.

Demirbaş, A. 2000b. Effect of lignin content on aqueous liquefaction products of biomass. Energy Conversion and Mgmt, Vol. 41, No. 15, pp. 1601-1607.

Demirbaş, A. & Celik, A. (2005a). Degradation of poplar and spruce wood chips usingalkaline glycerol. Energy Sources, Part A, Vol. 27, No.11, pp. 1073-1084.

Demirbaş, A.; Balat, M. & Bozbas, K. (2005b). Direct and catalytic liquefaction of woodspecies in aqueous solution. Energy Sources Vol.27, No.11, pp. 271-277.

Demirbaş, A. (2008). Liquefaction of biomass using glycerol. Energy Sources, Part A, Vol.30, No. 12, pp.1120-1126.

Diebold, J.P.; Czernik, S. (1997). Additives to lower and stabilize the viscosity of pyrolysis oils during storage. Energy & Fuels, Vol.11, pp. 1081-1091.

Ekaterini C.V. and Lemonidou, A.A. (2008). Hydrogen production via steam reforming of bio-oil components over calcium aluminate supported nickel and noble metal catalysts. Applied Catalysis A: General , Vol.351, No.1, pp. 111-121.

Elliott, D.C. ; Sealock, L.J. & Butner, R.S. (1988). Product Analysis from Direct Liquefaction of Several High-moisture Biomass Feedstocks, In Pyrolysis Oils from Biomass: Producing, Analyzing, and Upgrading, pp.179-188. J. Soltes and T. A. Milne, eds. Denver, Colo.: Am. Chem. Soc.

Fini, E.H. ; Yang S. ; Xiu, S. & Shahbazi, A. (2010). Characterization and Application of Manure-Based Bio-binder in Asphalt Industry. Transportation Research Board, 89th Annual Meeting January 13-17, 2010, Washington, D.C.pp.14

Gallivan, R.M. & Matschei, P.K. (1980). Fraction of oil obtained by pyrolysis of lignocellulosic materials to recover a phenolic fraction for use in making phenolic fraction for use in making phenol-formaldehyde resins. US Patent No. 42009647, American Can Co.

Gan, J.; Yuan, W.; Nelson, N.O.& Agudelo, S. C. (2010). Hydrothermal Conversion of Corn Cobs and Crude Glycerol. Biological Engineering, Vol. 2, No. 4, pp. 197-210.

Garcia-Perez, M.; Shen, J.; Wang, X. & Li, C. (2010). Production and fuel properties of fast pyrolysis oil/ bio-diesel blends. Fuel Processing Technology, Vol. 91, No. 3, pp. 296-305

Gharieb, H.K.; Faramawy, S. & Zaki, N.N. (1995). Liquefaction of cellulosic waste V. Water formation and evaluation of pyrolytic char as a byproduct of pyrolysis reaction. Fuel Sci. Technol. Int. 13 (7) (1995), pp. 895–909.

Goudriaan, F., Beld B.van de, Boerefijn F.R., Bos G.M., Naber J.E., Wal S.van der, and Zeevalkink, J.A. (2000). Thermal efficiency of the HTU process for biomass liquefaction. In Proceedings of the Progress in Thermochemical Biomass Conversion Conference, pp.1312-1325.

Gupta, R. (2008). Alkaline Pretreatment OF Biomass For Ethanol Production And Understanding The Factors Influencing The Cellulose Hydrolysis. Ph.D. Dissertation, Auburn University

He, R.; Ye, X.P.; Harte, F.& English, B. (2009). Effects of high-pressure homogenization on physicochemical properties and storage stability of switchgrass bio-oil. Fuel Processing Technology, Vol.90, pp. 415-421.

Heitz, M.; Brown, A. & Chornet, E. (1994). Solvent effects on liquefaction: Solubilizationprofiles of a Canadian prototype wood, Populus deltoides, in the presence of differentsolvents. Canadian J. Chem. Eng, Vol. 72, No. 6, pp. 1021-1027.

Hilten,R.N.; Bibens, B.P.; Kastner, J.R. & Das, K.C. (2010). In-Line Esterification of Pyrolysis Vapor with Ethanol Improves Bio-oil Quality. Energy Fuels, Vol. 24, pp. 673–682

Hill, J.; Nelson, E.; Tilman, D.; Polasky, S. & Tiffany, D. (2006). Environmental, economic and energetic costs and benefits of biodiesel and ethanol biofuels. Proc Natl Acad Sci USA. 103: 11206–11210

Huang, H.; Ramaswamy, S.; Tschirnera, U.W. & Ramaraob, B.V. (2008). A review of separation technologies in current and future biorefineries. Separation and Purification Technology. 62: 1-21

Huber G. W. & Dumesic, J. A. (2006). An overview of aqueous-phase catalytic processes for production of hydrogen and alkanes in a biorefinery. Catalysis Today, Vol. 111, No. 1-2, pp. 119-132.

Ikura, M.; Stanciulescu, M.; Hogan, E. (2003). Emulsification of pyrolysis derived bio-oil in diesel fuel Biomass Bioenergy, Vol. 3, pp. 221– 232.

Ikura, M.; Slamak, M.; Sawatzky, H. (1998). Pyrolysis liquid-in-diesel Oil Microemulsions. U.S.Pat. 5,820,640.

Itoh S., A. Suzuki, T. Nakamura, and S.Yokoyama. (1994). Production of heavy oil from sewage sludge by direct thermochemical liquefaction. Desalination. 98:127-133.

Jiang, X., and Ellis, N. (2010). Upgrading Bio-oil through Emulsification with Biodesel : Mixture Production. Energy & Fuels, Vol.24, pp.1358-1364.

Karagöz, S.; Bhaskar, T.; Muto, A.; Sakata, Y.; Oshiki, T. & Kishimoto, T. (2005). Low temperature catalytic hydrothermal treatment of wood biomass: analysis of liquid products. Chemical Engineering Journal, Vol. 108, pp.127-137.

Kim, Y.; Mosier, N.S. & Ladisch, M.R. (2008). Effect of Liquid Hot Water Pretreatment on Switchgrass Hydrolysis. The 2008 AIChE Annual Meeting. Philadelphia PA Klass, D.L. (1998). Biomass for renewable energy, fuels, and chemicals. San Diego, CA: Academic Press.

Knappert, D.R.; Grethlein, H.E. & Converse, A.O. (1981). Partial acid hydrolysis of poplar wood as a pretreatment for enzymatic hydrolysis. Biotechnology Bioengineering. 11:67-77

Kücük, M. M. (2005). Delignification of biomass using alkaline glycerol. Energy Sources, Part AVol., 27, No. 13, pp. 1245-1255.

Kranich, W.L. (1984). Conversion of sewage sludge to oil by hydroliquefaction. U.S. Enviromental Protection Agency. EPA-600/2-84-010. Cincinnati,OH Laser, M.; Schulman, D.; Allen, S.G.; Lichwa, J.; Antal, Jr. M.J. & Lynd, L.R. (2002). A comparison of liquid hot water and steam pretreatments of sugar cane bagasse for bioconversion to ethanol. Bioresource Technology. 81(1): 33-44

Lau, M. W.; Gunawan, C.; Balan, V. & Dale, B. E. (2010). Comparing the fermentation performance of Escherichia coli KO11, Saccharomyces cerevisiae 424A(LNH-ST) and Zymomonas mobilis AX101 for cellulosic ethanol production. Biotechnology for Biofuels. 3:11

Lee, D.; Yu, A.H.C.; Wong, K.K.Y. & Saddler, J.R. (1994). Evaluation of the enzymatic susceptibility of cellulosic substrates using specific hydrolysis rates and enzyme adsorption. Applied Biochemistry and Biotechnology. 45/45: 407–415

Li, D.; Zhao, Y.; Yao, F. & Guo, Q. (2009). Green Solvent for Flash Pyrolysis Oil SeparationEnergy & Fuels, Vol.23, pp. 3337–3338

Li, Y.; Ruan, R.; Chen, P.L.; Liu, Z.; Pan, X.; Lin, X.; Liu, Y.; Mok, C.K. & Yang, T. (2004). Enzymatic hydrolysis of corn stover pretreated by combined dilute alkaline treatment and homogenization. Transactions of the ASAE, 47: 821-825

Liu, Z. & Zhang, F. (2008). Effects of Various Solvents on the Liquefaction of Biomass to Produce Fuels and Chemical Feedstocks. Energy Conversion and Management, Vol.49, pp.3498-3504.

Lima, P.; Bonarini, A. & Mataric, M. (2004). Application of Machine Learning, InTech, ISBN 978-953-7619-34-3, Vienna, Austria

Lloyd T.A. & Wyman C.E. (2005). Combined sugar yields for dilute sulfuric acid pretreatment of corn stover followed by enzymatic hydrolysis of the remaining solids. Bioresour Technol. 96(18):1967-77

Lu, Q.; Li, W.Z. & Zhu, X.F. (2009). Overview of fuel properties of biomass fast pyrolysis oils. Energy Conversion and Management, Vol.50, No.5, pp.1376-1383.

Lynd, L.R.; Elander, R.T & Wyman, C.E. (1996). Likely features and costs of mature biomass ethanol technology. Applied Biochemistry and Biotechnology. 57/58: 741–761

Mazaheri, H. ; Lee, K.T. ; Bhatia, S. & Mohamed A.R. (2010). Sub/supercritical Liquefaction of Oil Palm Fruit Press Fiber for the Production of Bio-oil: Effect of solvents. Bioresource Technology. Vol.101, No.19, pp.7641-7647.

Mahfud, F.; Meli_an-Cabrera, I.; Manurung, R.; Heeres, H. (2007). Biomass to fuels: Upgrading of flash pyrolysis oil by reactive distillation using a high boiling alcohol and acid catalysts. Process Saf. Environ. Prot. Vol. 85, No.5, 466–472.

Minowa, T.; Kondo, T. & Sudirjo, S. T. (1998). Thermochemical liquefaction ofIndonesian biomass residues. Biomass and Bioenergy, Vol. 14, No.5-6, pp. 517-524.

Minami, E., and Saka S. (2003). Comparison of the Decomposition Behaviors of Hard Wood and Soft Wood in Supercritical Methanol. J Wood Sci. Vol. 49, pp. 73–78.

Minami, E., and Saka S. (2005). Decomposition Behavior of Woody Biomass in Water-added Supercritical Methanol. J Wood Sci. Vol. 51, pp. 395–400.

Minowa T., Murakami M., Dote Y., Ogi T., and Yokoyama S. (1995). Oil production from garbage by thermochemical liquefaction. Biomass and Bioenergy. 8(2):117-120.

Mohan,D; Pittman, C.U. & Steele, P.H. (2006). Pyrolysis of Wood/Biomass for Bio-oil: A Critical Review. Energy & Fuels, Vol. 20, 848-889.

Mosier, N.; Hendricksona, R.; Ho, N.; Sedlaka, M. & Ladisch, M.R. (2005). Optimization of pH controlled liquid hot water pretreatment of corn stover. Bioresource Technology 96: 1986-1993

Mosier, N.; Wyman, C.; Dale, B.; Elander, R,; Lee, Y.Y.; Holtzapple, M. & Ladisch, M. (2005). Features of promising technologies for pretreatment of lignocellulosic biomass. Bioresour Technol. 96(6):673-86

Nava, R. ; Pawelec, B. ; Castañoc, P. ; Álvarez-Galvánb, M.C. ; Loricerab, C.V. & Fierrob, J.L.G. (2009). Upgrading of bio-liquids on different mesoporous silica-supported CoMo catalysts. Applied Catalysis B : Enviromental, Vol. 92, pp. 154-167.

Oasmaa, A. & Czernik, S. (1999). Fuel Oil Quality of Biomass Pyrolysis Oils - State of the Art for the End Users. Energy & Fuels, Vol.13, pp.914-921.

Oasmaa, A.; Kuoppala, E.; Selin, J.F.; Gust, S. & Solantausta, Y. (2004). Fast pyrolysis of forestry residue and pine. 4. Improvement of the product quality by solvent addition. Energy & Fuels, Vol. 18 No.5, pp. 1578–1583.

Ocfmia, K. S.; Zhang, Y. & Funk, T. (2006). Hydrothermal processing of swine manure into oil using a continuous reaction system: Development and testing. Transaction of the ASABE, Vol.49, No. 2, pp. 533-541.

Ogi, T.; Minowa, T.; Dote, Y. & Yokoyama, S. Y. (1994). Characterization of oil produced byhe direct liquefaction of Japanese oak in an aqueous 2-propanol solvent system. Biomassand Bioenergy, Vol.7, No. 1-6: 193-199.

Perlack R.D.; Wright, L.L.; Turhollow, A.F.; Graham, R.L., Stokes B. J. & Erbach, D.C. (2005). Biomass as feedstock for a bioenergy and bioproducts industry: The technical feasibility of a billion-ton annual supply. Sponsored by USDOE and USDA2005.

Peng, J.; Chen, P.; Lou, H.; Zheng, X. (2009). Catalytic upgrading of biooil by HZSM-5 in sub- and super-critical ethanol. Bioresour. Technol. Vol.100, 3415–3418.

Peng, J.; Chen, P.; Lou, H.; Zheng, X. M. (2008). Upgrading of bio-oil over aluminum silicate in supercritical ethanol. Energy & Fuels, Vol.22, pp. 3489–3492.

Radlein, D.J.; Piskorz, J. & Majerski, P. (1996). Method of Upgrading Biomass Pyrolysis Liquids for use as Fuels and as Sources of Chemicals Reaction with Alcohol, in Patent, E. (ed.), Patent: CA2165858.

Ross, A.B.; Biller, P.; Kubacki, M.L.; Lin, H.; Lea-Langton, A. & Jones, J.M.. (2010). Hydrothermal processing of microalgae using alkali and organic acids. Fuel, Vol. 89, pp. 2234–2243.

Smith, W.J. & Consultancy, T. (2007). Mapping the Development of UK Biorefinery Complexes, NNFCC.

Suzuki A., Nakamura T., and Yokoyama S. (1990). Effect of operating parameters on thermochemical liquefaction of sewage sludge. Journal of Chemical Engineering of Japan. 23(1): 6-11.

Tang, Z.; Lu, Q.; Zhang, Y.; Zhu, X.F. & Guo, Q.X. (2009). One step bio-oil upgrading through hydrotreatment, esterification and cracking. Ind. Eng. Chem. Res., Vol. 48, pp. 6923-6929.

Tang, Y.; Yu, W.; Mo, L.; Lou, H. & Zheng, X. (2008). One-step hydrogenation-esterification of aldehyde and acid to ester over bifunctional Pt catalysts:A model reaction as novel route for catalytic upgrading of fast pyrolysis bio-oil. Energy Fuels, Vo. 22, No.5, pp.3484–3488.

Vasan, P. T.; Piriya, P. S.; Prabhu D. I. & Vennison, S. J. (2011). Cellulosic ethanol production by Zymomonas mobilis harboring an endoglucanase gene from Enterobacter cloacae. Bioresource Technolog. 102: 85–2589

Wang, D.; Czernik, S. & Montané, D. (1997). Biomass to hydrogen via pyrolysis and catalytic steam reforming of the pyrolysis oil and its fractions. Ind Eng Chem Res, Vol.36, pp. 1507–1518.

Wang, D.; Czermik, S.& Chornet, E. (1998). Production of hydrogen from biomass by catalytic steam reforming of fast pyrolytic oils. Energy Fuels, Vol. 12, pp. 19–24.

Wang, J.; Chang, J. and Fan, J. (2010). Upgrading of Bio-oil by Catalytic Esterification and Determination of Acid Number for Evaluating Esterification Degree. Energy & Fuels, Vol. 24, pp. 3251–3255.

Wyman C. E. (1999). BIOMASS ETHANOL: Technical Progress, Opportunities, and Commercial Challenges, Annual Review of Energy and the Environment. 24: 189- 226.

Xiong, W. M.; Zhu, M. Z.; Deng, L.; Fu, Y.; Guo, Q. X. (2009). Esterification of organic acid in bio-oil using acidic ionic liquid catalysts.Energy & Fuels, Vol. 23, pp. 2278–2283.

Xiu, S.; Shahbazi, A.; Shirley, V. & Cheng, D. (2010a). Hydrothermal pyrolysis of swine manure to bio-oil: Effects of operating parameters on products yield and characterization of bio-oil. Journal of Analytical and Applied Pyrolysis, Vol. 88, No.1, pp.73-79.

Xiu, S.; Shahbazi, A.; Wang, L. & Wallace, C. (2010b). Supercritical Ethanol Liquefaction of Swine Manure for Bio-Oils Production. American J. of Engineering and Applied Sciences, Vol.3, No. 2, pp. 494-500.

Xiu, S.; Shabazi, A.; Shirley, V. B.; Mims, M. R. & Wallace, C. W. (2010c). Effectiveness andMechanisms of crude glycerol on the biofuels production from swine manure through hydrothermal pyrolysis. J. Analytical and Applied Pyrolysis 87(2): 194-198.

Xu, C., and Etcheverry, T. (2008). Hydro-liquefaction of woody biomass in Sub- and Super- critical ethanol with iron-based catalysts. Fuel, Vol.87, pp.335-345.

Xu, J.; Jiang, J.; Sun, Y. & Lu, Y. (2008). Bio-oil upgrading by means of ethyl ester production in reactive distillation to remove water and to improve storage and fuel characteristics. Biomas s & Bioenergy, Vol.32, No.11, pp. 1056-1061.

Yang, B. & Wyman, C.E. (2004). Effect of xylan and lignin removal by batch and flow through pretreatment on the enzymatic digestibility of corn stover cellulose. Biotechnology and Bioengineering. 86(1): 88–95

Yang, Y. ; Gilbert, A. & Xu, C. (2009). Production of Bio-crude from Forestry Waste by Hydro-Liquefaction in Sub-/Super-Critical Methanol. AIChE Journal, Vol.55, No.3, pp.807-819.

Yang, Y.; Sharma-Shivappa, R., Burns, J. C. & Cheng, J. J. (2009). Dilute Acid Pretreatment of Oven-dried Switchgrass Germplasms for Bioethanol Production. Energy Fuels. 23 (7): 3759–3766.

Yaman, S. (2004). Pyrolysis of biomass to produce fuels and chemical feedstocks. Energy Conversion and Management,Vol. 45, Issue 5, March 2004, Pages 651-671

Yu, W.; Mo, L.; Lou, H.& Zheng, X. (2009). One-Step Hydrogenation-Esterification of Aldehyde and Acid to Ester over Bifunctional Pt Catalysts: A Model Reaction as Novel Route for Catalytic Upgrading of Fast Pyrolysis Bio-Oil.

Yuan X. Z., H. Li, G. M. Zeng, J. Y. Tong, W. Xie. (2007). Sub- and supercriticalliquefaction of rice straw in the presence of ethanol-water and 2-propanol-water mixture. Energy 32(11):2081-2088.

Zhang, B.; Shahbazi, A.;Wang, L.; Diallo, O. & Whitmore, A. (2010A). Alkali Pretreatment and Enzymatic Hydrolysis of Cattails from Constructed Wetlands. American Journal of Engineering and Applied Sciences. 3(2): 328-332

Zhang,B.; Shahbazi,A.; Wang, L.; Diallo O. & Whitmore A. (2010B). Hot-Water Pretreatment of Cattails for Extraction of Cellulose. Journal of Industrial Microbiology and Biotechnology. DOI: 10.1007/s10295-010-0847-x

Zhang, B.; Shahbazi, A.; Wang, L.; Diallo, O. & Whitmore, A. (2011). Dilute-sulfuric acid pretreatment and cellulose conversion of cattails. ASABE Annual Meeting, Louisville, Kentucky

Zhang, Q.; Chang, J.; T. Wang & Xu, Y. (2007). Review of biomass pyrolysis oil properties and upgrading research. Energy Conversion & Management, Vol.48, pp.87-92.

Zhao, Y.; Wang, Y.; Zhu, J.Y.; Ragauskas, A.& Deng, Y. (2008). Enhanced enzymatic hydrolysis of spruce by alkaline pretreatment at low temperature. Biotechnology and bioengineering. 99(6): 1320-1328

CHAPTER-8

UTILIZATION OF DIGESTERS' WASTE FOR BIOGAS PRODUCTION

1. INTRODUCTION: AN OVERVIEW

1.1. Is digestate, the waste produced by digesters, a highly effective organic fertilizer?

The general consensus among bio-power engineers and in literature is that the waste produced from digesters in biogas production is a highly effective fertilizer. Additionally, anaerobic digestion is considered to be a process that enhances the fertilizing properties of organic materials used in biogas production. These beliefs are seemingly derived from the observation that in the anaerobic stabilization of sludge, the ratio of organic to mineral components in the dry matter is roughly 2:1, but decreases to 1:1 following methanization. During the anaerobic digestion process, a portion of the organic dry matter in sludge is lost, resulting in a 40% reduction in the weight of the original dry matter. This reduction leads to an increase in the concentration of nutrients that were originally present. Anaerobic digestion primarily releases ammonium nitrogen from the original material, which mostly enriches the liquid phase due to its solubility. This process does not actually affect the levels of other nutrients (Straka 2006). The belief that waste from anaerobic digestion serves as a highly effective fertilizer is further supported by the observation of fertilized regions. The growths have a vibrant green color and possess a succulent texture. Their fresh appearance is a characteristic indication of mineral nitrogen, which leads to increased water retention by plants. Nevertheless, the dry matter content undergoes few changes, indicating that fertilization is ineffective. In order for organic matter to be classified as organic fertilizer, it must meet a fundamental requirement: it must be readily biodegradable by microbes, hence providing the essential energy for soil microorganisms.

1.2. Soil organic matter undergoing mineralization

This microbial transformation of organic matter in soil is mineralisation when organic carbon of organic substances is transformed to CO_2 and from mineralised organic matter those mineral nutrients are released that were already contained in organic matter in mineral (ionic) form and those that were in it in organic form. CO_2 is an important fertiliser in agriculture; it is the basic component for photosynthetic assimilation, for the formation of new organic matter produced by plants. As plants can take up only nutrients in mineral form (K+, NH4+, NO3-, Ca2+, Mg2+,H2PO4-, HPO42-, SO42-etc.) and nutrients in organic form (e.g. protein nitrogen, phosphorus of various organophosphates), it is not accessible to plants, and besides its main function – energy production for the soil microedaphon – the mineralization of organic matter in soil is an important source of mineral nutrients for plants. It is applicable solely on condition that organic matter in soil is easily mineralisable, i.e. degradable by soil microorganisms.

1.3. The benefits of using mineralising organic fertiliser for farmers: energy for soil microorganisms and release of mineral nutrients for plant nutrition

What are the key attributes we value in organic fertilizers? Are the benefits of increased energy and improved microbial activity in soil worth the cost of supplying mineral nutrients? Regrettably, simplistic economic viewpoints lead individuals who only scratch the surface to prioritize the benefits derived from mineral nutrients that are generated from organic matter. Calculating such a benefit is likewise straightforward. Quantifying the benefits derived from an enhanced microbial activity in soil is challenging and prone to significant inaccuracies. However, a competent manager will clearly prioritize such a benefit. The microbial activity of soil is a crucial factor in determining soil productivity. It has a significant impact on the physical properties of soil, as well as the levels of air and water content in the soil. Additionally, it has a role in the retention of nutrients in the soil for plant nutrition, as well as the potential loss of these nutrients by elution from the soil into groundwater. A biological factor is one of the primary components of the soil formation process. Without this factor, the material would not qualify as soil, but rather as a parent rock or a soil-forming substrate at best. Therefore, it may be stated that the major role of organic fertilizer is to stimulate microedaphon, rather than the release of mineral nutrients for plant utilization during the mineralisation process in the soil. However, this release of mineral nutrients does have a beneficial economic effect. The impact of mineral nutrients can be substituted by mineral fertilizers, but the energetic impact on soil microbial activity is vital and cannot be replaced.

1.4. What factors affect the quality of digestate as a fertilizer?

The digestate, the waste from digesters during biogas production, is composed of solid phase and liquid phase (fugate). We have demonstrated that the solid phase of the digestate is not an organic fertilizer because its organic matter is very stable and so it cannot be a relatively expeditious source of energy for the soil microedaphon (Kolář et al. 2008). Neither is it a mineral fertilizer because available nutrients of the original raw material and also nutrients released from it during anaerobic digestion passed to the liquid phase – fugate. The digestate, and naturally the fugate, have a low content of dry matter (fugate 0.8 – 3% by weight) and this is the reason why analytical data on the ones to tens of weight % of available nutrients given in dry matter foster an erroneous opinion in practice that these wastes are excellent fertilizers. In fact, fugates are mostly highly diluted solutions in which the content of the nutrients that are represented at the highest amount, mineral nitrogen, is only 0.04 – 0.4% by weight. The surplus of water during fertilization with this waste increases the elution of this nutrient in pervious soils while in less pervious soils the balance between water and air in the soil is impaired, which will have negative consequences. The quality of the digestate as an organic fertiliser (labile, not organic material that is hard to decompose) substantially influences not only the microbial decomposability of the input material but also the level of anaerobic digestion in the digester. In the past when the sludge digestion was carried out in municipal waste treatment plants in digesters at temperatures of 18°C-22°C (psychrophilic regime), the decomposability of the substrate after fermentation was still good, therefore the digested sludge was a good organic fertiliser. These days we work with less decomposable substrates in mesophilic ranges (around 40°C) or even in thermophilic conditions. The degree of decomposition of organic matter during fermentation is consequently high and the digestate as organic fertiliser is practically worthless.

1.5. An optimistic outlook - IFBB procedure

It would be ideal to realize biogas production from the liquid phase only – it would be possible to introduce high performance UASB (Upflow Anaerobic Sludge Blanket) digesters and to achieve the large saving of technological volumes but the concentration of substances in the liquid phase should have to be increased. The solid phase of substrates, which cannot be applied as an organic fertilizer after the fermentation process, would be used as biomass for the production of solid biofuels in the form of pellets or briquettes. But it would be necessary to reduce its chlorine content to avoid the generation of noxious dioxins and dibenzofurans during the burning of biofuel pellets or briquettes at low burning temperatures of household boilers and other low-capacity heating units.Wachendorf et al. (2007,2009) were interested in this idea and tried to solve this problem in a complex way by the hot-water extraction of the raw material (at temperatures of 5ºC, 60ºC and 80ºC) followed by the separation of the solid and liquid phase by means of mechanical dehydration when a screw press was used. This procedure is designated by the abbreviation IFBB (Integrated Generation of Solid Fuel and Biogas from Biomass). These researchers successfully reached the transfer ratio of crude fibre from original material (grass silage) to liquid phase only 0.18, which is desirable for biogas production, but for more easily available organic substances influencing biogas production, e.g. nitrogen-free extract, the ratio is 0.31. The transfer of potassium, magnesium and phosphorus to the liquid phase ranged from 0.52 to 0.85 of the amount in fresh matter, calcium transformation was lower, at the transfer ratio 0.44 – 0.48 (Wachendorf et al. 2009). Transformation to the liquid phase was highest in chlorine, 0.86 of the amount in original fresh matter, already at a low temperature (5ºC). The transfer of mineral nitrogen to the liquid phase before the process of anaerobic digestion is very low because there is a minute amount of mineral N in plant biomass and the major part of organic matter nitrogen is bound to low-soluble proteins of the cell walls. Nitrogen from these structures toughened up by lignin and polysaccharides is released just in the process of anaerobic digestion. Because in the IFBB process also organic nitrogen compounds (crude protein – nitrogen of acid detergent fibre ADF) are transferred to the liquid phase approximately at a ratio 0.40, the liquid phase, subjected to anaerobic digestion, is enriched with mineral nitrogen.

LikeWachendorf et al. (2009), we proceeded in the same way applying the IFBB system for the parallel production of biogas and solid biofuels from crops grown on arable land. The IFBB technological procedure is based on a high degree of cell wall maceration as a result of the axial pressure and abrasion induced with a screw press.

2. SIGNIFICANT ISSUES

2.1. The initial issue: organic matter of digestate is poorly degradable in soil, its labile fractions were already utilised in a digester

The point is that the digestate is not an organic fertiliser because its organic substance is poorly degradable. But its liquid fraction contains a small amount of mineral nutrients, mainly of nitrogen. The fugate (and also the digestate) can be considered as a very dilute mineral fertiliser, nitrogenous fertiliser. However, the agriculture sector is exposed worldwide to an enormous pressure on economic effectiveness while the costs of machinery, fuels and agricultural labour force are very high in relation to the price of agricultural products. Therefore the chemical industry helps farmers to save on transportation and application costs incurred by fertilisation when highly concentrated mineral fertilisers are produced. Even though they are substantially more expensive, from the aspect of cost accounting their use will finally pay off. Before the manufacture of town gas from coal using the ammonia water ended, farmers took the waste containing 1% of ammonia nitrogen only exceptionally even though it was practically free of charge.

With the current output of a biogas plant 526 kW (Chotýčany, South Bohemia) and daily dose of a substrate to the digester 46 t and practically identical production of digestate the daily production of mineral nitrogen is approximately 40 kg, which amounts to a relatively high value per year, almost 15 t of mineral nitrogen, but the dilution is unacceptable.

2.2. The secondissue: the digestate contains much water and therefore the solution with plant nutrients is very dilute.

If this waste is applied as a fertiliser, the water surplus increases the elution of this nutrient into the bottom soil in pervious soils. In impervious soils and in less pervious soils the imbalance between water and air in the soil is deteriorated with all adverse consequences: aerobiosis restriction, reduction in the count of soil microorganisms, denitrification and escape of valuable nitrogen in the form of N2or N-oxides into the atmosphere. Soil acidification takes place because organic substances are not mineralised under soil anaerobiosis and they putrefy at the simultaneous production of lower fatty acids. These soil processes result in a decrease in soil productivity. Currently, its probability is increasingly higher for these reasons:

- As a consequence of global acidification the frequency of abundant precipitation is higher in Europe throughout the year.
- As a result of rising prices of fuels, depreciation on farm machinery and human labour force farmers apply digestates or fugates in the closest proximity of a biogas plant. It causes the overirrigation of fertilised fields even though the supplied rate of nitrogen does not deviate from the required average.

The problem of an excessively high irrigation amount has generally been known since long: it occurred in Berlin and Wroclaw irrigation fields after irrigation with municipal waste water in the 19thand 20thcentury, in the former socialist countries after the application of agricultural and industrial waste waters and of slurry from litterless operations of animal production. Even though nobody surely casts doubt on the fertilising value of pig slurry or starch-factory effluents, total devastation of irrigated fields and almost complete loss of their potential soil productivity were quite normal phenomena (Stehlík 1988).

2.3. Key problems to address

A further part of this study should help solve these crucial problems:

- What is the rational utilisation of digestate and/or fugate and separated solid fraction of digestate in the agriculture sector that are generated by current biogas plants if we know that their utilisation as fertilisers is rather problematic?
- What are the prospects of utilisation of wastes from biogas production and what modifications in the technology of biogas production from agricultural wastes should be introduced?
- What problems should be solved by researchers so that the promising utilisation of wastes from biogas production could be realised?
- What is the optimum form of utilisation of wastes from biogas plants and why?

3. INFORMATION

3.1. Current optimum utilisation of digestate from biogas plants in the agriculture sector

3.1.1. Biodegradability (lability) and stability of organic matter

How many labile components of organic matter are lost during anaerobic digestion in a biogas plant can be demonstrated by determination of the degree of organic matter lability. For this purpose a number of methods can be used that are mostly based on resistance to oxidation or on resistance to hydrolysis.

Oxidation methods are based on oxidation with chemical oxidants, e.g. with a solution of K2Cr2O7in sulphuric acid at various concentrations – 6 M + 9 M + 12 M (Walkley 1947,Chan et al. 2001) or with a neutral solution of KMnO4at various concentrations (Blair et al. 1995,Tirol-Padre, Ladha 2004). The degree of organic matter lability is evaluated from the amount of oxidizable carbon in per cent of its total amount in particular variously aggressive oxidation environments or the reaction kinetics of the observed oxidation reaction is examined while its characteristic is the rate constant of the oxidation process.

In 2003 was proposed and tested the method to evaluate the kinetics of mineralisation of the degradable part of soil organic matter by the vacuum measurement of biochemical oxygen demand (BOD) of soil suspensions using an Oxi Top Control system of the WTW Merck Company, designed for the hydrochemical analysis of organically contaminated waters (Kolář et al. 2003). BOD on the particular days of incubation is obtained by these measurements whereas total limit BODtcan be determined from these data, and it is possible to calculate the rate constant K of biochemical oxidation of soil organic substances per 24 hours as the rate of stability of these substances. A dilution method is the conventional technique of measuring BOD and also rate constants. It was applied to determine the stability of soil organic substances but it was a time- and labour-consuming procedure. The Oxi Top Control method was used with vacuum measurement in vessels equipped with measuring heads with infrared interface indicator communicating with OC 100 or OC 110 controller while documentation is provided by the ACHAT OC programme communicating with the PC, and previously with the TD 100 thermal printer. Measuring heads will store in their memory up to 360 data sentences that can be represented graphically by the controller while it is also possible to measure through the glass or plastic door of the vessel thermostat directly on stirring platforms. The rate of biochemical oxidation of organic substances as the first-order reaction is proportionate to the residual concentration of yet unoxidised substances:

$$dy/dt = K (L - y) = KL_z \quad (1)$$

where:

L = total BOD

y= BOD at time t

Lz= residual BOD

k, K= rate constants

By integrating from 0 to t of the above relation the following equation is obtained:

$$L_z = L \,.\, e^{-Kt} = L \,.\, 10^{-kt} \quad (2)$$

L^z= L. e-Kt= L. 10-kt(2)

In general it applies for BOD at time t:

$$y = L\,(1 - 10\text{-}kt) \quad (3)$$

where:

y= BOD at time t

L= BODtotal

k= rate constant 24 hrs^{-1}

Used procedure is identical with the method of measurement recommended by the manufacturer in accordance with the Proposal for German Uniform Procedures DEV 46thBulletin 2000 – H 55, also published in the instructions for BOD (on CD-ROM) of WTW Merck Company.

The decomposition of organic matter is the first-order reaction. In these reactions the reaction rate at any instant is proportionate to the concentration of a reactant (see the basic equation dy/dt). Constant k is the specific reaction rate or rate constant and indicates the instantaneous reaction rate at the unit concentration of a reactant. The actual reaction rate is continually variable and equals the product of the rate constant and the instantaneous concentration. The relation of the reaction product expressed by BOD at time t (y) to t is the same as the relation of the reactant (L – y) at time t and therefore the equations are analogical.

$$(L - y) = L \, . \, e\text{-}kt \tag{4}$$

and

$$y = L\,(1 - e\text{-}kt) \tag{5}$$

If in the graph the residual concentration of carbon is plotted on the y-axis in a logarithmic scalelog (L – y)and the time in days from the beginning of experiment is plotted on the x-axis, we will obtain a straight line, the slope of which corresponds to the value -k/2.303.

The quantity of the labile fraction of organic matter can also be assessed by determination of soluble carbon compounds in hot water (Körschens et al. 1990,Schulz 1990) and their quality by determination of the rate constant of their biochemical oxidation (Kolář et al. 2003,2005a,b).

Hydrolytic methods are based on resistance of the organic matter different aggressive ways of hydrolysis that is realised at different temperature, time of action and concentration of hydrolytic agent, which is usually sulphuric acid. Among many variants of these methods the hydrolytic method according toRovira et Vallejo (2000,2002, 2007) inShirato et Yokozawa (2006) modification was found to be the best. This method yields three fractions: labile LP1, semi-labile LP2 and stable LP3. The per cent ratio of these three fractions, the sum of which is total carbon of the sample Ctot, provides a very reliable picture of the degree of organic matter lability.

Of course, there are a lot of methods based on the study of organic matter biodegradability in anaerobic conditions. First of all, it is the international standard ISO CD11734: Water quality – evaluation of the "ultimate" anaerobic biodegradability of organic compounds in digested sludge – Method by measurement of the biogas production, and particularly a very important paper using the Oxi Top Control measuring system manufactured by the German company MERCK for this purpose (Süssmuth et al. 1999).

Tests of methanogenic activity (Straka et al. 2003) and tests examining the activity of a microbial system (Zábranská et al. 1985a, b,1987) are methods that can describe the degree of organic matter lability in its ultimate effect. Our long-time work experiences in the evaluation of a huge amount of various analyses for the study of organic matter lability have brought about this substantial knowledge:

- The study of the ratio of organic matter labile fractions, i.e. of their quantity, is always incomplete. A more authentic picture of the situation can be obtained only if information on the quality of this labile fraction is added to quantitative data. Such a qualitative characteristic is acquired in the easiest way by the study of reaction kinetics of the oxidation process of this fraction. The process of biochemical oxidation and the calculation of its rate constant KBioare always more accurate that the calculation of its rate constant of oxidation by chemical oxidants KCHEM(Kolář et al. 2009a).

- It applies to current substrates for biogas production in biogas plants that with some scarce exceptions the degree of organic matter lability is very similar in both aerobic and anaerobic conditions. In other words: organic matter is or is not easily degradable regardless of the conditions concerned (Kolář et al. 2006).
- A comparison of various methods for determination of organic matter lability and its degradability in the anaerobic environment of biogas plant digesters and also for determination of digestate degradability after its application to the soil showed that hydrolytic methods are the best techniques. They are relatively expeditious, cheap, sample homogenisation and weighing are easy, and the results correlate very closely with methods determining the biodegradability of organic matter directly. E.g. with the exception of difficult weighing of a very small sample and mainly its homogenisation the Oxi Top Control Merck system is absolutely perfect and highly productive – it allows to measure in a comfortable way simultaneously up to 360 experimental treatments and to assess the results continually using the measuring heads of bottles with infrared transmitters, receiving controller and special ACHAT OC programme for processing on the PC including the graph construction. But its price is high, in the CR about 4 million Kč for the complex equipment. Hydrolytic methods require only a small amount of these costs and are quite satisfactory for practical operations (Kolář et al. 2008). However, for scientific purposes we should prefer the methods that determine anaerobic degradability of organic matter, designated by DC.

The substrate production of methane VCH_4S[the volume of produced methane (VCH_4c) after the subtraction of endogenous production of methane (VCH_4e) by the inocula] was determined by an Oxi Top Control Merck measuring system.

The calculation is based on this equation of state:

$$n = p \times V/RT \tag{6}$$

where:

n= number of gas moles

V= volume [ml]

P= pressure [hPa]

T= temperature [°K]

R= gas constant 8.134 J/mol °K

and the number of CO_2 and CH_4 moles in the gaseous phase of fermentation vessels is calculated:

$$n_{CO2\ g\ CH4} = (\Delta p \times Vg/RT) \times 10^{-4} \tag{7}$$

$$\Delta p = p_1 - p_0 \tag{8}$$

where: p_0= initial pressure

Fermentation at 35° C and continuous agitation of vessels in a thermostat lasts for 60 days, the pressure range of measuring heads is 500 – 1 350 kPa and the time interval of measuring pressure changes is 4.5 min. Anaerobic fermentation is terminated by the injection of 1 ml of 19% HCl with a syringe through the rubber closure of the vessel to the substrate. As a result of acidification CO_2 is displaced from the liquid phase of the fermentation vessel. The process is terminated after 4 hours. The number of CO_2 moles is calculated from the liquid phase:

$$nCO2\,l = \{[p2\,(Vg - VHCl) - p1 \times Vg]/RT\} \times 10^{-4} \tag{9}$$

The injection of 1 ml of 30% KOH into the rubber container in the second tube of the fermentation vessel follows. The sorption of CO_2 from the gaseous phase of the vessel is terminated after 24 hours and the total number of CO_2 moles in gaseous and liquid phases is calculated from a drop in the pressure in the vessel:

$$n_{CO2\,l,\,CO2\,g} = \{[p_3\,(V_g - V_{HCl} - V_{KOH}) - p_2\,(V_g - V_{HCl})]/RT\} \times 10^{-4} \tag{10}$$

where:

D_p = difference in pressures hPa

V_g= the volume of the gas space of the fermentation vessel ml

p_1= gas pressure before HCl application hPa

p_2= gas pressure before KOH application hPa

p_3= gas pressure after KOH application hPa

R = gas constant = 8.134 J/mol °K

T = absolute temperature = 273.15 + X °C

V_{HCl}= the volume of added HCl ml

V_{KOH}= the volume of added KOH ml

Based on the results, it is easy to calculate the number of CO_2 moles in the gaseous phase and by the subtraction from nCO_2 g $CH_4$the number of moles of produced methane:

$$n_{CH4} = (n_{CO2\,g\,CH4} + n_{CO2\,l}) - n_{CO2\,l\,CO2\,g} \tag{11}$$

The total number of moles of the gases of transported carbon:

$$n_{CO2\,g\,CH4} + n_{CO2\,l} = n_{total} \tag{12}$$

Baumann's solution A + B in deionised water of pH = 7.0 is used as a liquid medium (Süssmuth et al. 1999).

The standard addition of the inoculum corresponds roughly to an amount of 0.3% by volume (aqueous sludge from the anaerobic tank of the digester). Instead of Baumann's solution it is possible to use a ready-made nutrient salt of the MERCK Company for this system.

The operation of the Oxi Top Control measuring system was described in detail bySüssmuth et al. (1999).

Methane yield was calculated from the substrate production of methane $VCH_4S$by division by the initial quantity of the added substrate:

$$Y_{CH4g} = \frac{(V_{CH4C} - V_{CH4e})}{S} = \frac{V_{CH4S}}{S}[l/g] \tag{13}$$

where:

VCH_4C= methane yield of C-source

VCH_4e= methane yield of the added inoculum

S = substrate quantity at the beginning [g]

Lord's test and other methods suitable for few-element sets and based on the R range of parallel determinations were used for the mathematical and statistical evaluation of analytical results including the computation of the interval of reliability.

Anaerobic degradability is given by the equation:

$$D_c = \frac{C_g}{C_s} . 100 \tag{14}$$

where:

C_s= total C content in the sample

C_g= C content in methane released during the measurement of anaerobic degradability

The value of Cgis computed from the substrate production of methane VCH_{4S}:

$$C_g = \frac{12\ p\ V_{CH4S}}{RT} \tag{15}$$

(because 1 mol CH_4contains 12 g C)

where:

K = temperature (°K)

R = gas constant

P= pressure

VCH_4S= the volume of produced methane after the subtraction of endogenous production by the inoculum from total production

This method, which determines organic matter lability in anaerobic conditions, is so exact that it allows to investigate e.g. the digestive tract of ruminants as an enzymatic bioreactor and to acquire information on its activity, on feed utilisation or digestibility and on the influence of various external factors on the digestion of these animals (Kolář et al. 2010a) or to determine the share of particular animal species in the production of greenhouse gasses (Kolář et al. 2009b).

At the end of this subchapter dealing with the degree of organic matter lability and its changes after fermentation in a biogas plant these experimental data are presented: A mixture of pig slurry and primary (raw) sludge from the sedimentation stage of a municipal waste water treatment plant at a 1 : 1 volume ratio was treated in an experimental unit of anaerobic digestion operating as a simple periodically filled Batch-system with mechanical agitation, heating tubes with circulating heated medium at a mesophilic temperature (40°C) and low organic load of the digester (2.2 kg org. dry matter/m3) and 28-day fermentation. Acid hydrolysis of sludge, slurry and their mixture was done before and after anaerobic fermentation. The hydrolysis of samples was performed with the dry matter of examined sludge and its mixture with pig slurry including the liquid fraction after screening the material through a 250-µm mesh sieve. The method of hydrolysis according toRovira and Vallejo (2000,2002) as modified byShirato and Yokozawa (2006): 300 mg of homogenised sample is hydrolysed with 20 ml of 2.5 M $H_2SO_4$for 30 min at 105ºC in a pyrex tube. The hydrolysate is centrifuged and decanted, the residues are washed with 25 ml water and the wash water is added to the hydrolysate. This hydrolysate is used to determine Labile Pool I (LP I).

The washed residue is dried at 60ºC and hydrolysed with 2 ml of 13 M H_2SO_4overnight at room temperature and continuous shaking. Such an amount of water is added that the concentration of the acid will be 1 M, and the sample is hydrolysed for 3 hours at 105ºC at intermittent shaking. The hydrolysate is isolated by centrifugation and decantation, the residue is washed again with 25 ml of water and the wash water is added to the hydrolysate. This hydrolysate is used for the determination of Labile Pool II (LP II). The residue from this hydrolysis is dried at 60ºC and Recalcitrant Pool (RP) is determined from this fraction. C_{tot} is determined in all three fractions.

Degradability of organic matter of the test materials was studied by modified methods ofLeblanc et al. (2006) used to examine the decomposition of green mulch from Inga samanensis and Inga edulis leaves. These authors conducted their study in outdoor conditions (average annual temperature 25.1ºC) and we had to modify their method in the cold climate of this country. At first, the liquid phase of sludge, slurry and mixture was separated by centrifugation; the solid phase was washed with hot water several times and separated from the solid phase again. By this procedure we tried to separate the solid phase from the liquid one, which contains water-soluble organic substances and mineral nutrients. Solid phases of tested organic materials were mixed with sandy-loamy Cambisol at a 3:1 weight ratio to provide for inoculation with soil microorganisms and volume ventilation of samples with air. After wetting to 50% of water retention capacity the mixtures at an amount of 50 g were put onto flat PE dishes 25 x 25 cm in size. The material was spread across the surface of the dish. Cultivation was run in a wet thermostat at 25ºC, and in the period of 2 – 20 weeks dishes were sampled in 14-day intervals as subsamples from each of the four experimental treatments. The agrochemical analysis of the used topsoil proved that the content of available nutrients P, K, Ca and Mg according to Mehlich III is in the category "high" and pKKCl= 6.3. After drying at 60°C for 72 hours the content of lipids, crude protein, hemicelluloses, cellulose, lignin, total nitrogen and hot-water-insoluble dry matter was determined in the dish contents. After twenty weeks of incubation organic substances were determined in the dish contents by fractionation into 4 degrees of lability according toChan et al. (2001).

The content of hemicelluloses was calculated from a difference between the values of neutral detergent fibre (NDF) and acid detergent fibre (ADF), lignin was calculated from ADF by subtracting the result after lignin oxidation with KMnO4. Because ADF contains lignin, cellulose and mineral fraction, it was possible to determine the cellulose content by ashing the residue in a muffle furnace and by determination of mineral fraction. These methods were described byVan Soest (1963), modifications used by Columbian authors (Leblanc et al. 2006) were reported byLópez et al. (1992).

Ion exchange capacity [mmol.chem.eq./kg] was determined in dry matter of the examined materials according toGillman (1979), buffering capacity was determined in samples induced into the H+-cycle with HCl diluted with water at 1 : 1 and washed with water until the reaction to Cl-disappears. In the medium of 0.2 M KCl the samples were titrated to pH = 7 with 0.1 M NaOH and buffering capacity was calculated from its consumption.

Table.1 shows the analyses of a mixture of pig slurry and primary sludge used in the experiment. Obviously, compared to the values reported in literature our experimental materials had a somewhat lower content of organic substances in dry matter, and perhaps this is the reason why anaerobic fermentation reduced the content of organic substances by 39% only although the usual reduction by 45 – 65% for primary sludge was expected as reported in literature (Pitter 1981) and by 40 – 50% for pig slurry (Stehlík 1988). As a result of the organic dry matter reduction the content of nutrients in sludge after anaerobic fermentation is higher, nitrogen content is lower by about 20%. In this process organic nitrogen is converted to $(NH_4)^2 CO_3$, which partly decomposes into $NH_3 + H_2O + CO_2$ and partly passes into the sludge liquor.Roschke (2003) reported that up to 70% of total nitrogen might pass to the ammonium form at 54% degradation of organic substances of dry matter. Even though concentrations of the other nutrients in dry matter of the aerobically stabilised sludge increased as a result of the organic dry matter reduction, their content in the sludge liquor also increased (Tab. 2).

Table 1. The analysis of experimental pig slurry and primary sludge, mixture of pig slurry and primary sludge before methanisation in a digester and after methanisation in % of dry matter (pig slurry and primary sludge were mixed for anaerobic digestion at a 1:1 volume ratio). (Sample size n = 6, interval of reliability of the mean for a significance level = 0.05)

Organic substances		Pig slurry	Primary sludge	Mixture of slurry and sludge before methanisation	Mixture of slurry and sludge after methanisation
		65.1 2.6	62.7 2.4	64.1 2.4	36.9 1.5
Total nutrients	N	6.2 0.2	2.6 0.1	3.9 0.2	3.1 0.1
	P	1.6 0.1	0.7 0.0	1.1 0.0	1.3 0.1
	K	2.3 0.1	0.2 0.0	1.2 0.0	1.2 0.0
	Ca	2.8 0.1	2.6 0.1	2.5 0.1	2.8 0.1

Taking into account that the amount of water-soluble nutrients in the sludge liquor and organic forms of N and P dispersed in the sludge liquor in the form of colloid sol (but it is a very low amount) is related not only to the composition of the substrate but also to technological conditions of anaerobic digestion, digester load and operating temperature, it is evident that the liquid fraction of anaerobically stabilised sludge contains a certain amount of mineral nutrients, approximately 1 kg N/m3, besides the others, although differences in the concentration of P and K in the liquid fraction before and after fermentation are generally negligible.

Table 2. The analysis of the liquid fraction (sludge liquor) of a mixture of pig slurry and primary sludge from a waste water treatment plant (1 : 1) before fermentation and after fermentation in mg/l. The values A and B express % in the liquid phase of the total amount of sludge before and after fermentation (Sample size n = 5, interval of reliability of the mean for a significance level = 0.05)

	% A	% B	mg/l Before fermentation	mg/l After fermentation
Total N	8.40	55.20	246.2 14.7	994.7 59.6
Ammonia N	52.60	90.80	153.7 8.4	907.2 48.2
Total P	12.20	25.30	134.5 8.7	176.3 11.6
Total K	19.90	28.10	172.9 10.4	184.1 11.0

It is a very low amount, and there arises a question whether the influence of the liquid fraction on vegetation is given by the effect of nutrients or water itself, particularly in drier conditions. After anaerobic digestion the solid phase of sludge still contains a high amount of proteins and other sources of organic nitrogen that could be a potential pool of mineral nitrogen if the degradation of sludge after fermentation in soil is satisfactory.

Table 3. Proportions of the three pools of carbon in experimental materials, as determined by the acid hydrolysis method of Rovira and Vallejo (2002),(Sample size n = 4, interval of reliability of the mean for a significance level = 0.05),(Materials including the liquid fraction were used)

Material	Proportion		
	LP I	LP II	RP
Primary sewage sludge	68 5	23 2	9 1
Pig slurry	59 5	15 2	26 2
Mixture of primary sludge and pig slurry at a 1:1 volume ratio	63 5	20 2	17 1
Mixture of primary sludge and pig slurry at a 1:1 volume ratio after methanisation	18 2	16 1	66 5

The results of hydrolysis in Table 3 prove that pig slurry has 59% of its total carbon in LP I, which indicates great lability, corresponding to the hydrolysability of cereals and grasses according to Shirato and Yokozawa (2006). Primary sewage sludge is still better from this aspect, having almost 70% C in LP I. The degree of lability of the sludge and slurry mixture is relatively high and corresponds to the component ratio. After methanisation carbon content in LP I of the sludge and slurry mixture decreases to less than a third of the original amount and carbon of non-hydrolysable matters increases even almost four times in the RP fraction. The sum of LP I and LP II, i.e. the labile, degradable fraction of carbon compounds of the sludge and pig slurry mixture, was reduced by anaerobic digestion from 83% to 34%, that means approximately by 50%. These are enormous differences and they prove that mainly very labile organic substances are heavily destroyed by the anaerobic process even though a reduction in the content of organic substances during anaerobic fermentation is lower (by 39% in our experiment).

Table 4. The content of selected organic substances (%) and ion exchange and buffering capacity of the solid phase of primary sludge (A), pig slurry (B), sludge and pig slurry mixture at a 1:1 ratio before fermentation (C) and after fermentation (D) before and after 20 weeks of incubation with sandy-loamy Cambisol topsoil at a 3:1 ratio at 25°C in dry matter(Sample size n = 4 /hot-water-soluble dry matter n = 7/, interval of reliability of the mean for a significance level = 0.05)

	I Before incubation (25° C)				II After incubation (25°C, 20 weeks)			
	A	B	C	D	A	B	C	D
Lipids (petroleum ether extractable compounds) %	8.60 0.69	14.27 1.14	10.82 0.86	2.01 0.15	7.97 0.65	13.50 1.09	10.39 0,85	2.08 0,17
Proteins (Berstein) %	13.43 1.30	17.95 1.62	15.31 1.60	8.50 0.93	11.81 1.20	16.10 1.53	13.89 1.42	8.50 0.98
Hemicelluloses %	1.82 0.19	5.03 0.73	3.32 0.61	0.70 0.60	1.43 0.11	4.23 0.51	2.89 0.30	0.69 0.10
Cellulose %	7.45 0.92	11.18 1.33	9.61 1.05	6.03 0.95	5.42 0.82	9.27 0.98	7.96 0.94	6.05 0.83
Lignins %	4.84 0.62	5.16 0.84	4.99 0.75	5.18 0.92	4.83 0.91	5.18 1.07	4.98 0.84	5.20 0.91
Total N %	1.59 0.06	2.70 0.11	2.29 0.10	1.07 0.04	1.51 0.06	2.50 0.11	2.14 0.09	1.08 0.05
Hot-water insoluble dry matter %	98.25 2.94	98.26 2.95	98.25 2.95	98.23 2.92	89.05 2.67	85.17 2.60	87.26 2.58	98.20 2.93
Ion exchange capacity mmol chem. eq./kg	48 3	55 3	53 3	145 9	50 3	61 4	55 4	168 10
Buffering capacitymmol chem. eq./kg	62 4	69 4	65 4	157 9	65 4	72 4	70 4	179 11

Tab. 4shows the analysis of raw materials (sludge and pig slurry) and their mixture before and after anaerobic fermentation while Tab. 5 shows the analysis of their liquid fraction. The same results (Tab. 4)are provided by the incubation of the solid phase of sludge, pig slurry and their mixture before and after anaerobic fermentation when incubated with soil at 25°C and by the contents of lipids, crude protein, hemicelluloses, cellulose, lignin, total nitrogen and hot-water-insoluble dry matter; the same explicit conclusion can be drawn from the results of the fractionation of organic matter lability of the experimental treatments after 20-week incubation with soil according toChan et al. (2001) shown inTab. 5. A comparison of the results in Tab. 3 and 5 indicates that as a result of the activity of microorganisms of the added soil in incubation hardly hydrolysable organic matter was also degraded – differences between the most stable fractions F 3 and F 4 in Tab. 5 are larger by about 73% after anaerobic fermentation while in the course of acid chemical hydrolysis the content of non-hydrolysable fraction was worsened by anaerobic fermentation because it increased by about 290%. But it is a matter of fact that the soil microorganisms are not able to stimulate the anaerobically fermented sludge to degradation as proved by more than ¾ of total carbon in fraction 4.

Table 5. The fractionation of organic carbon (g/kg) of primary sludge, pig slurry, and sludge and slurry mixture at a 1:1 ratio before fermentation (A) and after fermentation (B) in a mixture with sandy-loamy Cambisol (3 : 1) in dry matter after 20 weeks of incubation at 25°C by the modified Walkley-Black method according to Chan et al. (2001) with a change in H_2SO_4concentration. (The values given in brackets are % of the C fraction in total dry matter carbon) (Sample size n = 5, interval of reliability of the mean for a significance level = 0.05).

	Unfermented primary sludge	Unfermented pig slurry	Mixture A	Mixture B	Soil only
Fraction 1 (12 N H2SO4)	59.84 7.18 (32.00)	55.38 6.52 (28.40)	54.09 6.50 (30.05)	2.65 0.30 (2.60)	1.30 0.17 (7.22)
Fraction 2 (18 N - 12 N H2SO4)	42.45 5.13 (22.70)	35.76 4.26 (18.34)	34.22 4.10 (19.01)	9.28 1.10 (9.07)	0.80 0.09 (4.44)
Fraction 3 (24 N - 18 N H2SO4)	27.34 3.28 (14.62)	20.18 2.53 (10.35)	20.30 2.42 (11.28)	11.13 1.33 (10.91)	3.70 0.44 (20.56)
Fraction 4(TOC = 24 N H2SO4)	57.37 6.85 (30.68)	83.67 10.01 (42.91)	71.39 8.55 (39.66)	78.97 9.40 (77.42)	1.22 1.42 (67.78)

The table results document that 20-week incubation decreased more or less the per cent content of examined organic substances except lignin (total N 5 – 8%, cellulose 17 – 25%, hemicellulose 13–22%, proteins 9–12%, lipids 4–7%, and the content of hot-water-insoluble dry matter by 10 – 15%) factually in all experimental treatments except the treatment of the anaerobically fermented mixture of primary sludge and pig slurry where a reduction in these matters is low or nil. Hence, primary sludge, pig slurry and their mixture can be considered as organic fertilisers but only before anaerobic fermentation. We recorded a substantially lower degree of degradation of selected organic substances in sludge, pig slurry and their mixture during incubation with 25% of sandy-loamy soil (5 – 25%) than didLeblanc et al. (2006) with phytomass of Inga samanensis and Inga edulis leaves, who reported about 50% degradation of total mass, hemicelluloses and nitrogen in mass. We are convinced that it is caused by a very different content of hemicelluloses in our materials compared to the materials used by the above-mentioned authors. No easily degradable hemicelluloses are present in sewage sludge or in pig slurry any longer, and obviously, only more stable forms pass through the digestive tracts of animals and humans. It is also interesting that after anaerobic fermentation and after 20-week aerobic cultivation at 25°C only the compounds (lipids

+ proteins + hemicelluloses in mixture II D account roughly for 11%) that could be considered as labile remained in the mixture of slurry and sludge. These are apparently their more stable forms as confirmed by the results in Tab. 5 which illustrate that to approximately 11% of organic carbon compounds it is necessary to add the % proportions of the first and second fraction on the basis of oxidisability according toChan et al. (2001). Literary sources report that the sum of lipids, proteins and hemicelluloses in the anaerobically stabilised sludge from municipal waste water treatment plants amounts to 13% – 39.6% of dry matter, so it is quite a general phenomenon.

The ion exchange capacity of sludge, pig slurry and their mixture before fermentation, before incubation and after incubation is very low and does not reach the values that are typical of sandy soil. It is increased by anaerobic fermentation along with incubation markedly but practically little significantly to the level typical of medium-textured soils. The same relations were observed for buffering capacity, which is not surprising. The results document that degradability of the organic part of anaerobically stabilised sludge worsened substantially and that it cannot be improved very markedly by the use of soil microorganisms and soil.

We have to draw a surprising conclusion that sludge as a waste from the processes of anaerobic digestion is a mineral rather than organic fertiliser and that from the aspect of its use as organic fertiliser it is a material of much lower quality than the original materials. We cannot speak about any improvement of the organic material by anaerobic digestion at all. Their liquid phase, rather than the solid one, can be considered as a fertiliser. If it is taken as a fertiliser in general terms, we do not protest because besides the slightly higher content of mineral nutrients available to plants (mostly nitrogen) it has the higher ion exchange capacity and higher buffering capacity than the material before anaerobic fermentation, but this increase is practically little significant.

3.1.2. Digestate composting

3.1.2.1. compost?

Just as in the assessment of digestate, the daily practice has greatly simplified the issue by not distinguishing between the main functions of mineral and organic fertilizers. This simplification has also resulted in an illogical situation regarding composting and the use of composts. Compost, in numerous nations, is commonly defined as partially decomposed organic matter, primarily derived from biodegradable trash, that contains a limited quantity of mineral nutrients and water. The primary criterion, typically established by a standard, is the specified nutrient composition, a minimum quantity of dry matter, the absence of harmful substances, and the complete decomposition of the original organic material to the extent that its source cannot be determined. These composts, known as 'pseudo' composts, are frequently provided to farmers at a much reduced price. This is because the expenses associated with their manufacturing are typically covered by producers of biodegradable waste who are seeking to dispose of challenging material.

The producers of these composts frequently question why farmers are reluctant to purchase them, despite their very affordable price. The minimum yield effect of fertilisation with these composts is attributed to their low nutrient content, which necessitates the application of tens of tons per 1 ha (10,000 m2). This, in turn, leads to increased transportation and handling expenses. Compared to the practice of using "green manure," which involves ploughing down fresh green matter from plants like clover, lucerne, stubble catch crops, and crops specifically grown for green manure, such as mustard and some varieties of rape, there are no benefits to be gained from fertilizing with these artificial composts. Soil microorganisms have a high level of efficiency in decomposing organic matter. They achieve this by maintaining a C : N ratio of 15 – 25 : 1. This process is comparable to composting, when biodegradable material is broken down in a compost pile. This benefits farmers by helping to dispose of waste. What are the characteristics of authentic compost? The definition clearly states that compost is a degraded and partially humified organomineral material, in which a portion of the organic component is stabilized by the mineral colloid fraction. It exhibits a high ion-exchange capacity, a high buffering capacity, and is resistant to rapid mineralization. It is evident that the absence of any reference of nutrition in this text has been seen by the reader. Indeed, they can be found in the compost, although their quantity may vary, which is of negligible significance. The compost must effectively retain nutrients in the soil through ion-exchange reactions and prevent their leaching from the topsoil and subsoil layers into the underlying soil or groundwater. This applies to nutrients derived from the compost, mineral fertilizers, or the natural soil substrate during the soil formation process. When producing "genuine" compost, it is important to make sure that the organic matter in the composted mixture is transformed through processes such as decomposition, mineralization, exothermic oxidation, and a synthetic process called humification. Humification produces complex compounds like humic acids, fulvic acids, and humins, which are important components of soil humus. It is important to distinguish between the phrases "humus" and "primary soil organic matter" as they are distinct combinations of substances with different properties. Humus is distinguished by its elevated ion-exchange capacity and extremely gradual mineralization (the half-life of humic acids' mineralization under soil conditions is 3,000 - 6,000 years!). Primary organic matter, albeit fully degraded but not humified, has contrasting characteristics. Occasionally, it may possess a substantial sorption capacity but lacks an ion-exchange capacity. The elevated ion-exchange capacity of humified organic matter gives rise to two further highly significant phenomena: The substantial surface pressures exerted by humus colloids in soil result in their interaction with equally reactive mineral colloids. These mineral colloids are silicate-based soil particles that are smaller than 0.001 mm in size. The term used to refer to these particles in the field of pedology is "physical clay". As particle size decreases, the specific surface area increases, resulting in more surface activity. Clay-humus aggregates are created, which are adsorption complexes. These complexes are the basic units of soil microaggregates that are well-aerated, mechanically stable, and elastic. These microaggregates can then combine to form larger macroaggregates, resulting in a structured and well-aerated soil. This soil contains a sufficient amount of capillary, semi-capillary, and non-capillary pores, allowing it to effectively handle precipitation water. During dry periods, capillary pores draw water upward from the lower layers of soil. Conversely, during rainy periods, non-capillary pores facilitate the movement of water in the opposite direction. This method fulfills the fundamental prerequisite for soil productivity. The concentration of nutrients in the soil solution, and consequently in the soil, is generally less significant compared to other factors. Another significant phenomenon associated with the ion-exchange properties of compost or soil is buffering capacity, which refers to the ability to tolerate changes in pH. Soils typically experience acidification due to multiple factors, including acid rains, electrolytic dissociation of acid fertilizers, and the high absorption of nutrients by plants from the soil solution. Plants maintain electroneutrality by using H^+ ions, which are created when water dissociates, to balance the intake of nutritional cations. This ensures that the overall electric charge remains constant. Otherwise, every plant would possess an electrical charge similar to that of an electrical capacitor. The humus or clay or clay-humus ion exchanger in compost or soil functions similarly to other ion exchangers. It operates in the same manner as plants during nutrient uptake. When there is an excess of any ion in the environment, such as H^+ in an acidifying soil, the plant binds to this H^+ and exchanges it for another cation that it had previously bound. The H^+ ion is obstructed in this manner, resulting in no alteration in the pH of the soil. Soils with a high buffering capacity are characterized by a significant amount of mineral or organic colloid fraction. This is commonly found in heavy-textured soils and organic soils with a high degree of humification of soil organic matter. As previously mentioned, it is clear that soils that should be fertilized with authentic composts are primarily light sandy and sandy-loam soils. These soils have rapid mineralization processes due to high aeration, causing the organic matter of potentially applied organic fertilizers to essentially "burn." Mineral nutrients are liberated from an organic fertilizer, but shortly thereafter, there is a deficiency of essential organic material in the soil. The soil microedaphon lacks adequate energy due to insufficient decomposition of organic materials, resulting in a limited ion-exchange capacity and a failure to undergo humification. This type of soil has poor water retention capacity, causing rainfall to rapidly wash away nutrients from the soil. The productivity of these soils can be significantly improved only with the application of authentic composts. Their clay-stabilized organic matter is resistant to excessive oxygen and remains decomposable, allowing it to sustain the necessary microbial activity in the soil.

3.1.2.2. What is the process for producing "authentic" compost?

Modern production of industrial composts is based on an idea that the compost is a substrate for plants with nutrient content. This is the reason why attention is mainly paid to the mechanical treatment of organic material – grinding, crushing and homogenisation. A homogenised blend, enriched with nutrients, applied water and/or compost additives, is subjected to fast fermentation. It is turned at the same time and homogenised again. The turning ensures a new supply of oxygen and if the compost has a sufficient amount of easily degradable organic matter, the temperature during composting increases up to 50 – 60°C, which allows a desirable breakdown of particles of the original organic material. The product acquires a dark colour, it is loose, often has a pleasant earthy smell while the odour of the original organic material is not perceptible any more. Farm sludge is often added to the compost formula as a nitrogen source or the improper C to N ratio is adjusted by the addition of mineral nitrogenous fertilisers. Slurry and liquid manure are used as an N and water source and sometimes limestone is added to prevent acidification. The aeration of the fermented pile of materials is provided by the addition of inert coarse-grained materials, mainly of wood chips, crushed straw, rubble, undecomposable organic waste and other materials available from local sources, whereas the use of horizontal and vertical ventilation systems is less frequent. It is often the type of "aeration" additive which explicitly shows that the compost producer prefers waste processing to the interest of future users of their products, farmers and productivity of their soils. The ion-exchange capacity of these composts is about 40 – 80 mmol chem. eq. 1000 g-1and it is very low. It characterises a light, little fertile sandy soil.

How is the real "genuine" compost produced? The following principles should be observed:

- Organic material of the compost formula should have a high degree of lability. If the compost producer does not have a sufficient amount of such very easily degradable organic material, its lability should be enhanced by saccharidic waste.
- **The C :** N ratio should be adjusted to the value 10 – 15 : 1, not to total C and total N, but to the value of Chwsand Nhws(hot water extractable carbon and nitrogen). Obviously, it is not worth adding to the compost a nitrogen source e.g. in waste polyamide because this nitrogen is not accessible. It is a flagrant example but we have detected many times that the C : N ratios are completely different from those the compost producers suppose them to be.
- The compost formula should have a high proportion of buffering agent. It should always be ground limestone or dolomite, it should never be burnt or slaked lime. Do not economize on this additive very much. It will be utilised excellently after the application of this compost to soils.

- Stabilisation of organic matter should be ensured by a sufficient amount of the clay mineral fraction. It must not be applied in lumps, but in the form of clay slurry, clay water suspension, used also for the watering of the blend of compost materials. Concrete mixers are ideal equipment for the preparation of clay slurry.
- The compost blend should be inoculated by healthy fertile topsoil. Soil microorganisms are adapted in a different way than the microorganisms of the intestinal tract of animals. Therefore slurry and liquid manure are sources of water and nitrogen but they are not a suitable inoculant even though they are often recommended in literature for this purpose.
- The basic requirement is to reach a high temperature (55 – 60°C) during composting and to maintain the second phase of temperature (40 – 50°C) for a sufficiently long time. This process will be successful only at a sufficiently high amount of highly labile organic matter in the compost formula, at a correct C : N ratio, at a correct water to air ratio in the pile (the moisture during fermentation should be maintained in the range of 50 – 60% of water-retention capacity) and at a reduction in heat losses. Heat losses of the compost into the atmosphere through the pile surface are relatively small. The highest quantity of heat is lost by conducting the heat through the concrete or the frozen ground of the compost pile, and mainly by an aerating system if it is installed.
- Humification processes, formation of humus acids and humins or their precursors at least, occur rather in later stages of fermentation and so we should accept that the good compost cannot be produced by short-term fermentation. Old gardeners fermented composts for 10 – 12 years, but their composts reached the ion-exchange capacity of 300 – 400 mmol chem. eq. 1000 g^{-1}.

3.1.2.3. What is the utilization of the digestate in compost manufacturing?

If besides decomposing exothermic processes synthetic endothermic processes are also to take place in compost when high-molecular humus substances (fulvic acids, humic acids and humins) are formed, these conditions must be fulfilled: very favourable conditions for the microflora development must exist in compost, and minimum losses and the highest production of heat must be ensured. For this purpose it is necessary to use a high admixture of buffering additive (limestone) in the compost formula, sufficient amount of very labile organic matter, thermal insulation of the base of fermented material because the heat transfer coefficient does not have the highest value for transfer from the composted pile into the atmosphere but mainly into solid especially moist materials, i.e. into concrete, moist or frozen earth, clay, bricks, etc. At a sufficient amount of labile fractions of organic matter the maximum heat production can be achieved only by a sufficient supply of air oxygen. Beware of this! The ventilation through vertical and horizontal pipes provides sufficient air for aerobic processes in the fermented material but at the same time the ventilation is so efficient that a considerable portion of reaction heat is removed, the material is cooled down and the onset of synthetic reactions with the formation of humus substances does not occur at all.

When sufficiently frequently turning the fermented material, the safest method of compost aeration and ventilation is the addition of coarse-grained material while inert material such as wood chips, chaff and similar materials can be used. It is however problematic because inert material in the fermented blend naturally decreases the concentration of the labile fraction of organic matter, which slows down the reaction rate of aerobic biochemical reactions and also the depth of fermentation is reduced in this way. It mainly has an impact on the synthetic part of reactions and on the formation of humus substances while the influence on decomposing reactions is smaller.

It would be ideal if during compost fermentation in a microbially highly active environment the inert aeration material were able not only to allow the access of air oxygen into the fermented material but also to decompose itself at least partly and to provide additional energy to biochemical processes in the pile in this way.

These requirements are excellently met by the solid fraction of digestate from biogas plants. It aerates the compost and although it lost labile fractions of organic matter in biogas plant digesters, it is capable of further decomposition in a microbially active environment. It releases not only energy but also other mineral nutrients. So this waste is perfectly utilised in this way. The average microbial activity of even very fertile, microbially active soils is not efficient enough for the decomposition of this stable organic material when the solid phase of digestate is used as an organic fertiliser. The decomposition rate is slow, especially in subsequent years, and therefore the resultant effect of the solid fraction of digestate as an organic fertiliser is hardly noticeable. The combination of anaerobic decomposition in the biogas plant digester and aerobic decomposition in compost could seem paradoxical, and some agrochemists do think so. The preceding exposition has shown that it is not nonsense.

Now let us answer the question: what dose of the solid fraction of digestate should be used in the compost formula? It depends on many factors: on the amount of the labile fraction of organic component and mainly on the degree of its lability (which can be determined in a reliable way by the above-mentioned method according toRovira and Vallejo 2002,2007,Shirato and Yokozawa 2006), on the aeration and porosity of materials used in the compost formula, on the number of turnings, on prevailing outdoor temperature, water content, degree of homogenisation and on other technological parameters.

In general: the higher the amount of the labile component of organic matter and the higher its lability (e.g. the content of saccharides and other easily degradable substances), the higher the portion of the solid fraction of digestate that can be used.

3.1.2.3 Now, the author presents concise evidence from their own research:

The basic compost blend was composed of 65% fresh clover-grass matter from mechanically mown lawns, 10% ground dolomite, 2% clay in the form of clay suspension, 20% solid phase of digestate (obtained by centrifugation with fugate separation) or 20% crushed wood chips and 3% PK fertilisers. The C : N ratio in the form of Chws: Nhws(hot-water-soluble forms) was 15 : 1, nitrogen was applied in NH4NO3in sprinkling water that was used at the beginning of fermentation at an amount of 70% of the beforehand determined water-retention capacity of the bulk compost blend. Inoculation was done by a suspension of healthy topsoil in sprinkling water. Fermentation was run in a composter in the months of April – November, and the perfectly homogenised material was turned six times in total. Water loss was checked once a fortnight and water was replenished according to the increasing water-retention capacity to 60%. The formation, amount and quality of formed humus substances were determined not only by their isolation and measurement but also by their specific manifestation, which is the ion-exchange capacity of the material. The original particles of composted materials were not noticeable in either compost (with the solid part of digestate and with wood chips), in both cases the dark coloured loose material with pleasant earthy smell was produced. Tab. 6 shows the analyses of composted materials and composts. The digestate was from a biogas plant where a mix of cattle slurry, maize silage and grass haylage is processed as a substrate. The material in which the aeration additive was polystyrene beads was used as compost for comparison.

Table 6. The content of fulvic acid carbon (CFA), humic acid carbon (CHA), their ratio and ion-exchange capacity T of the solid phase of digestate and wood chips at the beginning of fermentation and of composts with polystyrene (PS), wood chips and solid phase of digestate

	Solid phase of digestate	Wood chips	Compost		
			PS	Wood chips	Solid phase of digestate
CFAmg.kg-1	0	0	38	84	178
CHAmg.kg-1	0	0	15	20	62
CHA: CFA	-	-	0,39	0,24	0,35
Ion-exchange capacity T mmol chem. eq.kg-1	51	12	72	64	224

The results document that the ion-exchange capacity, and hence the capacity of retaining nutrients in soil and protecting them from elution after the application of such compost, increased very significantly only in the digestate-containing compost. The ion-exchange capacity of this compost corresponds to the ion-exchange capacity of heavier-textured humus soil, of very good quality from the aspect of soil sorption. The compost with wood chips produced in the same way does not practically differ from the compost with polystyrene but it does not have any humic acids and the ion-exchange capacity of these composts is on the level of light sandy soil with minimum sorption and ion-exchange properties. However, the total content of humus acids in the compost with the solid phase of digestate is very small and does not correspond to the reached value of the ion-exchange capacity of this compost. Obviously, precursors of humus acids that were formed during the fermentation of this compost already participate in the ion exchange. Humus acids would probably be formed from them in a subsequent longer time period of their microbial transformation. If only humus acids were present in composting products, at the detected low concentration of CFA+ CHA the T value of the compost with the solid phase of digestate would be higher only by 1 – 1.2 mmol.kg-1than in the compost with polystyrene or wood chips. Because it is more than a triple, other substances obviously participate in the ion exchange.

3.1.3. Utilization of digestate to enhance the quality of soils with high clay content

Optimum values of reduced bulk density Orfor soils are around 1.2 g.cm-3, but more important is the minimum value of bulk density for the restriction of root growth which is about 1.7 – 1.8 g.cm-3 for light soils and only 1.40 – 1.45 g.cm-3 for heavy-textured clay soils. Bulk density Oris a crucial parameter for the assessment of the soil compaction rate as an important negative factor of soil productivity. Bulk density of topsoil in the range of 0.95 – 1.15 g.cm^{-3} shows loose topsoil while the value 1.25 g.cm^{-3} indicates heavily compacted topsoil.

Another important value of soil is soil aeration VZ. It is expressed in volume % as the difference between porosity Poand momentous soil moisture $W_{obj.}$

$$V_z = P_o - W_{obj.} \qquad (16)$$

Optimum aeration e.g. for grasslands is 10% by volume, for soils for barley growing it is already as much as 24% by volume. Soil porosity Pois the sum of all pores in volume per cent, in topsoils it is around 55%, in subsoil it decreases to 45 – 35%. Sandy soils have on average P = 42% by vol., out of this 30% are large pores and 5% are fine pores while heavy-textured clay soils have the average porosity of 48% by vol., out of this only 8% are large pores and 30% are fine pores. Fine pores are capillary and large pores are non-capillary ones. Cereals should be grown in soils with 60 – 70% of capillary pores out of total porosity and 30 – 40% of non-capillary pores. Forage crops and vegetables require the soils with 75 – 85% of capillary pores and only 15 – 25% of non-capillary pores out of total porosity.

Ploughing resistance P is also significant. It is a specific resistance that must be overcome during cutting into and turning over the soil layer. It is expressed by the drawbar pull measured dynamometrically on the coupling hook of a tractor. It is related to the texture and moisture of soil, to its content of organic substances and ploughing depth. Ploughing resistance for light soils is 2 – 4 $t.m^{-2}$, for heavy-textured soils it is 6 – 8 $t.m^{-2}$. The units kp.dm-2are also used. For sandy soils the ploughing resistance of 25 – 28 $kp.dm^{-2}$ is usual, for clay soils it is 70 $kp.dm^{-2}$.

Hence heavy-textured soils are more responsive to the higher reduced bulk density of soil when roots develop poorly, they need more non-capillary pores to allow for the better infiltration of precipitation water, they also need higher aeration because they are mostly too moist and many aerobic processes including the microbial activity take place with difficulty. Of course, the high ploughing resistance is not desirable either for the economics of soil cultivation or for the production process of any crop. Therefore it is necessary to improve heavy-textured soils and the question is how. Organic fertilisers are not sufficient; peat was used previously but now it is banned to use it for the reason of the peat bog conservation, and synthetic soil amendments (Krilium, Flotal etc.) are currently too costly for the agriculture sector. An excellent material for the improvement of heavy-textured soils is the solid phase of digestate if ploughed down at higher doses than those used for the application of farmyard manure or compost, i.e. 100 – 150 t.ha-1. Even though we cannot expect any great release of mineral nutrients from organic matter of the solid phase of digestate due to high stability of this material, the improvement and aeration of heavy-textured soil with better conditions for the microbial activity of soil and undisturbed root growth often bring about a higher yield effect than is the yield effect of nutrients from high-quality organic fertilisers as shown by the results of this field trial:

When we still believed that the solid phase of digestate was an organic fertiliser, we laid out an exact field trial on a heavier-textured, loamy-clay soil with medium to good reserve of available nutrients. The trial had two treatments: the one treatment was fertilisation with the solid phase of digestate only (after fugate centrifugation) and the other treatment was the application of only mineral fertilisers in the form of pure salts at such a dose that the level of these easily available nutrients to plants was the same as the amount of unavailable or little available nutrients in the treatment fertilised with digestate. We wanted to find out from the yield of the grown crop what amount of mineral nutrients would be released from the digestate in comparison with completely available nutrients in the first year and in subsequent years of the crop rotation: early potatoes – winter barley – red clover – oats. We intended to compare the digestate with other organic fertilisers, e.g. farmyard manure which in the first year mineralises about a half of its nutrients bound in organic matter. But the result we obtained was surprising: in the first year the yield of early potatoes was higher by 12% in the digestate treatment although nobody could doubt that this treatment had a lower amount of nutrients than the variant fertilised with pure salts. The only explanation is that the higher yield effect in the digestate treatment was not caused by the higher input of nutrients but by the improvement in physical properties of heavy-textured soil that surely occurred as seen in Tab. 7. The favourable effect of the heavy-textured soil improvement on yield was positively reflected in subsequent years also in other crops of the crop rotation that were fertilised in both treatments in the same way, i.e. mineral fertilisers were applied. We drew a conclusion that in practice the yield effect is often ascribed to digestate nutrients although it is caused by better soil aeration and better root growth due to soil loosening after the application of digestate.

Table 7. Bulk density Or, aeration Vz, total porosity Po, proportion of large pores in total porosity and ploughing resistance P in a heavy-textured clay-loamy soil and after its improvement with the dose of 150 t-ha-1of the digestate solid phase

	Clay-loamy soils	
	initial	improved by digestate
Reduced bulk density Org.cm-3	1.43	1.38
Soil aeration Vz% by vol.	18.5	22.4
Total porosity Po% by vol.	43.9	43.8
Proportion of largepores in total porosity %	22.7	28.1
Ploughing resistance P kp.dm-2	63	50

3.2. Application of digestate through modification of standard biogas producing technology

Perspective utilisation of digestate is connected with envisaged modifications of the technology of biogas production in agricultural biogas plants. These plants have digesters for the solid phase only or the most frequent are liquid (suspension) digesters. These are digesters without partition wall where the biomass of microorganisms is carried by the processed substrate. In reactor systems for the technological processing of waste from chemical and food technologies and from the technology of municipal and industrial waste water treatment those digesters are preferred where the biomass of functional microorganisms is fixed onto a solid carrier or onto partition walls of apparatuses. It is often granulated and is maintained in the digester as a suspended sludge cloud. These reactors may be affected by short-circuiting and therefore they are sensitive to the particle size of the processed substrate but they withstand a much higher organic load than the digesters without partition wall. Of course, the reactor is smaller, cheaper and more efficient.

Hence a perspective modification of the biogas production technology in agricultural biogas plants is gradual transition to the procedures of anaerobic digestion that are currently used in industrial plants for the treatment of organic waste water. The promising utilisation of digestate from such digesters is mainly the manufacture of solid fuels in the form of pellets that are prepared from the solid phase of agricultural waste before the proper aerobic digestion of the material for a biogas plant. The first proposal of this type is the IFBB procedure, the principle of which was explained in Chapter 1.4. The liquid phase from the preparation of processed material, which is destined for anaerobic fermentation in digesters with partition wall, could be used as a liquid or suspension fertiliser but researchers would have to solve the cheap method of nutrient concentration in this waste. The current price of Diesel fuel, machinery and human labour and low purchase prices of agricultural products do not allow the application of highly diluted fertilisers and in fact handling of water. The problem is that a small biogas plant is only scarcely profitable. Hence economic reasons favour large-capacity plants with the volume of digesters 5 000 – 10 000 m3. In such large plants the reactors with partition wall would be unjustifiably expensive and therefore in these large-capacity facilities for biogas production it is necessary to use reactors without partition wall. The utilisation of their digestate should be based on this scheme: separation of digestate – concentration of fugate and its utilisation as a liquid mineral nitrogenous fertiliser. The solid phase of digestate should be used as an inert aeration component in compost production and as a material for the improvement and aeration of heavy-textured soils.

In any case, researchers must resolve a cheap method of nutrient concentration in fugate. A number of different reactors are available for small to medium-sized biogas plants with the treatment of material according to IFBB that were developed on a research basis mainly in the sixties to the nineties of the twentieth century. At first, these were reactors with suspension biomass, e.g. mixing contact anaerobic reactor (ACR – AG), its innovation was a membrane anaerobic reactor system (MARS) and sequencing batch reactors (SBR). Then reactors with immobilised biomass were developed that are divided into reactors with biomass on the surface of inert material and reactors with aggregated (granulated) biomass.

The former group is divided into upflow reactors and downflow reactors. Reactors with a mobile filling are the third variant. The latter group is divided into reactors with the internal separator of biogas and biomass, reactors with the external separator of biomass and reactors with partitions.

Further development brought about biofilm reactors where the biomass of microorganisms is fixed onto a solid carrier. These reactors are considered as facilities with the highest operating stability, very resistant to the fluctuation of operating conditions. But they do not usually allow for such a high load as reactors with suspension biomass. The oldest reactor of this series was an upflow anaerobic filter (UAF) reactor from 1967, then a downflow stationary fixed film reactor (DSFF) and downflow reactor with filling in bulk followed. Great progress was made by designing an anaerobic rotating biological contactor (ARBC) and fluidized bed reactor (FBR) in the eighties of the last century. A similar type of reactor, expanded bed reactor (EBR), also designated by AAFEB (anaerobic attached film expanded bed), is suitable to be operated at low temperatures. The detention time is only several hours and the portion of residual organic impurities is practically the same as in modern aerobic systems for the treatment of organically contaminated waters.

Further advance was the development of reactors with aggregated biomass. The most important representative of this group of digesters is an upflow anaerobic sludge blanket (UASB) reactor. It is a reactor with sludge bed and internal separator of microorganism biomass. The biggest reactor of this type (5 000 m3) processes waste water from the manufacture of starch in the Netherlands, it withstands the load of 12.7 kg chemical oxygen demand (COD) per 1 m3/day, 74% of organic matter is degraded and the detention time is 33 hours only. Besides the UASB reactor these reactors belong to this group: hybrid upflow bed filter (UBF) reactor, anaerobic baffled reactor (ABR), expanded granular sludge bed (EGSB) reactor, internal circulation (IC) reactor and upflow staged sludge bed reactor (USBB), often also called biogas tower reactor (BTR), and other design models of the UASB reactor.

At the end of this chapter it is to note that modern anaerobic reactors have almost amazing outputs – unfortunately, the more perfect the reactor, the more expensive, and also their advantage over huge digesters without partition wall we have got accustomed to in biogas plants is gradually disappearing. The selection of modern anaerobic reactors is also more difficult than the selection of conventional technology of reactors without partition wall, because they are mostly rather specific to the substrate to be processed. They also have higher demands on processing, attendance and checks.

The perspective possibility of using modern anaerobic reactors for biogas production in smaller plants and the simultaneous solution to the use of the digestate solid phase as a raw material for the production of solid pelleted biofuels initiated our study of the IFBB procedure (Chap. 1.4.) for the substrate commonly used in biogas plants in the CR. The results of our experimental work are presented below:

The IFBB technological procedure is based on a high degree of cell wall maceration as a result of the axial pressure and abrasion induced with a screw press.Reulein et al. (2007) used this procedure for dehydration of various field crops; it is also known from the technologies of processing rapeseed, sugar beet and leguminous crops for the production of protein concentrates (Telek and Graham 1983,Rass 2001) and in biorefineries for the extraction of lactic acid and amino acids (Mandl et al. 2006).

The basic substrate contained 37.5% by weight of cattle slurry and 62.5% by weight of solid substrates, i.e. a mixture of chopped maize silage and grass haylage of particle size max. 40 mm mixed at a 4.75 : 1 ratio, i.e. 51.6% of silage and 10% of haylage. In total, the substrate accounted for 19.3% of dry matter. This substrate at 15°C is designated by A. A portion of this substrate was mixed with water at a weight ratio of 1 : 5, put into a thermostat with a propeller stirrer at 15°C and intensively stirred for 15 minutes. Analogically, the other portion was also mixed with water at a substrate to water ratio of 1: 5 and put into a thermostat at a temperature of 60°C with 15-minute intensive stirring again. The sample of the substrate with water 15°C was designated by B, the sample with water 60°C was designated by C. The liquid phase from substrate A was separated by centrifugation while the liquid phases from substrate B and C were separated in a laboratory screw press for the pressing of fruits and vegetables. The separated liquid phases of substrates A, B and C were diluted with water to obtain a unit volume and the analytical results were recalculated to a transfer ratio in the liquid phase in relation to the content of particular nutrients in dry matter of the original substrate mixture.

The experiments conducted in an experimental unit of anaerobic digestion and in an equipment for IFBB made it possible to determine the content of mineral nutrients in substrate A after 42-day anaerobic digestion in mesophilic conditions (40°C), in the liquid phase of substrate A after anaerobic digestion, in the liquid phase of substrate B and C after recalculation to the dry matter content and concentration corresponding to substrate A, also after the process of anaerobic digestion under the same conditions (42 days, 40°C). The above recalculations enable to clearly show the advantages of the IFBB process in nutrient transfer from solid to liquid phase when substrate A and 5 times diluted substrates B and C are compared, but they may unfortunately evoke a distorted idea about the real concentration of nutrients in liquid phases. It is to recall that IFBB increases the mass flow and transfer to the liquid phase but with regard to the 5-fold dilution the nutrient concentration in liquid waste for fertilization continues to decrease. This is the reason why the table below shows the original, not recalculated concentrations in the fugate of fermented substrate A and in the fermented liquid phases of the same substrate in IFBB conditions designated by B and C, which document considerable dilution of these potential mineral fertilizers.

The solid phases of substrates A, B and C after anaerobic digestion were subjected to determination of organic matter hydrolysability in sulphur acid solutions according toRovira and Vallejo (2000,2002) as modified by Shirata and Yokozawa (2006); we already used this method to evaluate the degradability of a substrate composed of pig slurry and sludge from a municipal waste water treatment plant (Kolář et al. 2008).

Table 8 documents that the IFBB procedure proposed by German authors for grass haylage is applicable to the typical substrate of Czech biogas plants, to a mixture of cattle slurry, maize silage and grass haylage. In agreement with German experience the observed transfer ratios are markedly higher at 60°C compared to 15°C of hydrothermal conditions but the value of transfer ratios to the liquid phase is generally lower in our experiments. We ascribe this fact to the properties of the material and also to the achieved axial force of the used press that was apparently lower even though the same perforation size of the conical part of the press (1.5 mm) and slope of the body (1 : 7.5) were used.

Table 8. Dry matter content in the fresh mass of used materials and their chemical composition in % dry matter. The transfer ratio of mass flow to the liquid phase from the fresh mass of substrate not diluted with water at 15°C (A), diluted with water at a 1:5 ratio at 15°C (B) and diluted with water at a 1:5 ratio at 60°C (C). Liquid phase A was separated by centrifugation, liquid phases B and C with a screw press.(Sample size n = 5, reliability interval of the mean for a significance level = 0.05)

	Cattle slurry	Maize silage	Grass haylage	Substrate	Transfer ratio to liquid phase		
					A	B	C
Dry matter	6.4	28.9	18.7	19.3	0.06 0.01	0.18 0.04	0.20 0.03
N-compounds (N x 6.25)	25.6	11.5	7.4	16.3	0.05 0.01	0.20 0.04	0.26 0.05
Digestible nitrogen compounds	-	6.2	3.8	7.3	-	-	-
Nitrogen-free extract	-	52.8	48.6	49.9	0.30 0.03	0.45 0.05	0.48 0.05
Crude fibre	-	25.7	29.8	18.0	0.01 0.00	0.10 0.00	0.10 0.00
Fat	-	4.8	1.5	2.8	-	-	-
Organic substances	76.4	94.8	87.3	87.0	-	-	-
Mineral N (N - NH^{4+}, NO^{3-})	2.4	0.1	0.1	1.0	0.74 0.05	0.89 0.06	0.95 0.06
P	1.3	0.2	0.3	0.6	0.40 0.05	0.52 0.07	0.65 0.08
K	5.3	1.4	1.7	2.9	0.57 0.04	0.60 0.04	0.79 0.05
Ca	1.3	0.4	0.6	0.8	0.31 0.06	0.38 0.08	0.46 0.08
Mg	0.5	0.2	0.3	0.3	0.38 0.07	0.43 0.08	0.55 0.07
Na	0.1	0.1	0.1	0.1	0.70 0.08	0.77 0.04	0.80 0.08
Cl	0.3	0.2	0.2	0.2	0.77 0.06	0.85 0.05	0.85 0.06

The results in Table 8 illustrate that the separation of the liquid and solid phase of the substrate that has not been subjected to anaerobic digestion yet by means of centrifugation only is rather imperfect from the aspect of the mass flow of components. The IFBB system (water dilution, intensive stirring at a temperature of 60ºC and subsequent separation of the liquid and solid phase with a screw press) increases the transfer of organic and mineral substances into the liquid phase by about 15 – 20%, and it is also true of the saccharidic nitrogen-free extract and organic nitrogen compounds. This fact documents that the liquid phase has a higher amount of active, well-degradable organic material for anaerobic digestion, and so it is possible to expect not only the higher production of biogas but also more mineral nitrogen in the liquid after anaerobic digestion.

The high mass flow of alkaline metals and chlorine into the liquid phase, and on the contrary, the low transfer of calcium confirm the opinion of German researchers (Wachendorf et al. 2007,2009) that the IFBB procedure largely increases the quality of biomass solid phase as a material for the production of solid fuels: the production of polychlorinated dioxins and dibenzofurans is reduced, waste gases will be less corrosive and the temperature of ash fusion will be higher.

Nitrogen compounds of the substrate dry matter account for 16.3%, i.e. these nitrogen organic compounds contain 2.6% of nitrogen in dry matter (Table 8). The content of mineral nitrogen in the substrate before anaerobic digestion was 1%. If the digestate contains 2.26% of mineral nitrogen in the same dry matter after anaerobic digestion, it is to state that during anaerobic digestion about a half of organic nitrogen mineralized and enriched the original 1% content of the substrate before the fermentation process. But the dry matter content decreased in the course of fermentation, and therefore the concentration of all nutrients in digestate apparently increased contrary to the original substrate. In our experiment the concentration of substrate dry matter decreased from 19.3% to 13.3% by weight during anaerobic digestion. The content of mineral nitrogen amounting to 3.28% at this dry matter content corresponds to 2.26% of min. N at the original dry matter content of 19.3% by weight (Table 9). The contents of the nutrients P, K, Ca and Mg in digestate dry matter after anaerobic digestion (Table 9) are apparently substantially higher than before fermentation. However, anaerobic digestion did not actually bring about any increase in the content of these nutrients, and the increased concentrations completely correspond to a reduction in dry matter content, 19.3 : 13.3 = 1.45.

It is not a new fact, but Table 9 shows how this mineral nitrogen is transferred to the liquid phase of substrates B and C compared to substrate A. Obviously, the liquid phase of substrate B has a higher amount of mineral nitrogen than that of substrate A, and so the effects of the screw press, which already before anaerobic digestion enriched the liquid substrate B with colloidal solutions (sols) of nitrogen organic compounds from the crushed cell walls of the plant material that provided further mineral nitrogen during fermentation, were significantly positive at the same temperature. It was still more evident in the liquid phase of substrate C while a conclusion can be drawn that a higher temperature contributes to a higher extraction of insoluble or partly soluble nitrogen organic compounds from which further mineral nitrogen is released after subsequent fermentation.

Table 9. Contents of mineral nutrients after anaerobic digestion (42 days, 40°C) in digestate (substrate A), in its liquid phase and in fermented liquids from IFBB in % dry matter by weight

	Substrate A	Liquid phase of substrate A	Liquid phase of substrate after recalculation to dry matter content and concentration of substrate A	
			B	C
N	3.28	2.43	2.92	3.11
P	0.87	0.35	0.45	0.56
K	4.20	2.39	2.52	3.32
Ca	1.16	0.25	0.30	0.36
Mg	0.43	0.11	0.13	0.16

Table 10 documents the original (before their recalculation) concentrations of mineral nutrients in the liquid phase of substrates A, B and C. These results indicate that liquid phase A can be considered as a highly diluted mineral fertilizer. Even though the IFBB process increases the concentration of nutrients (nitrogen) in the liquid phase before and after fermentation (liquid phase B and C), the dilution is very high. The recommended dilution with water, used byWachendorf et al. (2009) and also in our experiments, produces liquid wastes diluted to such an extent after anaerobic digestion that they are practically hardly utilizable as a solution of mineral nutrients. Fugates are still rather problematic as mineral fertilizers, especially for applications in humid years and to soils with low microbial activity and consequently slow immobilization of mineral nitrogen, and naturally they are hardly applicable to pervious soils.

Table 10. Contents of mineral nutrients after anaerobic digestion (42 days, 40°C) in the liquid phase of digestate A and in liquid phases B, C with the application of IFBB in % by weight of solutions that should be used for mineral fertilization(Sample size n = 4, reliability interval of the mean for a significance level = 0.05)

	Liquid phase		
	A	B	C
N	0.32 0.03	0.09 0.01	0.10 0.01
P	0.05 0.00	0.01 0.00	0.02 0.10
K	0.31 0.04	0.08 0.01	0.11 0.15
Ca	0.03 0.00	0.01 0.00	0.01 0.00
Mg	0.01 0.00	0.00 0.00	0.00 0.00

(Note: Statistical evaluation of this recalculation table is based on original data in Table 10.)

Table 11 shows the results of hydrolytic experiments with solid phases of substrates A, B and C. They confirm the previously observed fact in the work with the substrate consisting of a mixture of pig slurry and primary sludge from a municipal waste water treatment plant that the solid phases of wastes from anaerobic digestion cannot be efficient as mineral fertilizers because of their very low degradability (Kolář et al. 2008). The IFBB process, which enriches the liquid phases with organic, easily degradable substances and improves biogas yields during the anaerobic degradation of only liquid phases, further depletes of these substances the solid phases of substrates and impairs their quality as organic fertilizers, even though it is not the case of an increase in the resistant component but only in worse hydrolysable LP II.

Table 11. Proportions of the three pools of carbon in the solid phase of substrate A before anaerobic digestion (A1), after anaerobic digestion (A2) and in the solid phase of substrate A1after IFBB procedure before anaerobic digestion at 15°C (B) and at 60°C (C) as determined by the acid hydrolysis method ofRovira and Vallejo (2002).(Sample size n = 4, reliability interval of the mean for a significance level = 0.05).

Solid phase of substrate	Proportion		
	LP I	LP II	RP
A1	43 8	41 7	16 2
A2	22 4	20 3	58 8
B	39 6	44 6	17 3
C	31 6	47 8	16 2

Note: Description of fractions according to the method ofRovira, Vallejo 2002:

- LP I = (labile pool I) = the reserve of very labile, easily hydrolysable organic substances expressed as % of the total amount of organic matter in a sample
- LP II = (labile pool II) = the reserve of intermediately labile, less easily hydrolysable organic substances in %
- RP = (recalcitrant pool) = the reserve of hydrolysis resistant, very hardly degradable organic substances in %

Therefore, it can be stated: We conducted a suitability assessment of the Integrated Generation of Solid Fuel and Biogas from Biomass (IFBB) technique for using ensiled grass matter as a substrate in new Czech biogas plants. The substrate typically consists of a mixture of bovine slurry, maize silage, and grass haylage. The agricultural chemical worth of the liquid phase derived from a biodigester was also assessed. Based on our findings, we determined that this method is appropriate for the substrate that was tested and enhances the agrochemical quality of a byproduct obtained from biogas production. Chlorine can be transferred to the liquid phase, allowing the solid phase to be used as a material for producing solid biofuels. This reduces the risk of generating polychlorinated dioxins and dibenzofurans during combustion. Nevertheless, the mineral nutrient concentration in the liquid phase significantly decreases during the IFBB operation due to dilution with water after anaerobic digestion, rendering its volume value insignificant.

REFERENCES

Blair G. J., Lefroy R. D. B., Lisle L. (1995): Soil carbon fractions based on their degree of oxidation and the development of a carbon management index for agricultural systems. Australian Journal of Agricultural Research, 46, 1459-1466.

Gillman G. P. (1979): A proposed method for the measurement of exchange properties of highly weathered soils. Australian Journal of Soil Research, 17, 129-139.

Chan K. Y., Bowman A., Oates A., 2001: Oxidizible organic carbon fractions and soil quality changes in an oxic Paleustalf under different pasture leys. Soil Science, 166, (1): 61-67.

Kolář L., Klimeš F., Ledvina R., Kužel S. (2003): A method to determine mineralization kinetics of a decomposable part of soil organic matter in the soil. Plant, Soil and Environment, 49, (1): 8-11.

Kolář L., Kužel S., Hanušová A., Gergel J., Ledvina R., Šindelářová M., Silovská Š., Štindl P. (2005a): The use of Spectroquant Merck BOD photometric test to evaluate the stability of organic matters in soil. Plant, Soil and Environment, 51, (1): 46-50.

Kolář L., Klimeš F., Gergel J., Kužel S., Kobes M., Ledvina R., Šindelářová M. (2005b): Methods to evaluate substrate degradability in anaerobic digestion and biogas production. Plant, Soil and Environment, 51, (4): 173-178.

Kolář L., Ledvina R., Kužel S., Klimeš F., Štindl S. (2006): Soil Organic Matter and its Stability in Aerobic and Anaerobic Conditions. Soil ◀ Water Research, 1, (2): 57-64.

Kolář L., Kužel S., Peterka J., Štindl P., Plát V. (2008): Agrochemical value of organic matter of fermenter wastes in biogas production. Plant, Soil and Environment, 54, (8): 321-328.

Kolář L., Kužel S., Horáček J., Čechová V., Borová-Batt J., Peterka J. (2009a): Labile fractions of soil organic matter, their quantity and quality. Plant, Soil and Environment, 55, (6): 245–251.

Kolář L., Maršálek M., Frelich J., Kužel S., Zedníková J., Švecová M. (2009b): Changes in methane release from organic matter passing through the digestive tract of horses.Czech Journal of Animal Science, 54, 112–120.

Kolář L., Frelich J., Brouček J., Kužel S., Borová-Batt J., Peterka J., Šlachta M., Volfová K.,

Pezlarová J., Hřebečková J., Čechová V. (2010a): Anaerobic degradability of organic matter of cattle faeces and a possibility of its utilization. Czech Journal of Animal Science, 55, (12): 538-547.

Kolář L., Kužel S., Peterka J., Borová-Batt J. (2010b): Agrochemical value of the liquid phase of wastes from fermenters during biogas production. Plant, Soil and Environment, 56, (1): 23-27.

Körschens M., Schulz E., Böhm R. (1990): Heisswasserlöslicher C und N im Boden als Kriterium für das N-Nachlieferungsvermögen. Zentralblatt für Mikrobiologie, Jena, 145, (4): 305-311.

Leblanc H. A., Nygren P., McGraw R. L. (2006): Green mulch decomposition and nitrogen release from leaves of two Inga spp. in an organic alley-cropping practice in the humid tropics. Soil Biology Biochemistry, 38, (2): 349-358.

López F., Rodríguez G., Kass M. (1992): Manual de métodos rutinarios. Laboratorio de nutrición animal. CATIE, Turrialba, Costa Rica, 52 pp.

Mandl M., Graf N., Thaller A., Böchzelt H., Schnitzer H., Steinwender M., Wachlhoffer R., Fink R., Kromus S., Ringhofer J., Leitner E., Zentek J., Novalin S., Mihalyi B., Marini I., Neureiter M., Narodoslawsky M. (2006): Green bio-refinery – primary processing and utilization of fibres from green biomass. Series: Reports from energy and environmental research. Vol. 67, BMVIT, Vienna. Pitter P. (1981): Hydrochemie. SNTL Praha, 373 pp.

Rass M. (2001): Rheology of biogenic solids under compression using peeled rapeseed as an example. Ph.D. thesis.University of Essen, Germany.

Reulein J., Scheffer K., Stülpnagel R., Bühle L., Zerr W., Wachendorf M. (2007): Efficient utilization of biomass through mechanical dehydration of silages. In: Proceedings of Florence, Italy: ETA – Renewable Energies.

Roschke M. (2003): Verwertung der Faulsubstrate, in: Biogas in der Landwirtschaft-Leitfaden für Landwirte und Investoren im Land Brandenburg. Ministerium für Landwirtschaft, Umweltschutz und Raumordnung des Landes Brandenburg, Postdam, 29-33.

Rovira P., Vallejo V. R. (2000): Examination of thermal and acid hydrolysis procedures in characterization of soil organic matter. Communications in Soil Science and Plant Analysis, 31: 81-100.

Rovira P., Vallejo V. R. (2002): Labile and recalcitrant pools of carbon and nitrogen in organic matter decomposing at different depths in soil: an acid hydrolysis approach. Geoderma, 107: 109-141.

Rovira P., Vallejo V. R. (2007): Labile, recalcitrant and inert organic matter in Mediterranean forest soils. Soil Biology ◄ Biochemistry, 39, (1): 202-215.

Shirato Y., Yokozawa M. (2006): Acid hydrolysis to partition plant material into decomposable and resistant fractions for use in the Rothamsted carbon model. Soil Biology ◄ Biochemistry, 38: 812-816.

Schulz E. (1990): Die heisswasserextrahierbaze C-Fraktion als Kenngrösse zur Einschätzung des Versorgungszustands mit organischer Substanz. Tagung-Berichte, Akad. Landwirtsch. Wiss., Berlin, 295, 269-275.

Stehlík K. (1988): Use of waste water for irrigation. Praha, MZVž ČR, 437 pp. (in Czech).

Straka F., Dohányos M., Zábranská J., Dědek J., Malijevský A., Novák J., Oldřich J. (2003): Biogas. GAS Říčany, 1, 1-517 pp. (in Czech).

Straka F., Dohányos M., Zábranská J., Dědek J., Malijevský A., Novák J., Oldřich J. (2006): Biogas. GAS Říčany, 2, 1-706 pp. (in Czech).

Süssmuth R., Doser Ch., Lueders T. (1999): Determination of the biological biodegradability of organic substances under anaerobic conditions using the Oxi Top Control measuring system. Universität Hohenheim, Institut für Mikrobiologie, Wissenschaftlich- Technische Werkstäten GmbH & Co., KG Weilheim, Germany.

Telek I., Graham H. D. (1983): Leaf protein concentrates. Westport, CT: AVI Publishing Company.

Tirol-Padre A., Ladha J. K. (2004): Assessing the reliability of permanganate-oxidizable carbon as an index of soil labile carbon. Soil Science Society of America Journal, 68, (3): 969-978.

Van Soest P. J. (1963): Use of detergents in the analysis of fibrous feeds. II. A rapid method for the determination of fiber and lignin. Journal of the Association of Official Analytical Chemists, 46, 829-835.

Wachendorf M., Fricke T., Grass R., Stülpnagel R. (2007): Ein neues Konzept für die bioenergetische Nutzung von Grünlandbiomasse. In: Wrage N., Isselstein J.: Mitteilungen der Arbeitsgemeinschaft Grünland und Futterbau, 8, 165-168.

Wachendorf M., Richter F., Fricke T., Grass R., Neff R. (2009): Utilization of semi-natural grassland through integrated generation of solid fuel and biogas from biomass. Grass and Forage Science, 64, 132-143.

Walkley A. (1947): A critical examination of a rapid method for determining organic carbon in soils. Soil Science, 63, 251-263.

Zábranská J., Dohányos M., Gran P. (1985a): Direct measurement of the Hydrogenase activity of Anaerobic Microorganisms. Watter Science Technology, 17, 303-306.

Zábranská J., Schneiderová K., Dohányos M. (1985b): Relation of coenzyme F 420 to the activity of methanogenic microorganisms. Biotechnology Letters, 7, 547-552.

Zábranská J., Dohányos M. (1987): Methods of Investigation of the Metabolism of Anaerobic Microorganisms. Acta hydrochimica hydrobiologica, 15, 43-55.

CHAPTER-10

PRODUCTION AND ASSESSMENT OF BIODIESEL

1. INTRODUCTION: AN OVERVIEW

Currently, fossil fuels are the most widely utilized energy sources globally; yet, there are some issues associated with their use. Firstly, the frequent increase in price renders petroleum economically unsustainable. Furthermore, the combustion of petrochemical sources results in the release of very hazardous pollutants that pose a significant risk to human health. One such pollutant is carbon dioxide, which is the primary contributor to global warming. Furthermore, fossil fuels are finite resources, meaning they have a limited lifespan. Vegetable oils are becoming increasingly popular as a viable substitute for gasoline due to their renewable properties and positive impact on the environment (Ferella et al., 2010).

Although vegetable oils have numerous benefits, they also have certain drawbacks when used as fuel. The utilization of vegetable oils in internal combustion engines is hindered due to their higher viscosity compared to diesel fuel and lower volatility. Consequently, they do not combust fully and result in the accumulation of deposits in the fuel injectors of diesel engines. Specialized literature outlines five methods for mitigating the aforementioned issues: combining vegetable oil with diesel, thermal cracking (pyrolysis), microemulsions, esterification, and transesterification (Ma & Hanna, 1999). The most preferred methods for producing biodiesel are currently esterification and transesterification processes (Janaun & Ellis, 2010).

Biodiesel, which refers to the straightforward alkyl monoesters of lengthy chain fatty acids obtained from renewable feedstocks, is the optimal alternative to diesel fuel. Consequently, research on this biofuel is experiencing consistent growth worldwide. In Brazil, the primary area of study is the development of biodiesel through the utilization of ethanol, as this type of alcohol is generated in significant quantities inside the nation. Ethanolysis yields a biodiesel that is less environmentally harmful compared to biodiesel made using methyl alcohol, as ethanol is obtained from sugar cane or maize. Production in the remainder of the globe primarily occurs through the methyl route and utilizes heterogeneous catalysts (Pighinelli, 2010). Biodiesel has strong biodegradability in both freshwater and soil environments, with a significant portion being mineralized within a maximum of 28 days under either aerobic or anaerobic circumstances (Makareviciene & Janulis, 2003; Pasqualino et al., 2006; Zhang et al., 1998). The biofuel is considered carbon-free because the plants used to make it absorb more carbon than is emitted when the biofuel is burned (Antolin et al., 2002; Lang et al., 2001; Sharma et al., 2008; Vicente & Martìnez, 2004). In addition, the combustion of biodiesel in diesel engines leads to a decrease in the emissions of hydrocarbons, carbon monoxide, particulate matter, and sulfur dioxide. However, it should be noted that the emission of nitrogen oxides increases due to the oxygen content present in biodiesel (Canakci et al., 2006; Labeckas

& Slavinskas, 2006; Turrio-Baldassarry, 2004). The current market price of biodiesel remains high due to the production expenses associated with its key raw materials, which consist of vegetable or animal fats and oils. According to Pighinelli (2010), almost 80% of the overall expenses in biodiesel manufacturing are attributed to obtaining triglycerols. A recurring issue that has received much attention is the rivalry between "food production" and "energy production". According to certain studies, utilizing the available area for oilseed growing could potentially lead to food shortages. To lower production costs and enhance competitiveness with petroleum diesel, biodiesel producers should select raw materials that are no longer available in their region, like soybean in Brazil. Additionally, they should explore alternative crops such as non-edible oils like Crambe Abyssinica and Jatropha Curcas, as well as utilize waste frying oils (Marchetti et al., 2007). Biodiesel production is often regarded as a highly relevant and contemporary global issue. This chapter will cover the production process of biodiesel, the key factors influencing the chemical reactions, and the crucial considerations for ensuring the quality of biodiesel in relation to its production. Additionally, it will address certain post-production parameters.

2. PRODUCTION OF BIODIESEL

Transesterification is the predominant method employed for the manufacture of biodiesel, and it may be implemented on both small scales, such as in laboratories, and large scales, such as in industrial settings, where millions of gallons of biofuel are produced. Although esterification can also produce biodiesel and is advised for raw materials containing oils with high levels of free fatty acids, this technology is only used commercially in a limited number of businesses. Therefore, we have chosen to only discuss the transesterification process in this chapter. This part will provide an explanation of the mechanics involved in the transesterification reaction, along with an identification of all the process variables that can impact the yield of biodiesel. Additionally, key optimization studies will be provided.

2.1 Transesterification

Transesterification, often referred to as alcoholysis, is the chemical process in which an oil or fat reacts with an alcohol to produce esters and glycerol. In order to fully carry out a transesterification reaction, a precise molar ratio of 3:1 between alcohol and triglyceride is required. Practically, in order to achieve the highest possible yield of ester, it is necessary for this ratio to exceed the stoichiometric ratio. A catalyst is commonly employed to enhance both the rate of reaction and the yield. In order to favor the formation of products, an excess amount of alcohol is employed due to the reversible nature of the reaction (Ma & Hanna, 1999). Figure 1 displays the reaction.

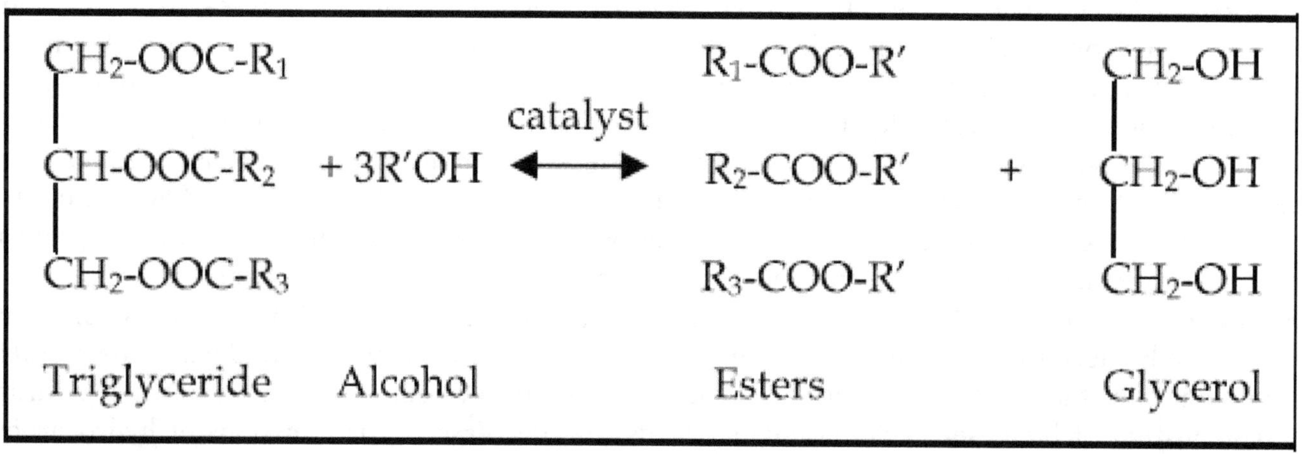

Fig. 1. Transesterification of triglycerides with alcohols.

The conventional technology of biodiesel production employs a basic homogeneous catalyst such as sodium or potassium hydroxides but when oil rich in free fatty acids (FFA) is used the basic catalyst and the FFA will interact to produce soap. This makes the amount of available catalyst for the transesterification reaction to be reduced and also complicates the down streaming separation and the biodiesel purification. Alternative processes for fatty acid ethyl ester (FAEE) production have been under development in order to employ different catalysts as heterogeneous ones (Marchetti & Errazu, 2011) such as metal oxides, metal complexes, active metals loaded on supports, zeolite, resins, membranes, and lipases (Kansedo et al., 2009).

Transesterification consists of a sequence of three consecutive reversible reactions, as can be seen on Figure 2. The first step is the conversion of triglycerides to diglycerides, followed by the conversion of diglycerides to monoglycerides, and finally monoglycerides to glycerol, yielding one ester molecule for each glyceride at each step.

1. Triglyceride (TG) + R'OH $\underset{K_4}{\overset{K_1}{\longleftrightarrow}}$ Diglyceride (DG) + R'COOR$_1$
2. Diglyceride (DG) + R'OH $\underset{K_5}{\overset{K_2}{\longleftrightarrow}}$ Monoglyceride (MG) + R'COOR$_2$
3. Monoglyceride (MG) + R'OH $\underset{K_6}{\overset{K_3}{\longleftrightarrow}}$ Glycerol (GL) + R'COOR$_3$

Fig. 2. The transesterification reactions of vegetable oil with alcohol to esters and glycerol (Freedman et al., 1986).

The main parameters affecting the transesterification reaction are molar ratio of vegetable oil to alcohol, catalyst type and amount, reaction time and temperature, the contents of free fatty acids (FFAs) and water in substrate oil (Freedman et al., 1984) and also the intensity of mixing during the chemical reaction.

2.2.1 The effects of moisture and free fatty acids

The starting materials used for alkali-catalyzed transesterification of glycerides must meet certain specifications. The presence of water during alkali catalyzed transesterification causes a partial reaction change to saponification, which produces soap. For that reason, the glycerides and alcohol must be substantially anhydrous (Wright et al., 1944). A small amount of soap favors the consumption of catalyst and reduces the catalytic efficiency, as well as causing an increase in viscosity, the formation of gels, and difficulty in achieving separation of glycerol. Ma et al. (1998) suggested that the free fatty acid content of the refined oil should be as low as possible, below 0.5%, and Feuge & Grose (1949) also stressed the importance of oils being dry and free of free fatty acids.

The use of alkali catalysts in the transesterification of used cooking oil is somewhat limited because the FFA in used cooking oil reacts with the most common alkaline catalysts (NaOH, KOH, and CH3ONa) and forms soap. Because water makes the reaction partially change to saponification, the alkali catalyst is consumed in producing soap and reduces catalyst efficiency. The soap causes an increase in viscosity, formation of gels which reduces ester yield and makes the separation of glycerol difficult. These two problems not withstanding, literature is replete with studies on the transesterification of waste cooking oil using alkaline catalyst (Marchetti & Errazu, 2010).

2.1.2 The impact of the molar ratio

One of the most important variables affecting the yield of ester is the molar ratio of alcohol to triglyceride. The stoichiometric ratio for transesterification requires three moles of alcohol and one mole of glyceride to yield three moles of fatty acid ester and one mole of glycerol. The molar ratio is associated with the type of catalyst used. For example, a reaction conduced with an acid catalyzed needed a 30:1 ratio of BuOH to soybean oil, while a alkali- catalyzed reaction required only a 6:1 ratio to achieve the same ester yield for a given reaction time (Freedman et al., 1986). Higher molar ratios result in greater ester conversion in a shorter time. During the ethanolysis of peanut oil with a molar ratio alcohol:oil of 6:1 the amount of glycerin liberated was more than did a 3:1 molar ratio (Feuge and Grose, 1949). In this point is important to consider the type of alcohol that is been used. This is because during ethanolysis, as this alcohol has chemical affinity for both glycerine and ester, the higher the molar ratio is more difficult to separate the both phases.

2.1.3 The impact of a catalyst

The catalysts for biodiesel can be separated into two major groups: homogeneous e heterogeneous. Homogeneous type forms a single phase mixture when added to oil and alcohol while the heterogeneous do not mix in the reaction medium. The group of homogeneous catalysts is divided into acid and basic and heterogeneous into metal oxides, metal complexes, active metals loaded on supports, zeolite, resins, membranes, and lipases The criterion for choosing which type of catalyst use should take into account firstly, the quality of raw material, but also the type of alcohol, the costs of the catalysts and technological route to be used for biodiesel production.

2.1.4 The impact of reaction time

The conversion rate of vegetable oils into biodiesel increases with reaction time. Freedman et al. (1984) transesterified peanut, cottonseed, sunflower and soybean oils under the condition of methanol to oil ratio of 6:1, 0.5% sodium methoxide catalyst and 60°C. After 1 minute, a yield of 80% of biodiesel was observed for soybean and sunflower oils, and after 60 minutes, the conversions were almost the same for all four oils.

During the transesterification of beef tallow with methanol the reaction was very slow during the first minute due to the mixing and dispersion of methanol into beef tallow. In the next five minutes, the reaction proceeded very fast (Ma et al., 1998). The production of beef tallow slowed down and reached the maximum value at about 15 min. The di- and monoglycerides increased at the beginning and then decreased. At the end, the amount of monoglycerides was higher than that of diglycerides.

2.1.5 The impact of the temperature of a chemical reaction

Transesterification can occur at different temperatures depending on the vegetable oil used, taking care not to exceed the boiling point of the alcohols used. In methanolysis of castor oil to methyl ricinoleate, with a molar ratio of 6:1-12:1 and 0.005-0.35% (by weight of oil) of

NaOH catalyst, the reaction proceeded most satisfactorily at 20-35°C (Smith, 1949). For the transesterification of refined soybean oil with methanol, molar ratio alcohol:oil of 6:1 and 1% NaOH of catalyst, three different temperatures were used (Freedman et al.,1984). After 0.1 h, ester yields were 94, 87 and 64% for 60, 45 and 32°C, respectively. After 1 h, ester formation was identical for the 60 and 45°C runs and only slightly lower for the 32°C run.

2.2 Biodiesel production optimization

In the world economic context, the optimization of industrial processes is a tool that has been applied frequently in order to reduce the consumption of raw materials, with the main objective of reducing production costs. Response surface methodology (RSM) is a powerful tool for the optimization of chemical reactions and/or industrial processes. The main advantages of this method include: (1) an understanding of how the test variables (process variables) affect the selected process response; (2) the determination of any possible interrelationship among the test variables; and (3) the characterization of the combined effect that all test variables may have on the process response (Myers & Montgomery, 1995). There are only few examples in the literature involving optimization of transesterification of vegetable oils to produce biodiesel, even knowing that the costs of biofuel production are high (Domingos et al., 2008).

Pinzi et al. (2011) evaluated together with the process variables, how raw material fatty acid composition affects the biodiesel production. The authors applied a factorial design to determine how the operation conditions affected the transesterification process (reaction temperature, initial catalyst concentration by oil mass and methanol concentration), while the reaction yield was considered as the response variable. The vegetable oils studied were maize (MME), sunflower (SFME), olive (OOME), coconut (CME), linseed (LME) and palm (PME). A different range of temperature (from 40oC to 60oC) was selected for coconut oil and palm oil, since these oils are rich in unsaturated fatty acids. Molar ratio methanol:oil ranged from 4.2:1 to 5.4:1 and catalyst concentration from 0.8 to 2.1% Optimized conditions for each of the raw materials are shown in Table 1.

A complementary study aimed to define the optimum time of reaction based on a study of reaction kinetics. During the transesterification reaction carried out at the optimized condition, samples were taken at 30s, 1, 2, 5, 10, 35, 60, 90 and 120 min. Kinetic curves showed that every transesterification reaction presents similar performance curve, with the exception of coconut that exhibits the lowest yield of fatty acid methyl esters (FAME). After 20-min reaction oils with longer fatty-acid chains (olive, corn, linseed and sunflower) achieve the optimal yield of FAME, but on the other hand palm oil and coconut oil show the best performance after 40-min reaction. The authors observed that the effect of catalyst concentration was influenced by fatty-acid composition. Vegetable oils composed by unsaturated fatty acids show a directly proportional dependence between the concentration of catalyst and yield, up to a maximum. Whereas, considering vegetable oils with (mono) saturated fatty acids, amounts of catalyst greater than the optimal value lead to soap production. Fatty acids chain length also seems to influence biodiesel conversion. Oils with longer fatty-acid chains need half of the reaction time requested by oils comprising shorter fatty-acid chains to achieve maximum yield.

Silva et al. (2011) discussed in their paper the production process optimization for biodiesel by transesterification of soybean oil with ethanol, where several parameters, including catalyst, alcohol/vegetal oil molar ratio, and temperature would influence the transesterification. The authors' main objective was to study how the process variables (ethanol-to-oil ratio, catalyst concentration, reaction time and temperature) affecting the yield of biodiesel and then optimize this process. The levels of process variables studied were: ethanol/oil ratio (i.e., 3:1, 6:1, 9:1, 12:1 and 15:1), catalyst concentration (0.1%, 0.5%, 0.9%, 1.3% and 1.7% w/v of NaOH), reaction time (40, 60, 80, 100, and 120 min) and temperature (40, 50, 60, 70 and, 80 °C). Optimum values of the process parameter for maximum efficiency (95% of ethyl esters) were molar ratio ethanol: soybean oil 9:1, catalyst concentration 1.3% w/v, temperature 40oC and reaction time 80 minutes. The analysis of the effects of process variables on yield in ethyl esters showed that the molar ratio, catalyst concentration and reaction time had a positive effect while the temperature had a negative effect. A positive effect means that the larger the values of process variables, the greater the yield of biodiesel.

Table 1. Optimization results according to Pinzi et al. (2011)

Factor	Optimum value
MME	Yield: 98.67 (wt%)
Reaction temperature (oC)	47.53
Catalyst concentration (wt%)	1.92
Methanol/oil (molar ratio)	5.4
SFME	Yield: 99.70 (wt%)
Reaction temperature (oC)	59.82
Catalyst concentration (wt%)	1.81
Methanol/oil (molar ratio)	5.4
OOME	Yield: 98.02 (wt%)
Reaction temperature (oC)	45
Catalyst concentration (wt%)	1.6
Methanol/oil (molar ratio)	6.03
CME	Yield: 90.01 (wt%)
Reaction temperature (oC)	60
Catalyst concentration (wt%)	1.7
Methanol/oil (molar ratio)	6.6
LME	Yield: 97.71 (wt%)
Reaction temperature (oC)	53
Catalyst concentration (wt%)	1.8
Methanol/oil (molar ratio)	6.02
PME	Yield: 98.91 (wt%)
Reaction temperature (oC)	65
Catalyst concentration (wt%)	1.81
Methanol/oil (molar ratio)	6.15

Another optimization study usedRaphanus sativus(L. Var) crude oil in ethanolysis with sodium hydroxide as catalyst. Three process variables were used to develop the experimental design: the ethanol:oil molar ratio (MR of 6:1 and 12:1), the catalyst concentration in relation to oil mass (C of 0.4 and 0.8 wt% NaOH) and the alcoholysis temperature (T of 45 and 65oC). This yield was expressed in relation to the oil mass used for ethanolysis, reason why some values were greater than 100%. Reaction temperature had no statistical significance over biodiesel yield. The highest biodiesel yield was 101.7% obtained at 65oC with a MR of 12:1 and 0.4 wt% of C. Nevertheless, when the alcoholysis temperature

was decreased to 45oC, phase separation improved and lower levels of soap accumulation were obtained in the ethyl ester phase. The authors recommend the following procedure for the ethanolysis ofRaphanus sativuscrude oil: MR of 11.7:1, NaOH concentration of 0.4 wt%, 45oC and vigorous agitation for 60 min as the first reaction stage, followed by a second stage in which MR and NaOH concentration can be reduced to 6:1 and 0.03 wt%, respectively (Domingos et al., 2008).

Heterogeneous catalysis optimization was studied by Marchetti & Errazu (2011). The reaction temperature's effects (30, 45 and 55oC), the initial amount of free fatty acid (2.8%, 9.9% and 19.5% w/w), the molar ratio of alcohol/oil (4.2:1, 5.01:1 and 6.1:1) and the type of catalyst (homogeneous - sulfuric acid or heterogeneous - Dowex monosphere 550A) over the main reaction are analyzed and their effects compared. Temperature and molar ratio had a positive effect over biodiesel production: when the temperature and molar ratio increase the final conversion increases as well. When the initial amount of free fatty acid was varied, experimental results show that the final conversion increases as the initial amount of free fatty acid increases. Therefore, this effect could also be seen on the total FAEE production since the final amount of biofuel will be produced from the triglycerides as well as from the fatty acids present in the reaction mixture. The last part of this paper, a comparative study was made between the production of esters using sulfuric acid and a base solid resin with ethanol anhydrous under similar operational conditions, such as T=55oC, initial amount of FFA=9.9% w/w, 2.2% w/w of each catalyst, and a molar ratio of alcohol/mixture of 6.1:1. Sulfuric acid reaches its final conversion in about 3 days time, while base solid resin reaches almost 100% in 70 hours.

As can be seen by the results showed earlier, biodiesel production is influenced by several process variables. The ideal combination of these variables will result in a higher yield in esters as well as a final product of higher quality. In addition, production costs could be reduced, including the industrial level. It was observed that depending on the feedstock, the type of alcohol and catalyst, the optimum conditions change. Another relevant point is related to the diversification of oil sources, which also help to reduce costs and to produce a quantity of fuel that meets global demand. In this scenario come acid oils, such as waste frying, and the heterogeneous catalysts, which are being actively researched.

3. ASSESSMENT OF BIODIESEL QUALITY

Various factors can affect the quality of biodiesel: The feedstock's quality. The composition of fatty acids in the original vegetable oil or animal fat. The production method and the accompanying materials utilized in this procedure. Parameters for post-production.

3.1 Production process factors

3.1.1 Reaction

The most important issue during biodiesel production is the completeness of the transesterification reaction. The triglycerides are converted to diglycerides, which in turn are converted to monoglycerides, and then to glycerol. Each step produces a molecule of a methyl or ethyl ester of a fatty acid. If the reaction is incomplete, then there will be triglycerides, diglycerides, and monoglycerides left in the reaction mixture. Each of these compounds still contains a glycerol molecule that has not been released. The glycerol portion of these compounds is referred to as bound glycerol. When the bound glycerol is added to the free glycerol, the sum is known as the total glycerol.

3.1.2 Free glycerol

Free glycerol refers to the amount of glycerol that is left in the finished biodiesel. Glycerol is essentially insoluble in biodiesel so almost all of the glycerol is easily removed by settling or centrifugation. Free glycerol may remain either as suspended droplets or as the very small amount that is dissolved in the biodiesel. Alcohol can act as co-solvent to increase the solubility of glycerol in the biodiesel. Most of this glycerol should be removed during the purification process. Water-washed fuel is generally very low in free glycerol, especially if hot water is used for washing. Distilled biodiesel tends to have a greater problem with free glycerol due to glycerol carry-over during distillation. Fuel with excessive free glycerol will usually have a problem with glycerol settling out in storage tanks, creating a very viscous mixture that can plug fuel filters and cause combustion problems in the engine.

3.1.3 Residual alcohol and catalyst

Since methanol or ethanol and the alkaline catalysts are more soluble in the polar glycerol phase, most will be removed when the glycerol is separated from the biodiesel.

However, the biodiesel typically contains 2-4% methanol after the separation, which may constitute as much as 40% of the excess methanol from the reaction. Most processors will recover this methanol using a vacuum stripping process. Any methanol remaining after this stripping process should be removed by the water washing process. Therefore, the residual alcohol level in the biodiesel should be very low. A specific value for the allowable alcohol level is specified in European biodiesel standards (0.2% in EN 14214), but is not included in the ASTM standard. Tests have shown that as little as 1% methanol in the biodiesel can lower the flashpoint of the biodiesel from 170ºC to less than 40ºC. Therefore, by including a flashpoint specification of 130ºC, the ASTM standard limits the amount of alcohol to a very low level (<0.1%). Residual alcohol left in the biodiesel will generally be too small to negatively impact the fuel's performance. However, lowering the flashpoint presents a potential safety hazard as the fuel may need to be treated more like gasoline, which also has a low flashpoint, than diesel fuel.

Most of the residual catalyst is removed with the glycerol phase. Like the alcohol, remaining catalyst should be removed during the water washing. Although a value for residual catalyst is not included in the ASTM standard, it will be limited by the specification on sulfated ash. Excessive ash in the fuel can lead to engine deposits and high abrasive wear levels. The European standard EN 14214 places limits on calcium and magnesium as well as the alkali metals sodium and potassium.

3.2 Post-production factors

3.2.1 Water and sediment

Water and sediment contamination are primarily concerns related to the maintenance and cleanliness of biodiesel production. Water can exist in two states: either as dissolved water or as suspended water droplets. Although biodiesel is commonly believed to be insoluble in water, it actually absorbs a significantly larger amount of water compared to diesel fuel. Biodiesel has the potential to contain up to 1500 parts per million (ppm) of dissolved water, whereas diesel fuel typically only contains around 50 ppm (Van Gerpen et al., 1997). The specifications for diesel fuel (ASTM D 975) and biodiesel (ASTM D 6751) both impose a maximum limit of 500 parts per million (ppm) for water content. Petroleum-based diesel fuel can accommodate a small quantity of water in suspension. Nevertheless, it is imperative to maintain the dryness of biodiesel. The presence of water at the bottom of diesel storage tanks is a challenging issue primarily caused by condensation. The presence of suspended water in fuel injection equipment poses a problem as it actively contributes to the corrosion of the closely fitting components inside the fuel injection system.

Water can also contribute to microbial growth in the fuel. This problem can occur in both biodiesel and conventional diesel fuel and can result in acidic fuel and sludges that will plug fuel filters. Sediment may consist of suspended rust and dirt particles or it may originate from the fuel as insoluble compounds formed during fuel oxidation. Some biodiesel users have noted that switching from petroleum-based diesel fuel to biodiesel causes an increase in sediment that comes from deposits on the walls of fuel tanks that had previously contained diesel fuel. Because its solvent properties are different from diesel fuel, biodiesel may loosen these sediments and cause fuel filter plugging during the transition period.

3.2.2 Storage stability

Storage stability pertains to the fuel's capacity to withstand chemical alterations when being stored for an extended period of time. Typically, these alterations involve oxidation resulting from exposure to atmospheric oxygen. The fatty acid makeup of biodiesel fuel plays a crucial role in regulating its resistance to oxidation. In general, the polyunsaturated fatty acids (C18:2, linoleic acid; C18:3 linolenic acid) are highly prone to oxidation. The alterations can be expedited by the existence of specific metals (including those composing the storage container) and illumination. Hydrolysis can occur if there is water. Oxidation of fuel typically leads to chemical reactions that generate hydroperoxides. These hydroperoxides can then give rise to short chain fatty acids, aldehydes, and ketones. Under favorable circumstances, the hydroperoxides can undergo polymerization. Thus, oxidation is commonly indicated by a rise in both the acid value and viscosity of the fuel. Frequently, these alterations are accompanied by a deepening of the biodiesel's hue from yellow to brown and the emergence of an odor resembling that of paint. In the presence of water, esters can undergo hydrolysis, resulting in the formation of long chain free fatty acids. This process also leads to an increase in the acid value. Currently, there is no universally recognized technique for quantifying the stability of biodiesel. The methodologies commonly employed for petroleum-based fuels, such as ASTM D 2274, have been proven to be unsuitable for biodiesel. Alternative methods, such as the Oil Stability Index or the Rancimat equipment, are commonly employed in the fats and oils business and appear to be more suited for analyzing biodiesel. Nevertheless, the engine sector lacks prior expertise in conducting these tests and lacks knowledge of acceptable values. Furthermore, the veracity of accelerated testing procedures has not been validated or linked to real engine issues. If the acid number, viscosity, or sediment content of biodiesel surpass the specified limits set by ASTM, it is not suitable for use as a transportation fuel. Food sector commonly use additives like BHT (butylated hydroxytoluene) and TBHQ (t-butylhydroquinone) to improve the shelf life of biofuels. Soybean oil-derived biodiesel inherently contains antioxidants, such as tocopherols (commonly known as vitamin E), which offer a degree of protection against oxidation. However, a portion of the tocopherols is lost during the refining process of the oil before biodiesel synthesis. Any fuel, including both diesel fuel and biodiesel, that will be stored for a duration exceeding 6 months should be treated with an antioxidant ingredient.

3.3 Quality control

All biodiesel production facilities should be equipped with a laboratory so that the quality of the final biodiesel product can be monitored. It is also important to monitor the quality of the feedstocks. One strategy used by many producers is to draw a sample of the oil (or alcohol) from each delivery and use that sample to produce biodiesel in the laboratory. This test can be fairly rapid (1 or 2 hours) and can indicate whether serious problems are likely in the plant. Measuring feedstock quality can usually be limited to acid value and water content. These are not too expensive and can be operated by less experienced technicians. To monitor the completeness of the reaction according to the total glycerol level specified in ASTM D 6751 requires the use of a gas chromatograph and a skilled operator. Large producers will find that having this equipment on-site is necessary. Commercial laboratories are available that can analyze the samples but there are costs and the time required may be several days. Smaller producers will need to use a more robust production process involving extra methanol or ethanol and probably multiple reaction steps. Then the product quality can be monitored through periodic testing by an outside laboratory.

Other possibilities for monitoring the transesterification reaction and assessing fuel quality are methods based on spectroscopy (such as near- infrared spectroscopy) or physical properties (such as viscometry). These methods are usually faster and easier to use than gas chromatography. However, some of them require extensive calibration. They also cannot accurately quantify glycerol at the low levels called for in the ASTM standard. To circumvent this, comparison to a reaction and product known to meet ASTM standards is needed.

4. STANDARDS FOR BIODIESEL

The main factor determining the quality of biodiesel is its compliance with the relevant standard. The technical criteria of biodiesel vary depending on the country or region of production. Biodiesel is subject to many quality standards, such as the European standard EN 14214 (Table 2), ASTM D6751 (Table 3), and other similar standards. The Comité Européen de Normalisation (CEN) established the European standard for Fatty Acid Methyl Esters (FAME) as vehicle fuel in 2003. This standard is identified by the number EN 14214. This standard establishes thresholds and techniques for measuring FAME, also referred to as biodiesel, which can be utilized either as an independent fuel or as an additive in diesel fuel. The CEN standard for diesel fuel, known as EN 590, mandates that any biodiesel added to the fuel must adhere to the EN 14214 norm. Currently, the European diesel fuel permits the blending of biodiesel up to a maximum of 5% by volume. Several EU nations have established national rules that permit the distribution of biodiesel as a separate fuel, particularly in Germany, for vehicles that have been specifically modified. The CEN is now examining a modified EN 590 specification for diesel fuel that will allow for a biodiesel blend of up to 7%. Currently, CEN is doing research to update the biodiesel standard EN 14214.

Table 2. Biodiesel Standard EN 14214 (Europe)

Property	Test method	Limits		Unit
		min	max	
Ester content	EN 14103	96.5		% (m/m)
Density, 15°C	EN ISO 3675 EN ISO 12185	860	900	kg/m³
Viscosity, 40°C	EN ISO 3104 EN ISO 3105	3.5	5.0	mm²/s
Flash point	EN ISO 3679	120		°C
Sulfur content	EN ISO 20846 EN ISO 20884		10.0	mg/kg
Carbon residue (10% dist. residue)	EN ISO 10370		0.30	% (m/m)
Cetane number	EN ISO 5165	51		
Sulfated ash	ISO 3987		0.02	% (m/m)
Water content	EN ISO 12937		500	mg/kg
Total contamination	EN 12662		24	mg/kg
Copper strip corrosion (3hr, 50°C)	EN ISO 2160	1		
Oxidative stability, 110°C	EN 14112	6.0		hr
Acid value	EN 14111		0.50	mg KOH/g
Iodine value	EN 14111		120	g iodine/100g
Linolenic acid content	EN 14103		12	% (m/m)
Content of FAME with ≥4 double bonds			1	% (m/m)
Methanol content	EN 14110		0.20	% (m/m)
Monoglyceride content	EN 14105		0.80	% (m/m)
Diglyceride content	EN 14105		0.20	% (m/m)
Triglyceride content	EN 14105		0.20	% (m/m)
Free glycerol	EN 14105, EN 14106		0.02	% (m/m)
Total glycerol	EN 14105		0.25	% (m/m)
Alkali metals (Na + K)	EN 14108, EN 14109		5.0	mg/kg
Earth alkali metal (Ca + Mg)	prEN 14538		5.0	mg/kg
Phosphorus content	EN 14107		10.0	mg/kg

The aim is to expand the types of oils that can be used as feedstock, while ensuring that the use of this product in blends or as a standalone fuel does not compromise the safety of cars. The European Commission has instructed CEN to update the EN 590 specification for diesel fuel to allow for a maximum blend of 10% biodiesel. The United States of America has opted to utilize the requirements established by

ASTM International for both traditional diesel fuel and biodiesel. The United States of America initiated specification work for biodiesel in 1993 within Committee D02 on Petroleum Products and Lubricants. The initial proposal at ASTM for biodiesel specifications was for B100, which is pure biodiesel, to be used as a standalone fuel. However, there was not enough data on the performance of biodiesel blends with more than 20% biodiesel and 80% conventional diesel (known as B20) to gain approval from the ASTM members. After 1994, the efforts within ASTM on biodiesel were mostly focused on establishing the specific characteristics of pure biodiesel that would make it a suitable fuel for existing diesel engines, specifically at the B20 level or lower. ASTM accepted a preliminary specification for B100 as a blend stock in 1999. The first complete specification was authorized in 2001 and made available for usage in 2002 under the name "ASTM D6751 Standard Specification for Biodiesel Fuel Blend Stock (B100) for Middle Distillate Fuels".

The approval process for D6751 followed the same idea as that used for the No. 1 and No. 2 grades of fuels under the standard specification, ASTM D975. If the parent fuels match their individual specifications, they can be mixed in any proportion and utilized in traditional diesel engines. If the parent fuels satisfied their respective criteria, there was no need for a distinct set of attributes for the completed blends of No. 1 and No. 2. The aforementioned restrictions apply to biodiesel as well. If biodiesel fulfills the specifications outlined in D6751 and conventional diesel meets the specifications in D975, the two can be mixed together and utilized in conventional engines. However, it is important to note that the maximum allowable biodiesel content in the final fuel blend is 20%.

Table 3. Biodiesel Standard ASTM D6751 (United States)

Property	Test method	Limits	Unit
Flash point (closed cup)	D 93	130.0 min.	°C
Water and sediment	D 2709	0.050 max.	% vol.
Kinematic viscosity, 40°C	D 445	1.9-6.0	mm²/s
Sulfated ash	D 874	0.020 max.	% mass
Sulfur	D 5453	0.0015 max or 0.05 max[a]	% mass
Copper strip corrosion	D 130	No. 3 max	
Cetane number	D 613	47 min	
Cloud point	D 2500	Report	°C
Carbon residue (100% sample)	D 4530	0.050 max	% mass
Acid number	D 664	0.80 max	mg KOH/g
Free glycerin	D 6584	0.020 max	% mass
Total glycerin	D 6584	0.240 max	% mass
Phosphorus content	D 4951	0.001 max	% mass
Distillation temperature, atmospheric equivalent temperature, 90% recovered	D 1160	360 max	°C

[a]The limits are for Grade S15 and Grade S500 biodiesel, respectively. S15 and S500 refer to maximum sulfur specifications (ppm).

While this mode of operation has served the US market well, there has been substantial effort since 2003 to develop and formally approve specifications for the finished blend of biodiesel and conventional diesel fuel. In addition, several improvements and changes to D6751 were also undertaken, some as a result of changes needed to secure approval of the finished blended biodiesel specifications. At the time of this report ballots to allow the formal acceptance of up to 5% biodiesel (B5) into the conventional diesel specifications for on/off road diesel fuel (ASTM D975) and fuel oil burning equipment (ASTM D396) and a new stand alone specification covering biodiesel blends between 6% and 20% have been approved through the Subcommittee level of Committee D02. In addition, a ballot to implement a new parameter in

D6751 to control the potential for filter clogging above the cloud point in B20 blends and lower has also passed the subcommittee and is on track for a June 2008 vote. Efforts to approve B100 and B99 as stand alone fuels have been discussed at ASTM, but have been put on hold in order to focus on the B5 and B6 to B20 blended fuel specification efforts.

This section describes the parameters of the specifications normally used in the biodiesel standards:

4.1 Ester content

This characteristic, like as distillation temperature, is crucial for identifying the existence of additional compounds and, in certain instances, for satisfying the legal criteria for biodiesel, specifically mono-alkyl esters. The low readings of pure biodiesel samples may be attributed to inadequate reaction conditions or the presence of different small components in the original fat or oil source. An elevated level of unsaponifiable compounds, such as sterols, residual alcohols, incomplete glycerides, and unseparated glycerol, can result in results that fall below the specified limit. Distilled methyl esters often have a larger ester concentration compared to undistilled ones, as most of these chemicals are eliminated during the distillation process of the final product (Mittelbach and Enzelsberger, 1999).

4.2 Density

Biodiesels typically exhibit greater densities compared to fossil diesel fuel. The values are contingent upon both the fatty acid composition and the purity of the substances. The density of a substance is directly proportional to its chain length, meaning that as the chain length decreases, the density increases. Additionally, the density also increases with an increasing number of double bonds. On the other hand, the presence of low density pollutants like methanol can decrease the density of the substance.

4.3 Viscosity

Biodiesel exhibits a higher kinematic viscosity compared to fossil fuel, and under some conditions, particularly at low temperatures, it can become highly viscous or even solidify. Increased viscosity has an impact on the rate of fluid flow and the properties of the fuel spray in the engine. Additionally, at low temperatures, it may potentially weaken the mechanical strength of the injection pump drive systems, especially when B100 diesel fuel is used independently.

4.4 Flash point

The flash point is a quantitative measure of the flammability of fuels, making it a crucial safety parameter for transportation and storage purposes. The flash point of petrol diesel fuel is approximately half that of biodiesels, making it a significant safety advantage for biodiesel. Pure biodiesels have a significantly higher flash point than the recommended limits, however this can fall quickly as the proportion of residual alcohol increases. Due to the strong correlation between these two factors, the flash point can serve as an indicator for the existence of methanol in biodiesel. The flash point serves as a criterion for classifying the transportation and storage of fuels, with varying thresholds across different regions. Therefore, achieving consistency in standards would likely necessitate a matching harmonization of regulations.

4.5 Sulfur

The presence of excessive levels of sulfur in fuels has been linked to adverse effects on both human health and the environment. This is why there is currently a tightening of national regulations to reduce sulfur content. Low sulfur fuels play a crucial role in facilitating the implementation of modern emissions control systems. Engines that run on high sulfur fuels release greater amounts of sulfur dioxide and particulate matter, which are considered to have a higher potential for causing genetic mutations. Furthermore, fuels that contain high levels of sulfur contribute to engine degradation and diminish the effectiveness and longevity of catalytic systems. Biodiesel fuels have long been acclaimed for their near absence of sulfur. The national criteria for biodiesel align with the statutory mandates for the highest permissible sulfur levels in conventional diesel fuel within the specific geographical area.

4.6 Carbon residue

Carbon residue refers to the quantity of carbon-based material that remains after a fuel sample is evaporated and subjected to pyrolysis under particular conditions. The word "carbon residue" is included in all three standards, despite the fact that this residue is not exclusively made up of carbon. This is owing to the widespread historical usage of the term. The metric functions as an indicator of the propensity of a fuel sample to generate deposits on injector tips and within the combustion chamber when utilized as automotive fuel. It is regarded as a crucial criterion for biodiesel quality, as it is connected to numerous other characteristics. Regarding biodiesel, the carbon residue is directly related to the quantities of glycerides, free fatty acids, soaps, and any leftover catalyst or impurities (Mittelbach 1996). Additionally, the parameter is affected by elevated levels of PUFA FAME and polymers (Mittelbach and Enzelsberger 1999). Carbon residue is restricted in biodiesel requirements due to these factors.

4.7 Cetane number

The cetane number of a fuel describes its propensity to combust under certain conditions of pressure and temperature. High cetane number is associated with rapid engine starting and smooth combustion. Low cetane causes deterioration in this behaviour and causes higher exhaust gas emissions of hydrocarbons and particulate. In general, biodiesel has slightly higher cetane numbers than fossil diesel. Cetane number increases with increasing length of both fatty acid chain and ester groups, while it is inversely related to the number of double bonds. The cetane number of diesel fuel in the EU is regulated at ≥51. The cetane number of diesel fuel in the USA is specified at ≥40. The cetane number of diesel fuel in Brazil is regulated and specified at ≥42.

4.8 Sulfated ash

Ash content describes the amount of inorganic contaminants such as abrasive solids and catalyst residues, and the concentration of soluble metal soaps contained in the fuel. These compounds are oxidized during the combustion process to form ash, which is connected with engine deposits and filter plugging (Mittelbach, 1996). For these reasons sulfated ash is limited in the fuel specifications.

4.9 Water content and sediment

The Brazilian and American standards combine water content and sediment in a single parameter, whereas the European standard treats water as a separate parameter with the sediment being treated by the Total Contamination property. Water is introduced into biodiesel during the final washing step of the production process and has to be reduced by drying. However, even very low water contents achieved directly after production do not guarantee that biodiesel fuels will still meet the specifications during combustion. As biodiesel is hygroscopic, it can absorb water in a concentration of up to 1000 ppm during storage. Once the solubility limit is exceeded (at about 1500 ppm of water in fuels containing 0.2% of methanol), water separates inside the storage tank and collects at the bottom (Mittelbach 1996). Free water promotes biological growth, so that sludge and slime formation thus induced may cause blockage of fuel filters and fuel lines. Moreover, high water contents are also associated with hydrolysis reactions, partly converting biodiesel to free fatty acids, also linked to fuel filter blocking. Finally, corrosion of chromium and zinc parts within the engine and injection systems have been reported (Kosmehl and Heinrich, 1997). Lower water concentrations, which pose no difficulties in pure biodiesel fuels, may become problematic in blends with fossil diesel, as here phase separation is likely to occur. For these reasons, maximum water content is contained in the standard specifications.

4.10 Total contamination

Total contamination is defined as the quota of insoluble material retained after filtration of a fuel sample under standardized conditions. It is limited to ≤ 24 mg/kg in the European specification for both biodiesel and fossil diesel fuels. The Brazilian and American biodiesel standards do not contain this parameter, as it is argued that fuels meeting the specifications regarding ash content will show sufficiently low values of total contamination as well. The total contamination has turned out to be an important quality criterion, as biodiesel with high concentration of insoluble impurities tend to cause blockage of fuel filters and injection pumps. High concentrations of soaps and sediments are mainly associated with these phenomena (Mittelbach, 2000).

4.11 Copper corrosion

This parameter characterizes the tendency of a fuel to cause corrosion to copper, zinc and bronze parts of the engine and the storage tank. A copper strip is heated to 50°C in a fuel bath for three hours, and then compared to standard strips to determine the degree of corrosion. This corrosion resulting from biodiesel might be induced by some sulfur compounds and by acids, so this parameter is correlated with acid number. Some experts consider that this parameter does not provide a useful description of the quality of the fuel, as the results are unlikely to give ratings higher than class 1.

4.12 Oxidation stability

Due to their chemical composition, biodiesel fuels are more sensitive to oxidative degradation than fossil diesel fuel. This is especially true for fuels with a high content of di - and higher unsaturated esters, as the methylene groups adjacent to double bonds have turned out to be particularly susceptible to radical attack as the first step of fuel oxidation (Dijkstra et al. 1995). The hydroperoxides so formed may polymerize with other free radicals to form insoluble sediments and gums, which are associated with fuel filter plugging and deposits within the injection system and the combustion chamber (Mittelbach & Gangl, 2001). Where the oxidative stability of biodiesel is considered insufficient, antioxidant additives might have to be added to ensure the fuel will still meet the specification.

4.13 Acid value

Acid value or neutralization number is a measure of free fatty acids contained in a fresh fuel sample and of free fatty acids and acids from degradation in aged samples. If mineral acids are used in the production process, their presence as acids in the finished fuels is also measured with the acid number. It is expressed in mg KOH required to neutralize 1g of biodiesel. It is influenced on the one hand by the type of feedstock used for fuel production and its degree of refinement. Acidity can on the other hand be generated during the production process. The parameter characterises the degree of fuel ageing during storage, as it increases gradually due to degradation of biodiesel. High fuel acidity has been discussed in the context of corrosion and the formation of deposits within the engine which is why it is limited in the biodiesel specifications of the three regions. It has been shown that free fatty acids as weak carboxylic acids pose far lower risks than strong mineral acids (Cvengros, 1998)

4.14 Iodine value, linolenic acid ester content and polyunsaturated

The iodine number is a quantitative indicator of the overall degree of unsaturation present in a combination of fatty acids. It is determined by the amount of iodine, in grams, that reacts with 100 grams of biodiesel. Engine manufacturers contend that fuels with elevated iodine numbers have a propensity to undergo polymerization and generate deposits on injector nozzles, piston rings, and piston ring grooves under high temperatures (Kosmehl and Heinrich 1997). In addition, it is believed that the presence of unsaturated esters in the engine oil might lead to the formation of high-molecular compounds that have a detrimental impact on the lubricating properties. This can ultimately result in engine damage, as stated by Schaefer et al in 1997. Nevertheless, the outcomes of many engine experiments suggest that polymerization events primarily occur in fatty acid esters that possess three or more double bonds (Worgetter et al., 1998; Prankl and Worgetter, 1996; Prankl et al., 1999). Seed oils that include three or more unsaturated esters make up only a small portion of their fatty acid composition. These oils are not considered suitable for use as feedstock according to some regional standards since they have a high iodine value. Several biodiesel specialists have proposed restricting the presence of linolenic acid methyl esters and polyunsaturated biodiesel, rather than focusing on the overall level of unsaturation as indicated by the iodine value.

4.15 Methanol or ethanol

Methanol (MeOH) or ethanol (EtOH) can cause fuel system corrosion, low lubricity, adverse affects on injectors due to its high volatility, and is harmful to some materials in fuel distribution and vehicle fuel systems. Both methanol and ethanol affect the flash point of esters. For these reasons, methanol and ethanol are controlled in the specification.

4.16 Mono, di and triglyceride

The EU standard specifies individual limit values for mono-, di- and triglyceride as well as a maximum value for total glycerol. The standards for Brazil and the USA do not provide explicit limits for the contents of partial acylglycerides. In common with the concentration of free glycerol, the amount of glycerides depends on the production process. Fuels out of specification with respect to these parameters are prone to deposit formation on injection nozzles, pistons and valves (Mittelbach et al. 1983).

4.17 Free glycerol

The content of free glycerol in biodiesel is dependent on the production process, and high values may stem from insufficient separation or washing of the ester product. The glycerol may separate in storage once its solvent methanol has evaporated. Free glycerol separates from the biodiesel and falls to the bottom of the storage or vehicle fuel tank, attracting other polar components such as water, monoglycerides and soaps. These can lodge in the vehicle fuel filter and can result in damage to the vehicle fuel injection system (Mittelbach 1996). High free glycerol levels can also cause injector coking. For these reasons free glycerol is limited in the specifications.

4.18 Total glycerol

Total glycerol is the sum of the concentrations of free glycerol and glycerol bound in the form of mono-, di- and triglycerides. The concentration depends on the production process.

Fuels out of specifications with respect to these parameters are prone to coking and may thus cause the formation of deposits on injector nozzles, pistons and valves (Mittelbach et al. 1983). For this reason total glycerol is limited in the specifications of the three regions.

4.19 Metals (Na+K) and (Ca+Mg)

Metal ions are introduced into the biodiesel fuel during the production process. Whereas alkali metals stem from catalyst residues, alkaline-earth metals may originate from hard washing water. Sodium and potassium are associated with the formation of ash within the engine, calcium soaps are responsible for injection pump sticking (Mittelbach 2000).These compounds are partially limited by the sulphated ash, however tighter controls are needed for vehicles with particulate traps. For this reason these substances are limited in the fuel specifications.

4.20 Phosphorus

Phosphorus in biodiesel stems from phospholipids (animal and vegetable material) and inorganic salts (used frying oil) contained in the feedstock. Phosphorus has a strongly negative impact on the long term activity of exhaust emission catalytic systems and for this reason its presence in biodiesel is limited by specification.

4.21 Distillation

This characteristic, such as the ester content, plays a crucial role in identifying the existence of additional compounds and, in certain instances, in complying with the legal definition of biodiesel, namely monoalkyl esters.

4.22 Cold climate operability

The low-temperature behavior of automotive diesel fuel is a crucial factor in determining its quality. If the fuel partially or completely solidifies, it can clog the fuel lines and filters, resulting in fuel starvation. This can lead to difficulties in starting the engine, driving, and even cause engine damage due to insufficient lubrication. The melting point of biodiesel products is determined by the length of the carbon chain and the number of double bonds. Biodiesel made from long chain saturated fatty acid esters tends to have poor performance at low temperatures.

5. CONCISE SUMMARY

Biodiesel is a significant emerging biofuel alternative. It can be derived from several sources such as vegetable oils or animal fats. Traditional processing methods utilize an alkali catalyzed process, which is not suitable for lower-cost feedstocks with significant levels of free fatty acids due to the generation of soap. Studies have demonstrated that pretreatment procedures with potent acid catalysts offer favorable conversion rates and produce final products of exceptional quality. These processes have also been expanded to enable the manufacture of biodiesel from feedstocks such as soapstock, which are commonly regarded as waste. Strict adherence to a quality standard is crucial for optimal fuel performance in the engine and is a prerequisite for the widespread adoption of biodiesel.

REFERENCES

Antolin, G.; Tinaut, F.V.; Briceno, Y.; Castano, V.; Perez, C. & Ramirez, A.L. Optimization of biodiesel production by sunflower oil transesterification.Bioresource Technology, Vol. 83 (2002), pp. 111–114, ISSN 0960-8524.

Canakci, M.; Erdil, A. & Arcaklioglu, E. Performance and exhaust emission of a biodiesel engine.Applied Energy, Vol. 83 (2006), pp. 594-605, ISSN 0306-2619.

Cvengros, J. Acidity and corrosiveness of methyl esters of vegetable oils.Fett/Lipid,Vol. 100, No 2 (1998), pp. 41-44, ISSN 1521-4133.

Dijkstra, A. J.; Maes, P. J.; Meert, D. & Meeussen, W. Interpreting the oxygen stability index.Oils-Fats-Lipids 1995 Proceedings of the World Congress of the International Society for Fat Research, pp. 629-637, The Hague, Netherland: P. J. Barnes & Associates, 1995.

Domingos, A. K.; Saad, E. B.; Wilhelm, H. M. & Ramos, L. P. Optimization of the ethanolysis ofRaphanus sativus(L. Var.) crude oil applying the response surface methodology.Bioresource Technology, Vol. 99 (2008), pp. 1837–1845, ISSN 0960-8524.

Ferella, F.; Di Celso, G. M.; De Michelis, I.; Stanisci, V. & Vegliò, F. Optimization of the transesterification reaction in biodiesel production.Fuel, Vol. 89 (2010), pp. 36 – 42, ISSN 0016-2361.

Feuge, R. O. & Grose, T.: Modification of vegetable oils. VII. Alkali catalyzed interesterilication of peanut oil with ethanol.Journal of the American Oil Chemists Society., Vol. 26 (1949), pp. 97-102, ISSN 1558-9331.

Freedman, B.; Butterfield, R.O. & Pryde, E.H. Transesterification kinetics of soybean oil.Journal of the American Oil Chemists Society,Vol. 63 (1986), pp. 1375–1380, ISSN 1558- 9331.

Freedman, B.; Pryde, E.H. & Mounts, T.L. Variables affecting the yield of fatty esters from transesterified vegetable oils.Journal of the American Oil Chemists Society, Vol. 61 (1984), pp. 1638–1643, ISSN 1558-9331.

Janaun, J. & Ellis, N. Perspectives on biodiesel as a sustainable fuel.Renewable and Sustainable Energy Reviews, Vol. 14 (2010), pp. 1312 – 1320, ISSN 1364-0321.

Kansedo, J.; Lee, K. T. & Bhatia, S. Biodiesel production from palm oil via heterogeneous transesterification. Biomass & Bioenergy, Vol. 33 (2 0 09), pp. 271–276, ISSN 0961- 9534.

Kosmehl, S. O. & Heinrich, H. (1997). The Automotive Industry's View on the Standards for Plant Oil-Based Fuels, In:Plant Oils as Fuels. Present State of Science and Future Developments, N. Martini & J. Schell (Eds.), 18-28, ISBN 3540647546 9783540647546, Berlin, Germany.

Labeckas, G.; Slavinskas S. The effect of rapeseed oil methyl ester on direct injection diesel engine performance and exhaust emissions.Energy Conversion Manage, Vol. 47 (2006), pp. 1954–1967, ISSN 0196-8904.

Lang, X.; Dalai, A.K.; Bakhshi, N.N.; Reaney, M.J. & Hertz, P.B. Preparation and characterization of bio-diesels from various bio-oils.Bioresource Technology, Vol. 80 (2001), pp. 53–62, ISSN 0960-8524.

Ma, F. & Hanna, M. A. Biodiesel production: a review.Bioresource Technology, Vol. 70 (1999), pp. 1–15, ISSN 0960-8524.

Ma, F., Clements, L. D. & Hanna, M. A.: Biodiesel fuel from animal fat. Ancillary studies on transesterification of beef tallow.Industrial Engineering Chemistry Research, Vol. 37 (1998), pp. 3768-3771, ISSN 1520-5045.

Makareviciene, V. & Janulis, P. Environmental effect of rapeseed oil ethyl ester.Renewable Energy, Vol. 28 (2003), pp. 2395–2403, ISSN 0960-1481.

Marchetti, J.M. & Errazu, A.F. Biodiesel production from acid oils and ethanol using a solid basic resin as catalyst.Biomass & Bioenergy, Vol. 34 (2 0 1 0), pp. 272–277, ISSN 0961-9534.

Marchetti , J.M., Miguel, V. U. & Errazu, A. F. Heterogeneous esterification of oil with high amount of free fatty acids.Fuel, Vol. 86 (2007), pp. 906–910, ISSN 0016-2361.

Mittelbach, M. Chemische und motortechnische Untersuchungen der Ursachen der Einspritzpumpenverklebung bei Biodieselbetrieb; Bund-Bundeslander- kooperations-projekt (2000).

Mittelbach, M. & Gangl, S. Long storage stability of biodiesel made from rapeseed and used frying oil. Journal of the American Oil Chemists Society, Vol. 78, No 6 (2001), pp. 573- 577, ISSN 1558-9331.

Mittelbach, M., & Enzelsberger, H. Transesterification of heated rapeseed oil for extending diesel fuel,Journal of the American Oil Chemists Society,Vol. 76 (1999), pp. 545–550, ISSN 1558-9331.

Mittelbach, M. Diesel fuel derived from vegetable oils, VI: specifications and quality control of biodiesel,Bioresource Technology, Vol. 56 (1996), pp. 7-11, ISSN 0960-8524.

Mittelbach, M., Worgetter, M.; Pernkopf, J. & Junek, H. Diesel fuel derived from vegetable oils: preparation and use of rape oil methyl ester.Energy in Agriculture,Vol. 2 (1983), pp. 369-384, ISSN 0167-5826.

Myers, R.H. & Montgomery, D.C. (1995).Response surface methodology: process and product optimization using designed experiments, John Wiley, Canada.

Pasqualino, J.C.; Montane, D. & Salvado, J. Synergic effects of biodiesel in the biodegradability of fossil-derived fuels.Biomass & Bioenergy, Vol. 30 (2006), pp. 874– 879, ISSN 0961-9534.

Pighinelli, A.L.M.T. Study of mechanical expeller and ethanolic transesterification of vegetable oils. PhD Thesis, School of Agricultural Engineering, State University of Campinas (UNICAMP), Campinas, 2010.

Pinzi, S.; Mata-Granados, J.M.; Lopez-Gimenez, F.J.; Luque de Castro, M.D. & Dorado, M.P. Influence of vegetable oils fatty-acid composition on biodiesel optimization.Bioresource Technology, Vol. 102 (2011), pp. 1059–1065, ISSN 0960-8524.

Prankl, H. & Worgetter, M.: Influence of the iodine number of biodiesel to the engine performance,Liquid Fuels and Industrial Products from Renewable Resources. Proceedings of the 3rdLiquid Fuel Conference, pp. 191-196, Nashville, Tennessee, USA, September 15-17, 1996.

Prankl, H.; Worgetter, M. & Rathbauer, J. Technical performance of vegetable oil methyl esters with a high iodine number.4thBiomass Conference of the Americas. Oakland, California, USA, 29 August to 2 September, 1999.

Schafer, A., D. Naber and M. Gairing. Biodiesel als alternativer Kraftstoff für Mercedes- Benz-Dieselmotoren. Mineralöltechnik,Vol. 43 (1997), pp. 1-32, ISSN: 0307-6490.

Sharma, Y.C.; Singh, B. & Upadhyay, S.N. Advancements in development and characterization of biodiesel: a review.Fuel, Vol. 87 (2008), pp. 2355–2373, ISSN: 0016-2361.

Silva, G. F.; Camargo, F.L. & Ferreira, A.L.O. Application of response surface methodology for optimization of biodiesel production by transesterification of soybean oil with ethanol. Fuel Processing Technology, Vol. 92 (2011), pp. 407–413, ISSN: 0378-3820.

Smith, M.K. Process of producing esters. US Patent 2 (1949), pp. 444-486.

Turrio-Baldassarry, L. Emissions comparison of urban bus engine fuelled with diesel oil and biodiesel blend.Science of the Total Environment. Vol. 327 (2004), pp. 147–162, ISSN: 0048-9697.

Van Gerpen, J.H.; Hammond, E.H.; Yu, L. & Monyem, A. Determining the influence of contaminants on biodiesel properties.Society of Automotive Engineers Technical Paper Series, Paper No. 971685, SAE, Warrendale, PA, USA, 1997.

Vicente, G.; Martìnez, M. & Aracil, J. Integrated biodiesel production: a comparison of different homogeneous catalysts systems.Bioresource Technology, Vol. 92 (2004), pp. 297–305, ISSN: 0960-8524.

Worgetter, M.; Prankl, H. & Rathbauer, J. Eigenschaften von Biodiesel.Landbauforschung Völkenrode. Sonderheft190 (Biodiesel- Optimierungspotentiale und Umwelteffekte) (1998), 31-43.

Wright, H. J., Segur, J. B., Clark, H. V., Coburn, S. K., Langdon, E. E. & DuPuis, R.N. A report on ester interchange.Oil Soap, Vol. 21(1944), pp. 145-148.

Zhang, X.; Peterson, C.; Reece, D.; Moller, G. & Haws, R. Biodegradability of biodiesel in aquatic environment. Transactions of The American Society of Agricultural Engineers, Vol. 41 (1998), pp. 1423–1430, ISSN 0001-2351.

CHAPTER- 11

BIOBUTANOL PRODUCTION AND UTILIZATION PERSPECTIVES

1. INTRODUCTION: AN OVERVIEW

In the present era, there is a growing demand for liquid fuels that may be used in transportation. As a result, there is a strong and extensive search for alternatives to fuels obtained from crude oil. Aside from bioethanol and ethyl or methyl esters of rapeseed oil, which are currently utilized as bio-components in European transportation fuels, researchers are exploring other alternatives. One of these options is biobutanol, which can be classified as a second-generation biofuel if it is derived from waste biomass or non-food agricultural products. While the production of this substance through biotechnology is more complex than the production of bioethanol, it offers several advantages for fuel preparation. These include a higher energy content, lower miscibility with water, lower vapor pressure, and lower corrosivity. Additionally, the producer of this substance, Clostridium bacteria, has the ability to ferment almost all available substrates. These factors may outweigh any drawbacks, making it a favorable option. The primary objective of this chapter is to provide a concise overview of the history of industrial biobutanol production. It aims to introduce the challenges associated with butanol formation by clostridia, including a brief description of different fermentation arrangements. Furthermore, it aims to present comprehensive fermentation data using lesser-known butanol producers, namely Clostridium pasteurianum NRRL B-592 and Clostridium beijerinckii CCM 6182. This text provides a concise summary of the utilization of biobutanol as a fuel for internal combustion engines. It focuses on the properties of biobutanol and its blends with petroleum fuels, as well as their emission characteristics. The emission measurement results obtained from three different passenger cars are used to illustrate these characteristics.

2. THEORETICAL FRAMEWORK

2.1. The historical development of industrial biobutanol synthesis.

The commencement of industrial acetone-butanol-ethanol (ABE) production through Clostridium fermentation is attributed to chemist Chaim Weizmann, who conducted his research at the University of Manchester in the United Kingdom. Weizmann aimed to synthesize rubber with butadiene or isoprene units using butanol or isoamyl alcohol as raw materials. He focused his efforts on identifying and isolating microbial organisms capable of producing butanol. Moreover, the advancement of the acetone-butanol process was expedited during World War I, since acetone derived from ABE fermentation of corn in Dorset, UK was utilized in the manufacturing of cordite. Nevertheless, in 1916, the German blockade impeded the provision of grain, prompting the relocation of production to Canada. Subsequently, with the United States joining the war, two distilleries in Terre Haute were repurposed for acetone manufacturing. Following the war, a consortium of American entrepreneurs acquired the Terre Haute facility and reinstated production in 1920. During this period, butanol gained recognition as a valuable solvent for automotive lacquers. Following the decrease in the price of molasses, new strains of solventogenic microorganisms were identified. In 1935, the first facility utilizing this feedstock was constructed at Bromborough, England, located near the port. By 1936, the Weizmann patent had expired, leading to the construction of new acetone-butanol facilities in several countries including the U.S.A., Japan, India, Australia, and South Africa. These plants often utilized molasses as the substrate. The Second World War further expedited the process of growth, with acetone emerging as the most essential product. The Bromborough facility underwent expansion, implementing a semi-continuous fermentation method that reduced fermentation time to 30-32 hours. Additionally, continuous distillation was successfully used. Following the war, almost two-thirds of the butanol production in the United States was achieved by fermentation. However, the growth of the petrochemical sector, along with the rising cost of molasses (which was being used for cattle feed), led to a progressive fall in industrial acetone-butanol fermentation. The majority of plants in Western countries had ceased operations by 1960, except for the Germiston factory in South Africa. This mill was able to continue production until 1983 due to the availability of inexpensive molasses and coal (Jones & Woods, 1986). Not only Western countries, but also the Soviet Union promoted the manufacturing of acetone and butanol. In the Dukshukino facility during the 1980s, the process was carried out in a semi-continuous manner using a multi-stage setup. This allowed for the combination of both saccharidic and starchy substrates, together with a small amount (up to 10%) of lignocellulosic hydrolyzate. Continuous distillation was also employed (Zverlov et al., 2006). The industrial manufacture of acetone and butanol through fermentation started in China around 1960. In the 1980s, there was a significant growth of this technique. The original batch fermentation method was modified into a semi-continuous 4-stage process. This modification resulted in a shorter fermentation cycle of 20 hours, a yield of around 35-37% from starch, and a productivity that was 2.3 times greater compared to the batch process (Chiao & Sun, 2007). In the late 20th century, it is likely that a majority of Chinese factories were shut down (Chiao & Sun, 2007). However, currently, China produces hundreds of thousands of tons of acetone and butanol per year through fermentation (Ni & Sun, 2009).

Industrial production of ABE in the former Czechoslovakia started with a slight delay comparing with other already mentioned countries. Bacterial cultures were isolated, selected and tested for many years by professor J. Dyr, head of the Department of Fermentation Technology of the Institute of Chemical Technology in Prague who lead a small research team and preparatory works for the plant design (Dyr & Protiva, 1958). Acetone - butanol plant was fully in operation from 1952 till 1965. The main raw materials were firstly potatoes which were later changed for rye. Various bacteria cultures (all were classified asClostridium acetobutylicum)were prepared for several main crops (potatoes, rye, molasses) which increased flexibility of the production. Annual production of solvents increased from year to year but did not exceed 1000 tons. Concentration of total solvents in the broth varied around 17–18 g.L-1. Process itself was run as batch, pH was never controlled, propagation ratio in large fermentation section was 1 : 35. The whole fermentation time was on average 36–38 h. Critical point for each fermentation was "break" in acidity after which started a strong evolution of gases and solvents. In case of potatoes and rye there were no nutrients supplied to the fermentation broth. The only process necessary for the pre-treatment of the raw materials of starch origin was their steaming under pressure in Henze cooker. Initial concentration of starch ranges from 4.5 to 5% wt. In spite of keeping all sanitary precaution (similarly today´s GMP) two types of unexpected failures occurred. Firstly it was contamination by bacteriophage (not possible to analyze it in those times) which appeared approx. three times during the lifetime and always was followed by a total sanitation and complete change of the producing strain. Secondly there appeared another unexpected event, i.e. a final turn to a complete acidification without initiation of solvent production indicated by a spore creation. This situation appeared in the range from 1 to 4% of the total number of batches.

2.2. The principle of acetone-butanol-ethanol (ABE) fermentation

The butanol production through acetone-butanol-ethanol (ABE) fermentation is an unique feature of some species of the genusClostridium; the most famous of them are strains ofC.acetobutylicum,C. beijerinckiiandC.saccharoperbutylacetonicumbut others with the same ability exist, too. Together with allClostridiumbacteria, solvent producers share some common characteristics like rod-shaped morphology, anaerobic metabolism, formation of heat resistant endospores, incapability of reduction of sulphate as a final electron acceptor and G+type of bacterial cell wall (Rainey et al., 2009).

ABE fermentation consists of two distinct phases, acidogenesis and solventogenesis. While the first one is coupled with growth of cells and production of butyric and acetic acids as main products the second one, started by medium acidification, can be characterized by initiation of sporulation and metabolic switch when usually part of formed acids together with sugar carbon source are metabolized to 1-butanol and acetone. The biphasic character of ABE fermentation coupled with alternation of symmetric and asymmetric cell division, first mentioned byClarke et al., (1988), is shown in Fig. 1. In the batch cultivation, first acidogenic phase is connected with internal energy generation and accumulation and also cells growth while second solventogenic phase is bound with energy consumption and sporulation. The tight connection of sporulation and solvents production was proved by finding a genespo0Aresponsible for both sporulation and solvent production initiation (Ravagnani et al., 2000). Metabolic pathway leading to solvents production and originating in Embden-Mayerhof- Parnas (EMP) glycolysis is shown in Fig.1, too. Pentoses unlike hexoses are converted to fructose-6-phosphate and glyceraldehyde-3-phosphate prior to their entrance to EMP metabolic pathway. Major products of the acidogenic phase - acetate, butyrate, CO_2 and H2are usually accompanied by small amounts of acetoin and lactate (not shown in Fig.1). The onset of solvents production is stimulated by accumulation of acids in cultivation medium together with pH drop. Butanol and acetone are formed partially from sugar source and partially by reutilization of the formed acids; and simultaneously a hydrogen production is reduced to a half in comparison with the acidogenic phase (Jones & Woods, 1986;Lipovsky et al., 2009). Functioning of all enzymes involved in the butanol formation has been reviewed, recently (Gheslaghi et al., 2009). Unfortunately, butanol is highly toxic to the clostridia and its stress effect causes complex response of the bacteria in which more than 200 genes regulating membrane composition, cell transport, sugar metabolism, ATP formation and other functionalities are involved and complicate any effort to increase butanol resistance (Tomas et al., 2004).

Solventogenic clostridia are known for their capabilities to utilize various mono-, di-, oligo- and polysaccharides like glucose, fructose, xylose, arabinose, lactose, saccharose, starch, pectin, inulin and others but usually the specific strain is not able to utilize efficiently all of named substrates. Although all genes of cellulosome were identified inC.acetobutylicumATCC 824 genome, the whole cellulosome is not functional what results in incapability of cellulose utilization (Lopez-Contreras et al., 2004). At first, starchy substrates like corn and potatoes were used for ABE fermentation but later blackstrap molasses became the preferential feedstock. Nowadays, a lot of researchers aim to use lignocellulosic hydrolyzates which, if available at a reasonable price and quality (no inhibitors), would be ideal feedstock for this process because clostridia can utilize diluted solutions of various hexoses, pentoses, disaccharides and oligosacharides efficiently.

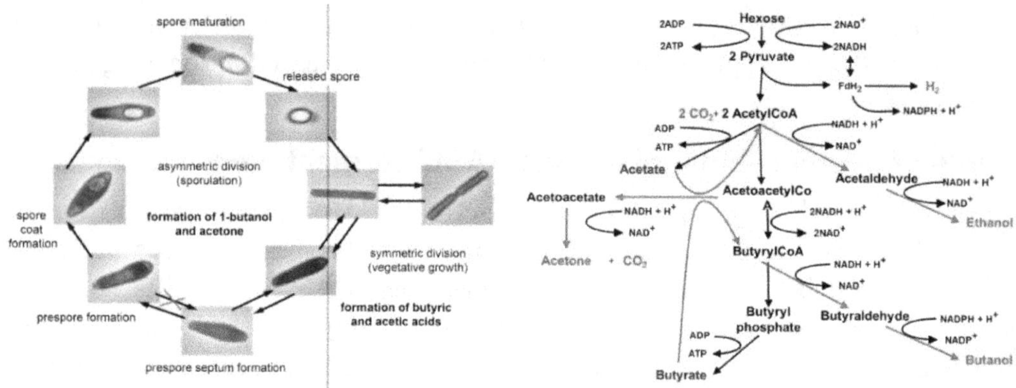

Figure 1. Life cycle of solventogenic clostridia and simplified metabolic scheme

2.2.1. Difficulties in the synthesis of butanol

Production of biobutanol by clostridia is not straightforward process and 1-butanol is neither a typical primary metabolite, the formation of which is connected with cells growth, nor a typical secondary metabolite like antibiotics or pigments. The metabolic switch from acido- to solventogenesis, regulation of which is usually connected with sporulation initiation, does not need to happen necessarily during the fermentation. Actually, when cells are well nourished and their growth rate approaches its maximum then cells reproduce and form only acids; this state has been many times observed in continuous cultivations (Ezeji et al., 2005) but sometimes it can occur even in batch cultivation as so-called "acid crash" (Maddox et al., 2000;Rychtera et al., 2010) which was generally ascribed to fast acetic and butyric acids formation. The proposed acid crash prevention was careful pH control or metabolism slowdown by lowering cultivation temperature (Maddox et al., 2000). However, very recently the novel possible explanation of this phenomenon has been revealed in intracellular accumulation of formic acid by C.acetobutylicumDSM 1731 (Wang et al., 2011). If acid crash is the phenomenon that usually happens at random in the particular fermentation, so-called strain degeneration is a more serious problem when the production culture loses either transiently or permanently its ability to undergo the metabolic shift and to produce solvents. The reliable prevention of the degeneration is maintaining the culture in the form of spore suspension (Kashket & Cao, 1995). A cause of degeneration was investigated in many laboratories using various clostridial strains and therefore also with different results. The degeneration of C.acetobutylicumATCC 824 is probably caused by loss of its mega plasmid containing genes for both sporulation and solvents production (Cornillot et al., 1997) but mechanism and reason of this degeneration were not offered by this study. Actually the authors (Cornillot et al., 1997) compared wild-type strain C.acetobutylicumATCC 824 with isolated degenerated mutants. It is questionable how often or under which conditions the degeneration of C.acetobutylicumATCC 824 happens because in the past, it was reported 218 passages of vegetative C.acetobutylicumATCC 824 cells did not almost influence their solvents formation (Hartmanis et al., 1986). The cells ofC.saccharoperbutylacetonicumN1-4 degenerated when quorum sensing mechanism in the population was impaired (Kosaka et al., 2007). The very detailed study of C.beijerinckiiNCIMB 8052 degeneration disclosed two different degeneration causes: involvement of global regulatory gene and defect in NADH generation (Kashket & Cao, 1995). It seems probable that degeneration has no single reason and if other strains were studied different reasons would be found.

ABE industrial fermentation was probably the first process that had to cope with bacteriophage infection of producing microorganism. The first severe bacteriophage attack was reported from Terre Haute plant in the U.S.A. in 1923 and the problems occurred at fermentation of corn by Clostridium acetobutylicum (the solvents yield was decreased by half for a year). From that time, Clostridiumstrains used for either starch or saccharose fermentations were attacked by various both lysogenic and lytic bacteriophages what was documented in the literature. The ABE plant in Germiston in South Africa faced to confirmed bacteriophage infection 4- times in its 46-year history (plus two unconfirmed cases). Till now, the best solution in battle against Clostridiumbacteriophages seems to be the prevention i.e. good process hygiene, sterilization, decontamination and disinfection (Jones et al., 2000).

Lactic acid bacteria represent the most common type of contamination having very similar requests for cultivation conditions (temperature, pH, anaerobiosis, composition of cultivation media) as clostridia and grow faster. These bacteria can cause not only losses in solvents yield but also can hamper the metabolic switch of clostridia because formed lactic acid over-acidifies the medium and poisons the clostridia in higher concentration. Other contaminants like Bacillusbacteria or yeast are encountered only scarcely (Beesch, 1953).

2.3. Innovative methods for the manufacture of biobutanol

In the past industrial applications, batch fermentation was a usual way how to produce biobutanol due to arrangement simplicity and attaining maximum biobutanol concentration, given by the used strain and cultivation medium, at the end of fermentation. Fed-batch fermentation can be regarded as modification of the batch process offering slight productivity increase by reduction of lag growth phase. However, taking into account possible industrial scale of the process, the preferential process arrangement is continuous ABE fermentation due to a lack of so called "dead" operation times. Nevertheless, its accomplishment in single bioreactor e.g. as chemostat is not usually easy because of biphasic process character when butanol production is not connected with growth directly (seeFig. 1). Theoretically, clostridial culture behaviour under chemostat cultivation conditions should follow an oscillation curve when acidogenesis is coupled with cell multiplication and decrease of substrate concentration. On the contrary, solventogenesis is coupled with decrease of specific growth rate due to sporulation what leads to cells wash-out and increase of substrate concentration in the medium. These two states should cycle regularly (Clarke et al., 1988) but in practice, irregular cycling with various depths of individual amplitudes is more probable as demonstrated several times (S.M.Lee et al., 2008). Moreover, chemostat cultivation conditions induce selection pressure on the microbial culture favouring non-sporulating, quickly multiplying cells what may cause culture degeneration i.e. the loss of the culture ability to produce solvents (Ezeji et al., 2005).

However, there are other options, tested in laboratory scale, how to arrange continuous ABE fermentation like multi-stage process splitting clostridial life cycle into at least two vessels, where first smaller bioreactor serves mainly for cells multiplication under higher dilution rate and in the second bigger bioreactor, actual solventogenesis takes place (Bahl et al., 1982). In addition, battery of bioreactors working in batch, fed-batch or semi-continuous regime ensuring continuous butanol output can also be considered continuous fermentation (Ni & Sun, 2009;Zverlov et al., 2006).

ABE fermentation in any regime can be combined with cells immobilization performed by different methods – entrapment in alginate (Largier et al., 1985), use of membrane bioreactor (Pierrot et al., 1986) or cells adsorption on porous material (S.Y.Lee et al., 2008;Napoli et al., 2010). Recently, final report of the US DOE grant (Ramey & Yang, 2004) has revealed a novel approach toward ABE fermentation. The principle of this solution is two step butanol production employing two microorganisms; at first Clostridium tyrobutyricumproduces mainly butyric acid which is consumed by second microorganism

Clostridium acetobutylicumand utilized for butanol production. The authors claimed they reached 50% yield of butyric acid in the first phase and 84% yield of butanol from butyrate. However, a pilot and a production plant planned for year 2005 have not been realized, yet. Nevertheless, this way of butanol production is still under research in U.S.A. (Hanno et al., 2010), focusing mainly on solventogenic clostridia that are capable of butyrate utilization for butanol production.

One of the main constraints of biotechnological butanol production is its low final concentration in fermented cultivation media caused by its severe toxicity toward producing cells. Average butanol concentration, stable reached in Germiston plant in South Africa, was 13 g.L-1(Westhuizen et al., 1982). Although higher butanol concentration (about 20 g.L-1) can be attained using e.g. mutant strainC. beijerinckiiBA101 (Qureshi & Blaschek, 2001a) cost of distillation separation is still high. Therefore efficient preconcentration methods applied either after the fermentation or more often during the fermentation are being searched now. Moreover, if such separation method is integrated with fermentation process it will increase amount of utilized substrate by alleviating product toxicity. Preferential separation methods in this context seem to be gas stripping (Ezeji et al., 2003), adsorption on zeolites or pervaporation (Oudshoorn et al., 2009).

3. BIOBUTANOL FERMENTATION EXPERIENCE AT ICT PRAGUE

The majority of the work was conducted using the strain Clostridium pasteurianum NRRL B-592, which differed significantly from the commonly used solvent-producing clostridia. One notable difference was the earlier commencement of solvent production, occurring during the exponential growth phase. The strain was selected based on its desirable characteristics, such as its ability to grow steadily and produce solvents, its resilience to slight changes in growing circumstances, and its resistance to strain degeneration. However, in certain instances, alternative solvent-producing strains, such as C.acetobutylicum DSM 1731 and C.beijerinckii CCM 6182, were also employed. The paper by Patakova et al., (2009 and 2011a) provides detailed information on the composition of growth media, maintenance of strains, description of cultivation methods, analytical methods employed, and expressions used to calculate fermentation parameters such as yield and productivity for batch, fed-batch, and continuous fermentations.

3.1. Methods of ABE study

Although fermentation control is a complex process, it mostly depends on a limited number of observable quantities such as pH or redox potential of the medium, as well as the concentrations of substrate(s), biomass, and metabolites determined offline. To enhance fermentation control and gain a deeper understanding of the process, fluorescent tagging of specific characteristics was utilized in conjunction with microscopy and flow cytometry. Flow cytometry, a high-throughput and multi-parametric technology capable of analyzing heterogeneous populations at the individual cell level, was recently employed for the first time to describe clostridial butanol fermentations. However, this study was conducted in a completely different environment (Tracy et al., 2008).

3.1.1. Use of fluorescent alternative of Gram staining for discrimination of acidogenic and solventogenic clostridial cells

The detailed description of the method development, particular application conditions and its use were published byLinhova et al., (2010a). The main idea of the staining is based on fact that clostridia are usually stained according to Gram as G+after germination from spores (motile, juvenile cells) and as G-when the cells started to sporulate. The change in Gram staining response corresponds to metabolic switch from acids to solvents formation and also with an alteration in a cell membrane composition i.e. thinning of peptidoglycan layer (Beveridge, 1990). Therefore the cells ofC.pasteurianumwere labelled with a combination of fluorescent probes, hexidium iodide (HI) and SYTO 13 that can be considered a fluorescent alternative of Gram staining. Cells of C.pasteurianumforming mainly acids fluoresced bright orange-red as G+bacteria and the solvent producing, sporulating cells exhibited green-yellow fluorescence as G-bacteria (see Fig.2). The red colour of labelled young cells was a result of a fact that green fluorescence of SYTO13 was quenched by that of HI while bright green-yellow colour of sporulating and/or old cells was caused by staining only by SYTO13 when HI did not permeate across the cell wall. Jones et al., (2008) used different combination of dyes (propidium iodide and SYTO 9) for labelling C.acetobutylicumATCC 824 during time course of batch cultivation but attained the same conclusion.

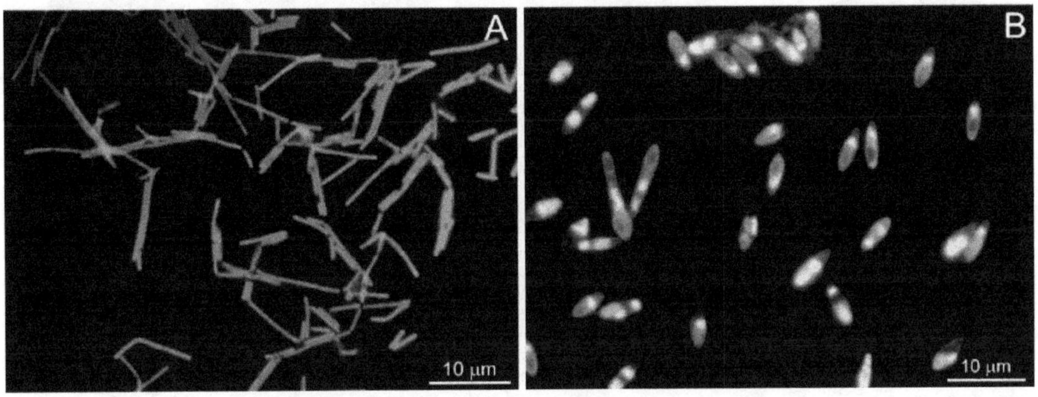

Figure 2. C.pasteurianumcells stained with hexidium iodide and SYTO 13 in acidogenic (A) and solventogenic (B) metabolic phases

Then,flowcytometryenablingquantificationoffluorescentintensitiesoflabelledclostridialpopulations was used for monitoring of physiological changes during fed-batch cultivation (Linhova et al., 2010a). For flow cytometry measurement, the cells were stained only by HI and the signal of fluorescent intensity acquired in a channel FL3 (red colour) was related to forward scatter signal (FSC) which corresponded to cell size in order to gain data independent on cell size. The data measured for C.pasteurianumwere compared with those for typical G+and G-bacteria i.e. for Bacillus megatheriumandEscherichia coliand there was a striking difference between the values of FL3/FSC for C.pasteurianumon one hand and those for B.megatherium and E.colion the other hand. While the values for B.megatherium(G+) and E.coli(G-) oscillated ±0.1 and ±0.2, respectively, in time course of 32 h in which they were sampled, the values forC. pasteurianum dropped from 3.1 to 0.8 during the cultivation. It was also evident that acidogenic phase had a very short duration and both metabolic phases overlapped. Further experiments are necessary to assess unambiguously the acquired data, however it is tempting to hypothesize that C.pasteurianumNRRL B-598 has a different pattern of acids and solvents formation when solvents production is connected rather with exponential growth phase than the well-known solventogenic strain C.acetobutylicumATCC 824 in which solvents production is generally assembled with stationary growth phase.

3.1.2. Use of flow cytometry for viability determination of clostridia

As to perform ABE fermentation means to handle clostridial population in different stages of the life cycle (see Fig. 1), determination of share of metabolically active i.e. vital cells in the population, is very important. Based on testing of various fluorescent viability probes with different principles of functioning, bisoxonol (BOX) was chosen as a convenient dye for C.pasteurianumviability determination (Linhova et al., 2010b). BOX stains depolarized cells with destroyed membrane potential i.e. nonviable cells. When the cells were fixed by 5 min boiling, whole population was labelled (Fig.3b)but in case of growing population (Fig.3a)most of cells remained non-stained. After optimization of staining conditions, flow cytometry was used for determination of culture viability (seeFig.4).

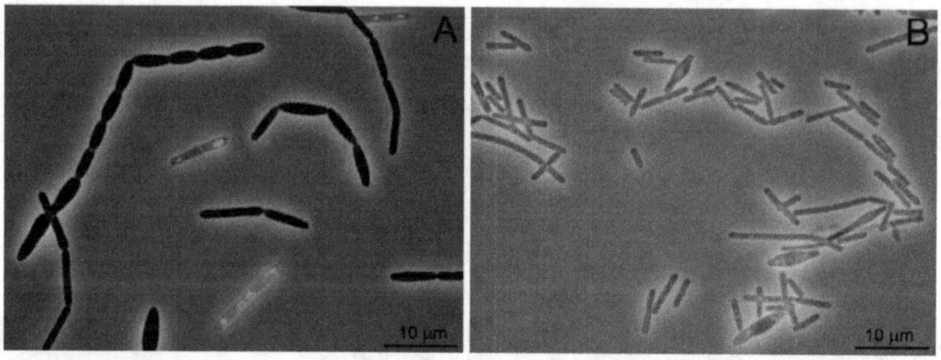

Figure 3. BOX stained viable (A) and fixed i.e. nonviable (B) cells ofC.pasteurianum

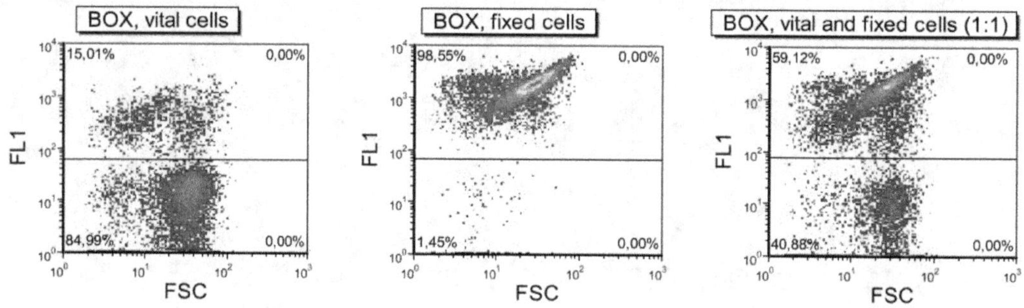

Figure 4. Dot-plot diagrams after BOX labelling ofC.pasteurianumpopulations of live (1), fixed (2) and mixture of live and fixed cells (3)(Population of viable cells in the left dot-plot diagram can be seen under the gate (in lower half of the diagram). In upper half of the left diagram, there are rests of cells after spores germination and sporulating cells, the share of which does not exceed 15%.)

Then the method was used for viability determination during batch cultivation (see Fig.5). Bioreactor was inoculated with spore suspension after heat shock that induced spores to grow and killed present vegetative cells. After the heat shock, the viability at the beginning of the fermentation was very low (Fig. 5B). In the exponential growth phase viability increased to ~78%, as expected. With glucose depletion (Fig. 5B) and reaching the highest concentration of 1-butanol (7.5 g.L-1seeFig.5A),the viability began to decrease. Relatively rapid viability decline at nutrient depletion conditions has already been observed by Novo et al., (1999) and Jepras et al., (1995) for S. aureus, E. coliand P. aeruginosa. They observed membrane potential decreased within a few minutes after removal of energy resources. Moreover, in our case, elevated 1-butanol concentration contributed to viability decline, too

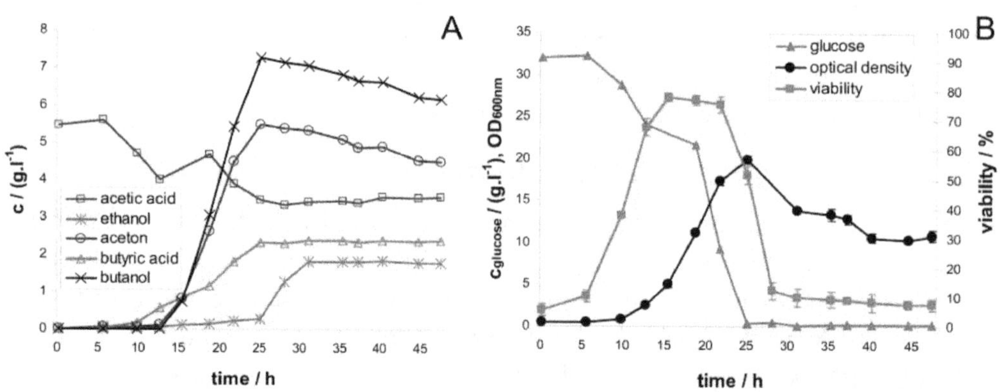

Figure 5. Comparison of viability determination with fermentation data during batch cultivation of C.pasteurianum

3.2. Feedstock comparison

Through screening in flask fermentations conducted within an anaerobic chamber, each feedstock (sugar beet, corn, and glucose) was paired with the suitable Clostridium strain based on their respective yield and productivity characteristics. Sugar beet is a long-standing crop cultivated in the Czech Republic for the past 160 years. It is known for its high productivity and can be utilized in the non-food sector for the manufacture of biofuels. Currently, sugar beet is being utilized for non-food purposes in the Czech Republic, specifically for the production of bioethanol at the Agroethanol TTD facility. Corn is mostly cultivated for cattle feed in CR. However, the amount of cattle herds decreases each. One of the key objectives of biofuel production is to support farmers and preserve fertile land areas. Corn can also be considered as an energy crop. Glucose was selected as the feedstock because it is considered a simple model of lignocellulosic material hydrolyzate, which may be used in the future. Table 1 provides a comparison of butanol production utilizing maize, sugar beet juice, and glucose, along with the relevant strains. Based on the stated parameters, sugar beet appears to be the preferred alternative. It is important to consider fermentation courses in all cases being compared. The fermentation of maize by C.acetobutylicum exhibited a biphasic behavior, resembling that described in textbooks. Initially, acids were produced, followed by a solventogenic phase accompanied by sporulation, during which the acids were reutilized. Nevertheless, the fermentation processes of sugar beet juice by C.beijerinckii and glucose by C.pasteurianum deviated from the expected pattern by initiating butanol production during the exponential growth phase (in both cases) and exhibiting minimal acid reutilization (in the case of C.pasteurianum).

Table 1. Comparison of bioreactor cultivations using different substrates and strains (Abbreviations B, ABE, YABE/S, YB/S, PABE stand for butanol, total solvents amount, yield of total solvents, yield of butanol and productivity of solvents formation.)

species	substrate	B ($g.L^{-1}$)	ABE ($g.L^{-1}$)	YABE /S(%)	YB/S (%)	PABE ($g.L^{-1}.h^{-1}$)
C.beijerinckii	sugar beet juice	11.6	16.2	37	26	0.40
C.acetobutylicum	corn	9.6	14.4	27	18	0.20
C.pasteurianum	glucose	7.3	11.8	35	18	0.23

Overall balances of mentioned fermentation courses can be expressed in form of equations (Eq. 1,Eq. 2,Eq. 3). Similar expression of products in numbers has already been published (seeEq. 4)byJones & Woods (1986) where this equation reflected average results achieved withC.acetobutylicumandC. beijerinckiistrains published in literature till 1986. In the equations (Eq. 1–Eq. 4), $C_{12}H_{22}O_{11}$, $C_6H_{12}O_6$, $C_4H_{10}O$, C_3H_6O, C_2H_6O, $C_2H_4O_2$, $C_4H_8O_2$ stand for saccharose, glucose, butanol, acetone, ethanol, butyric acid and acetic acid, respectively.

$$1.00C_{12}H_{22}O_{11} \rightarrow 1.22\ C_4H_{10}O + 0.60C_3H_6O + 0.04C_2H_6O + 0.25C_2H_4O_2 + 0.20C_4H_8O_2 + 1.60CO_2 + 0.80H_2 \quad (1)$$

$$1.00\ C_6H_{12}O_6 \rightarrow 0.42\ C_4H_{10}O + 0.21\ C_3H_6O + 0.04\ C_2H_6O + 0.12C_2H_4O_2 + 0.06C_4H_8O_2 + 0.58CO_2 + 0.36H_2 \quad (2)$$

$$1.00\ C_6H_{12}O_6 \rightarrow 0.54\ C_4H_{10}O + 0.40\ C_3H_6O + 0.02\ C_2H_6O + 0.19C_2H_4O_2 + 0.06C_4H_8O_2 + 6.77CO_2 + 3.98\ H_2 \quad (3)$$

Butanol production (Jones & Woods 1986):

$$1.00\ C_6H_{12}O_6 \rightarrow 0.56\ C_4H_{10}O + 0.22\ C_3H_6O + 0.07\ C_2H_6O + 0.14\ C_2H_4O_2 + 0.04\ C_4H_8O_2 + 2.21\ CO_2 + 1.35\ H_2 \quad (4)$$

Ratio of 1-butanol per unit of sugar (hexose) was the highest for saccharose (0.61) and the lowest for starch (0.42) but it can be stated the results were similar as presented by Jones and Woods (1986). The only exception was case of C.pasteurianum, in which remarkable amounts of carbon dioxide and hydrogen were produced not only in acidogenesis but throughout the whole fermentation period. Other experiences with the mentioned raw materials and also possible alternation of expensive but usual cultivation medium supplements, yeast extract or yeast autolysate, with cheap waste product of milk industry, whey protein concentrate, is presented in Patakova et al., (2009). Detailed description of the use of sugar beet juice as fermentation substrate for biobutanol production has been published, recently (Patakova et al., 2011b).

3.3. Impact of fermentation configuration on ABE fermentation

An overview of batch, fed-batch and two variants of continuous bioreactor fermentation experiments using glucose cultivation medium and the strain C.pasteurianumNRRL B-598 is presented inTable 2. Both batch and fed-batch cultivations were operated about 50h and a ratio of produced solvents (B:A:E) was about 2:1:0.1 in all cases. Batch cultivations were performed in media with initial glucose concentration 40 g.L-1and if usual total solvents yields referred in literature are about 30% (Ezeji et al., 2005;Shaheen et al., 2000) then similar solvents concentrations like those shown in Table 2were usually obtained. Therefore, higher initial glucose concentrations (60 and 80 g.L-1) were tested in flasks cultivations, however solvents concentrations remained either at the same level (for 60 g.L-1glucose) or they were lower (for 80 g.L-1glucose) in comparison with use of glucose concentration 40 g.L-1and significant portion of glucose stayed in media unconsumed what might indicate a phenomenon of substrate inhibition.

Consequently, fed-batch cultivations were employed (see Table 2) in which butanol and total ABE concentrations were moderately increased (about 10%) and lag growth phase was reduced to 50% i.e. 3 h (data not shown). Nevertheless, yield and productivity for both 1-butanol and total solvents remained almost the same as in case of batch cultivations. The reached maximal butanol concentration (8.3 g.L-1) is probably near the highest value tolerated by the used strain and a substantial improvement in an overall amount of produced butanol could be attained only by an integration of the cultivation with some on-line separation step.

C.pasteurianum (Abbreviations B, ABE, YABE/S, YB/S, PABEand D stand for butanol, total solvents amount, yield of total solvents, yield of butanol, productivity of solvents formation and dilution rate. Continuous cultivation proceeded as glucose-limitedaor glucose non-limitedbexperiments; values of yield and productivity were calculated in pseudo steady state. For detailed conditions of continuous fermentations see Patakova et al., 2011a.)

Table 2. Parameters of batch, fed-batch and continuous fermentations using

Cultivation	B ($g.L^{-1}$)	ABE ($g.L^{-1}$)	YABE/S(%)	B/A ratio	PABE(g.L-1.h-1)	D (h-1)
batch	7.3	11.8	35	2.0	0.23	-
fed-batch	8.3	12.3	23	2.2	0.25	-
continuousa	4.4	6.2	24	3.4	0.15	0.03
continuousb	4.0	5.9	20	1.8	0.20	0.07

Surprisingly, glucose-limited fermentation experiment showed superior results in comparison with glucose non-limited fermentation (see Table 2). The only exception was solvents productivity that was higher at the expense of unused substrate. In glucose non-limited continuous experiment, mutually adverse oscillations of butanol and glucose concentrations occurred unlike butanol concentration near constant value (pseudo steady state) achieved in glucose-limited fermentation. The glucose limitation is also believed to support long-term stability and to reduce strain degeneration (Fick et al., 1985). Fermentation courses in both cases were presented inPatakova et al., 2011a.

3.4. Comparing different strains

The fermentation process using C.acetobutylicum DSM 1731 and milled corn as a substrate was comparable to that described for C.acetobutylicum ATCC 824 (Lee S.Y. et al., 2008). This process exhibited distinct metabolic phases, involved the recycling of acids during solvent production, and resulted in the production of hydrogen, which reached its highest level during the acid production phase. Johnson et al. (1997) found that C.acetobutylicum DSM 1731 and C.acetobutylicum ATCC 824 have a DNA sequence similarity of 96%. The phenomenon known as the acid crash, which refers to the point where fermentation is completed in the acidogenic step, was occasionally observed without a clear explanation when employing this particular strain and milled maize as the substrate (Rychtera et al., 2010). Regrettably, the concentration of formic acid within the cells was not measured, hence it remains unverified if the acid crash in these instances was also attributed to formic acid (Wang et al., 2011). According to the Czech Collection of Microorganisms, the strain C.beijerinckii CCM 6218 is expected to be indistinguishable from the strain C.beijerinckii ATCC 17795. Interestingly, when the strain C.beijerinckii ATCC 17795 was examined for its ability to produce butanol using a growth medium containing molasses (Shaheen et al., 2000), it exhibited a low yield and maximum butanol production of 10% and 6.1 g.L-1, respectively. Furthermore, this strain, together with C.pasteurianum NRRL B-598, exhibited a distinct fermentation pattern compared to C.pasteurianum NRRL B-598. Moreover, the generation of butanol commenced during the exponential growth phase. The strain exhibited a greater metabolic rate for the substrate saccharose compared to the other studied strains, resulting in increased butanol productivity. The strain Clostridium pasteurianum NRRL B-598 utilized in this investigation exhibited notable variations in several physiological parameters compared to the species attributes documented in Bergey's Manual of Systematic Bacteriology (Rainey et al., 2009). While strains of the species C. pasteurianum are typically recognized for their ability to produce acetic and butyric acids or hydrogen (Rainey et al., 2009; Heyndrickx et al., 1991), the specific strain C. pasteurianum NRRL B-598 was mentioned in US Patent No 4539293 for its capacity to produce butanol when used in combination with another acidogenic strain, such as C. butylicum. Regrettably, the referenced patent does not include specific information regarding cultivation conditions, yields, solvent concentrations, and other relevant data.

3.5. Gas stripping is used to separate biobutanol from the fermentation medium

Gas stripping by nitrogen as a method potentially enabling both butanol preconcentration before final distillation and a way how to mitigate butanol toxicity during fermentation was studied separately from fermentation and stripping coefficient β, defined by equation (Eq. 5)was chosen as main criterion for stripping efficiency:

$$\beta = (-1/P_L)(dP_L/dt) \tag{5}$$

where PLis butanol concentration in a liquid phase.

Gas stripping of solvents from fermentation media is however only the first step towards isolation/ concentration of products (ABE), further steps consist in product change of state from the gas into liquid phases. There are several ways how to carry out this change of state but there are scarcely discussed.

Two of them i.e. application of low temperature (- 4ºC in condenser) and adsorption on charcoal followed by desorption by steam were tested. If low temperature was used for butanol conversion from the gas to liquid, average achieved preconcentration lay in the interval from 7 to 9. However, when the method of freezing was used then only 60% of solvents were captured in one gas cycle (probably due to insufficiency of freezing unit capacity) while at charcoal adsorption, 90% of solvents was captured. This affected stripping efficiency which was lowering gradually at freezing. On the contrary, main disadvantage of charcoal use was a gain of more diluted butanol solution (preconcentration from 2 to 4) after its displacement from charcoal by steam. Energy balance must be done for this process but it needs measurement in pilot scale.

Table 3.Comparison of butanol stripping from model solution and cultivation medium after fermentation

	Model solution ABE				Medium after fermentation			
	Initial conc. (g.L-1)	Final conc. (g.L-1)	Mean rate of stripping (g.L-1.h-1) (for 24 h)	Stripping coefficient (h-1) (for 24 h)	Initial conc. (g.L-1)	Final conc. (g.L-1)	Mean rate of stripping (g.L-1.h-1) (for 24 h)	Stripping coefficient (h-1) (for 24 h)
A	3.9	2.6	0.05	0.017	4.8	2.6	0.09	0.025
B	9.2	3.2	0.25	0.044	10.2	2.9	0.30	0.052
E	1.4	0.7	0.03	0.029	0.7	0.5	NA	NA

The profound influence of solution composition on stripping efficiency is shown in Table 3, where comparison of model (water) solution of solvents with medium after fermentation is provided. In this case, the stripping was carried out directly in the bioreactor (liquid volume 3L) using aeration ring as nitrogen distributor (flow rate 2 VVM). Schemes of stripping arrangements are provided in (Fribert et al., 2010).

Nevertheless, if summarized it can be stated that the mean rate of stripping for butanol and butanol preconcentrations achieved after application of freezing corresponded with already published values (Ezeji et al., 2003; Ezeji et al., 2005; Qureshi & Blaschek, 2001b). The mean butanol stripping rate exceeded the butanol productivity what indicated a potential successful integration of gas stripping with fermentation into one process.

4. BIOBUTANOL'S APPLICATION IN ROAD TRANSPORTATION

4.1. Perspectives of biobutanol use in road transport

The preferred use of biobutanol is the production of motor fuels for spark ignition engines by mixing with conventional gasoline; therefore biobutanol could become an option to bioethanol due to better potential in terms of its physico-chemical properties. Biobutanol concentration in fuel can reach up to 30% v/v without the need for engine modification. Since the butanol fuel contains oxygen atoms, the stoichiometric air/fuel ratio is smaller than for gasoline and more fuel could be injected to increase the engine power for the same amount of air induced. The oxygen content is supposed to improve combustion, therefore lower CO and HC emissions can be expected. Biobutanol and its mixtures can be used directly in the current gasoline supply system, such as transportation tanks and re-fuelling infrastructure. Biobutanol can be blended with gasoline without additional large-scale supply infrastructure, which is a big benefit as opposed to the bioethanol use. Finally biobutanol is non-poisonous and non-corrosive and it is easily biodegradable and does not cause risk of soil and water pollution.

4.2. Characteristics of Biobutanol-Gasoline Mixtures in Terms of their physical and Chemical Qualities

If compared to ethanol, biobutanol exhibits important advantages upon blending with gasoline. The mixtures have better phase stability in presence of water, low-temperature properties, oxidation stability during long-term storage, distillation characteristics and volatility with respect to possible air pollution.

The overview of the selected properties of butanol as compared to bioethanol and automotive gasoline meeting requirements of EN 228 is given inTable 4.

Table 4. Physico-chemical properties of alcohols and automotive gasoline (Wolf, 2007)

Parameter	Bioethanol	Biobutanol	Gasoline 95
Boiling Point (°C)	78.3	117.7	30-215
Density (kg.m-3)	794	809	720-750
Kinematic Viscosity (mm2.s-1)	1.5	3.6	0.4-0.8
Lower Heating Value (MJ.kg-1)	28.9	33.1	44.4
Heat of vaporisation (MJ.kg-1)	0.92	0.71	0.32
Research Octane Number RON	106-130	94	95
Motor Octane Number MON	89-103	80	85
Reid Vapour Pressure (kPa)	17	2.3	45-90
Stoichiometric air/fuel ratio	9	11.1	14.8
Oxygen Content (% w/w)	34.7	21.6	max 2.7

Due to the fact that oxygen content in biobutanol is lower than in ethanol, biobutanol can be added to the gasoline in higher concentrations with respect to EN 228 limit for the oxygen content in gasoline. Higher biobutanol content in gasoline does not require engine modification. The heating value (energy density) of biobutanol is close to that of gasoline, which has a positive effect on the fuel consumption.

Biobutanol has a slightly higher density compared to gasoline but the increase in density of biobutanol/ gasoline mixtures is so small that it does not cause problems with fulfilling limits for automotive gasoline containing up to 30% v/v biobutanol. Viscosity of biobutanol is significantly higher compared to the gasoline, which may affect engine fuel injection system at lower temperatures due to higher resistance to flow. However, this impact could be negligible for the blends with gasoline containing up to 30% v/v biobutanol.

4.2.1. Analysis of the fluctuation in the composition of biobutanol-gasoline mixtures

The vapour pressure of biobutanol and bioethanol is very low compared to gasoline. A disadvantage of the bioethanol use is a formation of volatile azeotropic mixtures of ethanol and hydrocarbons present in the gasoline which causes the increase in the vapour pressure of gasoline in the range of 6 - 8 kPa (Mužíková et al., 2009). The formation of azeotropes occurs in the concentration up to 10% v/v of biobutanol in gasoline but the highest increase in the vapour pressure is as low as 0.5 kPa at 5% v/v of biobutanol in gasoline. At higher biobutanol concentrations, another volatile and/or oxygen compound has to be added to compensate vapour pressure decrease and to keep good engine startability. The formation of azeotropes is also associated with decrease of the boiling points of the blends. While the addition of bioethanol influences negatively the distillation curve profile, biobutanol has minor effect on the distillation curve.

Because of the use of different gasolines in several European Union countries, the mixing of different oxygen compounds in the vehicle tank can occur, causing the simultaneous presence of ethers like MTBE and ETBE and other alcohols, especially ethanol, in combination with biobutanol in the gasoline. Ethers do not cause problems since their properties are close to hydrocarbons. They influence the vapour pressure of the butanol-gasoline blend proportionally according to the initial vapour pressure of pure components. On the contrary, bioethanol forms azeotropes, which can unpredictably change the vapour pressure of the mixture. The increase of vapour pressure depends on the final ethanol concentration in the mixture.

Biobutanol has significantly higher heat of vaporization than gasoline, which reduces the temperature of the air/fuel mixture and results in higher engine volumetric efficiency. At the same time it leads to lower compression temperature and longer ignition delay, which in turn may decrease the engine performance. The low vapour pressure and higher heat of vaporization is experienced to have a negative effect on the startability and cold start engine performance because of difficult fuel vaporization at low ambient temperatures (Xiaolong et al., 2009).

4.2.2. Water's impact on the stability of a phase

The water – fuel miscibility is very important factor for distribution of fuel blends. The content of small amounts of free water in the fuel is connected with the risk of corrosion problems, whereas larger amounts of water can impair fuel supply to the engine. Hydrocarbons in gasoline are very slightly miscible with water as opposed to alcohols. The solubility of water in petroleum gasoline is only 100 mg.kg-1, while bioethanol is completely miscible with water and solubility of water in biobutanol is 19.7% w/w. Bioethanol is very hygroscopic and its blends with gasoline are partially miscible with water depending on temperature and the ethanol concentration in the blend. Phase separation of water with bioethanol can occur at lower temperatures, which causes formation of the heterogeneous system composed of the hydrocarbon phase and water-ethanol phase. This fact is the reason why the bioethanol-gasoline blends cannot be distributed via common pipelines but only separately using tankers. Contrary to bioethanol, the ability of biobutanol to absorb significant amount of water is very low (Peng et al., 1996). Biobutanol has high affinity to hydrocarbons and the risk of potential phase separation is therefore minimized. Moreover, biobutanol remains in the hydrocarbon phase if the phase separation occurs. Biobutanol is not hygroscopic that is an important factor for the long-term storage of fuels. Accordingly, high stability of gasoline-biobutanol mixtures in the presence of water comparing with bioethanol was reported.

Ethers MTBE and/or ETBE can be added to the gasoline on purpose for increasing the octane number and oxygen content or they can be accidentally mixed in the fuel tank due to another type of gasoline fuelling. The presence of MTBE and ETBE slightly increases the miscibility of butanol-gasoline blend with water and decreases the temperature of the phase separation. Ethanol has the same behaviour, nevertheless due to the bioethanol ability to absorb humidity its presence in the blends is rather unfavourable.

4.2.3. Compatibility of biobutanol and its blends with different materials

Biobutanol is not as aggressive as the bioethanol with regard to the engine construction materials, sealants, and plastics. The fuel with 20% v/v of biobutanol has similar properties to the hydrocarbons in terms of swelling of polymers (Wolf, 2007).

The oxidation stability of biobutanol-gasoline blends may be compromised by potential impurities from biobutanol production (acetic and butyric acid, acetaldehyde and lower alcohols). The impurities in concentrations of 0.1% v/v in 10% v/v biobutanol-gasoline blends (which corresponds to 1% of impurities in biobutanol) can decrease the fuel oxidation stability by about 15%, therefore the purification with regard to the removal of fermentation by-products is very important step in biobutanol production.

The high boiling point of butanol may negatively influence its evaporation from engine oil after oil contamination caused by frequent cold starts. This phenomenon can occur especially at low ambient temperatures, when the fuel leaks into the engine oil through piston rings. In a normal engine operation biobutanol evaporates after the engine warm up and the motor oil additives are re-solved. However, the solubility of oil additives may be at risk in case of frequent cold starts and short routes in cold winter conditions.

4.3. Emission characteristics of butanol/gasoline blends in spark ignition engines

Besides the renewability of raw materials used for their production, alcohol fuels are reported to be advantageous over petroleum derived ones thanks to their better environmental characteristics. The oxygen contained in alcohol molecules is supposed to affect combustion process and cause soot and particulate reduction; some studies show that there is the potential for reduction of NOxemissions. While there was much information collected about the use and combustion behaviour of lower-molecular weight alcohols, such as methanol and ethanol, substantially less effort was yet put to the research of the properties of butanol (especially n-butanol as a product of fermentation during ABE process) upon their use in internal combustion engines.

For the evaluation of emission characteristics, it is very important to study combustion processes at different air/fuel ratio and thermodynamic conditions. The combustion of neat butanol as well as its mixtures with other fuels or chemicals was studied (Agathou & Kyritis, 2011;Broustail et al., 2011;Dagaut & Togbé, 2008;Sarathy et al., 2009) to obtain combustion velocities and kinetic data for modelling processes of butanol oxidation at the conditions of engine cylinder. However, it must be noted that real-world emissions level is affected by the interaction between fuel itself and the engine used, mainly its fuel injection system and engine control unit together with emission control systems – catalytic converters, particulate filters, exhaust recirculation etc.

Although butanol properties (boiling point, viscosity, octane number) predetermine it for the use in spark ignition engines as a partial substitute for conventional gasoline, a number of studies were carried out using butanol/diesel fuel mixtures in compression ignition engines. The addition of butanol (or other alcohols) significantly increases volatility and decrease lubricity of diesel fuel, which requires additional measurements for their use in today's diesel engines. Yao (Yao et al., 2010) studied emission characteristics of CI engine using diesel fuel containing 0 % to 15 % v/v n-butanol. By varying exhaust gas recirculation rates, they kept NOx emissions constant, while CO and PM (particulate matter) emissions significantly decreased with the concentration of n-butanol in the fuel. Rakopoulos et al. (Rakopoulos et al., 2010a) compared conventional diesel fuel, diesel fuel with 30% biodiesel (FAME), and biodiesel with 25 % n-butanol in turbo-charged CI truck engine; the experiments were focused on transient regimes causing temporary increase of pollutant emissions. Both FAME and butanol helped to improve the particulate emissions in the transient engine regimes, but in both cases the emissions of NOx increased. In stationary regimes at different engine speed and load, the authors (Rakopoulos et al., 2010b) determined emissions of all regulated pollutants. In all cases, the positive effect of butanol in diesel fuel was found on the emissions of particulates, NOx, and carbon dioxide, whereas hydrocarbon emissions slightly increased.

Much greater potential and possibility of utilization without necessity to solve technical problems has butanol used as a partial substitute of motor gasoline. The total miscibility with hydrocarbons, boiling point, flash point and other properties allow mixing butanol with gasoline in wide range of concentrations and combustion in common spark ignition engines. In comparison to other alcohols in the range of C1to C5mixed to gasoline in concentrations matching fuel oxygen content, butanol does not differ significantly in its effect on the emissions of regulated pollutants (Yacoub et al., 1998). The emissions of total hydrocarbons decrease, while significant increase takes place in the emissions of aldehydes, whose main constituent was formaldehyde.

One of the substantial drawbacks connected with the use of alcohols in SI engines is the problem of cold starts especially in winter conditions. Difficulties caused mainly by high heat of vaporization have to be eliminated by greater enrichment of air/fuel mixture in the period in which the engine heats up. This, on the other hand, can bring an increase in emissions of unburned or partially burned fuel due to near zero efficiency of catalytic converter in the early period after engine start.Irimescu (2010) modelled the situation for gasoline/butanol mixtures at different ambient temperatures and successfully verified the results with those obtained in experiments with a port injection engine.

The effect of butanol (or other alcohols) use in spark ignition engines depends also on the technique of fuel injection before its ignition in engine cylinder. Conventional way to prepare air/fuel mixture is the injection of fuel into the engine intake manifold, where it evaporates and the mixture is drawn to the cylinder in the suction cycle. Some engine manufacturers offer engines equipped with direct injection of fuel into the cylinder. Such engines allow the use of advanced techniques of emission control, such as lean (stratified) mixture combustion connected with the use of sequential injection. The direct injection engine was used by Cooney (Cooney et al., 2006), who investigated the effect of ethanol and butanol in blends with gasoline used in a series of engine tests conducted at varied loads. They reported the increase in engine efficiency at higher engine loads by a 4% with either 85 % n-butanol or 85 % ethanol. The efficiency is reported to be affected by lower octane number of n-butanol, even though knock combustion was not observed, and, on the other hand, by the higher flame speed of alcohols. Faster combustion can increase the efficiency if combustion timing was adjusted, while lower octane number should decrease it.

In contrast to modern engines of current passenger cars, there are still applications where carburetted engines or engines with open-loop control of fuel injection are used, without the ability to compensate for air-fuel ratio of specific fuels. In such cases, butanol blends result in approximately 50% enleanment connected with oxygen content in fuel, compared to ethanol. The authors evaluated the effect of the use of butanol-gasoline mixtures on pollutants emission of four different passenger cars equipped with spark ignition engines – from older Euro 2 vehicle to modern multipoint injection turbocharged one. As a baseline, unleaded gasoline with addition of 4 % ethanol was used. Mixtures containing butanol were prepared by addition of 10 %, 20 %, and 30 % pure synthetic n-butanol to the same gasoline. The properties of the mixtures were modified with small amounts of isooctane, toluene, and petroleum ether to keep their octane number and vapour pressure, which deteriorated by the addition of butanol. Four test vehicles manufactured by Skoda were used with different engine displacement, power, and technology level (see Table 5).

The emission tests were performed on a vehicle dynamometer according to ECE 83 emission test with the determination of CO, HC, and NOx emissions during two driving cycles. In addition to the measurement of regulated emissions, samples were taken during both phases of ECE 83 test for determination of individual hydrocarbons and aldehydes. Basic engine parameters were monitored during the tests using an engine diagnostic unit to detect possible abnormal operation states of engine control unit.

Table 5. Characteristics of vehicles used for emission tests

Vehicle type	Year of manufacture	Engine displacement [cm³]	Maximum power [kW]	Engine characteristics
Felicia Euro 2	1999	1289	50	Multi-point injection,four-cylinder
Fabia Euro4	2004	1198	47	Multi-point injection, three-cylinder
Octavia Euro4	2004	1781	110	Multi-point injection, 20V,five-cylinder, turbocharged

The inclusion of butanol in the gasoline had no impact on the regulated emissions, as observed in the ECE 83 test (Fig. 6). While there were notable variations in emission levels seen during certain ECE 83 test phases, the overall data indicates that all three cars experienced an increase in NOx emissions, while other controlled pollutants remained relatively stable. As anticipated, the utilization of butanol also resulted in a slight rise in aldehyde emissions, primarily composed of formaldehyde.

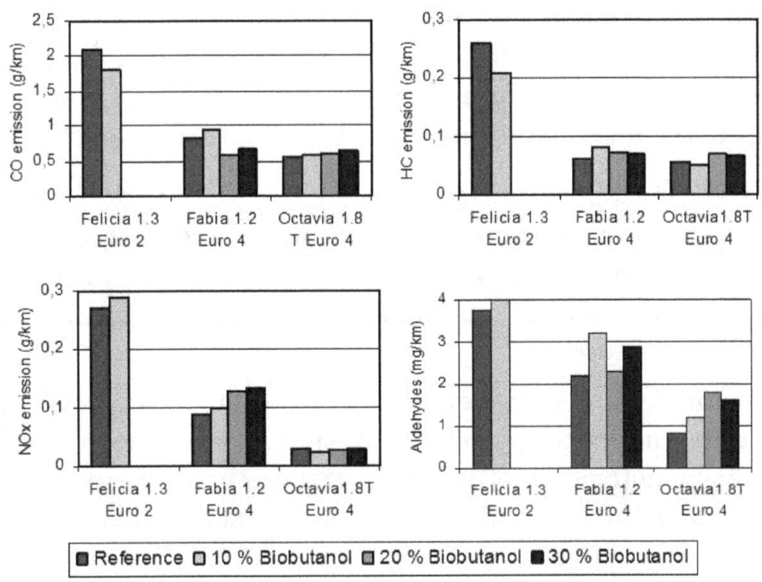

Figure 6. Effect of butanol in gasoline fuel on emissions of regulated pollutants (CO, HC, NOx) and aldehydes in ECE 83.03 emission test

5. CONCISE SUMMARY

The offered fermentation data holds significance across multiple domains:

- Fluorescence labeling and flow cytometry were found to be highly effective methods for assessing the physiological state of clostridial population during fermentation in a nearly real-time manner. Neither the fluorescence-based approach for discriminating acidogenic/ solventogenic state of individual cells nor the vitality staining using bisoxonol has yet been used on bacteria belonging to the genus Clostridium.
- The strain C.pasteurianum NRRL B-598 received the most significant focus due to its unprecedented level of detailed study. The comparison of three fermentation arrangements, namely batch, fed-batch, and continuous, provides a distinct dataset that is typically not accessible for the tested butanol producers. Due to its unique physiology compared to type C.pasteurianum strains and the observation of a very brief acidogenic metabolic phase and potential overlap of acidogenic and solventogenic phases as indicated by flow cytometry

analysis, additional inquiry is warranted to study the strain and its behavior. Furthermore, the strain can also be considered a highly promising hydrogen generator.

- The optimal fermentation conditions, resulting in a 37% yield of ABE and a productivity of 0.40 $g.L^{-1}.h^{-1}$, were obtained by utilizing sugar beet juice as the raw material and C.beijerinckii CCM 6182 as the microorganism. The sugar beet has the potential to become a substantial supply of sugar for non-food reasons, particularly in Europe and specifically in the Czech Republic. The strain C.beijerinckii CCM 6182, unlike its analog C.beijerinckii ATCC 17795, has not been extensively studied in terms of its abilities and fermentation characteristics. However, it has been observed that this strain exhibits similar behavior to C.pasteurianum NRRL B-598, specifically in terms of its favorable butanol production kinetics. One notable characteristic of this strain is the onset of butanol formation during the exponential growth phase.
- The initial studies on gas stripping as a possible method for concentrating and separating solvents from the fermented media have shown that this approach is feasible, under particular conditions. The gas stripping process should not have a negative impact on the fermentation process, and efforts should be made to minimize the transfer of solvents from the gas phase to the liquid phase in order to reduce costs. However, additional pilot trials are ideally required to fully evaluate the role of gas stripping in butanol production.

Regarding the utilization of biobutanol as a fuel for transportation, the following can be inferred:

- When compared to other bio-components used for mixing motor fuels, particularly bioethanol, biobutanol possesses highly desirable characteristics. These include a high energy content, low water solubility, complete miscibility with gasoline hydrocarbons, and suitable boiling point and vapour pressure.
- The utilization of gasoline with elevated levels (10% to 30% v/v) of butanol did not have any adverse impact on the operational characteristics of conventional spark ignition engines often found in passenger automobiles that are representative of the existing European vehicle fleet. Only a minor increase in NOx emissions and generation of aldehydes was seen during routine ECE 83 emission tests.

REFERENCES

Agathou, M.S. & and Kyritis, D.C. (2011) An experimental comparison of non-premixed biobutanol flames with the corresponding flames of ethanol and methane, Fuel, Vol. 90, No. 1, pp. 255-262, ISSN 0016-2361

Bahl, H.; Andersch, W. & Gottschalk, G. (1982) Continuous production of acetone and butanol by Clostridium acetobutylicum in two stage phosphate limited chemostat, European Journal of Applied Microbiology and Biotechnology, Vol.15, No.4, pp. 201-205, ISSN 0171-1741

Beesch, S.C. (1953) Acetone-butanol fermentation of starches, Applied Microbiology, Vol. 1, No. 2, pp. 85-96, ISSN 0099-2240

Beveridge, T.J. (1990) Mechanism of Gram variability in select bacteria, Journal of Bacteriology, Vol. 172, No. 3, pp. 1609-1620, ISSN 0021-9193

Broustail, G.; Seers, P.; Halter, F.; Moreac, G. & Mounaim-Rousselle, C. (2011) Experimental determination of laminar burning velocity for butanol and ethanol iso-octane blends, Fuel, Vol. 90, No. 1, pp. 1-6, ISSN 0016-2361

Chiao, J-S. & Sun, Z-H. (2007) History of acetone-butanol-ethanol fermentation industry in China: Development of continuous production technology, Journal of Molecular Microbiology and Biotechnology, Vol. 13, No. 1-3, pp. 12-14, ISSN 1464-1801

Clarke, K.G.; Hansford, G.S. & Jones, D.T. (1988) Nature and significance of oscillatory behavior during solvent production by Clostridium acetobutylicum in continuous culture, Biotechnology & Bioengineering, Vol. 32, No. 4, pp. 538-544, ISSN 0006-3592

Cooney, C.; Wallner, T; McConell, C.; Gillen, J.C.; Abell, C. & Miers, C.A. (2006) Effect of blending gasoline with ethanol and butanol on engine efficiency and emissions using a direct-injection, spark-ignition engine, In: Proceedings of the ASME internal combustion engine division 2009 spring technical conference ICES2009-76155, Milwaukee, Wisconsin, USA, 2006

Cornillot, E.; Nair, R.; Papoutsakis, E.T. & Soucaille, P. (1997) The genes for butanol and acetone formation in Clostridium acetobutylicum ATCC 824 reside on a large plasmid whose loss leads to degeneration of the strain, Journal of Bacteriology, Vol. 179, No. 17, pp. 5442-5447, ISSN 0021-9193

Dagaut, P. & Togbé, C. (2008) Oxidation kinetics of butanol-gasoline surrogate mixtures in a jet-stirred reactor: Experimental and modeling study, Fuel, Vol. 87, No. 15-16, pp. 3313-3321, ISSN 0016-2361

Dyr, J. & Protiva, J. (1958) Formation of neutral solvents in continuous fermentation by means of Clostridium acetobutylicum. In: Continuous Cultivation of Microorganisms A symposium, Málek, I., pp. 210-226, Publishing House of the Czechoslovak Academy of Sciences, Prague 1958.

Ezeji, T.C.; Qureshi, N. & Blaschek, H.P (2003) Production of acetone, butanol and ethanol by Clostridium beijerinckii BA101 and in situ recovery by gas stripping, World Journal of Microbiology and Biotechnology, Vol. 19, No. 6, pp. 595-603., ISSN 0959-3993

Ezeji, T.C.; Qureshi, N. & Blaschek, H.P. (2005) Industrially relevant fermentations. In Handbook on Clostridia, Dürre, P., pp. 797-812, CRC Press, ISBN 0-8493-1618-9, Boca Raton.

Fick, M.; Pierrot, P. & Engasser, J.M. (1985) Optimal conditions for long-term stability of acetone-butanol production by continuous cultures of Clostridium acetobutylicum, Biotechnology Letters, Vol. 7, No. 7, pp. 503-508, ISSN 0141-5492

Fribert, P.; Patakova, P.; Jahodova, L.; Lipovsky, J.; Linhova, M.; Rychtera, M. & Melzoch, K. (2010) Recovery of acetone, butanol and ethanol from fermentation broth by gas stripping, Proceedings of 19th International Congress of Chemical and Process Engineering, ISBN 978-80-02-02210-7, Prague, Czech Republic, September 2010

Gheslaghi, R.; Scharer, J.M.; Moo-Young, M. & Chou, C.P. (2009) Metabolic pathway of clostridia for producing butanol, Biotechnology Advances, Vol. 27, No. 6, pp. 764-781, ISSN 0734-9750

Hanno, R.; Qureshi, N.; Cotta, M. & Largus A. (2010) Mixed community bioreactors to convert lignocellulosic feedstock into the liquid biofuel butanol, February 27, 2011, Available from:http://www.ars.usda.gov/research/publications/publications.htm?seq_no_11 5=248960

Hartmanis, M.G.; Ahlman, H. & Gatenbeck, S. (1986) Stability of solvent formation in Clostridium acetobutylicum, Applied Microbiology and Biotechnology, Vol. 23, No. 5, pp. 369-371, ISSN 0175-7598

Heyndrickx, M.; De Vos, P. & De Ley, J. (1991) Fermentation characteristics of Clostridium pasteurianum LMG 3285 grown on glucose and mannitol, Journal of Applied Bacteriology, Vol. 70, No. 1, pp. 52-58, ISSN 1364-5072

Irimescu, A. (2010) Study of cold start air-fuel mixture parameters for spark ignition engines fueled with gasoline-isobutanol blends, International Communications in Heat and Mass Transfer, Vol. 37, No. 9, pp. 1203-1207, ISSN 0735-1933

Jepras, R.I.; Carter, J.; Pearson, S.C.; Paul, F.E. & Wilkinson, M.J. (1995) Development of a robust flow cytometric assay for determining numbers of viable bacteria, Applied and Environmental Microbiology, Vol. 61, No. 7, pp. 2696-2701, ISSN 0099-2240

Johnson, J.L., Toth, J., Santiwatanakul, S. & Chen, J.-S. (1997) Culture of Clostridium acetobutylicum from various collections comprise Clostridium acetobutylicum, Clostridium beijerinckii and two other distinct types based on DNA-DNA reassociation, International Journal of Systematic Bacteriology, Vol. 47, No. 2, pp. 420-424, ISSN 0020-7713

Jones, D.T.& Woods, D.R. (1986). Acetone-butanol fermentation revisited, Microbiological Rewievs, Vol. 50, No. 4, pp. 484-524, ISSN 0146-0749

Jones, D.T.; Shirley, M.; Wu, X. & Keis, S. (2000) Bacteriophage infections in the industrial acetone butanol (AB) fermentation process, Journal of Molecular Microbiology and Biotechnology, Vol. 2, No. 1, pp. 21-26, ISSN 1464-1801

Jones, S. W.; Paredes, C. J.; Tracy, B.; Cheng, N.; Sillers, R.; Sanger, R. S.& Papoutsakis, E. T. (2008) The transcriptional program underlying the physiology of clostridial sporulation, Genome Biology, Vol. 9, No. 7, pp. R114-R114.21, ISSN 1465-6914

Kashket, E.R. & Cao, Z.-Y. (1995) Clostridial strain degeneration, FEMS Microbiology Reviews, Vol. 17, No. 3, pp. 307-315, ISSN 0168-6445

Kosaka, T.; Nakayama, S.; Nakaya, K.; Yoshino, S. & Furukawa, S. (2007) Characterization of the sol operon in butanol-hyperproducing Clostridium saccharoperbutylacetonicum strain N1-4 and its degeneration mechanism, Bioscience, Biotechnology and Biochemistry, Vol. 71, No. 1, pp. 58-68, ISSN 0916-8451

Largier, S.T.; Long, S.; Santangelo, J.D.; Jones, D.T. & Woods, D.R. (1985) Immobilized Clostridium acetobutylicum P262 mutants for solvent production, Applied and Environmental Microbiology, Vol. 50, No. 2, pp. 477-481, ISSN 0099-2240

Lee, S.M.; Cho, M.O.; Park, C.H.; Chung, Y-C.; Kim, J.H.; Sang, B-I. & Um, Y. (2008) Continuous butanol production using suspended and immobilized Clostridium beijerinckii NCIMB 8052 with supplementary butyrate, Energy & Fuels, Vol. 22, No. 5, pp. 3459-3464, ISSN 0887-0624

Lee, S.Y.; Park, J.H.; Jang, S.H.; Nielsen, L.K.; Kim, J. & Jung, K.S. (2008) Fermentative butanol production by clostridia, Biotechnology and Bioengineering, Vol. 101, No. 2, pp. 209-228, ISSN 0006-3592

Linhova, M.; Patakova, P.; Lipovsky, J.; Fribert, P.; Paulova, L.; Rychtera, M. & Melzoch, K. (2010a) Development of flow cytometry technique for detection of thinning of peptidoglycan layer as a result of solvent production by Clostridium pasteurianum, Folia Microbiologica, Vol. 55, No. 4, pp. 340-344, ISSN 1874-9356

Linhova, M.; Branska, B.; Patakova, P.; Lipovsky, J.; Fribert, P.; Rychtera, M. & Melzoch, K.(2010b) Viability determination by flow cytometry during ABE fermentation,Proceedings of 19th International Congress of Chemical and Process Engineering, ISBN 978-80-02-02210-7, Prague, Czech Republic, September 2010

Lipovsky, J.; Patakova, P.; Rychtera, M.; Cizkova, H. & Melzoch, K. (2009) Prospects of butanolproduction from starch and cellulose materials, Chemicke Listy, Vol. 103, No. 6, pp. 479-483, ISSN 1213-7103

Lopez-Contreras, A.; Gabor, K.; Martens, A A.; Renckens, B. A. M.; Claassen, P. A. M.; van der Oost, J. & de Vos, W. M. (2004) Substrate induced production and secretion of cellulases by Clostridium acetobutylicum, Applied and Environmental Microbiology, Vol. 70, No. 9, pp. 5238-43, ISSN 0099-2240

Maddox, I.S.; Steiner, E.; Hirsch, S.; Wessner, S.; Gutierrez, N.A. ; Gapes, J.R. & Schuster, K.C.(2000) The cause of "acid crash" and "acidogenic fermentations" during the batch acetone-butanol-ethanol (ABE-) fermentation process, Journal of Molecular Microbiology and Biotechnology, Vol. 2, No. 1, pp. 95-100, ISSN 1464-1801

Mužíková, Z.; Pospíšil, M. & Sebor, G. (2009) Volatility and phase stability of petrol blends with ethanol, Fuel, Vol.88, No.8, pp. 1351-1356, ISSN: 0016-2361

Napoli, F.; Olivieri, G.; Russo, M.E.; Marzocchella, A. & Salatino, P. (2010) Butanol production by Clostridium acetobutylicum in a continuous packed bed bioreactor, Journal of Industrial Microbiology and Biotechnology, Vol. 37, No. 6, pp. 603-608, ISSN 1367-5435

Ni, Y. & Sun, Z. (2009) Recent progress on industrial fermentative production of acetonebutanol- ethanol by Clostridium acetobutylicum in China, Applied Microbiology and Biotechnology, Vol. 83, No. 3, pp. 415-423, ISSN 0175-7598

Novo, D.; Perlmutter, N.G.; Hunt, R.H. & Shapiro, H.M. (1999) Accurate flow cytometric membrane potential measurement in bacteria using diethyloxacarbocyanine and a ratiometric technique, Cytometry, Vol. 5, No. 1, pp. 55-63, ISSN 1552-4922

Oudshoorn, A.; Van der Wielen, L.A.M. & Straathof, A.J.J. (2009) Assessment of options for selective 1-butanol recovery from aqueous solution, Industrial & Engineering Chemical Research, Vol. 48, No. 15, pp. 7325-7336, ISSN 0888-5885

Patakova, P.; Lipovsky, J.; Cizkova, H.; Fortova, J.; Rychtera, M. & Melzoch, K. (2009) Exploitation of food feedstock and waste for production of biobutanol. Czech Journal of Food Sciences, Vol. 27, No. 4, pp. 276-283, ISSN 1212-1800

Patakova, P.; Lipovsky, J.; Paulova, L.; Linhova, M.; Fribert, P.; Rychtera, M. & Melzoch, K. (2011a) Continuous production of butanol by bacteria of genus Clostridium, Journal of Chemistry and Chemical Engineering, Vol. 5, No. 2, pp. 124-131, ISSN 1934-7375

Patakova, P.; Toure,S.S.M.; Simacek, P.; Cizkova, H.; Lipovsky, J.; Linhova, M.; Fribert, P., Rychtera, M. & Melzoch, K. (2011b) Exploitation of sugar beet for biobutanol production, Czech Sugar and Sugar Beet Journal, Vol. 127, No. 2, pp. 46-49, ISSN 1210-3306

Peng, Ch.; Lewis, C.K.; Stein, F.P. (1996) Water Solubilities in Blends of Gasoline and Oxygenates, Fluid Phase Equilibria, Vol.116 (1996), pp. 437-444, ISSN: 0016-2361

Pierrot, P.; Fick, M. & Engasser, J.M. (1986) Continuous acetone-butanol fermentation with high productivity by cell ultrafiltration and recycling, Biotechnology Letters, Vol. 8, No. 4, pp. 253-256, ISSN 0141-5492

Qureshi, N. & Blaschek, H.P. (2001a) Recent advances in ABE fermentation: hyper-butanol producing Clostridium beijerinckii BA101, Journal of Industrial Microbiology and Biotechnology, Vol. 27, No. 5, pp. 287-291., ISSN 1367-5435

Qureshi, N. & Blaschek, H.P. (2001b) Recovery of butanol from fermentation broth by gas stripping, Renewable Energy, Vol. 22, No. 4, pp. 557-564, ISSN 0960-1481

Rainey, F.A.; Hollen, B.J. & Small, A. (2009) Genus I. Clostridium Prazmowski 1880, 23AL. In Bergey`s manual of Systematic Bacteriology, The Firmicutes, Second Edition, Volume Three, De Vos, P.; Garrity, G.M.; Jones, D.; Krieg, N.R.; Ludwig, W.; Rainey, F.A.; Schleifer, K.-H. & Whitman, W.B, pp. 738-828, Springer, ISBN 978-0-387-95041-9, New York.

Rakopoulos, C.D.; Dimaratos, A.M.; Giakoumis, E.G. & Rakopoulos D.C. (2010a) Investigating the emissions during acceleration of a turbocharged diesel engine operating with biodiesel or n-butanol diesel fuel blends, Energy, Volume 35, No. 12, pp. 5173-5184, ISSN 0360-5442

Rakopoulos, D.C.; Rakopoulos, C.D.; Hountalas, D.T.; Kakaras, E.C.; Giakoumis, E.G. & Papagiannakis, R.G. (2010b) Investigation of the performance and emissions of bus engine operating on butanol/diesel fuel blends, Fuel, Vol. 89, No. 10, pp. 2781-2790,ISSN 0016-2361

Ramey, D. & Yang, S.-T. (2004) Production of butyric acid and butanol from biomass, In: Final report to the US Department of Energy DE-F-G02-00ER86106, February 27, 2011,

Ravagnani, A.; Jennert, K.C.B.; Steiner, E.; Grünberg, R.; Jefferies, J.R.; Wilkinson, S.R.; Young, D.I.; Tidswell, E.C.; Brown, D.P.; Youngman, P.; Morris, J.G.& Young, M. (2000) Spo A. directly controls the switch from acid to solvent production in solvent-forming clostridia, Molecular Microbiology, Vol. 37, No. 5, pp. 1172-1185, ISSN 0950-382X

Rychtera, M.; Patakova, P.; Fribert, P.; Lipovsky, J.; Linhova, M.; Paulova, L.; Melzoch, K.& Klefenz, H. (2010) Bottlenecks in bacterial production of fuel butanol, Proceedings of 19th International Congress of Chemical and Process Engineering, ISBN 978-80-02-02210-7,Prague, Czech Republic, September 2010

Sarathy, S.M.; Thomson, M.J.; Togbé, C.; Dagaut, P.; Halter, F. & Mounaim-Rousselle, C. (2009) An experimental and kinetic modeling study of n-butanol combustion, Combustion and Flame, Vol. 156, No. 4, pp. 852-864, ISSN 0010-2180

Shaheen, R.; Shirley, M. & Jones, D.T. (2000) Comparative fermentation studies of industrial strains belonging to four species of solvent-producing clostridia, Journal of Molecular Microbiology and Biotechnology, Vol. 2, No. 1, pp. 115-124, ISSN 1464-1801

Tomas, C.A.; Beamish, J.& Papoutsakis, E.T. (2004) Transcriptional analysis of butanol stress and tolerance in Clostridium acetobutylicum, Journal of Bacteriology, Vol. 186, No. 7, pp. 2006-2018, ISSN 0021-9193

Tracy, B.P.; Gaida, S.M.& Papoutsakis, E.T. (2008) Development and application of flowcytometric techniques for analyzing and sorting endospore-forming clostridia, Applied and Environmental Microbiology, Vol. 74, No. 24, pp. 7497-7506, ISSN 0099-2240

United States Patent No.4539293 (1985) Production of butanol by fermentation in the presence of cocultures of clostridium, Application 1983-10-05

Wang, S.; Zhang, Y.; Dong, H.; Mao, S.; Zhu, Y.; Wang, R.; Luan, G. & Li, Y. (2011) Formic acid triggers the acid crash of acetone-butanol-ethanol fermentation by Clostridium acetobutylicum, Applied and Environmental Microbiology, Vol. 77, No. 5, pp. 1674-1680,ISSN 0099-2240

Westhuizen van der, A.; Jones, D.T. & Woods, D.R. (1982) Autolytic activity and butanol tolerance of Clostridium acetobutylicum, Applied and Environmental Microbiology, Vol. 44,No.6, pp. 1277-1281, ISSN 0099-2240

Wolf, L. (2007) 1-Butanol as a Gasoline Blending Biocomponent, Mobile Sources Technical ReviewSubcommittee - March 28, 2007 Meeting, Aarlington, USA, available from http://www.epa.gov/oar/caaac/mstrs/March2007/Wolf.pdf

Yacoub, Y.; Bata, B. & Gautam, M. (1998) The performance and emission characteristics of C1- C5 alcohol-gasoline blends with matched oxygen content in a single-cylinder spark ignition engine, Proceedings of the Institution of Mechanical Engineers. Part A. Journal of Power and Energy, Vol. 212, No. 5, pp. 363-379, ISSN 0957-6509

Yao, M.; Wang H.; Zheng Z. & Yue, Y. (2010) Experimental study of n-butanol additive and multi-injection on HD diesel engine performance and emissions, Fuel, Vol. 89, No. 9, pp. 2191-2201, ISSN 0016-2361

Xiaolong, Y; Jing, Y.; Tieping, L. (2009) The Effect of an SI Engine Using Butanol-Gasoline Blended Fuel on Performance and Enviroment, ICEET '09 Proceedings of the 2009 International Conference on Energy and Environment Technology - Volume 01, pp. 402-405, ISBN: 978-0-7695-3819-8 IEEE Computer Society Washington, DC, USA, 2009

Zverlov, V.V.; Berezina, O.; Velikodvorskaya, G.A. & Schwartz, W.H. (2006) Bacterial acetone and butanol production by industrial fermentation in the Soviet Union: use of hydrolyzed agricultural waste for biorefinery, Applied Microbiology and Biotechnology, Vol. 71, No. 5, pp. 587-597, ISSN 0175-7598

CHAPTER-12

ADVANCING THE PROGRESS OF ALGAL BIOFUELS THROUGH THE ESTABLISHMENT OF A GENETIC FRAMEWORK

1. INTRODUCTION: AN OVERVIEW

According to Hoffert et al. (1998), it is expected that the worldwide energy demand would increase twofold in the next four decades. This allows for only a limited timeframe for a significant transformation in the underlying origins of worldwide energy. However, it is possible to meet our energy needs via alternate sources without hindering technological advancement, economic expansion, and political stability throughout this time. The urgency for clean and sustainable energy sources is heightened when considering the environmental repercussions associated with the release of carbon dioxide from fossil fuels. Increased electricity production will inevitably require a swift development of nuclear, wind, solar, and hydroelectric power sources. Despite the quick development of these energy sources, there are few options now available to support the continuous expansion of coal-based electricity until the middle of the century. Therefore, there is an urgent requirement for strategies to reduce carbon dioxide (CO2) emissions. Energy obtained from biomass offers a method to both capture CO2 and decrease the reliance on a fossil fuel-dependent infrastructure. Therefore, bioenergy possesses the benefit of being carbon neutral and will demonstrate its significance as a valuable resource in our collection of renewable energy alternatives. Ensuring carbon neutrality and utilizing energy efficiently are crucial not only in our daily lives, but also in the future generation of energy on a global scale. Biological systems achieve energy usage through a series of closely regulated metabolic networks, which are significantly more efficient than the internal combustion engine. The photosynthetic process is a highly significant and remarkable chemical mechanism for capturing energy. Although there have been significant advancements in photovoltaic technology in recent decades, the plants and algae that have evolved over billions of years offer a fully developed and efficient system for collecting solar energy.

1.1. An emphasis on biofuels

Biofuels, in essence, refer to chemically derived types of energy, such as hydrocarbons, that are derived from biological sources and may be used with the current infrastructure. Petroleum can be considered a biofuel since it consists of the fossilized remains of ancient biomass that has been extracted from the ground. In a contemporary setting, solar energy may be efficiently harnessed and transformed into liquid biofuel by converting photosynthetic biomass. This biofuel can then be used as a substitute for petroleum in transportation fuel. This notion is not novel; in fact, Rudolph Diesel originally conceived his piston-driven engine to operate on peanut oil, with the intention of enabling farmers to produce their own fuel (Nitschke and Wilson, 1965). Currently, biofuels are receiving significant attention as a valuable component of the renewable energy puzzle due to their smooth integration into the carbon cycle. Recent data indicates that the annual use of liquid transportation gasoline in the United States ranges from 130 to 140 billion gallons. It is expected that this demand will continue to rise to about 150 to 160 billion gallons per year, reaching its highest point around 2015 (U.S. Energy Information Administration, 2009). It seems probable that fuel usage will remain steady or gradually decline as we near the middle of the century. Even in the United States, where a faster shift towards hybrid or fuel-efficient automobiles is anticipated, there will still be a significant demand for liquid transportation fuels. While liquid fuels have a high energy density, the feasibility of using hydrogen for transportation appears to be a distant objective and may never be entirely practical. Presently, the manufacturing of hydrogen is a process that demands a significant amount of energy and necessitates a substantial investment in the infrastructure for distribution. On the other hand, plug-in electric vehicles face financial and environmental constraints due to their reliance on batteries and would necessitate a substantial period of time to replace the existing fleet of cars. Therefore, it is justified to place a greater emphasis on liquid biofuels for these specific reasons. Furthermore, petroleum is predominantly utilized for liquid transportation fuels worldwide. There are some situations where it is difficult to replace the dependence we have on liquid fuels, such as in the aircraft, trucking, and construction sectors. In addition, developing economies are expected to choose the most economical vehicle options, such as internal combustion engines, and will continue to rely on petroleum.

1.1.1. Ethanol is classified as a first-generation biofuel.

Ethanol possesses a lower energy density compared to other liquid fuels and is not fully compatible with the existing distribution infrastructure due to its hygroscopic nature, which can lead to the production of rust. Transporting ethanol produced in the Midwest to consumption sites on the East and West Coasts of the United States incurs significant expenses. In other regions, notably Brazil, where the manufacture of ethanol from sugar cane may be more cost-effective, there are still constraints, such as the substantial amount of energy required for distillation from the biologically produced dilute solutions. Utilizing food crops for biofuel production can potentially exert pressures on the global economy. The most significant concerns include the potential impact on food prices resulting from subsidized farming, the extensive use of fossil fuels in production, the degradation of agricultural land resources, and the potential changes in ecosystems owing to greater land use. The production of first-generation biofuels may lead to an increase in emissions of non-CO2 greenhouse gases. Applying nitrogen fertilizer to biofuel crops has the potential to enhance the emission of nitrous oxide, a greenhouse gas that is more potent than carbon dioxide. Additionally, it can lead to increased consumption of natural gas during the production of fertilizer. Although first-generation ethanol has sparked the biofuel revolution, it is evident that improved biofuels are necessary.

1.1.2. Advancements in biofuel technology

With the limitations of ethanol as a biofuel and the need to expand beyond food crop-based biofuel production, there is a pressing need for second- and third-generation biofuels. Since there appears to be little consensus on the meaning of second, third, and further generations of biofuels, we will adopt the term advanced ornext-generation biofuels. Next-generation biodiesel is a triglyceride derived fatty acid methyl ester (FAME), which does not originate from food crop sources. Additionally, there are clear limitations in some non-food crop terrestrial plant sources of triglycerides for biodiesel, such as palm oil and jatropha. Palm oil is rapidly becoming a major source of biodiesel to fulfill the European Union mandate of 10% liquid transportation biofuels by 2010 (European Parliament, 2009). Europe is a much larger consumer of diesel fuel for passenger vehicles with approximately 50% of the passenger cars sold in the EU currently having diesel engines (Smolinska, 2008). Unfortunately, the net result of the substantial increase in demand for palm oil has been an accelerated rate of palm plantation development and deforestation in tropical ecosystems.

Another promising group of next-generation biofuels are the alcohols with longer chains than ethanol, such as butanol and branched-chain alcohols. These fuels have a higher energy density than ethanol, do not absorb water as ethanol does, and possess very favorable combustion characteristics, such as high octane ratings. These alcohols are somewhat more unusual and rare in nature, but one example of a microorganism that is adept at producing these compounds isClostridium acetobutylicum.Regrettably, there are limitations on the use of this slow growing anaerobic organism for biofuel production. A search for other means for producing these promising longer and branched-chain alcohol biofuels in photosynthetic organisms is currently underway (Fortman et al., 2008).

While ethanol produced from cellulosic biomass is commonly touted as a promising advanced biofuel solution, the end product is still ethanol, which has all of the limitations stated above. There is strong motivation to move beyond ethanol, but what other means are available? Conversion of the sugars released by deconstruction of cellulosic biomass could easily be directed more usefully to one of the more desirable next-generation biofuels, such as microorganism-derived biodiesel or branched-chain alcohols. In a throwback to biofuels efforts of World War I, recently there has been renewed interest in the ability ofClostridialspecies to produce butanol and possibly other longer chain alcohol biofuels (Sillers et al., 2008). With the availability of genomes for these anaerobic bacteria, means to genetically enhance their productive capacities may be at hand. However, there are still a number of significant barriers to overcome for anaerobic fermentation to be a truly viable means of biofuel production. Alternatively, microbes such as bacteria and microalgae show promise as a renewable feedstock for a biofuels ranging from ethanol to biodiesel. The capacity of photosynthesis to capture solar energy is particularly attractive for producing renewable fuels because no intermediate chemical feedstock is required.

2. UTILIZING MICROALGAL BIOMASS FOR THE GENERATION OF BIOFUELS.

Algae are a diverse group of aquatic, photosynthetic organisms generally categorized as either macroalgae (i.e. seaweed) or microalgae, which are typically unicellular. Although the emerging field of algal biofuels remains in its infancy, microalgae have great potential to bring the promise of clean, sustainable fuel production before we must face the reality of fossil fuel depletion and exacerbated climate change. Algae are perhaps the most effective photosynthetic organisms for generating chemical energy from sunlight — the most abundant and renewable global energy source. It is believed that a large percentage of today's fossil fuels, particularly petroleum, originated as prehistoric algal blooms. As single-celled organisms, microalgae are capable of producing a large portion of their biomass as small molecule biofuel precursors since they lack macromolecular structural and vascular components needed to support and nourish terrestrial plants. As such, algae provide one of the most direct routes for the photosynthetic conversion of carbon dioxide and other organic substrates to biofuel. Moreover, the large surface area to volume ratio of these aquatic microorganisms is advantageous for absorption of nutrients and sunlight, which is reflected in the rapid growth rates observed in many species.

As aquatic organisms, microalgae offer many advantages over the terrestrial bioenergy crops with which they contend. Some of the most serious drawbacks of allocating portions of existing food crops to produce biofuels, particularly ethanol from corn and biodiesel from soy or rapeseed, are the obvious competition with food production and encouragement of subsidized operations. Both outcomes are coupled with severe economic ramifications. While cellulosic ethanol may avoid the food versus fuel controversy, this technology has yet to fully mature and will likely remain at the developmental stage for a number of years. In general, terrestrial crops have relatively long growing seasons and require arable land, oftentimes supplemented with costly fertilizers that can have harmful effects on the surrounding ecosystems. Additionally, there are greenhouse gases released in the process of generating fertilizer and harvesting terrestrial biomass. Furthermore, constant irrigation of these crops is yet another impediment, as this can be taxing on natural freshwater resources. While great strides are being made toward the optimization of cellulases for enzymatic degradation of lignocellulose, a significant amount of energy is still required to harvest and pre-treat (thermochemically breakdown) the cellulosic biomass, which constitutes an additional input of fossil fuel-derived energy.

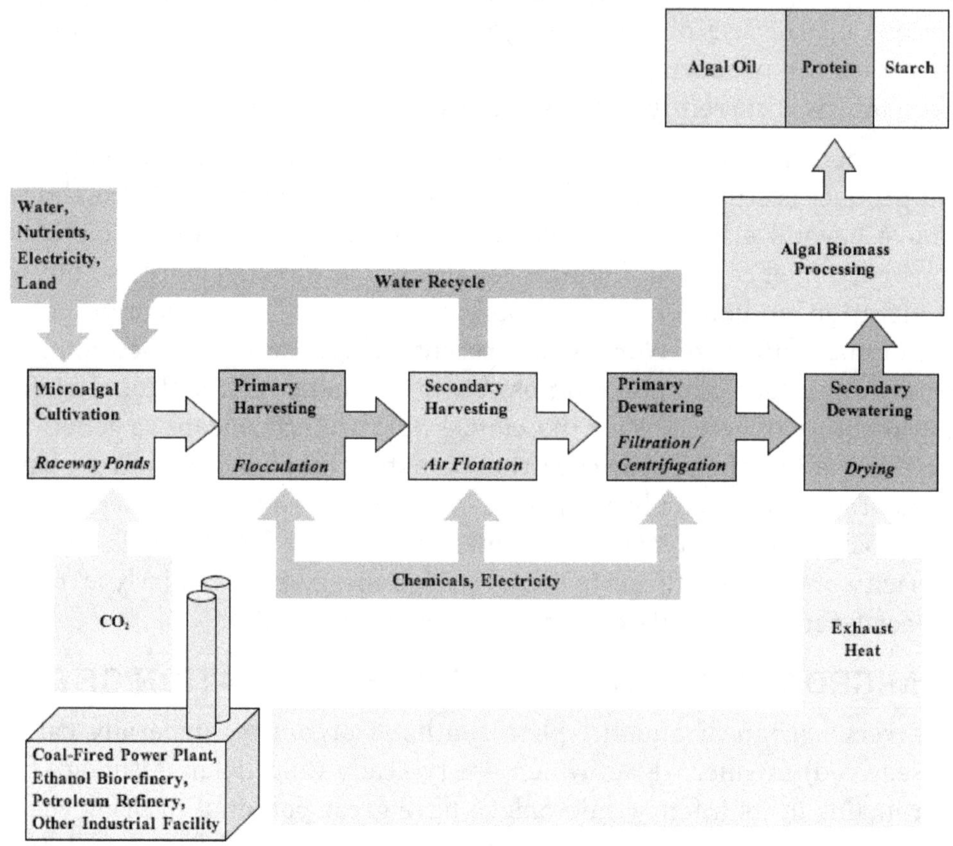

Figure 1. Algal process flow diagram with integrated industrial CO_2 sequestration.

Unlike terrestrial bioenergy crops, microalgae do not require fertile land or extensive irrigation and can be harvested continuously. Several species of algae provide an alternative to freshwater use by growing in brackish, sea, and even hypersaline water. Additionally, since algae consume carbon dioxide through the process of photosynthesis, large-scale cultivation can be used to remediate the CO_2 emissions from fossil fuel combustion (Benemann and Oswald, 1996) (Figure 1). Algal biomass also possesses secondary co-products such as antioxidant pigments, edible proteins, and nutraceutical oils that other alternative fuel crops lack (Spolaore et al., 2006). Lastly, since nearly all microalgae have a simple unicellular structure, algal biomass is devoid of lignocellulose. This strong structural polymer has proven to be a significant obstacle to releasing the energy trapped in terrestrial biomass. Not only do microalgae fully address each of the disadvantages of land-based biofuel crops, but they also are amenable to genetic engineering for the enhanced biosynthesis of a wide range of advanced biofuels and high-value added products.

Currently, three fundamental objectives remain critical to the implementation of economically- and technologically-feasible algal biofuel production: [1] increase of biological productivity through species selection and genetic engineering as well as optimization of culture conditions; [2] development of low-cost vessels for cultivation, whether they be closed photobioreactors or open pond systems; and [3] improvement of inexpensive downstream processing techniques for algal biomass, including harvesting, dewatering, and extraction of biofuel metabolites (Hejazi et al., 2004a; Shelef et al., 1984; Danquah et al., 2009). As with many novel sources of bioenergy, the complexity of the microalgal biofuel production process calls for a multidisciplinary approach in which biotechnological progress will be accompanied by advances in process engineering.

2.1. The commercial viability of microalgal biofuels

As the name 'microalgae' suggests, the relative size of these organisms may seem unsuitable for generating massive quantities of biofuel on a global scale; nonetheless, microalgae offer many advantageous qualities for biofuel production, especially when compared to terrestrial bioenergy crops. The basic principle of generating biofuel from microalgae is to exploit these cells as biological factories, whose lipid output can be as much as 70% of their total dry biomass, under optimal conditions. While many species of algae exhibit the natural capacity to produce abundant amounts of oil for conversion to biofuel, a major obstacle in the commercialization of such a process lies in the scalability. Many exciting breakthroughs in algal biotechnology have occurred on the lab bench, but mass cultivation of algae is still associated with some of the most challenging problems. A few areas of intense research focus include highly productive growth systems, temperature control, photooxidative stress tolerance, light intensity regulation, harvesting, and downstream processing. Whether it is through the manipulation of culture conditions or the application of mutagenesis and genetic engineering, the biological networks of these unicellular creatures can potentially be optimized to synthesize and/or secrete biofuel metabolites, particularly in the form of lipids and other hydrocarbons, in order to overcome some of the aforementioned stumbling blocks to large-scale cultivation.

It is difficult to convey a concise list of the ideal species for biofuel production because the organism must be paired not only with the climate in which it will be cultivated, but also the specific mechanism of cultivation and desired end products. Also, the lipid content can vary considerably depending on culture conditions. For example, nitrogen and silicon deprivation has shown to augment lipid accumulation in green algae and diatoms, respectively. As investigated by NREL's Aquatic Species Program, nutrient deprivation experiments and species collection and characterization efforts gave rise to an extensive list of microalgae with particularly high lipid contents (Sheehan et al., 1998).

There is, however, a serious caveat to high lipid accumulation in algae: the energy collected by the cell is partitioned into storage and, thus, made unavailable for immediate use. As a result, oleaginous species exhibit significantly slower growth rates than their more lean relatives. The fatty acid composition and growth characteristics of some of the more promising species are illustrated in Figure 2, where a clear balance can be seen between growth rate and lipid content. In this particular study, the algal species were cultivated first in airlift bioreactors, then in aerated polyethylene bags, and finally in outdoor raceway ponds for a period of four months. Subsequent biochemical evaluation found the neutral lipids to be predominantly C16and C18, which are ideal chain lengths for the composition of biodiesel (Gouveia et al., 2009).

Astonishingly, of the thousands of different species of algae, a mere fifteen organisms are commonly used for commercial applications (Raja et al., 2008) and only eight species' genomes have been sequenced

(Hallmann, 2007), but future bioprospecting endeavors paired with high-throughput screening are likely to discover more exemplary candidates for algal biofuel production.

2.2. Potential applications for genetic manipulation of algae

While microalgae are an abundant source of naturally oil-rich biomass, these cells can also act as biological factories designed to produce of a variety of promising biofuel precursors. By elucidating the complex metabolic networks involved in carbon utilization, there lies great opportunity for genetic and metabolic engineering of these organisms. Some of the major obstacles to metabolic engineering of algae stem from the lack of basic biological knowledge of these diverse creatures, including sparse genomic information and somewhat primitive methods of genetic transformation. As a result, the introduction of nuclear transgenes to microalgal cells relies on random chromosomal integration, which is highly susceptible to gene silencing; the subsequent recovery of stable transformants is limited to only a handful of species and is oftentimes irreproducible. Overcoming these biotechnological barriers, however, will present enormous opportunities to develop microalgae as versatile platform for biofuel production.

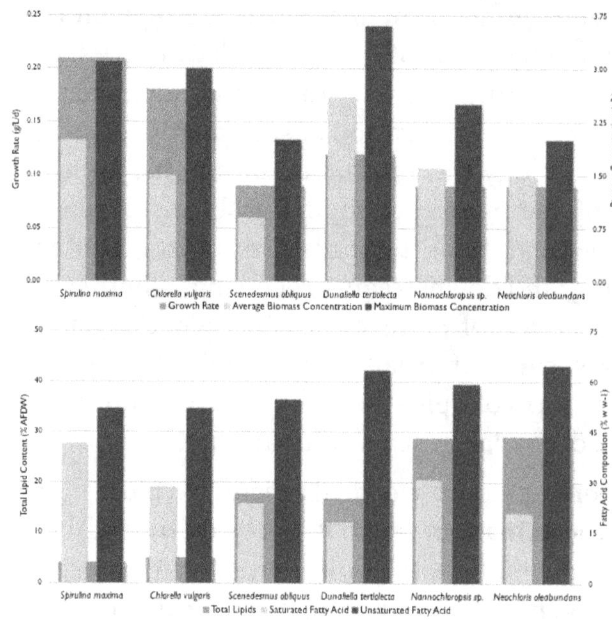

Figure 2. Growth characteristics and lipid profiles of commercially attractive microalgae.

A number of improvements in the productivity of green algae and diatoms would significantly enhance their capabilities as biofuel producers. Photoautotrophic algal growth rates and cell densities at commercial are low compared to microbial fermentation. Enhancement of growth through metabolic engineering with control of cell cycle would be a breakthrough for algal biofuels. Increasing biofuel feedstock production by improving the synthesis of biofuel precursors is imperative. Metabolic engineering of secretion pathways or developing means to readily strip hydrocarbons would allow the organism to survive while producing biofuel metabolites on a continue basis. This would reduce the amount of biomass that is produced per unit of biofuel, thus focusing resources on the primary product. Metabolic engineering might also increase the range of biofuel metabolites and other high-value added materials that can be synthesized. Furthermore, improved performance in a variety of photobioreactors and conditions is necessary. The development of organisms that can survive in environments that exclude invasive species and other contaminant microorganisms is another desirable attribute for open cultivation systems. Finally, the goal of designing suicide genes to prevent the unintended release of genetically modified organisms (GMOs) is an important consideration.

From a genetic engineering standpoint, there exist certain superior characteristics of eukaryotic algae as compared to other photosynthetic sources of biofuels. Focusing on green algae and diatoms will allow the transfer of well-established metabolic engineering approaches used in other eukaryotic systems. With eukaryotic algae there is the ability to perform both chloroplast and nuclear transformation, possibly increasing the complexity and range of metabolic engineering. Because eukaryotes have evolved a membrane bound secretory pathway, it is conceivable that eukaryotic algae could be genetically engineered to secrete lipid bodies into the media (Benning, 2008). For example, the algaBotryococcus brauniihas naturally evolved this mechanism for secreting hydrocarbons. If this biological feat could be achieved in other microalgal species, it would greatly simplify downstream processing by eliminating cell harvesting and lysis; thus, reducing the entire procedure to merely skimming the lipids from the top of the medium.

Many accomplishments have already been made in the field of microalgal bioengineering (Leon-Banares et al., 2004; Walker et al., 2005), the most relevant to biofuel production being increased photosynthetic efficiency and light penetration in C. reinhardtii (Mussgnug et al., 2007), but augmented lipid production through genetic alterations has yet to be achieved. Currently, C. reinhardtiiremains the workhorse of algal genetic engineering for its history as a model photosynthetic organism (Harris, 2001). The recently completed genome sequence of Dunaliella salinamay be a good starting point for genetic research of algal biofuel production; however, a single species cannot be expected to serve every application. With the limited availability of genomic data for microalgae (Hallmann, 2007), exploration of transgenic algae for bioenergy demands a genome project for a model biofuel production strain. Future efforts to probe the metabolic pathways of microalgae will likely employ technologies beyond genomic analysis, such as transcriptomics, metabolomics, proteomics, and lipidomics, to examine the broader biological landscape of algal metabolism (Jamers et al., 2009; Vemuri et al., 2005).

2.3. Large-scale cultivation of microalgae

The cultivation of macro- and microalgae is a long-standing and widely practiced method that has been used for almost 2,000 years. It yields a significant amount of biomass that is used for human nutrition, as well as for the production of commercially important biopolymers and specialty chemicals (Spolaore et al., 2006). For instance, the cultivation of the gelatinous cyanobacteria Nostoc in rice fields during 200 AD allowed a significant portion of the Chinese population to endure starvation (Qiu et al., 2002). Subsequently, the large-scale growth of microalgae has been used for commercial purposes, specifically for the manufacturing of either complete microalgae-based dietary supplements or extracts with health benefits, such as β-carotene, astaxanthin, and polyunsaturated fatty acids (e.g. DHA, omega-3). China, Japan, Australia, India, Israel, and the United States are prominent players in algal production on the global market.

2.3.1. Limitations on the growth of photoautotrophic algal biomass

In addition to certain biological limitations, several obstacles related to cultivation must be overcome to allow economical industrial scale-up of algal biofuel production. The conversion efficiency of solar energy to biomass by microalgae is governed, in part, by the inherent biological efficiency of photosynthesis, and largely by the effectiveness of light-transfer in liquid cultures. Some species of algae grown heterotrophically (i.e. supplemented with carbon sources other than CO_2 , such as sugars) can accumulate a greater amount of lipids (Wu et al., 2006); however, the costs associated with such cultivation may limit its applicability to biofuel production. The approach of heterotrophic algal biofuel production is the model for a number of algal biofuels start-up companies.

On the other hand, generating algal biomass for biofuels with energy directly from the sun rather than a chemical intermediate has its advantages. Microalgae essentially act as biological solar panels directly connected to biorefineries. Photoautotrophic cultivation has the added benefit of CO_2 sequestration from a point source. Although current commercial raceway ponds operate with areal productivities of 2-20 g m-2d-1, there remains much speculation regarding the maximum achievable algal biomass productivity. While heterotrophic modes of cultivation can yield very dense algal cell cultures, photoautotrophic cultures are not expected to exceed 60 g m-2d-1. Figure 3 shows a compilation of realistic areal productivities and theoretical projections for photoautotrophic algal cultivation in open ponds (Chisti, 2007; Weyer et al., 2008; Schenk et al., 2008; Wallace, 2008).

While a wide range of predictions has been made for the maximum attainable productivity of algal cultures, a straightforward analysis of energy transfer in algal biomass production reveals a few key bottlenecks. By following a single photon from its origin at the sun to the desired end product of algal oil, there are a number of unavoidable losses imposed on this conversion of sunlight by the inherent bioenergetics of cellular processes. Additional diversions of solar energy can be attributed to the algal growth system and can be minimized with proper design parameters.

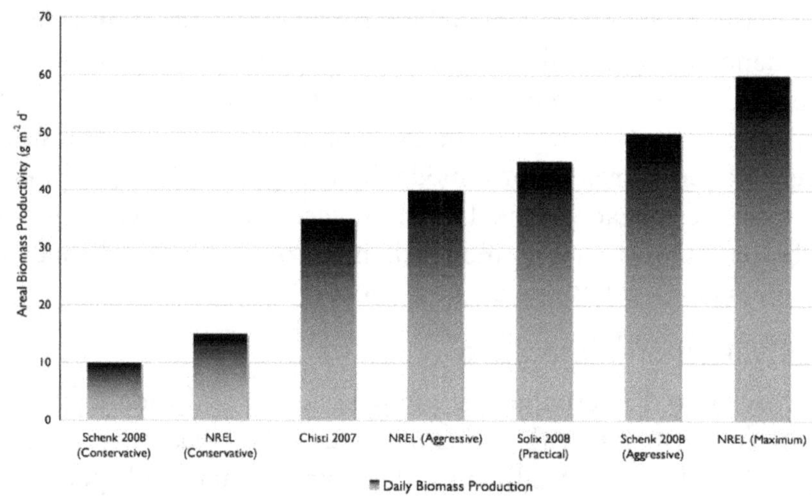

Figure 3. Projected algal biomass productivities for raceway ponds.

The first impediment that solar radiation faces when traveling to the Earth's surface is the local weather. As we know from our daily observations, cloudy skies can dramatically reduce the amount of light that reaches the ground. Additionally, while the equator receives high-intensity light year-round, solar irradiance diminishes as one travels away from the equator in latitude, thus near-equatorial zones are ideal for algal biomass production. Accordingly, Asia, Australia, and the United States are common sites for algal growth facilities. Figure 4 presents a map of solar data collected from 1990-2004 where black dots represent locations for which detailed weather analysis is available for algal production facilities (Weyer et al., 2008). In geographies that receive more exposure to sunlight, and accompanying high temperatures, evaporative water loss and cooling mechanisms become more important considerations.

Since there is little one can do to change the weather beyond choosing an adequate site for algal cultivation, the next constraint on solar energy collection comes from the limited spectrum of light that plants have adapted to utilize, deemed photosynthetically active radiation (PAR: λ = 400-700 nm), which accounts for only 45% of the total energy in the visible light spectrum (Weyer et al., 2008).

In conventional raceway ponds and photobioreactors, incident sunlight encounters billions of algal cells as it travels through the liquid culture — each cell absorbing some of the available energy. Thus, the transmission of light is severely inhibited by cell shading in these dense solutions. For example, the leaves of a tree have evolved to be essentially two-dimensional structures with only millimeter thicknesses; an algal culture volume can be thought of in a similar manner. In open ponds, only the cells on the surface are exposed to maximum sunlight, and those on the bottom of the 10-30 cm deep trough receive very little of this incoming energy. The advantage, though, of a liquid culture is that the shaded cells in the submerged regions can be recirculated to the surface periodically so that a large volume of biomass can be maintained. Additionally, advanced photobioreactor design can encourage optimal mixing patterns (Tredici and Zittelli, 1998).

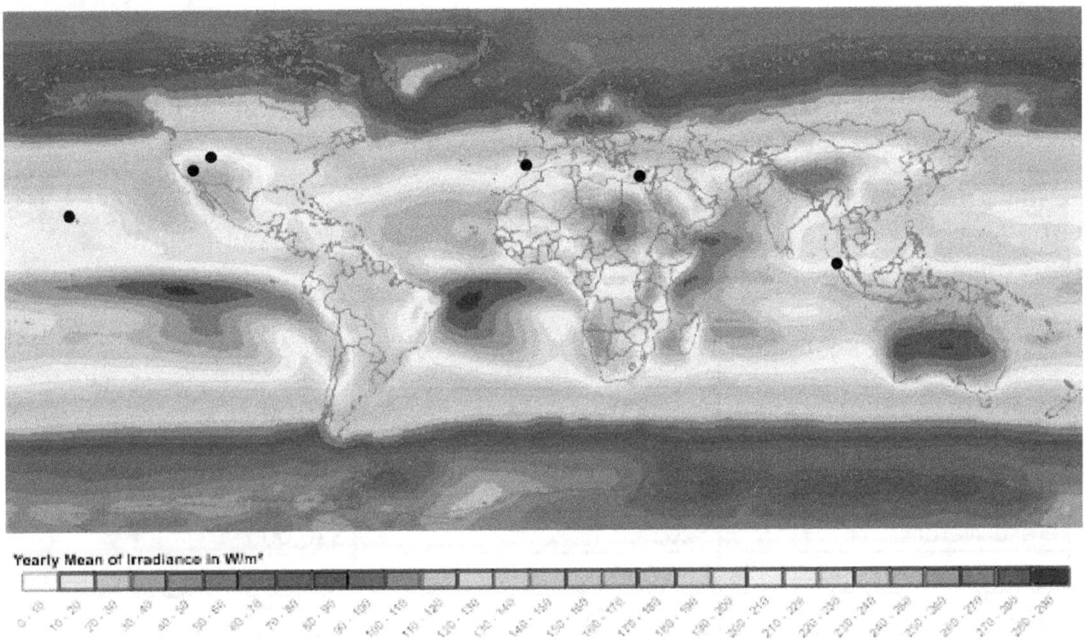

Figure 4. Global map of average annual solar radiation (Reprinted with permission from SoDa Services, Copyright Mines ParisTech / Armines 2006).

The next hurdle that usable photons must surmount is absorption by the molecular photosynthetic apparatus. As microalgae have adapted to survive in conditions of low light, they have evolved biochemical mechanisms that are incredibly adept at collecting light energy. Light-harvesting complexes, which surround the photosystems in the chloroplast membrane, act as antennae to gather and shuttle photons to the photosynthetic reaction center. One limitation to complete utilization of incident PAR lies in the maximum capacity of energy that each photosystem can handle. In fact, high-intensity light can quickly inundate the photosynthetic machinery, leading to the generation of free oxygen radicals, in a process called photoinhibition (Long et al., 1994). This obviously has a negative effect on the productivity of algal cultures by leading to cellular damage and premature population demise. Unlike the natural environmental conditions to which microalgae are accustomed, industrial cultivation strives for maximum collection of solar energy. In this case, algae can be exposed to high-intensity light for only a short period of time, on the scale of microseconds, allowing each photosystem to absorb light before becoming overwhelmed. Also at the biochemical level, there are inherent limitations to photosynthesis related to electron transfer. Estimates for the efficiency of photosynthesis correlate to roughly 25% energy conversion (Weyer et al., 2008). Optimization of photosynthesis is an ambitious genetic engineering goal and is likely to remain an intrinsic parameter of algal biomass production processes.

After the available sunlight has been harnessed by photosynthesis, there exist various levels of inefficiency related to the biological conversion of this energy to biomass. Principally, nearly 40% of the total energy is required to sustain basic cellular function and growth, leaving an estimated 60% of the total photosynthetically captured energy for biomass accumulation (Weyer et al., 2008). According recent analysis utilizing actual solar irradiance and weather data from various global locations (Figure 4), and taking each of the aforementioned assumptions into account, the projected range of algal biomass production is between 38 and 47 g m-2d-1(Weyer et al., 2008), which is in agreement other predictions (Figure 3). Current values for commercial biomass production in open ponds are typically 2-20 g m-2d-1, which provide sufficient profit margins for high-value products such as β-carotene, but are anticipated to meet the demands of cost-effective biofuel production in the near future. A new partnership between Seambiotic and the Israeli Electric Company has plans to produce algal biomass inexpensively for use as biofuel, with operating costs and profit margins listed in Table 1.

Table 1. Cost analysis of microalgal biomass production facilities in Israel (Reprinted with permission from Dr. Ami Ben-Amotz).

Annual Expenses(USD yr-1) 10-ha Raceway Farm	Nature Beta Technologies Commercial Plant	Seambiotic/Israeli Electric Pilot Plant
Manpower	$500,000 (20 Employees)	$120,00 (8 Employees)
Electricity	$180,000	$30,000
Nutrients	$36,000	$36,000
Land	$50,000	$10,000
Carbon Dioxide	$150,000	$5,000
Sea Water	$200,000	$5,000
Fresh Water	$20,000	$10,000
Miscellaneous Expenditures	$30,000	$20,000
Total	$1,166,000	$236,000
Revenue		
Biomass Production (yr-1)	70 tons (at 2 g m-2d-1)	700 tons (at 20 g m-2d-1)
Biomass Cost (USD kg-1)	$17.00	$0.34
Market Price (USD kg-1)	$4,000 (D. salinaβ-carotene)	< $0.50 (Potential Biofuels)

2.3.2. Raceway pond systems and photobioreactors are two types of systems used for cultivating algae

Although primitive in design, open raceway ponds are still the predominant algaculture system for high-value added products and have been used for decades (Benemann and Oswald, 1996;Sheehan et al., 1998). The fact that raceway ponds are uncomplicated makes them less productive and offers little control over the culture parameters; however, the low cost of this low-tech cultivation system allows it to compete with complex photobioreactors (Gordon et al., 2007). The surface-to-volume ratio and corresponding light penetration in open ponds are not ideal. As a result, ponds can only support low culture densities; however, the ease of scaling production to industrial proportions (> 1 million liters per acre) justifies their seemingly low efficiency. The use of raceway ponds is also vindicated by their uncomplicated design, which makes them readily available for implementation and relatively simple to clean and maintain. While open ponds are cost effective, they do have a large footprint and contamination by local algal species and threat of algal grazers pose serious risks. Invasion by indigenous microorganisms may be protected against by the use of greenhouse enclosures or by growing algae that can withstand hypersaline environments, such as Dunaliella salina. For the purpose of biofuel production, however, the low areal productivities of ponds alone may not be able to provide the necessary biomass feedstock economically. Commercial scale microalgal biofuel facilities will likely rely on integrated systems of high efficiency photobioreactors to provide a dense inoculum for readily scalable raceway ponds (Huntely et al., 2007).

While genetic engineering approaches may improve the photosynthetic and biosynthetic capabilities of microalgae, many innovative methods exist for optimizing photoautotrophic culture conditions to accomplish the same goal of increased yield (Muller-Feuga, 2004). The breadth of chemical engineering knowledge being applied to photobioreactor (PBR) design in order to enhance light and nutrient availability represents an important advance in the field (Tredici and Zittelli, 1998; Miron et al., 1999), particularly for modular and scalable reactors (Janssen et al., 2003; Hu et al., 1996); however, the cost of these technologies remains the ultimate constraint on feasibility.

Photobioreactors aim to optimize many, if not all, of the culture parameters crucial to microalgal growth. One condition achieved in PBRs, but not raceway ponds, is turbulent flow to produce enhanced mixing patterns. By simply increasing the Reynolds number of these systems, the resulting fluid dynamics have a positive effect on nutrient mass transfer, light absorption, and temperature control (Ugwu et al., 2008). However, highly turbulent flow within complex geometries comes at the expense of the cells' sensitivity to shear stress, which is one limitation of photobioreactor design. Another important consideration is the time scale over which the microalgae are transferred from the periphery of the bioreactor to the interior shaded region, as there exists a limit to the amount of light the cells can process before photoinhibiton occurs.

One of the most beneficial aspects of photobioreactors is the extended surface area achievable with tubular and flat-panel designs. This route to increased productivity takes advantage of allowing more algae to have contact with sunlight than an area of land would regularly allow with more basic cultivation systems; however, the additional cost of maintaining ideal temperatures and protecting these sometimes delicate devices from inclement weather pose some concern. Many of these nascent technologies depend entirely on the site of deployment and, as a result, require a great deal of customization (Janssen et al., 2003;Miron et al., 1999).

Some additional drawbacks of photobioreactors include the chemical gradients that can develop, particularly along tubular reactors. As a result of photosynthesis, a significant amount of oxygen can accumulate in these tubes and must be purged periodically. High concentrations of oxygen can both inhibit photosynthesis and, when combined with high irradiance, result in the formation of reactive oxygen species (ROS) (Tredici and Zittelli, 1998). These added difficulties contribute to the higher complexity and cost of photobioreactor operation. Some low-cost alternatives include vertical-column or hanging-bag bioreactors, which still provide a closed system for monocultures, with less control over culture parameters. These systems often rely on sparged air to provide both CO_2 and mixing force, as in airlift bioreactors. The ability to strike a balance between cost and productivity is the major challenge of microalgal cultivation, especially for applications that require closed cultivation, as is the case with genetically modified microalgae.

3. ADVANCINGTHECREATIONOFSELECTABLEMARKERSFORDUNALIELLASALINA

In the academic community, green microalgae serve as model organisms for photosynthetic research;Chlamydomonas reinhardtiiis the most reputable species for this work (Harris, 2001).Volvox carteri, a multicellular microalga, is another well-established model species for the elucidation of the genetic basis of cellular differentiation (Miller, 2002). In recent years, the green algaDunaliella salinahas been utilized to complement the study of photosynthesis, osmoregulation, carotenogenesis, and glycerol production (Jin et al., 2001; Liska et al., 2004; Thompson, 2005; Shaish et al., 1992; Chitlaru et al., 1991). Dunaliella salinais an attractive platform for both commercial and academic pursuits owing to its intriguing and advantageous abilities to survive in conditions of extreme salinity and produce significant amounts of β-carotene. Currently, the aspiration to genetically and metabolically engineer this organism in order to probe its biological networks and eventually enhance its productivity is an ambitious goal. Until recently, the stable expression of transgenes by this organism has been limited due to inexperience with genetic transformation and insufficient knowledge of the species' genome. While molecular methods of manipulation make C. reinhardtii and V. carteri experimentally tractable at many levels,there is a pressing need for the same tools to be developed for D. salina.

This section discusses attempts to genetically engineerD. salinathrough the development of selectable marker systems. The investigation includes detailed characterization of the growth response of D. salinato a number of antibiotics and herbicides commonly used for selection of microalgae, such as bleomycin, paromomycin, and phosphinothricin (PPT). Based on reported genetic sequence information for D. salina, promoter and 3'-UTR regions of highly active genes were selected as targets for genomic PCR, with the hopes of creating D. salina-specific plasmid transformation vectors. Although these efforts did not yield the intended results, this work establishes a foundation for genetic engineering of D. salina, which is expected to continue now that the sequenced genome has been made available (Smith et al. 2010).

3.1.Dunaliella salina as a host organism for biotechnological advancements

For many years, D. salina has been grown for its inherent capacity to synthesize β-carotene. The abundance of natural β-carotene in D. salina makes it highly lucrative to cultivate and process the biomass on a big scale, resulting in significant profitability (Ye et al., 2008). This particular green alga is well-suited for cultivation in outdoor ponds because it has the capacity to thrive in highly saline conditions, with salt concentrations up to eight times that of seawater. This characteristic significantly minimizes the risk of contamination by local microorganisms and eliminates the requirement for huge amounts of freshwater. Dunaliella spp. are comparable to Chlamydomonas spp. in their characteristics as elongated, single, flagellated cells measuring around 10 microns in size. However, D. salina and D. tertiolecta possess the ability to regulate osmotic pressure through a sophisticated system of ion channels, a flexible cell membrane that is not constrained by a cell wall, and the production of glycerol to counterbalance osmotic pressure (Goyal, 2007). Consequently, the absence of a strong cell wall makes it relatively easy to break down the algal biomass for subsequent processing. Moreover, the process of extracting specific compounds from microbial cells, known as "milking," has significantly advanced in recent years. During this procedure, the cells are exposed to a biocompatible organic solvent to facilitate the selective extraction of desired compounds into the solvent phase, while keeping the cells alive for further bioproduction. The effective demonstration of this procedure was using D. salina to extract β-carotene in a two-phase system (Hejazi et al., 2004b). The high market price of natural β-carotene is determined by the demand for this molecule and the continuous use of D. salina. However, the growing interest in biofuel production has led to a curiosity about the carotenogenesis pathways of D. salina (Lamers et al., 2004). Given that carotenoid compounds consist of elongated branched hydrocarbons, it is possible to modify the biosynthetic pathways of D. salina in order to generate hydrocarbons that are well-suited for usage as biofuel resembling gasoline. Having already created molecular biology tools for Dunaliella spp. (Polle et al., 2009), the sequencing and annotation of its 610 Mbp nuclear genome will enable more extensive genetic engineering efforts with this organism. During the experiments, only the genomes of chloroplast and mitochondria from D. salina CCAP 19/18 (GenBank GQ250046, GQ250045) were made available. Due to its distinct biotechnological applicability and extensive track record in mass production, D. salina is an optimal candidate for future advancements as a microalgae that produces biofuel.

3.2. Manipulation of the genetic makeup of D. salina

Owing to the attractiveness of D. salina for biotechnology, there is a renewed interest in engineering this organism. Publications have reported the genetic transformation ofD. salinaby both microparticle bombardment and electroporation (Geng et al., 2003; Tan et al., 2005). Some of the most impressive progress in the field has come from the Xue group at Zhengzhou University in China. With research covering optimization of transformation techniques, gene characterization, and enhanced gene expressing utilizing matrix attachment regions, their work provides important information and an exemplary research path to follow toward genetic engineering of D. salina (Wang et al., 2009; Lu et al., 2009; Jia et al., 2009a; Feng et al., 2009; Wang et al., 2007; Jia et al., 2009b; Liu et al., 2005; Jiang et al., 2003). The down-regulation of specific genes using RNAi in D. salina has also been reported (Jia et al., 2009a; Sun et al., 2008). These advances, however, are not readily reproducible and represent solitary accomplishments with an alga that has otherwise been difficult to transform.

3.2.1. Specific chemicals and genes that provide resistance to antibiotics

The bleomycin family of glycopeptide antibiotics is toxic to a wide range of organisms with as intercalator functionality able to cleave DNA. Bleomycin-resistance, attributed to theblegene, is an ideal selectable marker as the BLE protein acts in stoichiometric equivalent. Occurring as a dimer, each protein has a strong affinity for binding and inactivating two molecules of bleomycin. Therefore, the level of expression of this exogenous gene can be directly correlated with antibiotic tolerance observed phenotypically. As such, it has been developed as a selectable marker system for nuclear transformation of C. reinhardtii and V. carteri and is likely to have similar applicability in other algae (Lumbreras et al., 1998;Hallmann et al., 1999). Since the protein must enter the nucleus, it requires high levels of expression to be active; thus, inherently selecting for clonal isolates with high-level expression. Commercial forms of bleomycin include bleocinTM and zeocinTM, which are known to affect mammalian, insect, yeast, bacterial, and plant cells.

Paromomycin has been used extensively as a selective agent for the genetic engineering of microorganisms as well as an antibiotic for the treatment of bacterial infections in humans. Paromomycin acts by binding to the 16S ribosomal RNA of microbes, effectively interfering with protein synthesis. The aminoglycoside 3'-phosphotransferase geneaph from Streptomyces rimosus already exists in a codon bias similar to that of Chlamydomonasand confers resistance to paromomycin, kanamycin, and neomycin. Paromomycin is highly toxic to green algae and has been used as an effective selective agent in bothC. reinhardtii and V. carteri (Sizovaet al., 2001; Jakobiak et al., 2004).

Phosphinothricin (PPT), also known as glufosinate-ammonium (GLA) and bialaphos, is a natural amino acid that competitively inhibits glutamine synthesis in plants and animals (Hoerlein, 1994). For this reason, it is used in a number of herbicides, including Basta®and RoundUp®. A number of agricultural crops have been engineered by to resist PPT through introduction of thebar (bialaphos resistance) gene from Streptomyces hygroscopicus, which encodes a phosphinothricin acetyl transferase (Thompson et al., 1987). In the academic community, PPT is used as a selective agent for work with Arabidopsis and thebargene is carried on pGR117 for Agrobacterium-mediated transformation of A. thaliana (Akama et al., 1995).

3.2.2. Genetic regulatory factors of D. salina that originate from within the organism itself.

Prior to the availability a fully sequenced genome, the genetics of D. salinawere explored for useful elements such as regulatory sequences of highly expressed genes. Highly active endogenous promoter and 3'-untranslated region (UTR) pairs are of particular interest and significance to expressing transgenes inDunaliella. Recent publications describing the use of the actin,rbcS, carbonic anhydrase, and ubiquitin promoters (Jiang et al., 2005; Walker et al., 2004; Chen et al., 2009) and nitA 3'-UTR (Li et al., 2007; Xie et al., 2007) are the basis for many of the pioneering attempts to genetically engineer D. salina, including the work presented hereafter.

3.3. Experimental procedures and techniques

3.3.1. Cultivation of microalgae cells

D. salinastrains CCAP 19/18 and UTEX 1644 were obtained from the Culture Collection of Algae and Protozoa (UK) and the Culture Collection of Algae at University of Texas at Austin, respectively, and maintained on sterile agar plates (1.5% w/w) containing 1 M NaCl Dunaliella medium (Weldy et al., 2007). Cells were cultivated photoautotrophically in 1-L glass Fernbach flasks at 27º C (± 1) using 1 M NaCl Dunaliella medium (DM). Each axenic batch culture was inoculated with 10 ml of exponentially growing cells (1 × 106cells ml-1), constantly stirred, bubbled with sterile air, and illuminated with cool-white fluorescent bulbs at an intensity of 80 μE m-2s-1.

3.3.2. Generation of dosage response curves

In order to test the efficacy of the antibiotics bleocinTM (EMD Biosciences), paromomycin (MP Biomedicals), and the PPT-containing (200 g L-1) commercial herbicide Basta® (Bayer Crop Sciences) on D. salinaCCAP 19/18, algal cells were grown in the presence of various concentrations of each selective agent (Table 3)using 100-ml stirrer flasks under the same conditions mentioned previously, with the exception of aeration. Each culture was inoculated with 10 ml of exponentially growing cells (1 × 105cells ml-1) and the cell density was measured with a Zeiss Axiovert 100 inverted light microscope using a hemocytometer over a one week period. Additionally, cells were plated on agar to analyze viability on solid medium. Measurements were taken in duplicate and experiments were repeated two times.

3.3.3. Molecular cloning and construction of vectors

All plasmids were prepared and isolated from bacterial hosts using the QIAPrep Kit (Qiagen) following the supplier's protocol.E. coliDH5α cells (Invitrogen) were grown axenically either on LB agar or in LB medium containing pertinent antibiotic selective pressure (ampicillin or kanamycin) at 37º C. Genomic DNA was extracted fromD. salinausing standard CTAB protocol adapted fromVolvox carteri. PCR was performed using the MJ Mini (Bio-Rad) to amplify the actin (GenBank: AF541875) and rubisco small subunit (GenBank: AY960592) promoters and nitA 3'-UTR (GenBank: EF156403) fromD. salinaUTEX 1644 genomic DNA. Likewise, the genesbleandbarwere amplified from the plasmids pSP124 and pGR117, respectively. All primers employed in this study are listed below in Table 2. Each PCR product was subsequently ligated into the subcloning plasmid pGEM®-T Easy (Promega) using T4 DNA Ligase and its corresponding buffer (Invitrogen). Sequencing of the genetic fragments derived from PCR was performed at the UMBC Biological Sciences Dept. DNA sequencing facility using BigDye®(Applied Biosystems).

Table 2. Primers used to amplify genes and promoters for vector development.

Target Sequence	Primers (including NotI, SmaI, HindIII, and XhoI for subcloning)
actin promoter	5'-AATAATAGCGGCCGCCACGGCTCACCATCTTGTTT-3' 5'-AATAATACCCGGGTTGATCTCTCTGTCACCCCT-3'
rbcS2 promoter	5'-AATAATAAGCGGCCGCAGACATGAACCTATA-3' 5'-AATAATAACCCGGGAGGTCTTGGCAATGA-3'
bar	5'-AATAATACCCGGGATGAGCCCAGAACGACGCCC-3' 5'-AATAATAAAGCTTTCAGATTTCGGTGACGGGCA-3'
ble	5'-AATAATACCCGGGATGTTCTTTACTTTTTTACA-3' 5'-AATAATAAAGCTTCTAGAGTGGGTCGACGTCGG-3'
aphVIII	5'-AATAATACCCGGGCGAAGCATGGACGATGCGTT-3' 5'-AATAATAAAGCTTTCAGAAGAACTCGTCCAACA-3'
nitA 3'-UTR	5'-AATAATAAAGCTTGCGGGGTCAGCAGGAGCGAC-3' 5'-AATAATACTCGAGTCGATCAGCCTTTGCAATCC-3'

3.3.4. Manipulation of the genetic material in D. salina

Electroporation: A population of D. salina CCAP 19/18 cells was obtained from a 250-ml culture during its exponential growth phase, with a concentration of approximately 1 × 106 cells ml-1. The cells were then washed multiple times with a pre-electroporation buffer (consisting of 0.2 M mannitol and 0.2 M sorbitol) to eliminate any remaining salts. Subsequently, the cells were suspended in an electroporation buffer (containing 0.08 M KCl, 0.005 M CaCl2, 0.01 M HEPES, 0.2 M mannitol, and 0.2 M sorbitol) to achieve a final density of 8 × 107 cells ml-1 (Sun et al., 2005). Subsequently, 500 µl of electroporation samples were created in cuvettes with a 0.4 cm electrode gap (Bio-Rad). These samples consisted of either 4 × 107 or 1 × 106 cells ml-1, 20 mg ml-1 of pSP124, and 2 mg ml-1 of herring or salmon sperm DNA (Sigma). Following a 10-minute period of incubation on ice, the process of electroporation was carried out utilizing the Gene Pulser® II with Capacitance Extender Plus (Bio-Rad). Two electroporation conditions were evaluated, with the first condition having a capacitance of 500 µF and a voltage of 400 V (equivalent to 1 kV cm-1), and the second condition having a capacitance of 25 µF and a voltage of 1.6 kV (equivalent to 4 kV cm-1). The resistance of each sample was 50 Ω. After the electric pulse, the cells were promptly supplied with DM and then left to recuperate in darkness for 12 hours at room temperature. To identify possible transformants, the cells were placed on DM plates containing 1.5% agar and 4 mg/L of bleocin. The cells were observed for a week.

Microparticle Bombardment

Spherical gold particles with a diameter of less than 10 µm (Aldrich) were created by washing and suspending them in sterile dH2O many times until a concentration of 50 mg ml-1 was reached. Twenty shots from the microparticle gun involved the ethanol precipitation of roughly 20 µg of DNA onto 12.5 mg of gold particles (250 µl) for a duration of one hour at -80º C. After a brief centrifugation of the gold solution at 14,000 revolutions per minute (RPM), the resulting particle was rinsed with 70% ethanol, centrifuged twice, and ultimately resuspended in 78% ethanol. The resuspended solution was then stored on ice for later use with the transformation gun. Prior to transformation, a 1-L culture of D. salina CCAP 19/18 in exponential phase (approximately 1 × 106 cells ml^{-1}) was harvested using a Sorvall® RC-5B Refrigerated Superspeed Centrifuge, Du Pont Instruments, at a speed of 5,000 RPM for 10 minutes at a temperature of 25º C (± 5). The harvested cells were then suspended in 5 ml of fresh DM. The process of cell bombardment and recovery was conducted in the same manner as previously described for the transformation of Volvox, as outlined by Schiedlmeier et al. in 1994. Subsequently, each filter paper was immersed in 25 ml of DM and left to recover for two days without antibiotic selective pressure, following the previously mentioned culture conditions, except for aeration. Transformed colonies were then chosen based on the minimum inhibitory concentration (M.I.C.) of the suitable agent on solid medium.

3.4. Results

3.4.1. Assessment of the lowest amounts that prevent the growth of microorganisms

The presence of 0.25 mg L-1 of bleocin in the liquid culture greatly inhibited the growth rate of D. salina CCAP 19/18. When exposed to concentrations of 0.5 and 1.0 mg bleocin L-1, the proliferation of cells was entirely halted. However, the cells were able to remain alive for more than a week. Due to the limited cell division, these cells showed signs of stress such as low chlorophyll coloring and decreased motility. Significantly, in the present investigation, doses of bleocin at 2.0 mg L-1 and above were shown to ultimately eliminate the algal cultures, although a few viable cells remained for a period of three to five days. Comparable outcomes were noted when D. salina was grown on solid medium with similar levels of bleocin. However, the growth rate on agar plates was predictably considerably lower. A considerable proportion of D. salina colonies demonstrated the ability to endure on solid medium with a concentration of 2.0 mg bleocin L-1 for approximately one month. However, concentrations of 4.0 mg bleocin L-1 and above resulted in cell death within one week. Figure 5 illustrates the growth patterns of liquid cultures with extended viability levels of 0.25, 0.5, and 1 mg L-1. The results indicate that the minimum inhibitory concentration (M.I.C.) of bleocin for this particular microalgal strain is 2.0 mg L-1 when grown in liquid culture and 4.0 mg L-1 when grown on solid medium. Both selection criteria necessitate a minimum of one week of exposure to the corresponding minimum inhibitory concentration (M.I.C.) of bleocin. The herbicide Basta®, which contains PPT as its active ingredient, demonstrated superior potency and rapid selectivity compared to other studied agents. Concentrations of PPT at 1 mg L-1 and above exhibited cytotoxic effects on D. salina cells, leading to their death within a few hours. When exposed to a concentration of 0.5 mg PPT L-1, the growth rate was almost non-existent, and these cultures could not be maintained for more than three days. Finally, despite the presence of 0.25 mg PPT L-1, which caused a decrease in the rate of cell division, the cultures remained alive and functional for more than a week. The elevated concentration of PPT resulted in observable symptoms of toxicity, including reduced movement and heightened carotenoid coloring in the algae. Dunaliella salina exhibited considerably higher tolerance to PPT when grown on solid medium compared to liquid culture. Cells remained viable on agar plates with concentrations of 0.25, 0.5, and 1.0 mg L-1 for a duration exceeding one month. The suppression of cell development was observed both in terms of the relative number of surviving cells, which decreased as the concentration of PPT increased, and the extended survivability of these slower developing colonies due to their gradual depletion of available nutrients. Following a period of one month, the cells exhibited a high concentration of β-carotene and were visibly characterized by an orange hue. Figure 6 displays a direct comparison of D. salina spot tested on a 1 M NaCl medium without any selective agent added (on the left) and with 0.25 mg PPT L-1 (on the right). After one week, it was evident that the PPT concentration of 2.0 mg L-1 surpassed the threshold for total cell killing on solid medium.

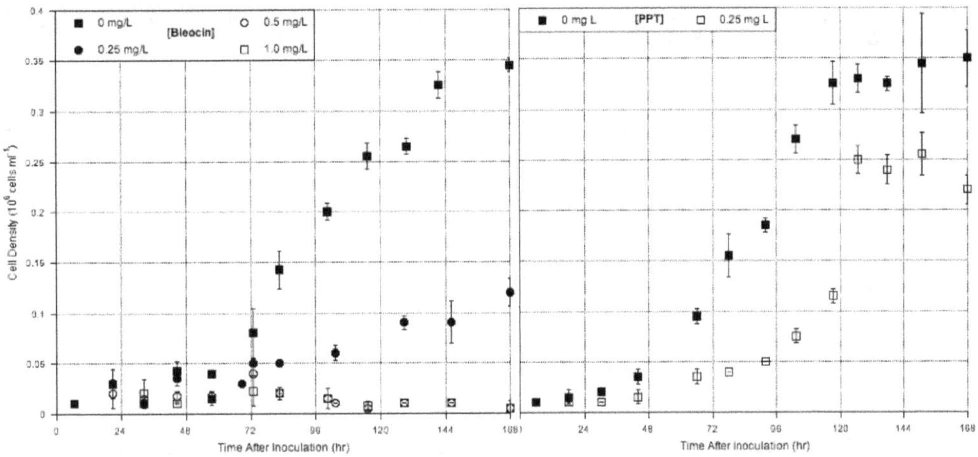

Figure 5. Dosage response curves forD. salinaCCAP 19/18.

While paromomycin is toxic to related microalgae,D. salinaCCAP 19/18 was found to be remarkably insensitive to this antibiotic. In the presence of concentrations as high as 400 mg paromomycin L-1, the algal cells were able to proliferate vigorously. Consequently, paromomycin was deemed unsuitable as a selective agent for the genetic transformation ofD. salina.

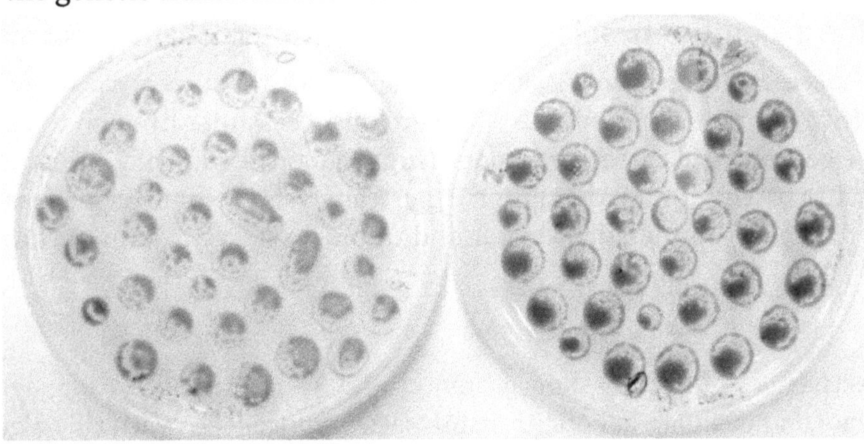

Figure 6. Stress-induced accumulation of β-carotene byD. salina.

Table 3. D. salinaCCAP 19/18 growth response to antibiotic and herbicide exposure characterized as either normal (+), inhibited (–), or negligible (0). Concentrations that engendered death of the entire algal population within one week are denoted with an ×.

[Bleocin] mg L^{-1}	Growth Rate (Liquid)	Cell Viability (Solid)
4.0	×	×
3.0	×	–
2.0	×	+
1.0	0	+
0.5	0	+
0.25	–	+
0	+	+
[Phosphinothricin] mg L-1	Growth Rate (Liquid)	Cell Viability (Solid)

[Bleocin] mg L^{-1}	Growth Rate (Liquid)	Cell Viability (Solid)
4.0	×	×
3.0	×	×
2.0	×	×
1.0	×	+
0.5	×	+
0.25	–	+
0	+	+
[Paromomycin] mg L-1	Growth Rate (Liquid)	Cell Viability (Solid)
400	+	+
300	+	+
200	+	+
100	+	+
50	+	+
25	+	+
0	+	+

3.4.2 Examining the genome of D. salina to identify promoters that are consistently active

Despite others' success with genomic PCR ofDunaliella-specific nucleotide sequences, our efforts to obtain theactinandrbcS2promoters andnitA3'-UTR were unsuccessful. Although the products from these PCR attempts appeared to be the correct fragment length, upon sequencing, it became clear that these were not the targets that we set out to amplify. BLAST analysis of some sequences recovered showed relevant homology to genes fromDunaliella viridisandArabidopsis lyrata, but not the targeted promoters specific toD. salina.

3.5. Discussion

The lack of effectiveness of the D. salina promoter and 3'-UTR amplification significantly hindered our attempts to create and evaluate genetic transformation approaches with this alga. The unsuccessful efforts to enhance the actin, rbcS2, and nitA regulatory elements give rise to doubts regarding the accuracy of the sequences recorded in GenBank. Another potential factor contributing to this lack of success could be attributed to the particular strain of D. salina utilized as the source for these sequences. The previous research on these promoters and 3'-UTR has exclusively utilized the UTEX 1644 D. salina strain. It is possible that the UTEX 1664 sample sent to us was either mistaken or contaminated. Moreover, the authenticity of the published sequence information attributed to UTEX 1644 seems questionable. Based on our observations, the CCAP 19/18 strain consistently demonstrated the ability to synthesize β–carotene when cells gathered and dried on the inner surface of the culture flasks, unlike UTEX 1644. To definitively determine the identity of each strain, it is necessary to do genomic analysis of the 18S rRNA. This approach has been successfully applied to many species of Dunaliella, including the UTEX 1664 and CCAP 19/18 strains (Olmos et al., 2000; Polle et al., 2008). Because we were unable to create expression vectors specifically for Dunaliella, we were only able to genetically modify D. salina using the C. reinhardtii bleomycin-resistance plasmid, pSP124. Research conducted by Walker et al. (2004) has provided evidence that genetic regulatory sequences from D. salina show activity in C. reinhardtii. Therefore, it is plausible to assume that the promoters of C. reinhardtii could also be effective in D. salina. Regrettably, despite conducting multiple experiments involving electroporation and microparticle bombardment, no viable transformants were successfully obtained following the selection process on bleocin plates.

It was hypothesized that the gold microparticles would exert a significant force of impact on D. salina due to its absence of a cell wall. However, electroporation should have been a more suitable method. Testing several voltage levels (4 and 1 kV cm-1) and cell densities (4 × 107 or 1 × 106 cells ml-1) for electroporation in order to transform cells and express the ble gene temporarily was found to be unsatisfactory. Through all techniques of transformation, we observed that control samples remained viable following the treatment, indicating that the electrical pulse did not cause cell death. In the absence of endogenous D. salina promoters, it is not possible to ascertain whether the lack of transgenic expression was due to inadequate transformation conditions or inactive promoters. Despite the challenges faced during the molecular research involving D. salina, tests related to antibiotic and herbicidal tolerance produced findings that will enhance the microbiological knowledge of D. salina and assist in future genetic modification of the organism.

4. CONCISE SUMMARY

Our method of genetically modifying a Dunaliella salina aims to pave the way for future endeavors in genetic engineering of this organism and maybe serve as a model for genetic exploration of other unique algae species. Through our dose response studies, we successfully reduced the already limited selection of medicines suitable for D. salina to only two options: the antibiotic bleomycin and the herbicide phosphinothricin. Furthermore, we were able to determine the minimum inhibitory concentrations of both substances in both solid and liquid mediums. Due to a lack of adequate sequence data, we were unable to create the intended D. salina transformation vectors. Additionally, attempts to use existing Chlamydomonas vectors for transformation were unsuccessful. Dunaliella salina is considered a fragile organism since it lacks a cell wall. As a result, existing transformation techniques may be too aggressive for this organism. It is also plausible to assume that gene silencing is a problem in D. salina. In addition to improving transformation protocols suitable for this alga, there is significant interest in developing molecular techniques to enhance the stable integration and expression of transgenes in Dunaliella species. While there have been significant achievements in the field of D. salina molecular biology, further research is needed to explore the genetic aspects of this alga. The recent release of the nuclear genome of D. salina CCAP 19/18, along with its chloroplast and mitochondrial genomes, is expected to significantly promote the advancement of genetic and metabolic engineering in this organism. Microalgal cultivation systems, similar to computers, require the combination of the algal organism being cultivated and the vessel used to accumulate the cells. Although it is possible for a single software to be compatible with many hardware devices, often they are developed and designed in conjunction with each other. The same principle applies to algae culture systems. Regardless of whether the growing environment is a raceway pond or a photobioreactor, there are countless potential algae species that may be developed. As microalgal biotechnology progresses towards modified algae and high-performance photobioreactors (PBRs), the distinct characteristics of the organism will be combined with concerns for bioreactor design. As the computational capacity of microchips continues to grow and new versions of operating systems become more frequent, it is anticipated that algae species chosen or modified for high productivity would consistently require more from their cultivation systems, and vice versa.

REFERENCES

Agathou, M.S. & and Kyritis, D.C. (2011) An experimental comparison of non-premixed biobutanol flames with the corresponding flames of ethanol and methane, Fuel, Vol. 90, No. 1, pp. 255-262, ISSN 0016-2361

Bahl, H.; Andersch, W. & Gottschalk, G. (1982) Continuous production of acetone and butanol by Clostridium acetobutylicum in two stage phosphate limited chemostat, European Journal of Applied Microbiology and Biotechnology, Vol.15, No.4, pp. 201-205, ISSN 0171-1741

Beesch, S.C. (1953) Acetone-butanol fermentation of starches, Applied Microbiology, Vol. 1, No. 2, pp. 85-96, ISSN 0099-2240

Beveridge, T.J. (1990) Mechanism of Gram variability in select bacteria, Journal of Bacteriology, Vol. 172, No. 3, pp. 1609-1620, ISSN 0021-9193

Broustail, G.; Seers, P.; Halter, F.; Moreac, G. & Mounaim-Rousselle, C. (2011) Experimental determination of laminar burning velocity for butanol and ethanol iso-octane blends, Fuel, Vol. 90, No. 1, pp. 1-6, ISSN 0016-2361

Chiao, J-S. & Sun, Z-H. (2007) History of acetone-butanol-ethanol fermentation industry in China: Development of continuous production technology, Journal of Molecular Microbiology and Biotechnology, Vol. 13, No. 1-3, pp. 12-14, ISSN 1464-1801

Clarke, K.G.; Hansford, G.S. & Jones, D.T. (1988) Nature and significance of oscillatory behavior during solvent production by Clostridium acetobutylicum in continuous culture, Biotechnology & Bioengineering, Vol. 32, No. 4, pp. 538-544, ISSN 0006-3592

Cooney, C.; Wallner, T; McConell, C.; Gillen, J.C.; Abell, C. & Miers, C.A. (2006) Effect of blending gasoline with ethanol and butanol on engine efficiency and emissions using a direct-injection, spark-ignition engine, In: Proceedings of the ASME internal combustion engine division 2009 spring technical conference ICES2009-76155, Milwaukee, Wisconsin, USA, 2006

Cornillot, E.; Nair, R.; Papoutsakis, E.T. & Soucaille, P. (1997) The genes for butanol and acetone formation in Clostridium acetobutylicum ATCC 824 reside on a large plasmid whose loss leads to degeneration of the strain, Journal of Bacteriology, Vol. 179, No. 17, pp. 5442-5447, ISSN 0021-9193

Dagaut, P. & Togbé, C. (2008) Oxidation kinetics of butanol-gasoline surrogate mixtures in a jet-stirred reactor: Experimental and modeling study, Fuel, Vol. 87, No. 15-16, pp. 3313-3321, ISSN 0016-2361

Dyr, J. & Protiva, J. (1958) Formation of neutral solvents in continuous fermentation by means of Clostridium acetobutylicum. In: Continuous Cultivation of Microorganisms A symposium, Málek, I., pp. 210-226, Publishing House of the Czechoslovak Academy of Sciences, Prague 1958.

Ezeji, T.C.; Qureshi, N. & Blaschek, H.P (2003) Production of acetone, butanol and ethanol by Clostridium beijerinckii BA101 and in situ recovery by gas stripping, World Journal of Microbiology and Biotechnology, Vol. 19, No. 6, pp. 595-603., ISSN 0959-3993

Ezeji, T.C.; Qureshi, N. & Blaschek, H.P. (2005) Industrially relevant fermentations. In Handbook on Clostridia, Dürre, P., pp. 797-812, CRC Press, ISBN 0-8493-1618-9, Boca Raton.

Fick, M.; Pierrot, P. & Engasser, J.M. (1985) Optimal conditions for long-term stability of acetone-butanol production by continuous cultures of Clostridium acetobutylicum, Biotechnology Letters, Vol. 7, No. 7, pp. 503-508, ISSN 0141-5492.

Fribert, P.; Patakova, P.; Jahodova, L.; Lipovsky, J.; Linhova, M.; Rychtera, M. & Melzoch, K. (2010) Recovery of acetone, butanol and ethanol from fermentation broth by gas stripping, Proceedings of 19th International Congress of Chemical and Process Engineering, ISBN 978-80-02-02210-7, Prague, Czech Republic, September 2010

Gheslaghi, R.; Scharer, J.M.; Moo-Young, M. & Chou, C.P. (2009) Metabolic pathway of clostridia for producing butanol, Biotechnology Advances, Vol. 27, No. 6, pp. 764-781, ISSN 0734-9750

Hanno, R.; Qureshi, N.; Cotta, M. & Largus A. (2010) Mixed community bioreactors to convert lignocellulosic feedstock into the liquid biofuel butanol, February 27, 2011, Available from:http://www.ars.usda.gov/research/publications/publications.htm?seq_no_11 5=248960

Hartmanis, M.G.; Ahlman, H. & Gatenbeck, S. (1986) Stability of solvent formation in Clostridium acetobutylicum, Applied Microbiology and Biotechnology, Vol. 23, No. 5, pp. 369-371, ISSN 0175-7598

Heyndrickx, M.; De Vos, P. & De Ley, J. (1991) Fermentation characteristics of Clostridium pasteurianum LMG 3285 grown on glucose and mannitol, Journal of Applied Bacteriology, Vol. 70, No. 1, pp. 52-58, ISSN 1364-5072

Irimescu, A. (2010) Study of cold start air-fuel mixture parameters for spark ignition engines fueled with gasoline-isobutanol blends, International Communications in Heat and Mass Transfer, Vol. 37, No. 9, pp. 1203-1207, ISSN 0735-1933

Jepras, R.I.; Carter, J.; Pearson, S.C.; Paul, F.E. & Wilkinson, M.J. (1995) Development of a robust flow cytometric assay for determining numbers of viable bacteria, Applied and Environmental Microbiology, Vol. 61, No. 7, pp. 2696-2701, ISSN 0099-2240

Johnson, J.L., Toth, J., Santiwatanakul, S. & Chen, J.-S. (1997) Culture of Clostridium acetobutylicum from various collections comprise Clostridium acetobutylicum, Clostridium beijerinckii and two other distinct types based on DNA-DNA reassociation, International Journal of Systematic Bacteriology, Vol. 47, No. 2, pp. 420-424, ISSN 0020-7713

Jones, D.T.& Woods, D.R. (1986). Acetone-butanol fermentation revisited, Microbiological Rewievs, Vol. 50, No. 4, pp. 484-524, ISSN 0146-0749 Jones, D.T.; Shirley, M.; Wu, X. & Keis, S. (2000) Bacteriophage infections in the industrial acetone butanol (AB) fermentation process, Journal of Molecular Microbiology and Biotechnology, Vol. 2, No. 1, pp. 21-26, ISSN 1464-1801

Jones, S. W.; Paredes, C. J.; Tracy, B.; Cheng, N.; Sillers, R.; Sanger, R. S.& Papoutsakis, E. T. (2008) The transcriptional program underlying the physiology of clostridial sporulation, Genome Biology, Vol. 9, No. 7, pp. R114-R114.21, ISSN 1465-6914

Kashket, E.R. & Cao, Z.-Y. (1995) Clostridial strain degeneration, FEMS Microbiology Reviews, Vol. 17, No. 3, pp. 307-315, ISSN 0168-6445

Kosaka, T.; Nakayama, S.; Nakaya, K.; Yoshino, S. & Furukawa, S. (2007) Characterization of the sol operon in butanol-hyperproducing Clostridium saccharoperbutylacetonicum strain N1-4 and its degeneration mechanism, Bioscience, Biotechnology and Biochemistry, Vol. 71, No. 1, pp. 58-68, ISSN 0916-8451

Largier, S.T.; Long, S.; Santangelo, J.D.; Jones, D.T. & Woods, D.R. (1985) Immobilized Clostridium acetobutylicum P262 mutants for solvent production, Applied and Environmental Microbiology, Vol. 50, No. 2, pp. 477-481, ISSN 0099-2240

Lee, S.M.; Cho, M.O.; Park, C.H.; Chung, Y-C.; Kim, J.H.; Sang, B-I. & Um, Y. (2008) Continuous butanol production using suspended and immobilized Clostridium beijerinckii NCIMB 8052 with supplementary butyrate, Energy & Fuels, Vol. 22, No. 5, pp. 3459-3464, ISSN 0887-0624

Lee, S.Y.; Park, J.H.; Jang, S.H.; Nielsen, L.K.; Kim, J. & Jung, K.S. (2008) Fermentative butanol production by clostridia, Biotechnology and Bioengineering, Vol. 101, No. 2, pp. 209-228, ISSN 0006-3592

Linhova, M.; Patakova, P.; Lipovsky, J.; Fribert, P.; Paulova, L.; Rychtera, M. & Melzoch, K. (2010a) Development of flow cytometry technique for detection of thinning of peptidoglycan layer as a result of solvent production by Clostridium pasteurianum, Folia Microbiologica, Vol. 55, No. 4, pp. 340-344, ISSN 1874-9356

Linhova, M.; Branska, B.; Patakova, P.; Lipovsky, J.; Fribert, P.; Rychtera, M. & Melzoch, K. (2010b) Viability determination by flow cytometry during ABE fermentation, Proceedings of 19th International Congress of Chemical and Process Engineering, ISBN 978- 80-02-02210-7, Prague, Czech Republic, September 2010

Lipovsky, J.; Patakova, P.; Rychtera, M.; Cizkova, H. & Melzoch, K. (2009) Prospects of butanol production from starch and cellulose materials, Chemicke Listy, Vol. 103, No. 6, pp. 479-483, ISSN 1213-7103

Lopez-Contreras, A.; Gabor, K.; Martens, A A.; Renckens, B. A. M.; Claassen, P. A. M.; van der Oost, J. & de Vos, W. M. (2004) Substrate induced production and secretion of cellulases by Clostridium acetobutylicum, Applied and Environmental Microbiology, Vol. 70, No. 9, pp. 5238-43, ISSN 0099-2240

Maddox, I.S.; Steiner, E.; Hirsch, S.; Wessner, S.; Gutierrez, N.A. ; Gapes, J.R. & Schuster, K.C. (2000) The cause of “acid crash” and “acidogenic fermentations” during the batch acetone-butanol-ethanol (ABE-) fermentation process, Journal of Molecular Microbiology and Biotechnology, Vol. 2, No. 1, pp. 95-100, ISSN 1464-1801

Mužíková, Z.; Pospíšil, M. & Sebor, G. (2009) Volatility and phase stability of petrol blends with ethanol, Fuel, Vol.88, No.8, pp. 1351-1356, ISSN: 0016-2361

Napoli, F.; Olivieri, G.; Russo, M.E.; Marzocchella, A. & Salatino, P. (2010) Butanol production by Clostridium acetobutylicum in a continuous packed bed bioreactor, Journal of Industrial Microbiology and Biotechnology, Vol. 37, No. 6, pp. 603-608, ISSN 1367-5435

Ni, Y. & Sun, Z. (2009) Recent progress on industrial fermentative production of acetonebutanol- ethanol by Clostridium acetobutylicum in China, Applied Microbiology and Biotechnology, Vol. 83, No. 3, pp. 415-423, ISSN 0175-7598

Novo, D.; Perlmutter, N.G.; Hunt, R.H. & Shapiro, H.M. (1999) Accurate flow cytometric membrane potential measurement in bacteria using diethyloxacarbocyanine and a ratiometric technique, Cytometry, Vol. 5, No. 1, pp. 55-63, ISSN 1552-4922

Oudshoorn, A.; Van der Wielen, L.A.M. & Straathof, A.J.J. (2009) Assessment of options for selective 1-butanol recovery from aqueous solution, Industrial & Engineering Chemical Research, Vol. 48, No. 15, pp. 7325-7336, ISSN 0888-5885

Patakova, P.; Lipovsky, J.; Cizkova, H.; Fortova, J.; Rychtera, M. & Melzoch, K. (2009) Exploitation of food feedstock and waste for production of biobutanol. Czech Journal of Food Sciences, Vol. 27, No. 4, pp. 276-283, ISSN 1212-1800

Patakova, P.; Lipovsky, J.; Paulova, L.; Linhova, M.; Fribert, P.; Rychtera, M. & Melzoch, K. (2011a) Continuous production of butanol by bacteria of genus Clostridium, Journal of Chemistry and Chemical Engineering, Vol. 5, No. 2, pp. 124-131, ISSN 1934-7375

Patakova, P.; Toure,S.S.M.; Simacek, P.; Cizkova, H.; Lipovsky, J.; Linhova, M.; Fribert, P., Rychtera, M. & Melzoch, K. (2011b) Exploitation of sugar beet for biobutanol production, Czech Sugar and Sugar Beet Journal, Vol. 127, No. 2, pp. 46-49, ISSN 1210- 3306

Peng, Ch.; Lewis, C.K.; Stein, F.P. (1996) Water Solubilities in Blends of Gasoline and Oxygenates, Fluid Phase Equilibria, Vol.116 (1996), pp. 437-444, ISSN: 0016-2361

Pierrot, P.; Fick, M. & Engasser, J.M. (1986) Continuous acetone-butanol fermentation with high productivity by cell ultrafiltration and recycling, Biotechnology Letters, Vol. 8, No. 4, pp. 253-256, ISSN 0141-5492

Qureshi, N. & Blaschek, H.P. (2001a) Recent advances in ABE fermentation: hyper-butanol producing Clostridium beijerinckii BA101, Journal of Industrial Microbiology and Biotechnology, Vol. 27, No. 5, pp. 287-291., ISSN 1367-5435

Qureshi, N. & Blaschek, H.P. (2001b) Recovery of butanol from fermentation broth by gas stripping, Renewable Energy, Vol. 22, No. 4, pp. 557-564, ISSN 0960-1481

Rainey, F.A.; Hollen, B.J. & Small, A. (2009) Genus I. Clostridium Prazmowski 1880, 23AL. In Bergey`s manual of Systematic Bacteriology, The Firmicutes, Second Edition, Volume Three, De Vos, P.; Garrity, G.M.; Jones, D.; Krieg, N.R.; Ludwig, W.; Rainey, F.A.; Schleifer, K.-H. & Whitman, W.B, pp. 738-828, Springer, ISBN 978-0-387-95041-9, New York.

Rakopoulos, C.D.; Dimaratos, A.M.; Giakoumis, E.G. & Rakopoulos D.C. (2010a) Investigating the emissions during acceleration of a turbocharged diesel engine operating with biodiesel or n-butanol diesel fuel blends, Energy, Volume 35, No. 12, pp. 5173-5184, ISSN 0360-5442

Rakopoulos, D.C.; Rakopoulos, C.D.; Hountalas, D.T.; Kakaras, E.C.; Giakoumis, E.G. & Papagiannakis, R.G. (2010b) Investigation of the performance and emissions of bus engine operating on butanol/diesel fuel blends, Fuel, Vol. 89, No. 10, pp. 2781-2790, ISSN 0016-2361

Ramey, D. & Yang, S.-T. (2004) Production of butyric acid and butanol from biomass, In: Final report to the US Department of Energy DE-F-G02-00ER86106, February 27, 2011, Available from: http://74.125.155.132/scholar?q=cache:BqatEryJKOwJ:scholar.google.com/+author:

Ravagnani, A.; Jennert, K.C.B.; Steiner, E.; Grünberg, R.; Jefferies, J.R.; Wilkinson, S.R.; Young, D.I.; Tidswell, E.C.; Brown, D.P.; Youngman, P.; Morris, J.G.& Young, M. (2000) Spo0A directly controls the switch from acid to solvent production in solvent-forming clostridia, Molecular Microbiology, Vol. 37, No. 5, pp. 1172-1185, ISSN 0950-382X

Rychtera, M.; Patakova, P.; Fribert, P.; Lipovsky, J.; Linhova, M.; Paulova, L.; Melzoch, K.& Klefenz, H. (2010) Bottlenecks in bacterial production of fuel butanol, Proceedings of 19th International Congress of Chemical and Process Engineering, ISBN 978-80-02-02210-7, Prague, Czech Republic, September 2010

Sarathy, S.M.; Thomson, M.J.; Togbé, C.; Dagaut, P.; Halter, F. & Mounaim-Rousselle, C. (2009) An experimental and kinetic modeling study of n-butanol combustion, Combustion and Flame, Vol. 156, No. 4, pp. 852-864, ISSN 0010-2180

Shaheen, R.; Shirley, M. & Jones, D.T. (2000) Comparative fermentation studies of industrial strains belonging to four species of solvent-producing clostridia, Journal of Molecular Microbiology and Biotechnology, Vol. 2, No. 1, pp. 115-124, ISSN 1464-1801

Tomas, C.A.; Beamish, J.& Papoutsakis, E.T. (2004) Transcriptional analysis of butanol stress and tolerance in Clostridium acetobutylicum, Journal of Bacteriology, Vol. 186, No. 7, pp. 2006-2018, ISSN 0021-9193

Tracy, B.P.; Gaida, S.M.& Papoutsakis, E.T. (2008) Development and application of flowcytometric techniques for analyzing and sorting endospore-forming clostridia, Applied and Environmental Microbiology, Vol. 74, No. 24, pp. 7497-7506, ISSN 0099-2240

United States Patent No.4539293 (1985) Production of butanol by fermentation in the presence of cocultures of clostridium, Application 1983-10-05

Wang, S.; Zhang, Y.; Dong, H.; Mao, S.; Zhu, Y.; Wang, R.; Luan, G. & Li, Y. (2011) Formic acid triggers the acid crash of acetone-butanol-ethanol fermentation by Clostridium acetobutylicum, Applied and Environmental Microbiology, Vol. 77, No. 5, pp. 1674-1680, ISSN 0099-2240

Westhuizen van der, A.; Jones, D.T. & Woods, D.R. (1982) Autolytic activity and butanol tolerance of Clostridium acetobutylicum, Applied and Environmental Microbiology, Vol. 44, No.6, pp. 1277-1281, ISSN 0099-2240

Wolf, L. (2007) 1-Butanol as a Gasoline Blending Biocomponent, Mobile Sources Technical Review Subcommittee - March 28, 2007 Meeting, Aarlington, USA, available from http://www.epa.gov/oar/caaac/mstrs/March2007/Wolf.pdf

Yacoub, Y.; Bata, B. & Gautam, M. (1998) The performance and emission characteristics of C1-C5 alcohol-gasoline blends with matched oxygen content in a single-cylinder spark ignition engine, Proceedings of the Institution of Mechanical Engineers. Part A. Journal of Power and Energy, Vol. 212, No. 5, pp. 363-379, ISSN 0957-6509

Yao, M.; Wang H.; Zheng Z. & Yue, Y. (2010) Experimental study of n-butanol additive and multi-injection on HD diesel engine performance and emissions, Fuel, Vol. 89, No. 9, pp. 2191-2201, ISSN 0016-2361

Xiaolong, Y; Jing, Y.; Tieping, L. (2009) The Effect of an SI Engine Using Butanol-Gasoline Blended Fuel on Performance and Enviroment, ICEET '09 Proceedings of the 2009 International Conference on Energy and Environment Technology - Volume 01, pp. 402-405, ISBN: 978-0-7695-3819-8 IEEE Computer Society Washington, DC, USA, 2009

Zverlov, V.V.; Berezina, O.; Velikodvorskaya, G.A. & Schwartz, W.H. (2006) Bacterial acetone and butanol production by industrial fermentation in the Soviet Union: use of hydrolyzed agricultural waste for biorefinery, Applied Microbiology and Biotechnology, Vol. 71, No. 5, pp. 587-597, ISSN 0175-7598

CHAPTER-13

RHEOLOGICAL ANALYSIS OF BIO-OILS PRODUCED BY PILOT SCALE MICROWAVE ASSISTED PYROLYSIS

1. INTRODUCTION: AN OVERVIEW

Renewable energy is increasingly significant in resolving both environmental and economic challenges associated with the use of fossil fuels. Lignocellulosic materials are the most plentiful sustainable resources found on our planet (Lynd et al., 2005). Biomass can be used as a source of energy through either biochemical or thermochemical processes. Pretreatment of biomass is an essential initial step in the biochemical process, as it allows enzymes to access cellulose and hemicellulose by opening up the structure of the biomass cell wall. Pyrolysis, gasification, and combustion are the primary thermochemical methods used to extract energy from biomass. According to Yu et al. (2007), combustion can achieve an efficiency of around 30% at its highest. Gasification has garnered significant attention due to its superior efficiency when compared to combustion (Bridgwater, 2004). Wornat et al (1994) found that the combustion of bio-oils derived from biomass pyrolysis is more effective. Bio-oil has benefits in terms of storage and transport, as well as its versatility as an energy carrier and a source of chemicals (Bridgwater, 2004). The thermochemical process enables the conversion of biomass with low carbohydrate content or that cannot be fermented into alcohol fuels, hence enhancing the technological resilience of initiatives aimed at attaining the 30 x 30 objective. Pyrolysis is a thermally-driven reaction that involves the decomposition of a substance in the absence of oxygen. The initial stage of gasification and combustion involves the partial or complete oxidation of the substrate. In gasification, gas is the primary output, accounting for 85% of the total. On the other hand, in most forms of pyrolysis, bio-oil is the primary result, making for 70-80% of the total. The quantity of pyrolysis products, such as syngas/producer gas (a combination of CO and H2), bio-oil, and bio-char (charcoal), will differ based on the specific pyrolysis techniques used (conventional, fast, vacuum, flash, and ultra), the characteristics of the biomass (type of feedstock, moisture content, particle size), and the parameters of the reaction (heating rate, temperature, and residence time). In order to obtain bio-oil through fast pyrolysis, Bridgwater (2003) identified four crucial requirements: extremely rapid heating rates (1000°C/s), high rates of heat transfer (600-1000 W/cm2), short periods of time for the vapors to reside (usually less than 2 seconds), and quick cooling of both pyrolysis vapors and aerosols. Over the past two decades, various reactor designs have been investigated to fulfill the quick heat-transfer needs of the fast pyrolysis process, as the reactor is its fundamental component. To achieve exceptionally high levels of heating and heat transmission during pyrolysis, it is typically necessary to use a biomass feed that has been finely ground.

Microwave irradiation is a method of turning biomass into valuable goods and chemicals through pyrolysis. Microwave assisted pyrolysis (MAP) differs from other fast pyrolysis methods in that it does not necessitate extensive grinding of the material, allowing for the utilization of huge chunks of wood logs. Additionally, MAP has the capability to process mixed feedstocks such as municipal solid wastes. Microwave irradiation offers a distinct advantage over traditional heating due to its ability to rapidly heat the inside components. In addition, microwave energy deposition in the dielectric loss mode of heating can result in evenly distributed heating (Miura et al., 2000) and is simple to regulate. MAP offers several technical advantages compared to conventional pyrolysis. It eliminates the need for size reduction (powder form) and does not require agitation and fluidization. As a result, the pyrolytic gas and bio-oils produced are cleaner than those obtained from gasification and conventional pyrolysis. The lack of a carrier gas during fluidization leads to an increased heating value of the syngas generated. Due to the well-established nature of microwave heating, the cost of creating a microwave heating system for biomass pyrolysis would be relatively low. Both studies conducted by the International Energy Agency on Bioenergy and research conducted in Finland have shown that bio-oil is the most cost-effective liquid fuel derived from biomass. The desired quality of bio-oils would be determined by the specific intended application. The calorific value, viscosity, density, surface tension, and distillation characteristics play a crucial role in fuel applications (Garcia-Perez et al., 2006a). To accomplish the aforementioned features, it is necessary for bio-oil to possess (i) a low concentration of solid particles, (ii) a high level of uniformity and stability, and (iii) a relatively high flash point. Solantausta et al (1994) determined that the utilization of bio-oil in gas turbines can be enhanced through the optimization of its physical and chemical characteristics, such as ash content, alkali content, heating value, and viscosity. The viscosity of bio-oil has an impact on the spray pattern and the size of the droplets. The elevated viscosity of bio-oil leads to more pressure drops along the pipeline, necessitating greater effort from the fuel pump to sustain a consistent flow rate. In their study, Doll et colleagues (2008) formulated an equation that quantifies the automation properties of the fuel. This equation incorporates many parameters such as the characteristic number (K), density (ρ), Weber number (We), Reynolds number (Re), surface tension (γ), and kinematic viscosity (v). To achieve the desired low characteristic number (K), it is necessary to control two physical properties of the fuel: kinematic viscosity (v) and surface tension (γ). Furthermore, viscosity is regarded as the more significant of these two elements. The viscosity of bio-oils exhibits significant variation, ranging from 35 to 1000 cP at a temperature of 40 °C. This variation is influenced by factors such as the kind of feedstock used and the specific circumstances of the processing method (Bridgwater, 2004; Czernik & Bridgwater, 2004). Diebold (2002) states that an effective gathering of volatile elements during the manufacturing of bio-oil leads to a bio-oil that contains a higher proportion of low-molecular weight components, resulting in a reduced viscosity, improved solvency properties, and potentially enhanced storage characteristics. Typically, the generation of bio-oil via pyrolysis is a thermodynamically non-equilibrium procedure. This procedure necessitates a brief period of time spent in a region of elevated temperature, followed by a swift cooling process, resulting in the production of a bio-oil that is not in a state of equilibrium (Ringer et al., 2006). The bio-oil contains several reactive organic components that undergo interactions to reach a state of equilibrium during storage. The interactions lead to the creation of bigger molecules, which in turn raises the viscosity of the bio-oil (Diebold & Czernik, 1997; Ringer et al., 2006). Due to the significant oxygen content (40-50 wt %) and water content (15-30 wt %), as well as the low H/C ratios, bio-oils cannot be utilized as transportation fuels without undergoing previous upgrading. Another challenge is the reduced stability of bio-oils during storage, which is caused by the presence of unsaturated chemicals and their poor compatibility with traditional liquid fuels (Samolada et al., 2000). Catalytic biomass pyrolysis is a favorable method because it eliminates the need for expensive condensation and re-evaporation processes before upgrading bio-oil (Samolada et al., 2000; Lu et al., 2009a).

Multiple studies have demonstrated that the viscosity of bio-oil is contingent upon the specific feedstocks used, the type of pyrolyzer employed, and the conditions under which pyrolysis takes place. The quality of the bio-oil is mostly influenced by the type of feedstock, in addition to the postproduction processing procedures. To enhance comprehension of the impact of feedstock on product quality, it is necessary to conduct a comparative analysis of different feedstocks (Oasmaa et al., 2005a). Hence, the present investigation aimed to assess and evaluate the viscosity of bio-oils derived from various feedstocks using microwave pyrolysis, and to analyze their properties utilizing storage and loss moduli.

1.1. Importance of bio-oil viscosity

The viscosity of a bio-oil refers to its internal friction that opposes its flow. Viscosity is a crucial characteristic of gasoline that must be taken into account when designing and choosing equipment for handling, processing, and transportation. The viscosity of bio-oil has an impact on the functioning of fuel injection equipment, especially when the increase in viscosity hampers the flow of fuel at low temperatures. The quality and practical use of bio-oil as fuel are highly influenced by its viscosity and elemental compositions. Specifically, it is preferable to have lower viscosity and oxygen concentration (Ertas & Alma, 2010). Bio-oil typically exhibits more viscosity than both crude oil and diesel fuel, as supported by studies conducted by Onay & Kockar (2006), Parihar et al. (2007), and Pootakham & Kumar (2010a,b). Pootakham and Kumar (2010a) found that the loading equipment used for petroleum products, such as gasoline and diesel fuels, has a flow rate ranging from 0.9 to 1.3 m3/min. However, for bio-oil, the equipment can be operated at a lower flow rate of 0.6 m3/min and a safety operating pressure of 205 kPa (or 30 psi) to ensure safety, as stated by Jones and Pujadó (2006). Bio-oil exhibits higher viscosity than crude oil at ambient temperature. However, within the temperature range of 35-45ºC, the viscosity of bio-oil closely resembles that of crude oil (Bridgewater, 1999; Thamburaj, 2000; Pootakham & Kumar, 2010a,b). To facilitate the transportation of bio-oil through a pipeline, it is necessary to maintain the temperature of the pipeline within the range of 35–45ºC. This temperature range ensures that the viscosity of the bio-oil remains comparable to that of crude oil (Pootakham & Kumar, 2010a, b). Thangalazhy-Gopakumar et al (2010) found that the viscosity of bio-oil is higher compared to diesel (0.011 Pa.s) and gasoline (0.006 Pa.s). Typically, fuel with a high viscosity leads to inadequate atomization and incomplete combustion. This causes the injection nozzles and combustion chamber to accumulate excessive carbon deposits, and the lubricating oil to become contaminated with unburnt residues. The viscosity of the fuel has a direct impact on the atomization and mixing process in the combustion chamber. When it comes to fuel application, a lower viscosity makes it simpler to pump and atomize, resulting in the production of smaller droplets (Ji-Lu, 2008). Therefore, bio-oils in their unaltered state are not appropriate for utilization in contemporary diesel engines (Özaktas et al., 1997). Their high acidity, low thermal stability, low calorific value, high viscosity, and poor lubricating qualities restrict their application as transportation fuel (Garcia-Perez et al., 2006b). Oasmaa et al (2005) indicated that, in order to be suitable for engine use, the viscosity of the substance should fall between the 10-20 cSt range, while the solids content should be below 0.1 wt%. It is widely recognized that bio-oils are fundamentally distinct from petroleum fuels. It is imperative to define precise fuel standards for the commercial use of bio-oils as liquid fuels. The standards should encompass the essential features, such as viscosity, lubricity, homogeneity, stability, heating value, pH, water content, flash point, solids, and ash (Qiang et al., 2008). The viscosity of bio-oil is influenced by factors such as temperature, feedstock, water content, amount of collected light ends, and the degree of aging of the pyrolysis oil (Ji-Lu, 2008). As an illustration, bio-oil derived from P. indicus and F. mandshurica exhibited a kinetic viscosity ranging from 70 to 350 mPa s, while bio-oil obtained from rice straw had a minimum kinetic viscosity of approximately 5 to 10 mPa s. The lower viscosity of the bio-oil from rice straw can be attributed to its high water content (Luo et al., 2004). Water has both detrimental and beneficial impacts on the storage and usage of bio-oils. The adverse consequences include a decrease in heating values, the occurrence of phase separation, an increase in ignition latency, and a reduction in combustion rates and adiabatic flame temperatures during the combustion process. Moreover, it results in early evaporation and subsequent challenges in injecting the substance during the preheating phase. The beneficial impacts include viscosity reduction, atomization facilitation, and combustion-related pollutant emission reduction (Calabria et al., 2007). In addition, hydroxyl radicals derived from water can impede the creation of soot and also expedite its oxidation. Senso°z and Kaynar (2006) found that the viscosity of bio-oils is influenced by the length of the fatty acid chains and the amount of saturated bonds. The increased density of bio-oil compared to water indicates the presence of heavy components (Sensöz et al., 2006). The presence of lignin in the original feedstock has a beneficial effect on the molecular weight and viscosity of bio-oil (Fahmi et al., 2008). In a recent study, Ertas and Alma (2010) conducted a comparison between the average molecular weight and molecular weight distribution of bio-oil derived from laurel extraction residues (with values of 664 g mol/l and 1.52, respectively) and found that they were highly similar to the corresponding values for switchgrass bio-oil (658 g mol/l and 1.49, respectively) as reported by He et al. (2009a). The viscosity of bio-oil undergoes a rise during storage as a result of gradual polymerization and condensation processes. This increase in viscosity is further intensified by elevated temperatures. Inhibitors, such as hydroquinone, have a significant impact on the rate at which bio-oil viscosity increases. This is because the inhibitors suppress thermal polymerization processes, leading to a reduction in viscosity. (Ji-Lu, 2008). Garcia-Perez et al (2010) determined that the rise in viscosity of bio-oils is caused by the incorporation of lignin generated oligomers. Condensation reactions in bio-oil result in an increase in water content with time (Garcia-Perez et al., 2002). The instability can be ascribed to the existence of alkali metals in the ash, which are being transported by the char particles along with the vapors. According to Diebold (2002), these alkali metals operate as catalysts for polymerization events, leading to an increase in viscosity. Basic techniques like the addition of polar solvents, diesel, or other fuels can help mitigate certain undesirable features of bio-oil. Polar solvents, such as methanol or ethanol, can enhance the volatility and heating capacity while reducing the viscosity and acidity. According to Ji-Lu and Yong-Ping (2010), the inclusion of ethanol enhances the volatility, stability, and heating value, while reducing the viscosity, acidity, and corrosivity. According to Onay and Kockar (2006), the viscosity of bio-oil can be altered by blending it with diesel or other fuels. Various techniques, including emulsification, hydrotreating, and catalytic cracking, are being studied to enhance the fuel characteristics of bio-oils. However, these approaches are not discussed in detail in this chapter.

2. MATERIALS AND METHODS

2.1. Biomass selection

Aspen's low density, strength, and rigidity render it unsuitable for numerous structural applications (Mackes & Lynch, 2001). Nevertheless, aspen can also serve as a substitute for bedding material or as a source of feedstock for biofuel or bio-oil.The biomass yield of canola is influenced by irrigation and ranges from 5 to 10 tons per hectare (Enayati et al., 2009). As per the US Canola Association, canola was grown on 3.2 million hectares in 2009, yielding 16-32 million tons of canola straw. In the early stages of US history, maize cobs played a significant role as a fuel for heating residential dwellings, farm structures, and small enterprises. Currently, maize cobs are resurfacing as a promising source for direct combustion, gasification, and cellulosic ethanol production. This is because they provide several advantages, including as their compact and consistent size, high heat value, and low nitrogen and sulfur contents, which surpass those of other competing feedstocks. On a dry matter basis, the typical cob yield accounts for approximately 14% of the grain yield and makes up around 16% of the corn stover biomass in a field (Roth & Gustafson, 2010). Blaschek and Ezeji (2010) reported that 15% of corn stover consists of corn cob. According to several estimates from different sources, the annual production of corn stover ranges from 200 to 250 million dry tons (Sokhansanj et al., 2002; Kadam & McMillan, 2003). The projected annual production of corn cobs in the US is 30-40 million metric tons, based on the fact that stover makes about 15-16% of the total.

2.2. Bio-oil production

The bio-oils were generated by either batch or continuous microwave assisted pyrolysis (MAP) procedures conducted at the University of Minnesota (Figure 1). Approximately 250 grams of samples were placed in a 1000 milliliter quartz flask, which was then inserted into the microwave cavity of a Panasonic NN-SD797S. The power level was adjusted to 8, resulting in an output power of approximately 1000 W. Following approximately 12 minutes of microwave exposure, the volatile pyrolyzates were condensed using cool water at a temperature of approximately 4-5 °C. The bio-oil was the percentage obtained from bottles attached to the lower part of the condensers. The condensates that were attached to the inner surface of the tubes were subsequently cleansed with ethanol and concentrated at a temperature of 80°C using a vacuum rotovap (Buchi R-141, Flawil, Switzerland) until reaching a nearly constant weight. The resulting concentrate was then combined with the bio-oil.

Figure 1. Batch and continuous microwave assisted pyrolysis (Courtesy: Dr. R. Ruan, University of Minnesota, MN, USA)

As mentioned earlier aspen, canola, and corn cob were the feedstocks selected for the bio-oil production. Aspen pellets were used for the production of bio-oil in a batch MAP process. Canola compost pellets were also used in a batch MAP process. All corn cobs were ground to less than 1 cm in size before MAP. Corn cob 1 and corn cob 2 bio-oils were produced using a continuous MAP, where heavy fractions of the bio-oil were not collected. Corn cob 3 bio-oil was produced using a batch MAP, where heavy fractions of the bio-oil were collected. Corn cob 4 was the bio-oil produced in a batch MAP of corn cob pretreated with 4% sulfuric acid; the bio-oil contained more water and furfural than other bio-oil samples. Fig. 2 shows bio-oils from different feedstocks produced through MAP. Heavy fractions of the bio-oils were collected for all batch processes. The bio-oil separated into two phases: about top one-third of the bio-oil was in oily phase and the bottom two-thirds were in aqueous phase. The oily phase is a relatively stable, light emulsion containing water soluble chemicals and light oily components; whereas the aqueous phase is a large molecule oily mixture characterized by high viscosity and water insolubles (Yang et al., 2010). The aqueous phase yield varied between batch (aspen and canola >60%) and continuous (corn cobs 10-20%) production. The aqueous phase yield of corn cobs were in agreement with corn stover (Yang et al., 2010).

Figure 2. Samples of bio-oil produced through MAP from different feedstocks

2.3. Characterization of bio-oil

The viscosity measurement and dynamic rheology analysis were conducted using a Rheometer (ATS Rheosystems, Rheologica Instruments Inc, NJ) equipped with a cup and bob assembly. A volume of approximately 15 mL of the sample was placed in a cup, and a shear rate was applied by means of the bob. The viscosity of bio-oils was measured using a rheometer at a temperature of 20°C, with shear rates ranging from 1 to 200/s. The tension and viscosity were determined for the applied shear rate using the software. To determine the linear viscoelastic area, a stress sweep was conducted at a consistent frequency. In order to conduct a frequency sweep ranging from 1 to 100 Hz, the stress level chosen was from the linear region that was determined during the stress sweep test. A temperature sweep ranging from 20 to 100°C was conducted using a constant stress and frequency determined in the previous phases. The viscoelastic moduli, specifically the storage modulus (G') and the loss modulus (G"), were measured for all samples as a function of frequency. A logarithmic scale graph was used to determine the linear viscoelastic area.

3. RESULTS AND DISCUSSION

3.1. Influence of temperature on the viscosity of bio-oils derived from various feedstocks

The viscosities of bio-oils produced from different feedstocks were measured at 20, 40, 50, 60, 80, and 100ºC and are shown in Fig. 3. In general, the viscosity of bio-oil at 20ºC was higher than that of 100ºC, irrespective of the feedstocks and with or without catalyst. This observation was in agreement with viscosity of bio-oil from pine wood chips (Thangalazhy-Gopakumar et al., 2010). Bio-oils used in this study had a higher viscosity at lower shear rate and the viscosity decreased exponentially at higher shear rate (>10/s) and similar result was reported for bio-oil from pine wood chips (Thangalazhy-Gopakumar et al., 2010). All the bio-oils used in this study showed non-Newtonian behavior as evident from Fig. 1. As mentioned in the introduction section, viscosity plays an important role in atomization through influencing inertial and aerodynamic instabilities. The Sauter mean diameter (SMD) of spray increases with viscosity for Newtonian fluid, whereas elasticity or shear thinning behavior of non-Newtonian fluid would affect the SMD. Thus, it is important to examine the non-Newtonian or viscoelastic nature of bio-oil since it may exhibit these effects during the application. According toLu et al (2009b), most of the bio-oils behave as Newtonian fluids at temperatures lower than 80ºC, whereas all the bio-oils used in this study even at high temperature (100ºC) behaved as non-Newtonian fluid. A prevalent shear thinning behavior was observed at 50 and 80ºC byTzanetakis et al (2008) and similar behaviors were observed for all the bio-oils irrespective of the type of process (batch or continuous), with or without catalyst and kinds of feedstocks.

The viscosity of bio-oil between 20, 40, and 80ºC were statistically different for all the feedstocks. Although the viscosities of bio-oil from corn cob were different at 20ºC, the differences vanished at 80ºC. In general, the viscosity of bio-oils produced from different feedstocks decreased with an increase in temperature. Similar trends were reported for bio-oils produced from different feedstocks such as softwood bark (Boucher et al., 2000a, b), sugarcane bagasse (Garcia-Perez et al., 2002), rice husk (Zhang et al., 2006), switchgrass (Boateng et al., 2007), corn stover (Yu et al., 2007), hardwood (Tzanetakis et al., 2008), pine and oak wood and bark (Ingram et al., 2008), pine wood chips (Thangalazhy-Gopakumar et al., 2010), and rice husk (Ji-Lu & Yong-Pong, 2010). When temperature was increased from 20 to 40ºC, viscosity of bio-oil from canola showed a minimum decrease of 9% and bio-oil from corn cob 1 showed a maximum decrease of 25%. A further increase in temperature to 80ºC resulted in viscosity decrease of 26 and 52%, respectively, for the bio-oil produced from canola and corn cob 1. Bio-oil derived from hardwood showed a similar behavior; however, the decrease was seven fold (Tzanetakis et al., 2008). The bio-oil viscosity measured at 40ºC in this study was ten-fold lower than the viscosity (0.02 Pa.s) of the bio-oil produced from (heterotrophic) microalgae (Miao & Wu, 2004).

The viscosity of bio-oils produced from different feedstocks though MAP was lower than the light fuel viscosity of 4 cSt (Mohan et al., 2006), the heavy fuel oil viscosity of 50 cSt (Czernik and Bridgwater, 2004), the US #4 fuel oil viscosity of 5.5-24 cSt (Oasmaa et al., 2009), commercial automotive #2 diesel viscosity of 2-4.5 cSt (Islam et al., 2010), diesel viscosity of 0.011 Pa s (Thangalazhy-Gopakumar et al., 2010), and was higher than JP4 viscosity of 0.88 cSt (Chiaramonti et al., 2007) and gasoline viscosity of 0.006 Pa s (Thangalazhy-Gopakumar et al., 2010) at a temperature of 40ºC. Considering the viscosity criteria (15 cSt at 35-45ºC and 21.5 cSt at 30ºC) presented by researchers (Pootakham & Kumar 2010a; Islam et al., 2010) for loading/handling and pipe transportation, the bio-oils from different feedstocks produced through MAP can be easy to load using existing petroleum loading equipments and easy to transport through pipe also. According to ASTM burner fuel standard, the bio-oil can have a maximum viscosity of 125 cSt at 40ºC without filtering (Oasmaa et al., 2009). Considering this limit, the viscosity of the bio-oils used in this study had a much low viscosity and these bio-oils can be used as burner fuel.

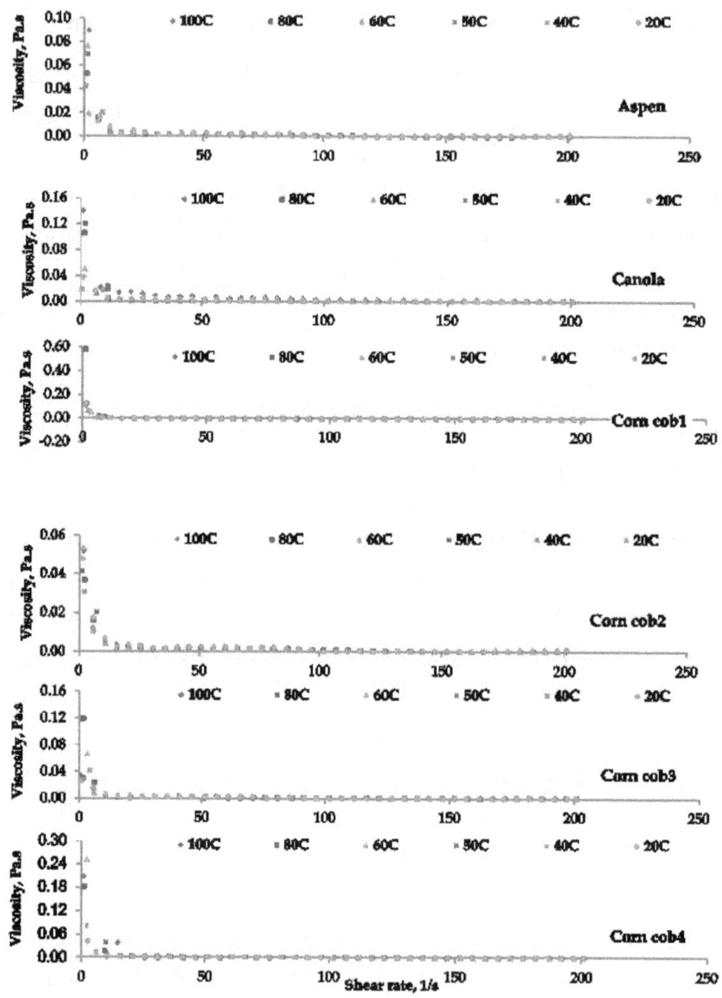

Figure 3. Viscosity of bio-oils produced from different feedstocks at indicated temperatures

Czernik and Bridgwater (2004) reported that the viscosity of bio-oil produced from wood would vary between 40 and 100 cP at 50ºC, whereas the viscosity of the heavy fuel oil is 180 cP. As evident from Fig. 3, the viscosity of bio-oils produced from different feedstocks through batch or continuous MAP with or without catalyst had a significantly lower viscosity (1.5-2.2 cP) than the viscosity values reported byCzernik and Bridgwater (2004) and the viscosity of bio-oil (33 cP) from hardwood at 50ºC (Tzanetakis et al., 2008). The viscosity of bio-oils used in this study was lower than that of bio-oil from sugarcane bagasse (12.1-28 cSt) measured at 60 and 80ºC (Das et al., 2004). This result indicates that the bio-oils produced through MAP can be easily atomized. One possible reason for low viscosity of the bio-oils in this study was the absence of agitation and fluidization in MAP resulted in a clearer bio-oil (free from fine char/ particles) than that of conventional pyrolysis.

3.2. Viscosity comparison across several feedstocks

As noted in Fig. 3, bio-oil from canola had lower viscosity than that of aspen. Among the bio-oils produced from corn cob, the viscosity of bio-oil with catalyst (corn cob 4) was lower than other bio-oils due to more water content. The viscosity of bio-oil produced from corn cob without catalyst was similar to that of liquid phase of canola and it was lower than that of liquid phase of aspen.

All the bio-oils from different feedstocks were behaved as a non-Newtonian fluid. Similar behaviors have been reported for bio-oils (upper layer) obtained from forest residues (Garcia-Perez et al., 2006a) and pine and oak bark (Ingram et al., 2008). However,Ingram et al (2008) reported Newtonian behavior at 25ºC for the bio-oils produced from pine and oak wood through auger reactor but at higher temperatures (50 and 80ºC) they showed mild shear thinning behaviors. Rheological data of shear rate and shear stress of the bio-oils were fitted according to the Power-Law model

$$\mu = k\gamma^{n-1} \tag{1}$$

where, μ is the viscosity (Pa-s), k is the consistency coefficient (Pa.sn), γ is the shear rate, 1/s and n is the flow behavior index of the fluid (dimensionless). The power law conststants for different bio-oils are presented in Table 1. The flow behavior indexesnless than 1 suggests that presence of the pseudoplastic behavior (shear thinning). A possible reason might be breakdown of (waxy) structure would result in low viscosity at high shear rate. In general, the values of flow behavior index are more reliable than that of consistency coefficient (Johnson, 1999).The deviation of flow behavior index from 'unity' indicates the degree of deviation from Newtonian behavior. For shear thinning, the index value can be anywhere between 0 and 1. The smaller the value of n, the greater is the degree of shear thinning (Chhabra & Richardson, 1999). Considering the above points, canola aqueous phase exhibited strong shear thinning behavior than that of the rest. In general, the bio-oils from corn cob approaches Newtonian behavior as the n values were close to unity.

Table 1. Power law constants of the bio-oils at different temperatures

Bio-oil	Power law constants	Temperature, ºC					
		20	40	50	60	80	100
Aspen	k	0.019	0.030	0.040	0.071	0.063	0.153
	n	0.456	0.229	0.126	0.003	0.031	0.298
Canola	k	0.065	0.105	0.094	0.093	0.128	0.223
	n	0.132	0.077	0.085	0.056	0.179	0.058
Corn cob 1	k	0.030	0.047	0.040	0.058	0.123	0.103
	n	0.289	0.107	0.144	0.034	0.165	0.175
Corn cob 2	k	0.061	0.097	0.039	0.046	0.067	0.103
	n	0.084	0.074	0.143	0.056	0.083	0.173
Corn cob 3	k	0.040	0.070	0.048	0.05	0.107	0.070
	n	0.250	0.010	0.095	0.08	0.131	0.093
Corn cob 4	k	0.059	0.066	0.089	0.091	0.046	0.048
	n	0.043	0.044	0.063	0.146	0.008	0.001

3.3. Comparing the measures of viscosity with the viscosities of bio-oil.

Accurate measurement of the viscosity of bio-oil/fuel is essential for the proper operation of fuel supply systems and atomisers. The viscosity of bio-oil can be measured according to the ASTM D 445 using the following equation

$$\eta = \pi \mathrm{Pr}^4 t / 8lv = \pi h \rho g r^4 t / 8lv \tag{2}$$

where η is the viscosity (dynes/cm2or poise), v is the volume of liquid (c.c.), t is the liquid flowing time (s), r is the radius of narrow tube (cm), l is the length of narrow tube (cm). This is most widely followed method, as evident from table 2. The viscosity of bio-oil can be measured using capillary or rotational viscometers and they are reported as kinematic (cSt) or dynamic viscosity (mPa.s). The kinematic viscosity of the bio-oil can be converted into dynamic viscosity if the density (kg/dm3) of bio-oil is known at a given temperature using the following formula

$$\eta(\mathrm{mPa.s}) = \eta\ (\mathrm{cSt}) * \rho\ (\mathrm{kg} / \mathrm{dm}^3) \tag{3}$$

Table 2. Comparison of viscosity measurement and bio-oil viscosity produced from various feedstocks through different pyrolysis

Bubbling fluidized bed pyrolyzer, 450-550°C	Guayule		60	2472.7		1.1362 g/ mL	Boateng et al., 2009
Auger reactor, 450°C	Pine wood	Stony Brook Model PDVa-100	50	55.2	Std	1.18 g/ mL	Hassan et al., 2009a
Bubbling fluidized bed fast pyrolysis, 500°C	Barley straw	Grabner Mini Visll automatic microviscometer	40	23.5			Mullen et al., 2010
	Barley hull		40	10.2			
Fast pyrolysis VTT, 520°C, 1-2s	Pine		20	100-200		1100-1300	Oasmaa et al., 2004
Fast pyrolysis VTT, 520°C, 1-2s	Forest residue		40	15-30			Oasmaa et al., 2005b
Fast pyrolysis VTT, 520°C, 1-2s	Forest residue	ASTM D445	20	100			Oasmaa & Kuoppala, 2003
		-do-	40	30			
Fluidized bed flash pyrolysis, 530°C	Wheat straw	ASTM D445	50	11	ASTM D 4052	1.186 kg/ dm³ at 15°C	Sipila et al., 1998
	Pine wood	-do-		46		1.266 kg/ dm³	
	Hard wood	-do-		30		1.233 kg/ dm³	
Fast pyrolysis, 420-540°C	Rice husk	Glass capillary viscometer	40	13.2	Densitometer	1.14 g/ mL at 30°C	Lu et al., 2008
Fast pyrolysis, 300-600°C	Sugarcane bagasse	Capillary viscometer		7.1 cP	Wt/ Volume	1167	Anto & Thomas, 2009
Attrition free pyrolysis, 340-540°C, 300-900s	Beach wood chips	Dynamic stress rheometer (SR 200), shear rate 3/ s	40	3.6-6.9 cP	Wt/ Volume	1200-1230 at room temp	Guillain et al., 2009
Ablative pyrolysis, 625°C, 130 m/ s	Oak wood	Brookfield digital viscometer model LVTD	40	159 cP		1.29 g/ cm³	Czernik et al., 1994
Slow pyrolysis, 650°C, 15°C/ min, 30 min	Rape seed cake	ASTM D88	50	42.6	ASTM 1744	900	Çulcuoglu et al., 2005
Slow pyrolysis, 600°C, 5°C/ min, 2 h	Oil palm empty fruit bunch	ASTM	50	13.52	ASTM	1.031 g/ cm³	Khor et al., 2009
Vacuum pyrolysis, 500°C, 14 kPa	Softwood bark	ASTM D445-88 Cannon Fenske upflow viscometer (modified) Bohlin rheometer (0.5 Pa)	25	128	ASTM D369	1.066 g/ mL at 20°C	Boucher et al., 2000a
		-do-	40	38		1.051 g/ mL	
		-do-	50	24		1.043 g/ mL	
		-do-	60	15		1.034 g/ mL	
		-do-	80	7.5			
Vacuum pyrolysis, 500°C, 14 kPa	Softwood bark	ASTM D445 Cannon Fenske upflow viscometer	30	85			Boucher et al., 2000b
		-do-	50	24			
		-do-	80	8			
Vacuum pyrolysis, 530°C, 40 kPa	Softwood bark	ASTM D445-88 Brookfield viscometer model LVDV III+	30	62	ASTM D369	1.188	Ba et al., 2004
		-do-	60	15			
Vortex reactor-NREL	Hybrid poplar	Brookfield viscometer #18 spindle, shear rate 79/ s	40	22 cP		1.2 g/ cc	Diebold& Czernik, 1997
	Bagasse		60	510.3		1.0904 g/ mL	
Auger reactor, 500°C, 5.9 min	Pine chips	Brookfield synchrometric LVT viscometer	25	125.6 cP	Densitometer	1.183	Garcia-Perez et al., 2007
	Pine pellets	-do-	25	44.8 cP		1.236	
Auger reactor, 450°C, 30 s	Pine wood	Brookfield viscometer LV-DVI+, Rheometer, TA instruments Model 1000N	50	60.9 cP	Pycnometer ASTM D4052	1.19 g/ cc	Ingram et al., 2008
	Oak wood	-do-	50	41.6 cP		1.20 g/ cc	
	Pine bark	-do-	80	70 cP		1.17 g/ cc	
	Oak bark	-do-	80	131 cP		1.20 g/ cc	
Auger reactor, 450°C	Pine wood	Stony Brook 200 58 viscometer		80.7	ASTM D4052	1.19 g/ mL	Bhattacharya et al., 2009
Auger reactor, 450°C	Pine wood	Stony Brook viscometer model PDVa-100	40	30			Hassan et al., 2009b
	Pine bark	-do-		137.9			
	Cottonwood	-do-		77			
	Cottonwood bark	-do-		472.7			
Microwave, 600W, 2450Hz, 40 min	Corn stover	ASTM D445 Rotational viscometer Brookfield DV-E	20	1270 cP		1.25 g/ mL	Yu et al., 2007
		-do-	40	185 cP			
		-do-	50	60 cP			
		-do-	80	34 cP			
Fast pyrolysis 530 °C, 12 kPa	Sugarcane bagasse	ASTM D445-88 Cannon Fenske upflow viscometer	20	116.5	ASTM 369 Picnometer	1211	GarciaPerez et al., 2002
			40	26.7		1195	
			60	11.2		1180	
			80	5.4		1160	
Slow pyrolysis 500°C, 11°C/ min	Olive waste	ASTM D445-88	50	51	ASTM D1298	1070 at 15°C	SensÕz et al., 2006
Fixed bed reactor 500°C, 40°C/ min, 30 min	Rapeseed	ASTM D445-88	50	43	ASTM D1298	918 at 30°C	SensÕz et al., 2000
500°C, 50°C/ min, N_2 flow rate of 100 cm³/ min	Safflower seed press cake	ASTM D445-88	50	225	ASTM D4052	1079 at 15°C	SensÕz & Angin, 2008

ASTM D445 specifies that the kinematic viscosity of conventional fuels, which are Newtonian fluids, is normally determined. In their comprehensive research, Leroy et al (1988) extensively examined the rheological properties of various bio-oils derived from wood. They found that these bio-oils displayed a predominantly Newtonian behavior within the shear rate range of 1 to 1000/s. On the contrary, the bio-oils utilized in this investigation exhibited shear thinning characteristics. In the case of a Newtonian fluid, the viscosity remains consistent as the shear rate increases. Radovanovic et al (2000) described a method for quantifying the viscosity of bio-oil using a falling ball viscometer. Osamaa et al (2009) suggested the use of Cannon-Fenske viscometer tubes instead of Ubbelohde tubes for obtaining more precise results with dark colored liquids. This is because the flow direction in Cannon-Fenske tubes is more suited for accurate measurements. If the bio-oil appears visually uniform, there is no need to prefilter the sample. Prior to sampling, it is crucial to remove any air bubbles and allow for a 15-minute equilibration period in order to accurately determine viscosity at a specific temperature. Table 2 presents a comparison of viscosity measurement and viscosity of bio-oils produced from various feedstocks using different pyrolysis reactors. The measurement of bio-oil density and the corresponding density values are provided to facilitate the conversion of viscosity units from kinematic to dynamic. As per ASTM D445-88, viscosity measurements should be taken at temperatures of 20 and 40ºC. However, the table shows that the viscosity values were reported at various temperatures ranging from 20 to 100ºC. The bio-oils employed in this investigation exhibited decreased viscosity compared to the viscosities stated in the table, regardless of temperature.

3.4. Stress and frequency sweep

Typically, the linear boundaries of viscoelastic behavior are established by defining the stress range in which G' and G" remain constant and unaffected by stress. The graph in Figure 4 illustrates the storage and loss moduli of bio-oil derived from various feedstocks, as they vary with shear stress. The bio-oils displayed linear areas in Fig. 4, with shear stress ranging from 10 to 100 Pa. A shear stress of 55 Pa was chosen for the frequency and temperature sweep. As shown in the figure, the storage modulus (G') was consistently higher than the loss modulus for all feedstocks. However, Ba et al (2004) found that the bio-oil produced from softwood bark through vacuum pyrolysis had a dominant behavior of the loss modulus. Garcia-Perez et al (2008) found that the storage modulus of the bio-oil produced from mallee woody biomass was lower than the loss modulus. In their investigation, Ba et al (2004) discovered four distinct zones, including two plateaus, for the variable G'. However, this particular study did not detect the same phenomenon. The bio-oils derived from corn cob (3 and 4) generated in a batch MAP exhibited a higher storage modulus compared to the bio-oils derived from corn cob produced in continuous MAP, as shown in Figure 5. The loss modulus of the bio-oil obtained from batch MAP (corn cob 3 and 4) was lower compared to the bio-oil obtained from continuous MAP of corn cobs and batch MAP of other feedstocks. The storage and loss moduli exhibited comparable values for the bio-oils derived from corn cob in a batch MAP, regardless of the presence or absence of a catalyst (Fig. 5).

The frequency sweep was conducted in the range of 0.1-100 Hz and found the linear region between 1 and 100 Hz. Accordingly, the frequency sweep experiments were repeated and identified the linear region between 10 and 100 Hz as shown in Fig. 3. As G' approaches a slope of more than 2, which confirms the existence of linear viscoelastic region (Tzanetakis et al., 2008). Similarly, G" approaches a slope of more than 2 (less than slope of G'), which is also consistent with linear viscoelastic behavior. A frequency of 50 Hz was selected from the linear range depicted in Fig. 4 for temperature sweep. The frequency sweep was also confirmed that the storage modulus was predominant than that of respective loss modulus of each bio-oil.Tzanetakis et al (2008) reported that the loss modulus was ten times higher than that of storage modulus for the bio-oil produced from hardwood, and it was an opposite observation. Similar to stress sweep, the bio-oil produced from corn cob 3 and 4 in a batch MAP had lower loss modulus than the bio-oil from other feedstocks including corn cob. The storage and loss moduli were overlapping between the bio-oils at a low frequency (10Hz) and a clear difference was observed as frequency increases. The bio-oil from aspen and canola had a higher storage and loss moduli than that of bio-oils from corn cob.

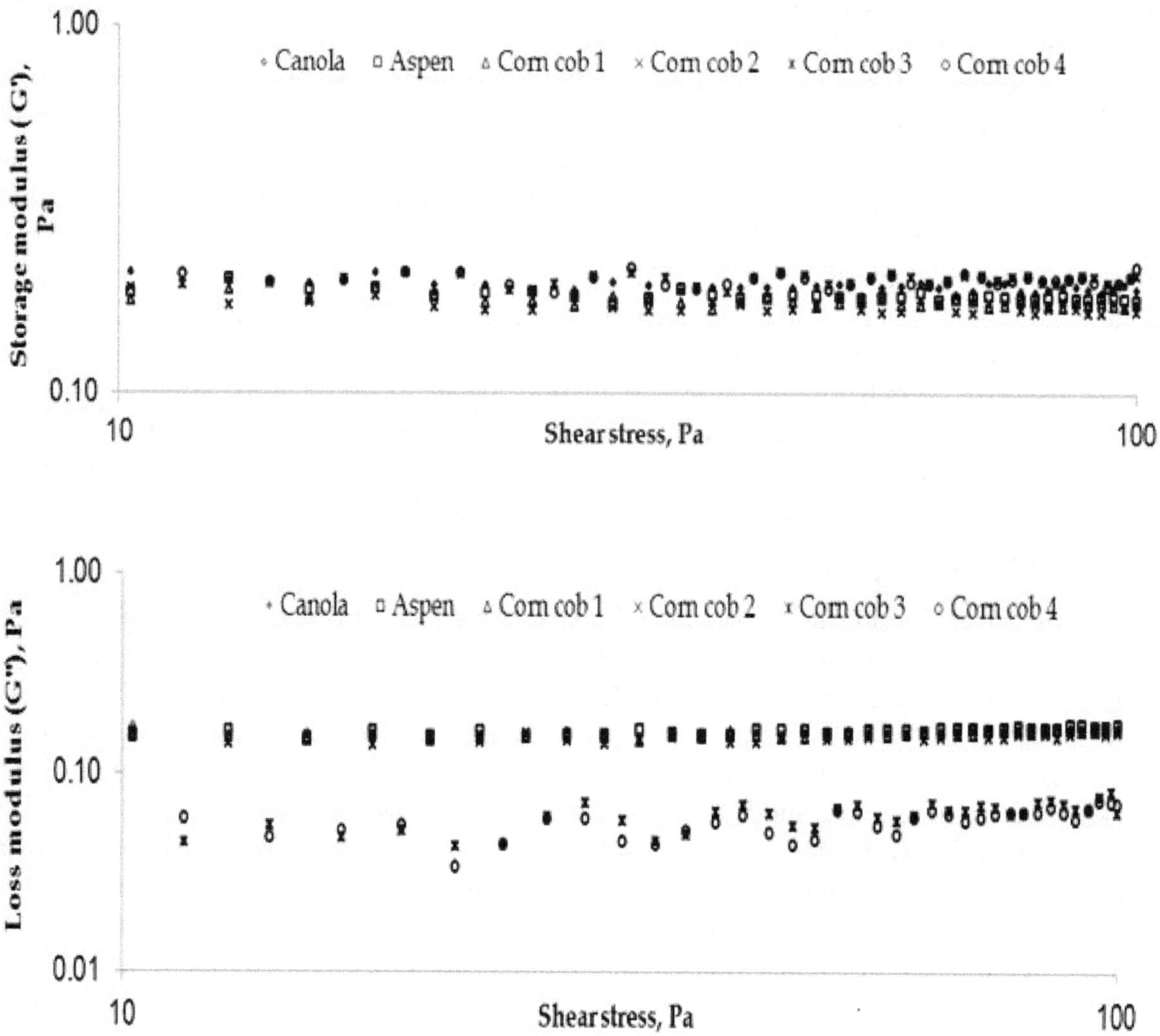

Figure 4. Stress sweep of bio-oils from produced from different feedstocks

3.5. Temperature sweep

The bio-oils were subjected to a temperature sweep using a shear stress of 55 Pa and a frequency of 50 Hz. The temperature was increasing at a constant pace of 10ºC per minute. Figure 6 displays the trend of storage and loss modulii for all the bio-oils. The modulus of bio-oil derived from canola exhibited a declining pattern as the temperature surpassed 80ºC. Furthermore, the temperature sweep analysis reaffirmed that the storage modulus of the bio-oils derived from various feedstocks was more prominent than the loss modulus. Generally, bio-oils are considered to be Newtonian fluids, as stated by Oasmaa & Peacocke (2001) and Oasmaa et al. (2003). However, according to Fig. 6, the storage and loss moduli lines indicate that bio-oils behave as non-Newtonian fluids regardless of the feedstocks.

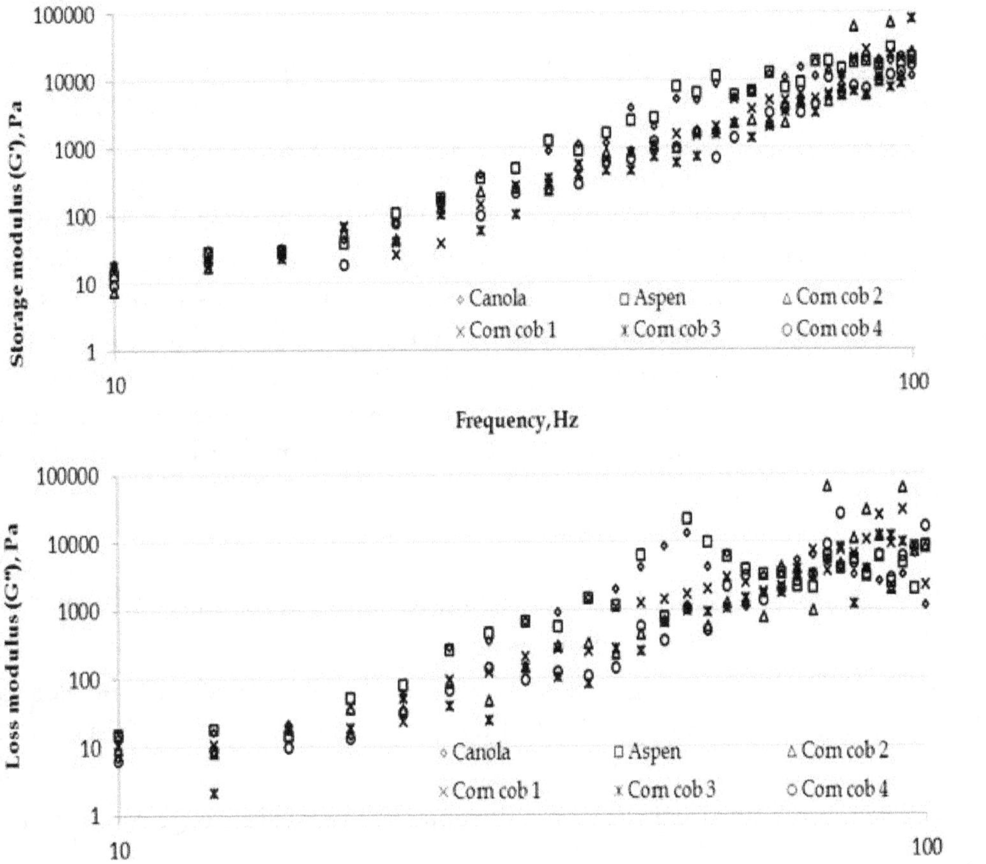

Figure 5. Frequency sweep of bio-oils from produced from different feedstocks

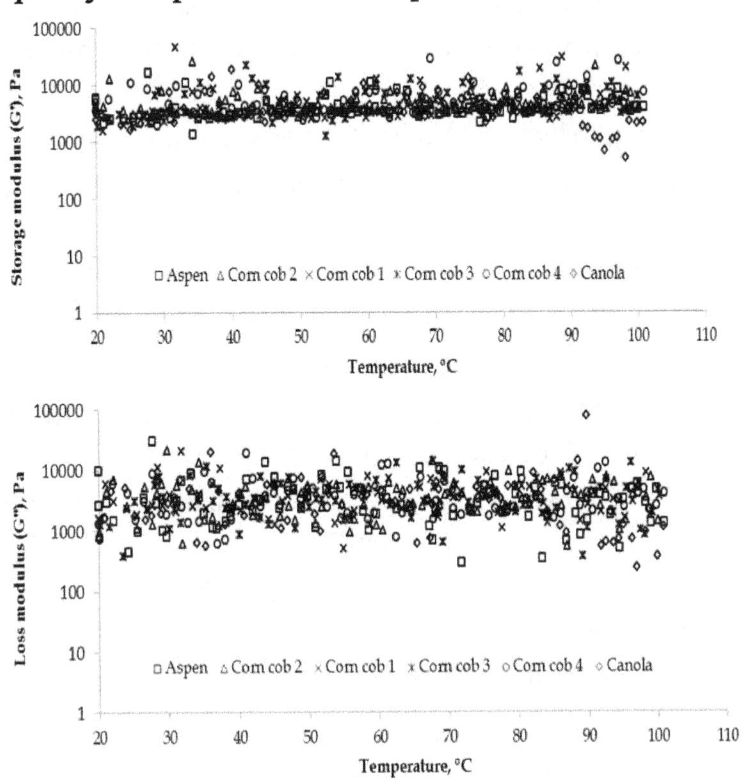

Figure 6. Temperature sweep of bio-oils produced from different feedstocks

4. CONCISE SUMMARY

Viscosity is a crucial factor when dealing with bio-oils. A comprehensive rheological analysis was conducted on bio-oils derived from aspen, canola, and maize cobs using MAP. The findings from the basic viscosity and dynamic rheology tests showed that the bio-oils examined in this research exhibited characteristics of non-Newtonian fluids. The bio-oils exhibited viscosities ranging from 0.00272 to 0.00196 Pa.s at 20ºC and from 0.00236 to 0.00163 Pa.s at 40ºC. The viscosities of these bio-oils are significantly lower compared to the viscosities of bio-oils obtained from various feedstocks as documented in earlier research. Furthermore, the bio-oils utilized in this research had a viscosity that was lower than that of light and heavy fuel oil, the US #4 fuel oil, and the ASTM burner fuel standard, suggesting a wide range of possible uses. The viscosity range of bio-oils suggests that they possess favorable handling and processing characteristics. However, it is important to note that viscosity alone does not determine the suitability of bio-oil for a particular application. Hence, it is necessary to examine other aspects in order to evaluate the appropriateness of these bio-oils.

REFERENCES

Akama K., Puchta H., & Hohn B. 1995. Efficient Agrobacterium-mediated transformation of Arabidopsis thaliana using the bar gene as selectable marker. Plant Cell Rep 14:450- 545.

Benemann J.R. & Oswald W.J. 1996. Algal mass culture systems. In Systems and Economic Analysis of Microalgae Ponds for Conversion of CO2 to Biomass. US Department of Energy, Pittsburgh, PA, pp. 42-65.

Benning C. 2008. A role for lipid trafficking in chloroplast biogenesis. Progess in Lipid Research 47:381-389.

Chen T., Liu H., Lu P., & Xue L. 2009. Construction of Dunaliella salina heterotrophic expression vectors and identification of heterotrophically transformed algal strains. Chinese Journal of Biotechnology 25:392-398.

Chisti Y. 2007. Biodiesel from microalgae. Biotechnology Advances. 25:294-306.

Chitlaru E. & Pick U. 1991. Regulation of glycerol synthesis in response to osmotic changes in Dunaliella Plant Physiology 96:50-60.

Danquah M.K., Gladman B., Moheimani N., & Forde G.M. 2009. Microalgal growth characteristics and subsequent influence on dewatering efficiency. Chemical Engineering Journal 151:73-78.

European Parliament. 2009. Directive 2009/28/EC, L 140. pp. 16-59

Feng S., Xue L., Liu H. and Lu P. 2009. Improvement of efficiency of genetic transformation for Dunaliella salina by glass beads method. Molecular Biology Reports 36:1433-1439.

Fortman J.L., Chhabra S., Mukhopadhyay A., Chou W., et al. 2008. Biofuel alternatives to ethanol: pumping the microbial well. Trends in Biotechnology 26:375-381.

Geng D., Wang Y., & Wang P. 2003. Stable expression of hepatitis B surface antigen gene in Dunaliella salina. J Appl Phycol 15:451-456.

Gordon J.M. & Polle J.E.W. 2007. Ultrahigh bioproductivity from algae. Appl Microbiol Biotechnol 76:969-975.

Gouveia L. and Oliveira A.C. 2009 Microalgae as raw material for biofuels production. J Ind Microbial Biotechnol 36:269-274.

Goyal A. 2007. Osmoregulation in Dunaliella, Part II: Photosynthesis and starch contribute carbon for glycerol synthesis during a salt stress in Dunaliella tertiolecta. Plant Physiology and Biochemistry 45:705-710.

Hallmann A. 2007. Algal Transgenics and Biotechnology. Transgenic Plant Journal 1:81-98.

Hallmann A. & Rappel A. 1999. Genetic engineering of the multicellular green alga Volvox: a modified and multiplied bacterial antibiotic resistance gene as a dominant selectable marker. The Plant Journal 17:99-109.

Harris E.H. 2001. Chlamydomonas as a Model Organism. Annu Rev Plant Physiol Plant Mol Biol 52:363-406.

Hejazi M.A., Holwerda E. & Wijffels R.H. 2004a. Milking microalga Dunaliella salina for β-carotene production in two phase bioreactors. Biotechnol Bioeng 85:475-481.

Hejazi M.A., Kleinegris D. & Wijffels R.H. 2004b. Mechanisms of extraction of β-carotene from Dunaliella salina in two-phase bioreactors. Biotechnol Bioeng 88:593-600.

Hoerlein G. 1994. Glufosinate (phosphinothricin), a natural amino acid with unexpected herbicidal properties. Rev Environ Contam Toxicol 138:73-145.

Hoffert M.I., Caldeira K., Jain A.K., Haites E.F., et al. 1998. Energy Implications of future stabilization of atmospheric CO2 content. Nature 395:881-884.

Hu Q., Guterman H. & Richmond A. 1996. A flat inclined modular photobioreactor for outdoor mass cultivation of photoautotrophs. Biotechnol Bioeng 51:51-60.

Huntely M.E. & Redalje D.G. 2007. CO2 mitigation and renewable oil from photosynthetic microbes: a new appraisal. Mitigation and Adaptation Strategies for Global Change 12:573-608.

Jakobiak T., Mages W., Scarf B., Babinger P., et al. 2004. The bacterial paromomycin resistance gene, aphH, as a dominant selectable marker in Volvox carteri. Protist 155:381-393.

Jamers A., Blust R., & Coen W.D. 2009. Omics in algae: Paving the way for a systems biological understanding of algal phenomena? Aquatic Toxicology 92:114.

Janssen M., Tramper J., Mur L.R., & Wijffels R.H. 2003. Enclosed outdoor photobioreactors: light regime, photosynthetic efficiency, scale-up, and future prospects. Biotechnol Bioeng 81:193-210.

Jia Y., Xue L., & H. Liu J.L. 2009a. Characterization of the glyceraldehyde-3-phosphate dehydrogenase (GAPDH) dene from the halotolerant alga Dunaliella salina and inhibition of its expression by RNAi. Current Microbiology 58:426-431.

Jia Y., Xue L., Li J., & Liu H. 2009b. Isolation and proteomic analysis of the halotolerant alga Dunaliella salina flagella using shotgun strategy. Journal 2009.

Jiang G.-Z., Lu Y.-M., Niu X.-L., & Xue L.-X. 2005. The actin gene promoter-driven bar as a dominant selectable marker for nuclear transformation of Dunaliella salina. Acta Cenetica Sinica 32:424-433.

Jiang G.Z., Nu X.L., Lu Y.M., Xie H., et al. 2003. Cloning and characterization of Hsp70a cDNA fragment of Dunaliella salina. Yi Chuan 25: 573-576.

Jin E.S., Polle E.W., & Melis A. 2001. Involvement of zeaxanthin and of the Cbr protein in the repair of photosystem II from photoinhibition in the green alga Dunaliella salina. Biochim Biophys Acta 1506:244-259.

Lamers P.P., Janssen M., Vos R.C.H.D., Binoand R.J., et al. 2004. Exploring and exploiting carotenoid accumulation in Dunaliella salina for cell-factory applications. Trends Biotechnol 26:631-638.

Leon-Banares R., Gonzalez-Ballester D., Galvan A., & Fernandez E. 2004. Transgenic microalgae as green cell factories. Trends in Biotechnology 22:45-52.

Li J., Xue L., Yan H., Wang L., et al. 2007. The nitrate reductase gene-switch: a system for regulated expression in transformed cells of Dunaliella salina. Gene 403:132-142.

Liska A.J., Shevchenko A., Pick U., & Katz A. 2004. Enhanced photosynthesis and redox energy production contribute to salinity tolerance in Dunaliella as revealed by homology-based proteomics. Plant Physiology 136:2806-2817.

Liu H.T., Zang W.D., Lu Z.M., Wang N., et al. 2005. Cloning and analysis of psaB cDNA of Dunaliella salina. Chinese Journal of Biotechnology 21:643-645.

Long S.P., Humphries S., & Falkowski P.G. 1994. Photoinhibition of photosynthesis in nature. Annu Rev Plant Physiol Plant Mol Biol 45:633-662.

Lu P., Yan H., Li J., Liu H., et al. 2009. An optimal electroporation system for Dunaliella salina. Chinese Journal of Biotechnology 25:520-525.

Lumbreras V., Stevens D.R., & Purton S. 1998. Efficient foreign gene expression in Chlamydomonas reinhardtii mediated by an endogenous intron. Plant J 14:441- 447.

Miller S.M. 2002. Taming the fierce roller: an “enhanced” understanding of cellular differentiation in Volvox. Bioessays 24:3-7.

Miron A.S., Gomez A.C., Camacho F.G., Grima E.M., et al. 1999. Comparative evaluation of compact photobioreactors for large-scale monoculture of microalgae. Journal of Biotechnology 70:249-270.

Muller-Feuga A. 2004. Microalgae for Aquaculture: The current global situation and future trends, Blackwell Science. Look Back at the U.S. Department of Energy’s Aquatic Species Program: Biodiesel from Microalgae. TP- 580-24190. National Renewable Energy Lab, Golden, Colorado.

Schiedlmeier B., Schmitt R., Muller W., Kirk M.M., Gruber H., Mages W., & Kirk D.L. 1994. Nuclear transformation of Volvox carteri. Proc Natl Acad Sci USA, 91:5080-5084.

Shelef G., Sukenik A., & Green M. 1984. Microalgae harvesting and processing: a literature review. Solar Energy Research Institute, pp. 1-65.

Schenk P.M., Thomas-Hall S.R., Stephens E., Marx U.C., Mussgnug J.H., Posten C., Kruse O., Hankamer B. 2008. Seconds generation biofuels: high-efficiency microalgae for biodiesel production. Bioenergy Research. 1:20-43.

Sillers R., Show A., Tracy B., & Papoutsakis E.T. 2008. Metabolic engineering of the nonsporulating, non-olventogenic Clostridium acetobutylicum strain M5 to produce butanol without acetone demonstrate the robustness of the acid-formation pathways and the importance of the electron balance. Metabolic Engineering 10:321.

Sizova I., Fuhrmann M., & Hegemann P. 2001. A Streptomyces rimosus aphVIII gene coding for a new type phosphotransferase provides stable antibiotic resistance to Chlamydomonas reinhardtii. Gene 277:221-229.

Smith D.R., Lee R.W., Cushman J.C., Magnuson J.K., Tran D., & Polle J.E.W. 2010. The Dunaliella salina organelle genomes: large sequences, inflated with intronic and intergenic DNA. BMC Plant Biology, 10:83.

Smolinska J. 2008. EU Economic Report. ACEA - European Automobile Manufacturers' Association.

Spolaore P., Joannis-Cassan C., Duran E., & Isambert A. 2006. Commercial applications of microalgae. Journal of Bioscience and Bioengineering 101:87-96.

Sun G., Zhang X., Sui Z., & Mao Y. 2008. Inhibition of pds gene expression via the RNA interference approach in Dunaliella salina (Chlorophyta). Mar Biotechnol 10:219-226.

Sun Y., Yang Z., Gao X., Li Q., et al. 2005. Expression of foreign genes in Dunaliella by electroporation. Mol Biotechnol 30:185-192.

Tan C., Quin S., Zhang Q., Jiang P., et al. 2005. Establishment of a micro-particle bombardment transformation system for Dunaliella salina. J Microbiol 43:361-365.

Thompson C.J., Movva N.R., Tizard R., Crameri R., et al. 1987. Characterization of the herbicide-resistance gene bar from Streptomyces hygroscopicus. EMBO J 6:2519-2523.

Thompson G.A. 2005. Mechanisms of osmoregulation in the green alga Dunaliella salina. Journal of Experimental Zoology 268:127-132.

Tredici M.R. and Zittelli G.C. 1998. Efficiency of sunlight utilization: tubular versus flat photobioreactors. Biotechnol Bioeng 57:187-197.

Ugwu C.U., Aoyagi H., & Uchiyama H. 2008. Photobioreactors for mass cultivation of algae. Bioresource Technology 99:4021-4028.

U.S. Energy Information Administration. Annual Energy Outlook 2009 with Projections to 2030. DOE/EIA-0383(2009).

Vemuri G.N. & Aristidou A.A. 2005. Metabolic engineering in -omics era: elucidating and modulating regulatory networks. Microbiology and Molecular Biology Reviews 69:197-216.

Walker T.L., Becker D.K., & Collet C. 2004. Characterisation of the Dunaliella tertiolecta RbcS genes and their promoter activity in Chlamydomonas reinhardtii. Plant Cell Rep 23:727-735.

Walker T.L., Collet C., & Purton S. 2005. Algal transgenics in the genomic era. Journal of Phycology 41:1077-1093.

Wallace R., Ringer M., Pienkos P. 2008. National Renewable Energy Laoratory (NREL) Report to Congress: Algal Biofuels.

Wang T., Xue L., Hou W., Yang B., et al. 2007. Increased expression of transgene in stably transformed cells of Dunaliella salina by matrix attachment regions. Applied Microbiology and Biotechnology 76:651-657.

Wang T., Xue L., Xiang J., Li J., et al. 2009. Cloning and characterization of the 14-3-3- protein gene from the halotolerant alga Dunaliella salina. Molecular Biology Reports 36:207-214.

Weldy C.S. & Huesemann M. 2007. Lipid production by Dunaliella salina in batch culture: effects of nitrogen limitation and light intensity. U.S. Department of Energy Journal of Undergraduate Research 7:115-122.

Weyer K., Bush D., Darzins A., & Wilson D. October 24, 2008. Theoretical Maxium Algal Oil Production. Algae Biomass Summit Solix.

Wu Q. & Miao X. 2006. Biodiesel production from heterotrophic microalgal oil. Bioresource Technology 97:841-846.

Xie H., Xu P., Jia Y., Li J., et al. 2007. Cloning and heterologous expression of nitrate reductase genes from Dunaliella salina. J Appl Phycol 19:497-504.

Ye Z.-W., Jiang J.-G., & Wu G.-H. 2008. Biosynthesis and regulation of carotenoids in Dunaliella: Progresses and prospects. Biotechnol Advances 26:352-360.

CHAPTER-14

SIMULTANEOUS PRODUCTION OF BIOETHANOL AND POWER

1. INTRODUCTION: AN OVERVIEW

In recent times, there has been a growing interest in using biomass as a fuel source in many countries as a means to mitigate the effects of global warming, which is mostly produced by the burning of fossil fuels. (Mousdale, 2010). Many academics anticipate that bioethanol has the potential to serve as a viable alternative to petroleum. Indeed, bioethanol has minimal energy and exergy loss during chemical reactions, saccharification, and fermentation for ethanol synthesis, as it is solely generated by energy conversion via chemical reactions (Cardona et al., 2010). Nevertheless, during the process of fermentation, the resulting product contains a significant quantity of water, hence hindering the ability to fully optimize the heat value of the product. Hence, it is necessary to separate the ethanol-water combination in order to get pure ethanol for use as fuel (Zamboni et al., 2009a, 2009b; Huang et al., 2008). Distillation is commonly employed for the separation of this combination in practical applications (Fair 2008). Conventional distillation is widely recognized as a process that requires a significant amount of energy. Additionally, it is not possible to directly manufacture pure ethanol fuel from a distillation column due to the formation of an azeotropic mixture between ethanol and water. In order to separate pure ethanol from ethanol-water mixtures through distillation, the employment of an entrainer (azeotrope breaking agent) is required. This is because the azeotropic mixture, which vaporizes without any alteration in its composition, needs to be broken. Either benzene, cyclohexane, or isopropyl alcohol can serve as entrainers for the ethanol-water mixture. Consequently, a minimum of two separation units are necessary in order to obtain pure ethanol, resulting in additional energy consumption (Doherty & Knapp 2008). Approximately half of the thermal energy contained in bioethanol is estimated to be necessary for the process of distilling ethanol from the combination. In order to decrease energy usage in the process of bioethanol production, several researchers have suggested the use of membrane separations (Baker 2008, Wynn 2008) or pressure swing adsorption (PSA) (Modla & Lang, 2008) as alternatives to azeotropic distillation. These alternatives have proven successful in developing suitable membranes or sorbents to achieve an effective separation. However, in numerous instances, they have given minimal consideration to the comprehensive process scheme or have devised heat integration procedures relying on traditional heat recovery technologies, such as the widely recognized heat cascading use. Conventional heat recovery methods limit the ability to minimize the minimum energy consumption of the overall process due to constraints on changes to the state of the process stream (Hallale 2008, Kemp 2007). Furthermore, the majority of cost reduction assessments for bioethanol facilities have been carried out using these traditional techniques and technology. Therefore, the cost of bioethanol products continues to be elevated in comparison to fossil fuels.

In recent times, there has been a development in the notion of co-production of energy and products, which involves reevaluating the energy and production system with a focus on improving energy conversion efficiency and energy conservation. However, in order to achieve co-production, it is important to thoroughly examine and optimize the heat and power needed for manufacturing in every individual process. Thus, the researchers have devised a self-heat recuperation technique using exergy recuperation (Kansha et al. 2009) and implemented it in various chemical processes for co-production (Fushimi et al. 2011, Kansha et al. 2010a, 2010b, 2010c, 2011, Matsuda et al.2010). This chapter presents the introduction and application of self-heat recuperation technology in the separation processes of bioethanol production for co-production. Furthermore, this study examines the practicality and energy equilibrium of simultaneously producing bioethanol and power by biomass gasification, utilizing a self-heat recuperation process.

2. ENERGY EQUILIBRIUM IN THE PROCESS OF TRADITIONAL BIOETHANOL MANUFACTURING

It assumed that the amount of energy in feed stock wet biomass is 100 and that 50% of this energy consists of that from reactant sugars, such as starch, cellulose and others. Thus, the amount of energy of the original component of sugar (50) transfers to ethanol (46) and heat (4) through chemical reactions (saccharification and fermentation) with water. This energy is estimated from the following calculation; the caloric value of sugar is 685 kcal/mol, the caloric value of ethanol is 316 kcal/mol and 2 mol ethanol is produced from 1 mol sugar through the above reaction. The pure ethanol product is then separated by distillation and additional heat energy (23) is required for this distillation work when azeotropic distillation is used for the separation. Non-reactants contain a large amount of water, for which the higher heat value is almost equal to the evaporation heat, leading to a net heat value of 0. The above energy relation is shown in Fig. 1. Beyond this, some additional energy is required to produce heat energy from the wet biomass for distillation (23). This additional energy (15) is used to dry the wet biomass in a heater to produce dry biomass that is used as fuel for distillation. Figure 2 shows the total energy balance including this additional energy. It is noted that 50-80% moisture content in wet biomass is assumed in this energy analysis, because many types of wet biomass exist in this range, such as those that originate from ligneous, garbage and sludge. It can be seen from Fig. 2 that 138 units of energy in the wet biomass feed stock is required to produce 46 energy units of ethanol and that about 1/3 of the energy of the wet biomass can be utilized as bioethanol for fuel. Thus, 2/3 of the wet biomass feed stock energy is wasted. Even though this wasted heat energy could potentially be heat sources for other processes, the exergy ratio and temperature of the waste heats are quite low. Thus, it is difficult to achieve energy saving from this by heat integration technologies such as cascading utilization. In fact, the highest required temperature during bioethanol production is normally at the distillation column reboiler and this temperature is lower than 150°C. This heat is exhausted from the condenser at below 100°C. To utilize the biomass energy more effectively, it is clear that the energy consumption during distillation for separating water and product ethanol and for drying of the wet biomass must be reduced. When an integrated system of distillation and membrane separation processes are utilized to substitute for azeotropic distillation, the energy required can be decreased from 23 to 12 units (8: distillation, 4: membrane separation). However, the pressure difference for membrane separation requires electric power. If we assume that the power generation efficiency from dry biomass is 25% and 75% of the energy for the membrane separation process is provided by electricity, 35 energy units from wet biomass are required for distillation and dehydration by membrane separation.

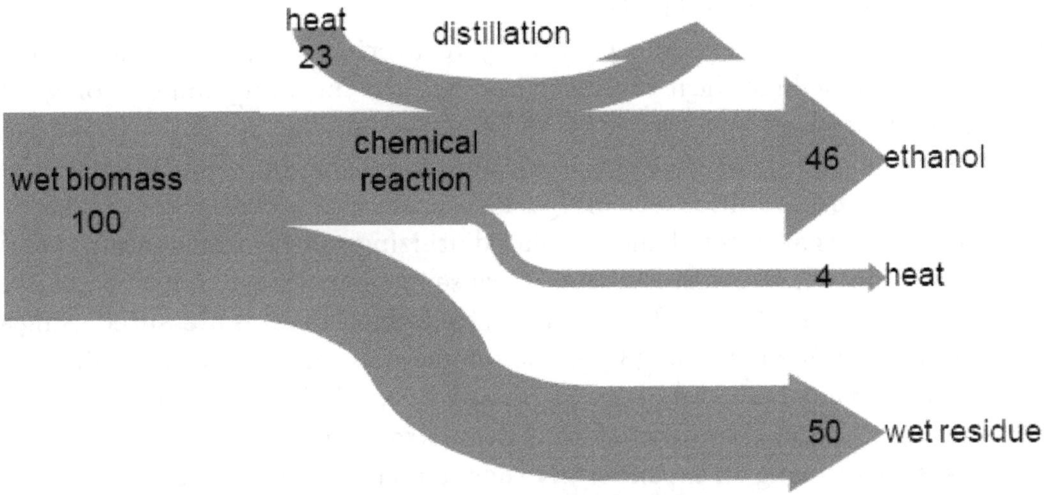

Fig. 1. Energy balance for bioethanol production

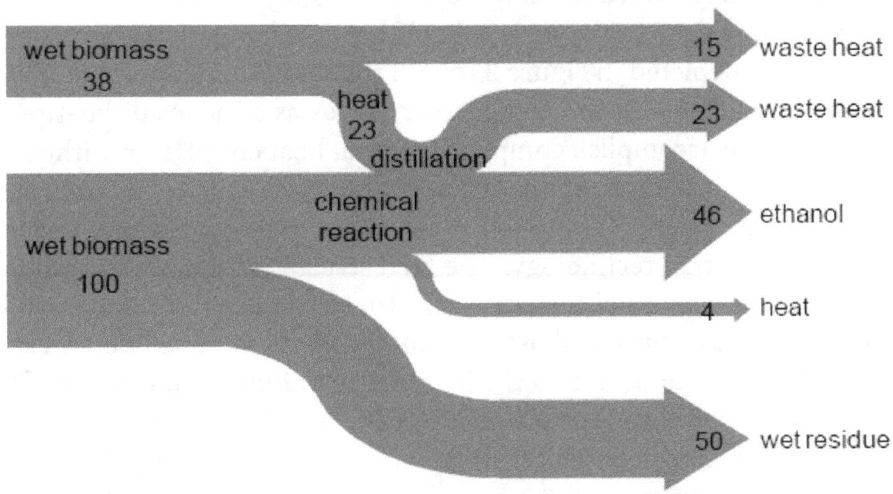

Fig. 2. Total energy balance for bioethanol production

3. SELF-HEAT RECUPERATION TECHNOLOGY AND SELF-HEAT RECUPERATIVE PROCESSES

The self-heat recuperation technique, as described by Kansha et al. in 2009, enables the recirculation of both latent heat and sensible heat in a process. This technology reduces energy consumption by utilizing compressors and self-heat exchangers that are based on exergy recuperation. This technology involves dividing a process unit based on its functions in order to achieve a balance between heating and cooling loads. This is done through the use of enthalpy and exergy analysis. Additionally, the cooling load is recovered by compressors and swapped with the heating load. Consequently, the heat of the process stream is efficiently circulated without any additional heat, leading to a significant reduction in energy consumption for the operation. By implementing this technology in both the distillation and dehydration processes, the energy balance for ethanol production can be substantially altered compared to the previously mentioned scenario. This section presents the design methodology for self-heat recuperative processes, starting with a simple thermal process. It then proceeds to introduce the application of self-heat recuperative processes in separation processes.

3.1 Self-heat recuperative thermal process

In order to decrease energy usage in a process by recovering heat, heating and cooling functions are typically combined to facilitate heat exchange between the incoming and outgoing streams, hence promoting heat circulation. A self-heat exchange system is a system that adopts such integration. In order to optimize the self-heat exchange load, a heat circulation module has been suggested for the heating and cooling operations of the process unit, as seen in Figure 3 (Kansha et al., 2009). Figure 3 (a) depicts a thermal process in which gas streams are heated and circulated utilizing self-heat recuperation technology. The feed stream is heated using a heat exchanger (1→2) to raise its temperature from the initial temperature, T0, to the desired temperature, T1. The effluent stream from the process is pressured using a compressor to recover the heat of the effluent stream (3→4). The temperature of the stream leaving the compressor is increased to T1' through adiabatic compression. Stream 4 is cooled using a heat exchanger to facilitate self-heat exchange, resulting in a transfer of heat from stream 4 to stream 5. The effluent stream is subsequently depressurized using an expander in order to reclaim a portion of the energy expended by the compressor. This results in optimal internal heat circulation achieved through self-heat recuperation. The effluent stream is ultimately cooled to temperature T0 using a cooler (6→7). It should be noted that the total heating duty is equivalent to the internal self-heat exchange load, QHX, and does not include any external heating load, as depicted in Figure 3 (b). In the scenario of optimum adiabatic compression and expansion, the work inputted into the compressor serves as a means of heat pumping. This allows for the effluent temperature to accomplish complete internal heat circulation without any loss of energy. Hence, the implementation of self-heat recuperation has the potential to significantly decrease energy usage. Figure 3 (c) depicts a thermal process that involves the circulation of heat in vapor/liquid streams, utilizing the self-heat recuperation technology. The feed stream is heated using a heat exchanger (1→2) to raise its temperature from the initial temperature, T0, to the desired temperature, T1. The outflow from the following procedure is compressed using a compressor (3→4). The latent heat can be transferred between the feed and effluent streams due to the increase in boiling temperature of the effluent stream to Tb' through compression. Therefore, the effluent stream is cooled by the heat exchanger by self-heat exchange (4→5) while recovering its heat. The effluent stream is subsequently depressurized using a valve (5→6) and ultimately cooled to temperature T0 using a cooler (6→7). This results in optimal internal heat circulation by self-heat recuperation, which is comparable to the gas stream scenario mentioned before. It is important to observe that the total heating duty is equivalent to the internal self-heat exchange load, QHX, and does not include any external heating load, as depicted in Figure 3 (d). The sensible heat of the vapor and liquid in the incoming stream can be transferred to the sensible heat of the outgoing stream, while the heat required for vaporization in the incoming stream can be transferred to the heat released during condensation in the outgoing stream. Consequently, the self-heat exchange system in gas streams and/or vapor/liquid streams reduces the energy used by the heat circulation module to a fraction ranging from 1/22 to 1/2 of the original amount.

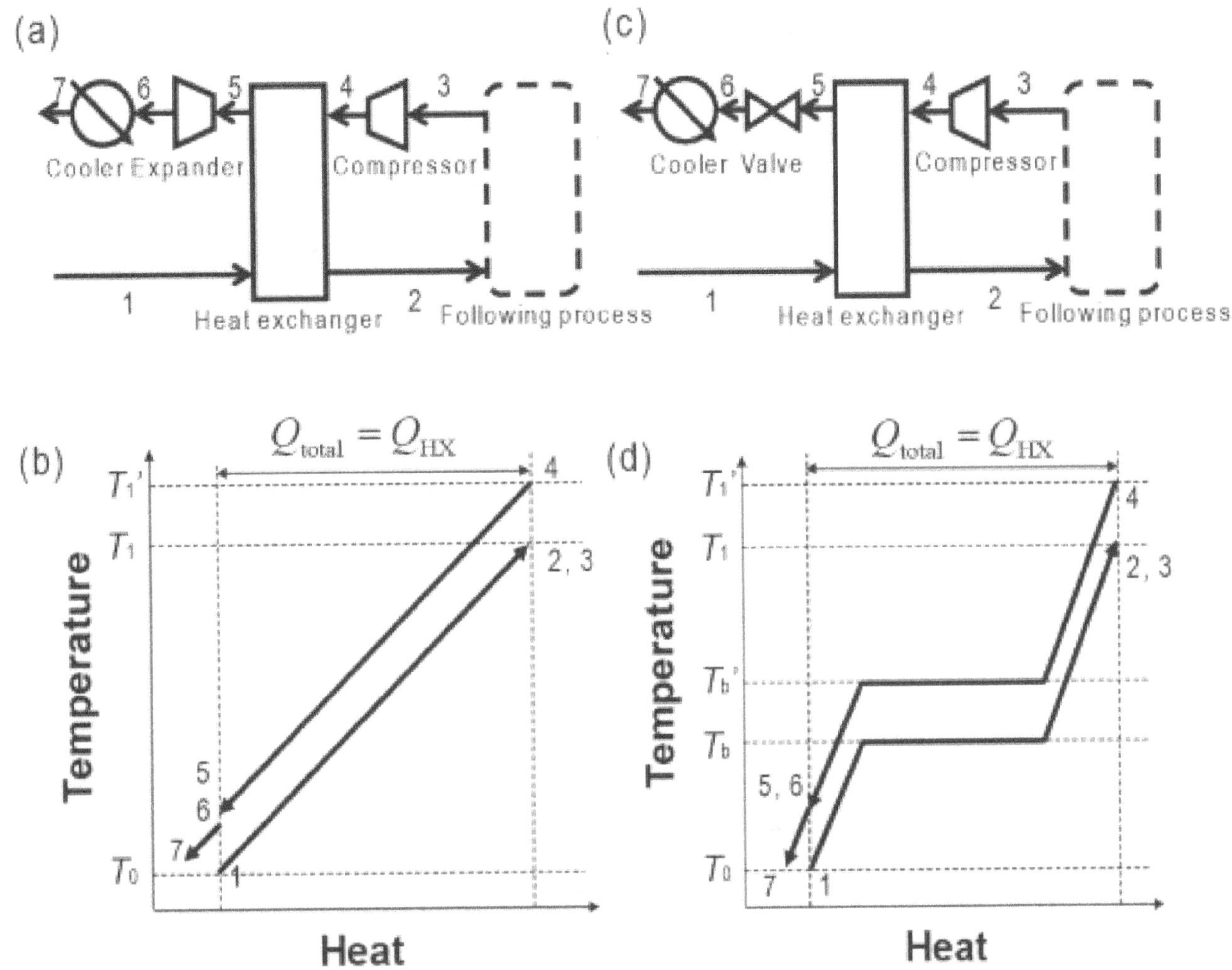

Fig. 3. Self-heat recuperative thermal process a) process flow of gas streams, b) temperature- heat diagram of gas streams, c) process flow of vapor/liquid streams, d) temperature-heat diagram of vapor/liquid streams

3.2 Process that involves the recovery and reuse of heat within a distillation system.

In order to decrease the amount of energy needed, a system has been designed that includes both the distillation column and the preheating section. This system expands the self-heat recuperative thermal process to distillation operations, as demonstrated in Figure 4 (Kansha et al., 2010a, 2010b). The distillation process can be categorized into two pieces: preheating and distillation. These portions are responsible for balancing the heating and cooling load through enthalpy and exergy analysis. The self-heat recuperation technique is utilized in both sections. In the preheating section, one of the streams originating from the distillation section is in the form of vapor, while the stream going to the distillation section consists of both vapor and liquid phases. This is done to ensure that the enthalpy of the incoming and outgoing streams in the section is balanced. The enthalpy of the feed and effluent streams in the preheating section is balanced in order to ensure that the enthalpy of the streams in the distillation section is automatically balanced as well. Therefore, the amount of heat required for the reboiler is the same as the amount of heat removed by the condenser in the distillation column. Thus, the sensible heat of the vapor and liquid in the incoming streams can be transferred to the sensible heat of the outgoing streams, while the heat required for vaporization can be transferred to the heat released during condensation in each segment.

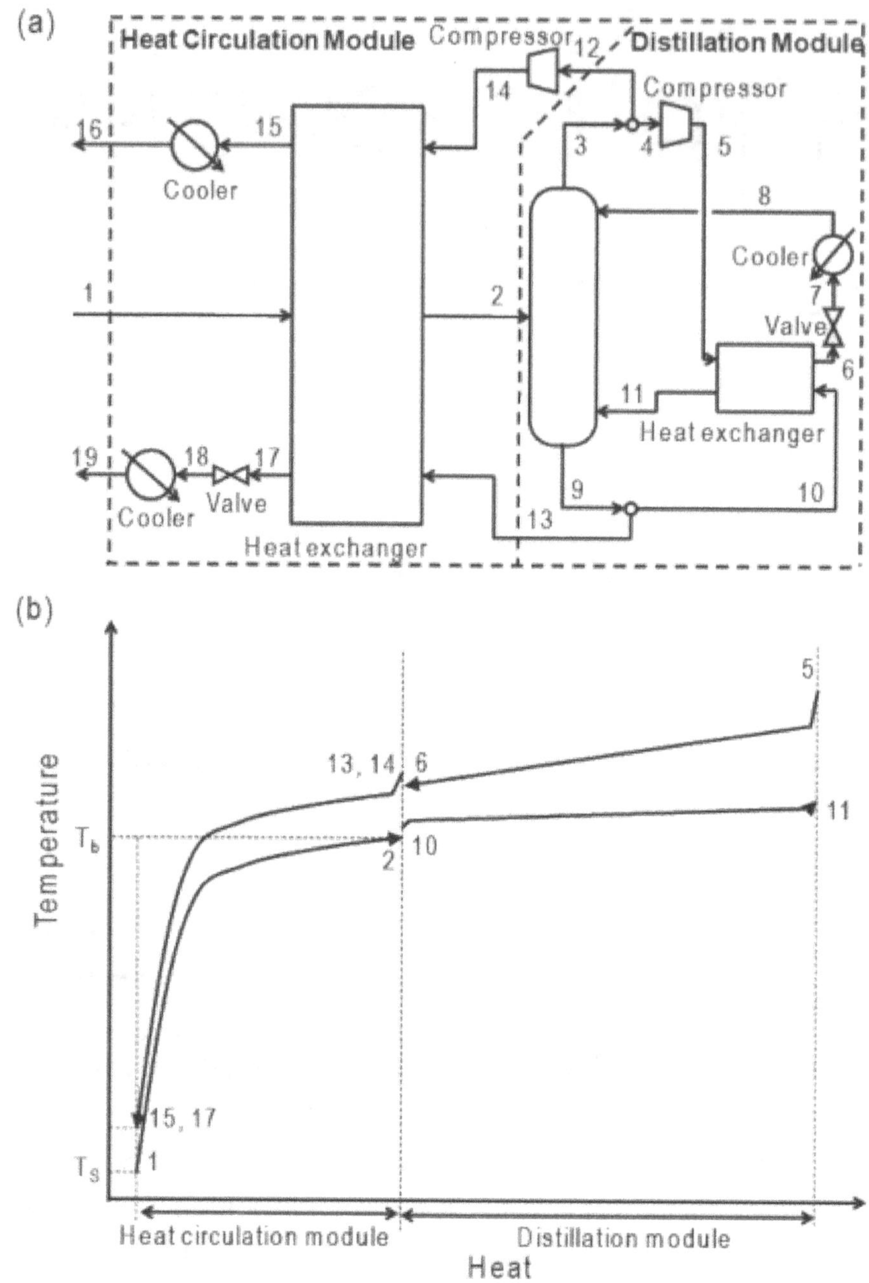

Fig. 4. Self-heat recuperative distillation process a) process flow diagram, b) temperature- heat diagram

Figure 4 (a) shows the structure of a self-heat recuperative distillation process consisting of two standardized modules, namely, the heat circulation module and the distillation module. Note that in each module, the summation of the enthalpy of the feed streams and that of the effluent streams are equal. The feed stream in this integrated process module is represented as stream 1. This stream is heated to its boiling point by the two streams independently recuperating heat of the distillate (12) and bottoms (13) by the heat exchanger (1→2). A distillation column separates the distillate (3) and bottoms (9) from stream 2. The distillate (3) is divided into two streams (4, 12). Stream 4 is compressed adiabatically by a compressor and cooled down by the heat exchanger (2). The pressure and temperature of stream 6 are adjusted by a valve and a cooler (6→7→8), and stream 8 is then fed into the distillation column as a reflux stream. Simultaneously, the bottoms (9) is divided into two streams (10, 13). Stream 10 is heated by the heat exchanger and fed to the distillation column (10→11). Streams 12 and 13 are the effluent streams from the distillation module and return to the heat circulation module. In addition, the cooling duty of the cooler in the distillation module is equal to the compression work of the compressor in the distillation module because of the enthalpy balance in the distillation module.

The effluent stream (12) from the distillation module is compressed adiabatically by a compressor (12→14). Streams 13 and 14 are successively cooled by a heat exchanger. The pressure of stream 17 is adjusted to standard pressure by a valve (17→18), and the effluents are finally cooled to standard temperature by coolers (15→16, 18→19). The sum of the cooling duties of the coolers is equal to the compression work of the compressor in the heat circulation module. Streams 16 and 19 are the products.

Figure 4 (b) shows the temperature and heat diagram for the self-heat recuperative distillation process. In this figure, each number corresponds to the stream numbers in Figure 4 (a), andTsandTbare the standard temperature and the boiling temperature of the feed stream, respectively. Both the sensible heat and the latent heat of the feed stream are subsequently exchanged with the sensible and latent heat of effluents in heat exchanger 1. The vaporization heat of the bottoms from the distillation column is exchanged with the condensation heat of the distillate from the distillation column in the distillation module. The heat of streams 4 and 12 are recuperated by the compressors and exchanged with the heat in the module. It can be seen that all the self-heat is exchanged. As a result, the exergy loss of the heat exchangers can be minimized and the energy required by the distillation process is reduced to 1/6–1/8 of that required by the conventional heat exchanged distillation process.

3.1.2 Self-heat recuperative azeotropic distillation is a method used for removing water from a substance.

Conventional azeotropic distillation processes, which have one distillation column for dehydration to separate ethanol and another to separate water from their mixture, are divided into three modules. The sum of the feed enthalpy is made equal to that of the effluent stream enthalpy in each module to analyze the heating and cooling loads of all process streams by following self-heat recuperation technology. According to this analysis, the recovery streams are selected and the internal heat of the process stream in each module can be recovered and recirculated using a compressor and heat exchanger through self-heat recuperation technology.

Figure 5 a) shows the structure of the self-heat recuperative azeotropic distillation module (Kansha et al. 2010c), consisting of three modules, namely, the first distillation module, the heat circulation module, and the second distillation module. In this self-heat recuperative distillation module, stream 1 represents a feed stream of the ethanol-water azeotropic mixture and stream 2 represents an entrainer (benzene and cyclohexane) feed stream. These streams are fed into the distillation column of the first distillation module. The vapor stream from the first distillation process is compressed adiabatically by a compressor (4→5). Subsequently, stream 5 is cooled in a heat exchanger (5→6), and the pressure and temperature of stream 6 are adjusted by a valve and a cooler (6→7→8). The liquid stream (8) is divided into two streams (9 and 10) in a decanter. Stream 9 consists mainly of the entrainer, which is recycled to the feed benzene (3). The bottom (11) of the distillation column is divided into two streams (12 and 14). Stream 14 becomes a product stream (pure ethanol). Stream 12 is heated in the heat exchanger and fed into the distillation column. In the heat circulation module, the effluent stream (10) from the first distillation module is heated in a heat exchanger and fed to the distillation column in the second distillation module. At the same time, the recycled stream, which is the distillate stream from the second distillation module, is adiabatically compressed by a compressor (18→27) and cooled by exchanging heat in the heat exchanger (27→28). The pressure and temperature of stream 28 are adjusted by a valve and cooler (28→29→30) and stream 30 is fed into the distillation column of the first distillation module as the recycled stream. Next, in the second distillation module, the feed stream (15) is separated into the distillate (16) and the bottoms (17) by the distillation column. The vapor distillate (16) is divided into two streams (18 and 19) by a separator. Stream 18 is recycled to the heat circulation module, while stream 19 is adiabatically compressed (19→20) and exchanged with the heat in a heat exchanger (20→21). The temperature and pressure of stream 21 are adjusted by a valve and a cooler (21→22→23), and then the effluent stream is fed into the distillation column. Subsequently, the bottom stream (17) from the distillation column is divided into two streams (24 and 25). Stream 25 is the product water. The other stream (24) is vaporized in the heat exchanger and fed into the distillation column (26).

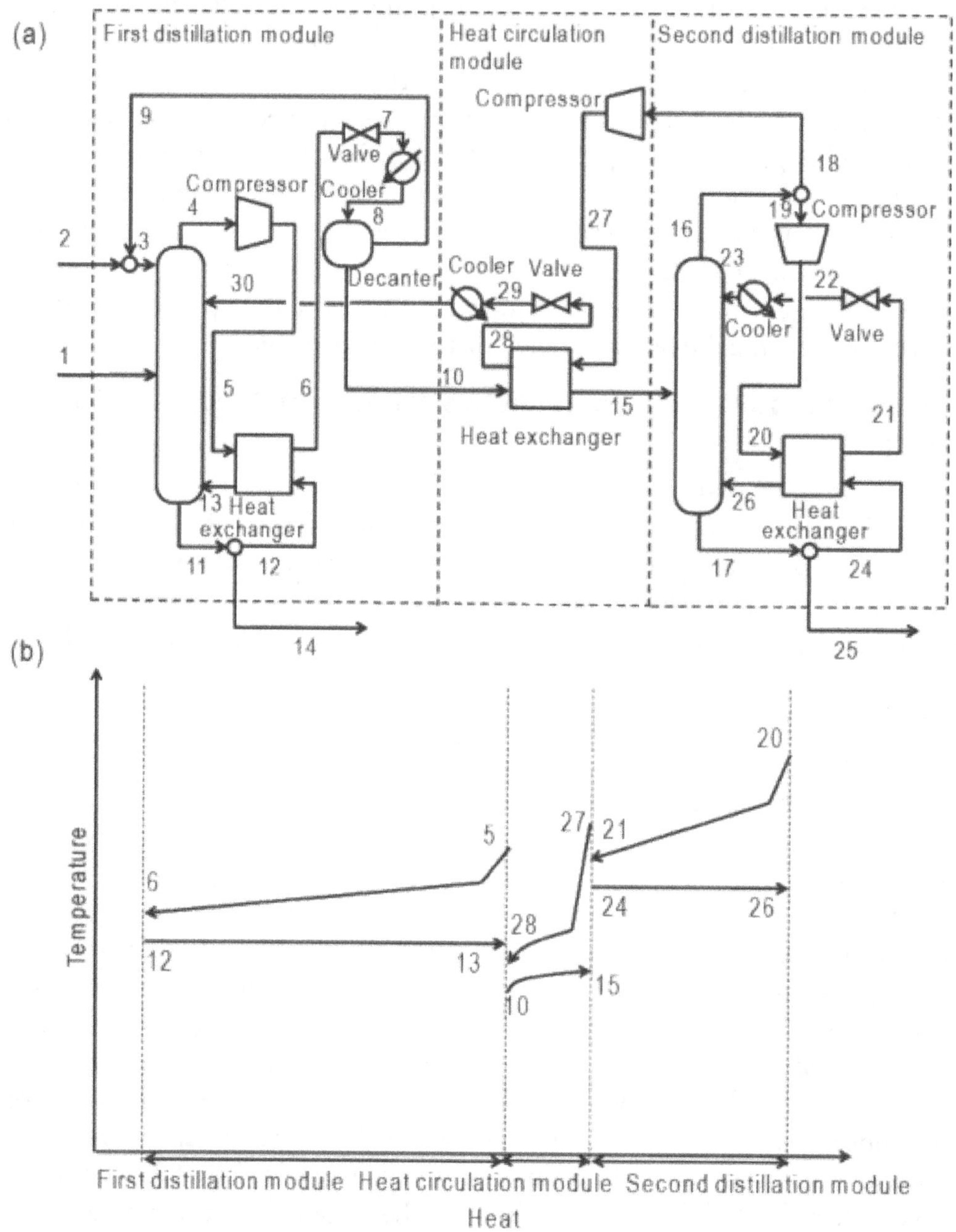

Fig. 5. Self-heat recuperative azeotropic distillation process for dehydration a) process flow diagram, b) temperature-heat diagram

Figure 5 b) shows a temperature–heat diagram for the self-heat recuperative distillation module for azeotropic distillation. Note that numbers beside the composite curve correspond to the stream numbers in Figure 5 a). It can be seen that the latent heats of the effluent streams are exchanged with those of the feed streams, as well as the sensible heats in each module, leading to minimization of the exergy loss in the heat exchangers. From this figure, it can be understood that all of process heat is recirculated without any heat addition and the total heating duty was covered by internal heat recovery. All of the compression work in each module was discarded into coolers in each module, because the sum of enthalpy in the feed streams was equal to that of the effluent streams in each module when using internal heat recovery. As this relationship indicates, the compression work was used for inducing heat recovery and circulation in each module and exhausted as low exergy heat. As a consequence, the energy required of the self-heat recuperative distillation module for azeotropic distillation is 1/8 of that of the conventional azeotropic distillation process.

3.1.3 Self-heat recuperative drying

Biomass resources typically have a significant moisture content, which results in increased transportation expenses, degradation during storage, and decreased thermal efficiency after conversion. Drying is an essential technique for effectively utilizing biomass (McCormick & Mujumdar 2008). Furthermore, the energy consumption in the production process is significantly increased by the need to dry biomass, as a result of the substantial quantity of energy required for water evaporation due to its high latent heat. Furthermore, this attribute of the drying process is identical to that of the heat and distillation processes. Recently, a drying process utilizing self-heat recuperation technology was proposed by Fushimi et al. in 2011. Figure 6 a) depicts a diagrammatic representation of a self-heat recuperative drying process. The moist sample is subjected to heating in a heat exchanger, transitioning from state 1 to state 2. The warmed moist sample and vapor are subsequently introduced into an evaporator (dryer) along with dry gas to facilitate evaporation (16). The energy required for evaporation is provided by highly heated steam and gas (7). The hot and arid specimen (3) is isolated and then cooled by the desiccated gas (15) (3→5). Once the unseparated sample has been removed to avoid its entry into the compressor, the evaporated steam and gas (4) undergo compression (7) through a compressor. The sensible and latent heats of the compressed steam and gas are transferred in the heat exchanger (7→8) and then directed to a condenser for the purpose of separating the water and gas. Subsequently, the water is drained (10). The valve and cooler are used to regulate the pressure and temperature of the drain water, increasing it from 10 to 12 and then to 14. Concurrently, a portion of the gas's pressure energy (9) is regained in an expander. The gas temperature is subsequently reduced using a cooler (13). The depleted gas can be reused as the gas supply (15). In order to utilize this gas as the dry gas feed, it is essential to introduce makeup gas to account for the loss caused by the dissolution of a tiny quantity of gas in the condenser water. In the context of a practical drying process, the sample that has been dried is promptly separated following the evaporation and then returned to the heat exchanger to make use of the heat. Nevertheless, in order to expedite the drying process (increase drying rate) and supply the necessary energy for drying, the evaporator is supplied with hot gas from the sample. An increase in gas flow rate results in a higher energy demand for compression due to two reasons: firstly, a bigger amount of gas needs to be compressed, and secondly, a lower partial pressure of steam necessitates a higher compression pressure for condensation. Therefore, it is necessary to optimize the gas flow rate.

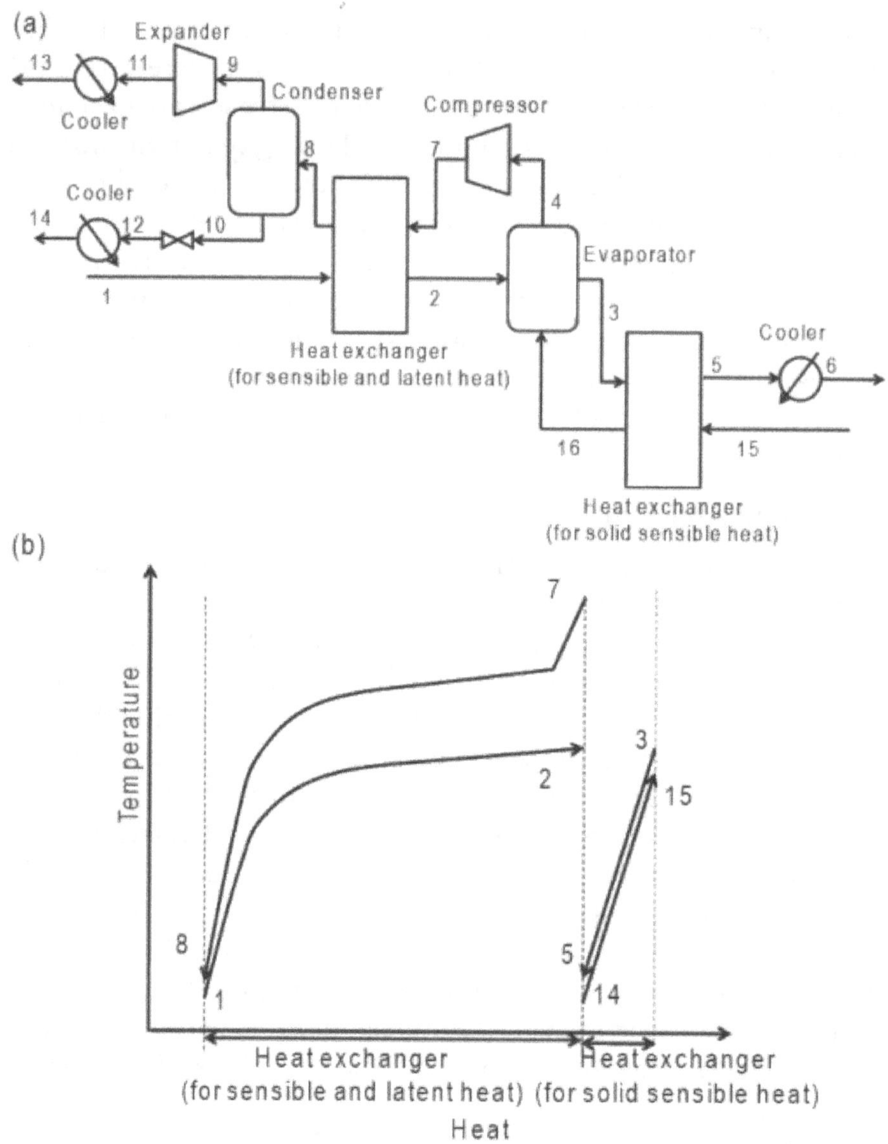

Fig. 6. Self-heat recuperative drying process for dehydration a) process flow diagram, b) temperature-heat diagram

Figure 6 b) shows a temperature-heat diagram of the self-heat recuperative drying process. Note that the numbers beside the composite curve in this temperature-heat diagram correspond to the stream numbers in Figure 6 a). It can be seen that the condensation heat of the steam in the effluent stream (7→8) is exchanged with the evaporation heat of the feed stream (1→2), as well as the sensible heats in a heat exchanger. At the same time, the heat of solid sample after evaporation is exchanged with the heat of the gas stream in the other heat exchanger and this heat is supplied to the feed solid sample. These result in the reduction of exergy loss in the heat exchangers. Based on this diagram, it is evident that all process heat is circulated without any more heat being added, and that the entire heating requirement is met through internal heat recovery. All the compression work performed in each module was rejected into coolers because to the fixed base conditions of the stream being set at standard settings. Therefore, in order to transfer heat inside the process stream and achieve self-heat recuperative drying, heat exchangers and a compressor are utilized. The energy consumption for this process is just 1/7 of that required for traditional heat recovered drying.

4. INTEGRATION WITH BIOMASS GASIFICATION

To adopt self-heat recuperative processes, it is necessary to generate power in substitution for heat energy. According to the energy balance shown in Figs. 1 and 2, much residue with insufficient heat value for utilization due to its high moisture content is produced during bioethanol production. By integrating the self-heat recuperative drying process with power generation, this wet biomass can be utilized for energy. In this section, an integrated system for self-heat recuperative bioethanol production with biomass gasification is introduced.

4.1 Biomass gasification and its impact on the system

One of the easiest ways to generate power from biomass is direct combustion of biomass in a boiler, wherein thermal energy is produced and power is generated from this thermal energy by using a steam turbine (boiler and turbine generator). However, energy conversion efficiency under this procedure is not good enough. To increase the conversion efficiency of energy from biomass to power, biomass gasification reaction is used. Gasification reactions can be divided into two mechanisms; pyrolysis and gasification by chemical reaction (partial oxidation, etc.) Biomass gasification normally passes through both of these. After passing through a series of gasification procedures, the gases are fed into a gas turbine, and then the power is generated. Gasification reactions are normally endothermic reactions, and must be provided with heat during reactions. However, the overall energy conversion efficiency will be increased compared with the boiler and turbine generator. In addition, a further increase in energy conversion efficiency, through a biomass-based integrated gasification combined cycle (IGCC) technology has been investigated (Bridgwater 1995).

It is currently assumed that the energy conversion efficiency of biomass through power generation and biomass gasification is 25%. The energy amount of the wet residue is 50 in Figs. 1 and 2. It assumed that half of the energy amount of this wet residue can be utilized for drying the biomass. According to the analysis of self-heat recuperative drying above, 1/8 of the amount of energy for water evaporation is required for power to dry this wet residue using self-heat recuperative drying. This means that power (8) can be generated and a part of this power (3) is used for drying, leading to 4% of the initial wet biomass being converted to power as net energy (5) from the wet residue as shown in Fig. 7.

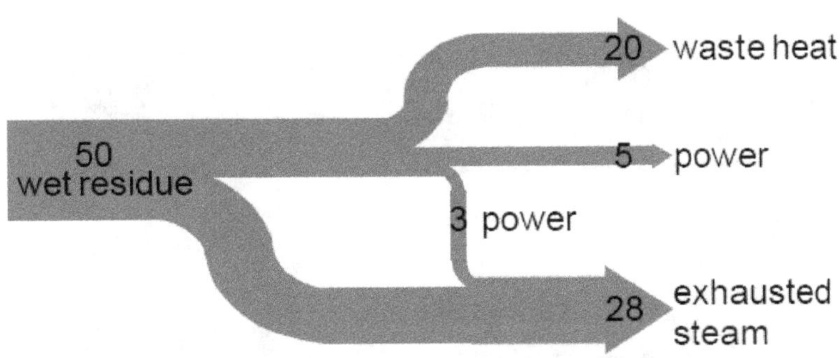

Fig. 7. Power generation from wet residue during bioethanol production

5. ENERGY BALANCE FOR SELF-HEAT RECUPERATIVE BIOETHANOL PRODUCTION

The same assumption as for section 2 is assumed; the amount of energy in the wet biomass feed stock is 100, 50% of the energy value of the wet biomass consists of the energy value of reactant sugars such as starch, cellulose and others, and the amount of energy of the original sugar component (50) transfers to ethanol (46) and heat (4) through chemical reactions (saccharification and fermentation) with water.

By applying the self-heat recuperative distillation and azeotropic distillation process to the distillation and dehydration process, the additional heat energy for distillation is converted to power. At the same time, the energy (23) in Figure 1 is reduced to 4. This value was estimated from the energy reduction results from the self-heat recuperative processes in section 3.

By integrating the aforementioned biomass gasification in section 4 with the self-heat recuperative processes introduced in section 3, bioethanol (46) and power (1) can be produced as co-products from wet biomass (100) during bioethanol production, as shown in Fig. 8. Wet residue (non-reactants contain a large amount of water, for which the higher heat value is almost equal to the required evaporation heat, leading to net heat value of 0) in Figs. 1 and 2 can be utilized as the energy supply. Thus, it can be understood that 46% of the energy of the wet biomass is transferred to the bioethanol and 1% of the energy to power.

Furthermore, the additional wet biomass (38) required to provide the distillation heat (23) is no longer necessary for this bioethanol production. Thus, power (4) can be generated from the additional wet biomass by using a self-heat recuperative drying process and biomass gasification, as shown in Fig. 9. As a result, 33% (= 46/138×100) of the energy of the wet biomass is transferred to bioethanol and 4% (= 5/138×100) is transferred to power for co- production. It can be said that this bioethanol production procedure achieves not only energy savings but also reduction of exergy dissipation for the whole process, leading to achievement of optimal co-production. In addition, substituting the azeotropic distillation process by dehydration uses a membrane separation. All of the self-heat recuperative processes and biomass gasification are applied to produce this energy. The energy required can be decreased to 4 as power, where the same assumptions as used for the results described above are used in the calculation, such that power generation efficiency from dry biomass is 25% and 75% of the energy required for the membrane separation process is provided by electricity. This value of power is the same as the energy required by applying self-heat recuperative processes to the distillation and dehydration processes. Although the energy required by membrane separation process is smaller than that of azeotropic distillation in the conventional processes, it becomes equal after applying the self-heat recuperative processes.

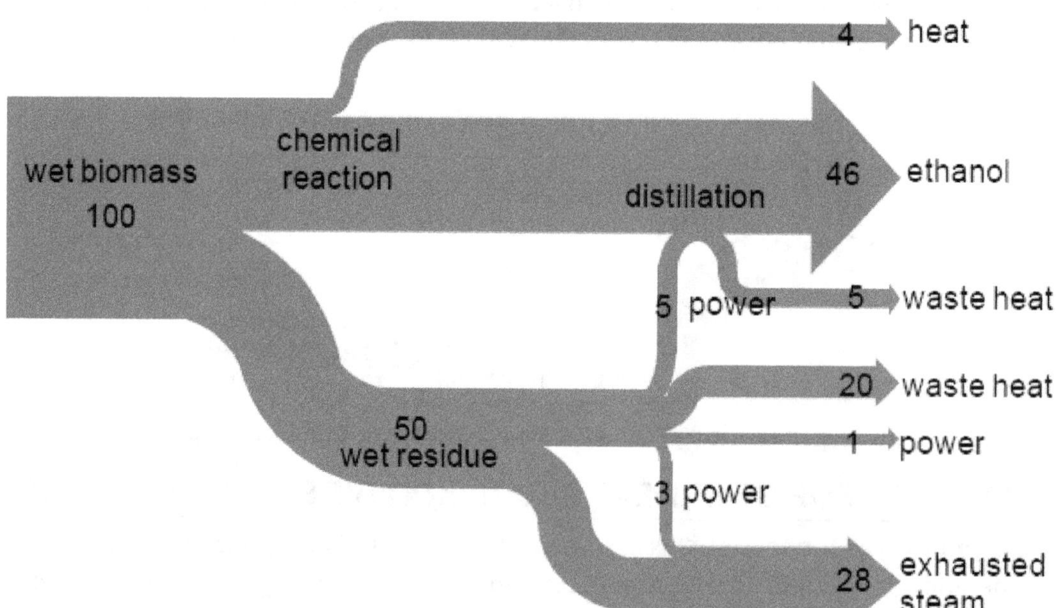

Fig. 8. Energy balance for bioethanol production with self-heat recuperation

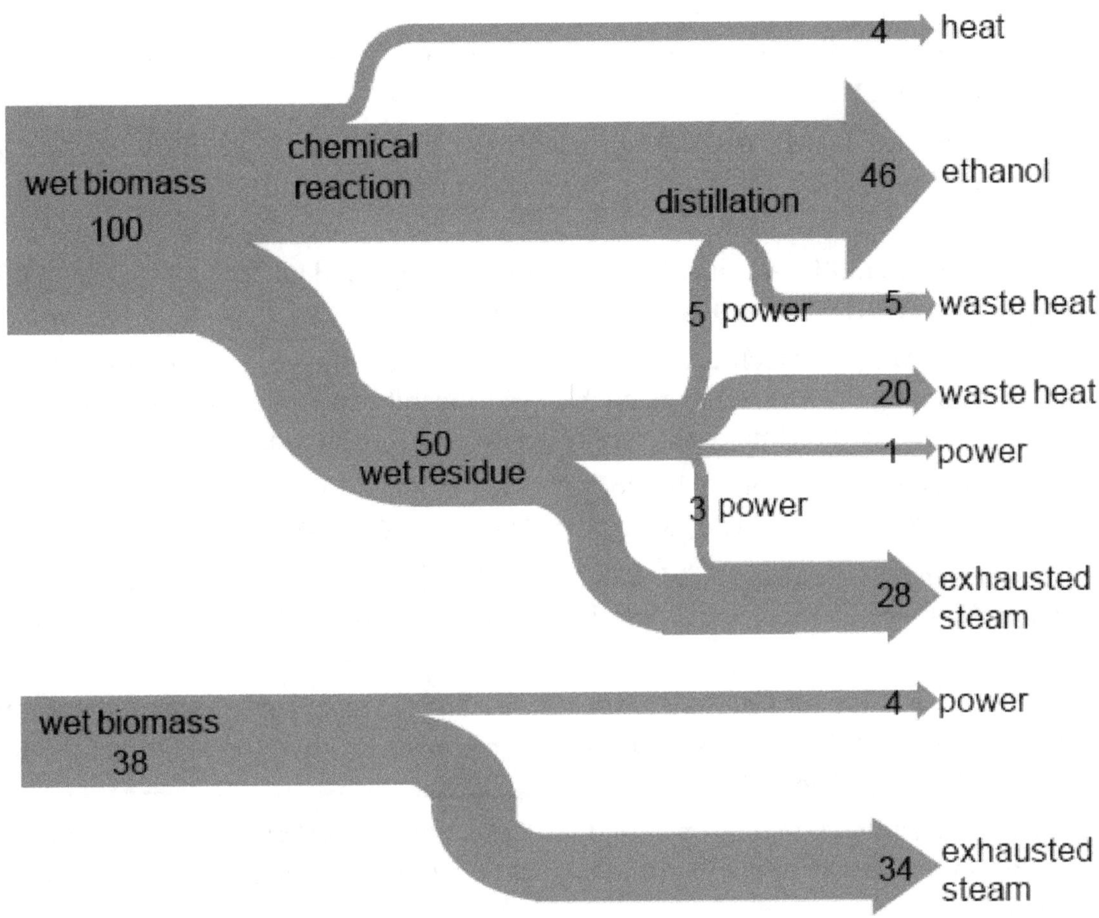

Fig. 9. Total energy balance for bioethanol production with self-heat recuperation

6. CONCISE SUMMARY

This chapter presents a recently developed self-heat recuperation technology and investigates the possibility of simultaneously producing bioethanol and power by combining self-heat recuperative processes with biomass gasification for power generation. The examination is conducted by analyzing energy balances. Upon analyzing the energy balance of conventional bioethanol production processes, it becomes evident that a significant amount of energy is utilized for the separation of water through distillation and drying. Consequently, the operational costs associated with bioethanol production are elevated, thereby restricting the potential societal benefits of bioethanol. By integrating self-heat recuperative methods into distillation, azeotropic distillation, and drying processes, both the energy demands and the amount of wasted residue can be significantly decreased. This is achieved by efficient heat circulation and the utilization of biomass gasification to convert the residue into a power source. Therefore, it has been demonstrated that the simultaneous generation of bioethanol and power is viable, allowing for the economic benefits of the bioethanol product. Ultimately, this technology is anticipated to facilitate the adoption of bioethanol and reduce worldwide CO_2 emissions.

REFERENCES

Baker, R.W. (2008). Membrane Technology, Introduction, In:Kirk-Othmer Separation Technology 2nd Ed. Vol. 2,A. Seidel, (Ed.), 446-502, John Wiley & Sons, ISBN 978-0- 470-12741-4, NJ, USA

Bridgwater, A.V. (1995). The technical and economic feasibility of biomass gasfication for generation, Fuel, Vol. 74, No. 5, pp. 631-653, ISSN 0016-2361

Cardona, C.A.; Sanchez, O.J. & Gutierrez, L.F. (2010).Process Synthesis for Fuel Ethanol Production, ISBN 978-1-4398-1597-7, FL, USA

Doherty, M.F. & Knapp, J.P. (2008). Distillation, Azeotropic and Extractive, In:Kirk-Othmer Separation Technology 2nd Ed. Vol. 1,A. Seidel, (Ed.), 918-984, John Wiley & Sons, ISBN 978-0-470-12741-4, NJ, USA

Fair J.R. (2008). Distillation, In:Kirk-Othmer Separation Technology 2nd Ed. Vol. 1,A. Seidel, (Ed.), 871-917, John Wiley & Sons, ISBN 978-0-470-12741-4, NJ, USA

Fushimi, C.; Kansha, Y.; Aziz, M.; Mochidzuki, K.; Kaneko, S.; Tsutsumi, A.; Matsumoto, K.; Yokohama, K.; Kosaka, K.; Kawamoto, N.; Oura, K.; Yamaguchi, Y. & Kinoshita, M. (2011). Novel drying process based on self-heat recuperation technology, Drying Technology, Vol. 29, No. 1, pp.105-110, ISSN 0737-3937

Hallale, N. (2008). Process Integration technology, In:Kirk-Othmer Separation Technology 2nd Ed. Vol. 2,A. Seidel, (Ed.), 837-871, John Wiley & Sons, ISBN 978-0-470-12741-4, NJ, USA

Huang, H.-J.; Ramaswamy, S.; Tschirner, U.W. & Ramarao, B.V. (2008). A review of separation technologies in current and future biorefineries, Separation and Purification Technology, Vol. 62, No. 1, pp.1-21, ISSN: 1383-5866

Kansha, Y.; Tsuru, N.; Sato, K.; Fushimi, C. & Tsutsumi, A. (2009). Self-heat recuperation technology for energy saving in chemical processes,Industrial and Engineering Chemistry Research, Vol. 48, No. 16, pp.7682-7686, ISSN 0888-5885

Kansha, Y.; Tsuru, N.; Fushimi, C.; Shimogawara, K. & Tsutsumi, A. (2010a). An innovative modularity of heat circulation for fractional distillation,Chemical Engineering Science, Vol. 65, No.1, pp.330-334, ISSN 0009-2509

Kansha, Y.; Tsuru, N.; Fushimi, C. & Tsutsumi, A. (2010b). Integrated process module for distillation processes based on self-heat recuperation technology,Journal of Chemical Engineering of Japan, Vol. 43, No. 6, pp. 502-507, ISSN 0021-9592

Kansha, Y.; Tsuru, N.; Fushimi, C. & Tsutsumi, A. (2010c). New design methodology based on self-heat recuperation for production by azeotropic distillation,Energy & Fuels, Vol. 24, No. 11, pp. 6099-6102, ISSN 0887-0624

Kansha, Y.; Kishimoto, A.; Nakagawa, T. & Tsutsumi, A (2011). A novel cryogenic air separation process based on self-heat recuperation,Separation and Purification Technology, Vol. 77, No. 3, pp. 389-396, ISSN 1383-5866

Kemp, I.C. (2007).Pinch Analysis and Process Integration A User Guide on Process Integration for the Efficient Use of Energy 2nd Ed., Elsevier, ISBN 13 978-0-75068-260-2, Oxford, UK

Matsuda, K.; Kawazuishi, K.; Hirochi, Y.; Sato, R.; Kansha, Y.; Fushimi, C.; Shikatani, Y.; Kunikiyo, H. & Tsutsumi, A. (2010). Advanced energy saving in the reaction section of the hydro-desulfurization process with self-heat recuperation technology, Applied Thermal Engineering, Vol. 30, No. 16, pp. 2300-2305, ISSN 1359-4311

McCormick, P.Y. & Mujumdar, A.S. (2008). Drying, In:Kirk-Othmer Separation Technology 2nd Ed. Vol. 1,A. Seidel, (Ed.), 984-1032, John Wiley & Sons, ISBN 978-0-470-12741-4, NJ, USA

Modla, G. & Lang P. (2008). Feasibility of new pressure swing batch distillation methods, Chemical Engineering Science, Vol. 63, No. 11, pp. 2856-2874, ISSN 0009-2509 Mousdale, D.M. (2010). Introduction to Biofules, CRC Press, ISBN 978-1-4398-1207-5, FL, USA Wynn, N.P. (2008). Pervaporation, In:Kirk-Othmer Separation Technology 2nd Ed. Vol. 2,A. Seidel, (Ed.), 533-550, John Wiley & Sons, ISBN 978-0-470-12741-4, NJ, USA

Zamboni, A.; Shah, N. & Bezzo, F. (2009a). Spatially Explicit Static Model for the Strategic Design of Future Bioethanol Production Systems. 1. Cost Minimization,Energy & Fuels, Vol. 23, No. 10, pp. 5121-5133, ISSN 0887-0624

Zamboni, A.; Shah, N. & Bezzo, F. (2009b). Spatially Explicit Static Model for the Strategic Design of Future Bioethanol Production Systems. 2. Multi-Objective Environmental Optimization,Energy & Fuels, Vol. 23, No. 10, pp. 5134-5143, ISSN 0887-0624

CHAPTER-15

CONVERSION OF NON-HOMOGENEOUS BIOMASS TO ULTRACLEAN SYNGAS AND CATALYTIC CONVERSION TO ETHANOL

1. INTRODUCTION: AN OVERVIEW

The development of biofuels is justified by the need to reduce greenhouse gas emissions, address growing energy prices, and provide security of supply. Nevertheless, the food prices documented in 2007 and 2008 had a significant impact on over 100 million individuals, leading to widespread undernourishment globally (Rastoin, 2008). The food crisis has arisen due to a combination of causes, including insufficient investment in agriculture, extensive speculation on agricultural commodities, and the conflict between biofuels and food production. According to estimates, the production of food will need to be increased by 50% by the year 2050 in order to provide for the 9 billion people residing on Earth (Rastoin, 2008). One of the difficulties of the next century is the recycling of carbon from residual trash to generate biofuels. Multiple firms are now working on creating technology that can turn leftover waste streams into syngas, which is then further processed into alcohols. "Green" ethanol is crucial for decreasing reliance on petroleum and offering environmental advantages, particularly in the fuel additive industry. Ethanol functions as both an oxygenate and an octane booster. The waste-to-syngas method offers an alternative solution to the food vs. fuel debate, while simultaneously reducing the amount of garbage in landfills and boosting the recovery of carbon. By employing this method, the amount of ethanol produced exceeds 350 liters per dry tonne of feedstock that is introduced into the gasifier (Enerkem's technology is used as an illustration). The residual heat, which is a byproduct of the process, is utilized within the process itself and can also be employed for external heating or cooling purposes. Enerkem Inc. is progressing the technique from small-scale testing to larger-scale testing, and ultimately to full-scale commercial use. Specifically, they are now constructing a facility in Edmonton, Alberta that will be capable of processing 12,500 kg/h of sorted and biotreated urban garbage. The economics of the process are advantageous at the mentioned commercial capacity due to the plant's modular architecture, moderate operational expenses, and the inclusion of a tipping fee for the residue that is fed into the gasifier. The initial section of this chapter will introduce the process of preparing feedstock, followed by the steps of gasification and gas conditioning. The performance of gasification for syngas production, which has the desired H2/CO ratio for downstream synthesis, is determined by the properties of the heterogeneous feedstock. The subsequent section of the chapter will focus on the process of methanol synthesis in a three-phase reactor utilizing syngas. The final section of the chapter will concentrate on the catalytic processes involved in the conversion of methanol into bio-ethanol.

2. SYNTHESIS GAS (SYNGAS) PRODUCTION BY GASIFICATION

2.1. Characteristics and composition of heterogeneous wastes as feedstock

Biomass refers to organic matter obtained from plants or animals that possesses potential chemical energy. An example of biomass is wood, which was historically the initial source of fire utilized by humans and continues to be employed for cooking and heating purposes. On a global level, biomass currently ranks as the fourth largest energy source, but it possesses the potential to surpass all others and claim the top spot. According to Prins et al. (2005), photosynthesis has the capacity to store biomass energy annually that is 5-8 times greater than the current global energy consumption. The fundamental process of photosynthesis involves the conversion of carbon dioxide and water into glucose and oxygen. This process is endothermic, meaning it requires an input of energy, which is provided by photons. Examples of biomass include agricultural wastes and forest industry byproducts such as branches, straw, stalks, and sawdust. An exemplary instance of residual agricultural biomass is the byproduct of sugar cane ethanol production in Brazil, which yields 280 kg of residual bagasse with a dry solids content of 50%. Lignocellulosic materials can be harvested and reclaimed due to their inherent energy value (Ballerini and Alazard-Toux, 2006). However, it is crucial to leave a portion of this material in situ as it contributes to maintaining soil fertility. It is crucial for governments and communities to recognize that hydrocarbon-based waste material represents an untapped energy resource that should be utilized. Waste can exist in either solid or liquid states. The options for disposal include landfilling, incineration, or conversion. Materials such as municipal solid waste, electrical transmission poles, railroad ties treated with creosote, sludge from wastewater treatment, and wood from construction and demolition operations containing paints and resins, all contain carbon that can be utilized in bio-refineries. Enerkem is currently constructing a bio-refinery in Edmonton, AB, which aims to valorize these materials. To evaluate the use of residual biomass or Municipal Solid Waste (MSW) as a raw material for producing bio-ethanol, it is necessary to have a fundamental knowledge of the composition of the raw material and the specific characteristics that determine its effectiveness as a feed in the gasifier. The key factors to consider are: moisture content, ash content, volatile matter content, elemental composition, and heating value. Biomass moisture content refers to the amount of water present in the material, expressed as a percentage of the substance's weight. This weight can be expressed either on a wet basis or on a dry basis. When the moisture content is measured on a "wet" basis, the weight of the water is stated as a percentage of the combined weight of the water, ash, and dry-ash free matter. In order to optimize the gasification reaction, it is occasionally essential to achieve a specific amount of dryness in the feedstock. Indeed, a greater amount of moisture is transported through increased oxygen consumption in order to maintain the optimal temperature in the gasifier. The temperature of gasification plays a critical role in determining the efficiency of the process. An optimal condition is present when there is precisely the necessary amount of oxygen required to fully carry out the gasification reaction. The biomass with 20% moisture has a temperature of approximately 660°C, while the biomass with 10% moisture has a temperature of around 695°C (Prins et al., 2005). According to van der Drift et al. (2001), increasing the amount of oxygen will result in an increase in the creation of carbon dioxide and a decrease in gasification efficiency. Consequently, the heating value of the synthesis gas will decrease. Insufficient oxygen will facilitate the reduction of carbon, resulting in an elevation in methane production.

The inorganic component, often known as the ash content, can be quantified using the same method as the moisture content. Typically, the ash content is measured based on the dry weight of the sample. Both the overall ash content and chemical makeup are crucial factors in the gasification process. The ash's behavior during combustion and gasification is influenced by its composition. For instance, the presence of liquefied ash can lead to complications in both combustion and gasification reactors. The issues might range from ash removal blockage caused by slagging ash to serious operational concerns in fluidized bed systems. Therefore, it is essential to measure the ash melting point. Volatile matter pertains to the portion of biomass that is liberated when the biomass is subjected to temperatures beyond its dehydration thresholds. Through the heating process, the biomass undergoes decomposition, resulting in the formation of volatile gasses and solid char. Biomass generally has a significant amount of volatile matter, reaching up to 80%, while coal has a relatively low volatile matter content, less than 20%. The elemental makeup of the ash-free organic component of residual biomass exhibits a high degree of uniformity. The primary constituents consist of carbon, oxygen, and hydrogen. Additionally, a minor amount of nitrogen and sulfur is often present in most biomass. Table 1 displays the basic makeup of biomass as determined from ultimate analysis.

Table 1. Elementary composition of residual biomass

Element	Wt% (dry basis)
Carbon	44 – 51
Hydrogen	5.5 - 6.7
Oxygen	41 – 50
Nitrogen	0.12 - 0.6
Sulfur	0 - 0.2

The heating value of a fuel is an indication of the energy chemically bound in the fuel with reference to a standardized environment. The standardization involves the temperature, state of water, and the combustion products. The calorific value is presented as the higher heating value and the lower heating value. The higher heating value represents the heat release per unit of mass when the material (at 25°C) is completely oxidized to carbon dioxide and water and then returned to 25°C. A calorimetric bomb is the standard instrument used to measure this value. The lower heating value is not taking into account the energy supplied by the condensation of water (latent heat of vaporization of water at 25°C which is 2440 kJ/kg). The water includes moisture from the feedstock and the product from the reaction between oxygen and hydrogen comprised in the raw material (Borman and Ragland, 1998). Basic and complementary information about biomass intended for gasification can be obtained via the proximate and ultimate analyses.

Proximate analysis measures moisture content, volatile matter, fixed carbon, ash content and calorific value. Ultimate analysis provides information about elementary composition of the biomass in weight percentage of carbon, hydrogen, oxygen, sulphur and nitrogen. The carbon to hydrogen ratio in the feedstock has a direct impact on the syngas, more particularly on the ratio of $H2/CO$ (Higman and van der Burgt, 2008).

Table 2 (depicted below) presents a comparison of different feedstock properties. Moisture content was provided after drying of feedstock and compositions are approximated (Ciferno and Marono, 2002).

Table 2. Properties of biomass feedstocks

Analysis		Sawdust	Bagasse	Switchgrass	Straw	Bituminous Coal	Dry Sewage (MSW)
Utlimate Analysis (% w/w) dry basis	C	50	48	43	43,5	61.5	20.5
	H	6,3	6.0	5.6	4.2	4.2	3.2
	O	43	42	46	40.3	6	17.5
	N	0,8	-	0.5	0.6	1.2	2.3
	S	0,03	-	0.1	0.2	5.1	0.6
Proximate Analysis (% w/w) dry basis	Moisture	7.8	1	8.4	7.6	8.7	4.7
	Volatiles	74	80	73	68.8	36.1	41.6
	Fixed Carbon	25.5	15	13.5	13.5	42	2.3
	Ash	0.03	4	4.5	10.1	21.9	56
	HHV (MJ/kg)	19.3	17	15.4	17	27	8

Note that higher calorific value formulas, refering to equation (Eq. 1), were developed from the composition of the feedstock. Dulong formula is one of them (Higman and van der Burgt, 2008). However, numerous correlations were developed and reported in literature. For instance, experiments on more than 200 species were performed at the Indian Institute of Technology (Bombay) and the following equation was derived from empirical data.

$$HHV\ (MJ/kg) = 0.3491C + 1.1783H - 0.1034\ O - 0.0211\ A + 0.1005S - 0.0151N \quad (1)$$

Where C is the weight fraction of carbon, H of hydrogen, O of oxygen, A of ash, S of sulfur and N of nitrogen. This correlation offers an average absolute error of 1.45 % (Channiwala and Parikh, 2002).

The inorganic component, often known as the ash content, can be quantified using the same method as the moisture content. Typically, the ash content is measured based on the dry weight of the sample. Both the overall ash content and chemical makeup are crucial factors in the gasification process. The ash's behavior during combustion and gasification is influenced by its composition. For instance, the presence of liquefied ash can lead to complications in both combustion and gasification reactors. The issues might range from ash removal blockage caused by slagging ash to serious operational concerns in fluidized bed systems. Therefore, it is essential to measure the ash melting point. Volatile matter pertains to the portion of biomass that is liberated when the biomass is subjected to temperatures beyond its dehydration thresholds. Through the heating process, the biomass undergoes decomposition, resulting in the formation of volatile gasses and solid char. Biomass generally has a significant amount of volatile matter, reaching up to 80%, while coal has a relatively low volatile matter content, less than 20%. The elemental makeup of the ash-free organic component of residual biomass exhibits a high degree of uniformity. The primary constituents consist of carbon, oxygen, and hydrogen. Additionally, a minor amount of nitrogen and sulfur is often present in most biomass. Table 1 displays the basic makeup of biomass as determined from ultimate analysis.

2.2. Production and purification of synthesis gas

Gasification is a process that generates a synthetic gas, also known as syngas, which is mostly made up of hydrogen, methane, and carbon monoxide (Marie-Rose et al., 2011, Villano et al., 2010). The process of gasification was initiated during the late 1800s, primarily for the purpose of producing town gas used for street lighting and cooking. Additional applications included thermal processes for heating, the manufacturing of raw materials for the chemical sector, and power generation. During the Second World War, when oil supplies were scarce, wood gasification was employed for transportation as well as for generating heat and power. The energy crisis of the 1970s, specifically the oil shocks of 1973 and 1979, served as a catalyst for the advancement of gasification technology (Higman and van der Burgt, 2008). The fundamental principle of gasification is to chemically react a carbon-containing feedstock with steam and oxygen, which can be air, enriched air, or pure oxygen. The chemical processes that occur most frequently during gasification are as follows (Higman and van der Burgt, 2008). Thermal decomposition, also known as pyrolysis, involves the breakdown of a substance by dehydration and breaking processes. This process produces gases, intermediate vapours, and carbon structures referred to as "char". The "char" undergoes partial oxidation, resulting in the production of carbon monoxide (CO) and carbon dioxide (CO2). This process generates heat to facilitate the otherwise endothermic reactions.

The partial oxidation

$$C + \frac{1}{2} O_2 = CO \qquad -111 \text{ MJ/kmol} \qquad (2)$$

$$CO + \frac{1}{2} O_2 = CO_2 \qquad -283 \text{ MJ/kmol} \qquad (3)$$

$$H_2 + \frac{1}{2} O_2 = H_2O \qquad -242 \text{ MJ/kmol} \qquad (4)$$

Steam-carbon, i.e. the water-gas reaction, that converts carbon structures into H2and CO.

The water gas reaction

$$C + H_2O = CO + H_2 \qquad 131 \text{ MJ/kmol} \qquad (5)$$

Steam reforming of intermediates formed by thermal decomposition.

The Steam methane reforming reaction

$$C_xH_y + xH_2O = x\,CO + (\tfrac{1}{2} y + x)\, H_2 \qquad (7)$$

The steam reforming reaction

$$C_xH_y + xH_2O = x\,CO + (\tfrac{1}{2} y + x)\, H_2 \qquad (7)$$

The dry reforming reaction

$$C_xH_y + xCO_2 = 2x\,CO + (\tfrac{1}{2} y + x)\, H_2 \qquad (8)$$

Conversion of Non-Homogeneous Biomass to Ultraclean Syngas and Catalytic Conversion to Ethanol. Reactions involving CO_2 and H_2 with carbon and with intermediates are kinetically slower than the steam induced reactions at the conditions used in gasifiers.

The Boudouard reaction

$$C + CO_2 = 2CO \qquad 172 \text{ MJ/kmol} \qquad (9)$$

The methanation reaction

$$C + 2H_2 = CH_4 \qquad -75 \text{ MJ/kmol} \qquad (10)$$

Water gas shift reactions that lead to a desired H2/CO ratio.

The Water gas shift reaction

$$CO + H_2O = CO_2 + H_2 \qquad -41 \text{ MJ/kmol} \qquad (11)$$

Steam is hence chemically involved in the gasification, but it also served as moderating agent to limit the temperature. Carbon dioxide could also be used to this purpose. Other elements in feedstocks such as sulphur, nitrogen and chlorine will lead to different molecular species contaminating the syngas such as H2S, COS, NH3,HCN and HCl.

Different types and configurations of reactors can be used to produce syngas: fixed bed, entrained flow gasifier and bubbling fluidized bed. Fixed bed gasifier (essentially slowly moving-beds) can be updraft or downdraft.

The Updraft fixed bed gasifier is probably the oldest technology. Biomass or coal are fed from the top and move slowly by gravity. A grate supports the material while oxidizing the gas in its way upwards. The synthesis gas is withdrawn at the top of the reactor and ash from the bottom. There is a combustion zone in the bottom where char combusts in contact of oxygen to form CO_2 . Char reacts with carbon dioxide to form two moles of carbon monoxide (Eq. 9). A zone of pyrolysis take place a little higher in the fixed bed because all oxygen has been consumed at this level. Finally, hot synthesis gas dries the feedstock at its entrance (Ciferno and Marono, 2002,Paes, 2005). This simple proven process has as drawback the production of a tar-rich syngas. Downdraft fixed bed gasifier employs co-current flow of oxidant and feedstock. This configuration has low tar formation as an outcome. Nevertheless, a part of the carbon remains unconverted (Ciferno and Marono, 2002).

Entrained flow gasifier utilizes a blast of oxidant in a co-current arrangement with the feed grounded to ensure carry-over. A high temperature is necessary because of the short residency time, counted in seconds. More oxygen is thus required to reach higher temperature (Higman and van der Burgt, 2008). With this type of gasifier, the particles entrained are passed to a cyclone and a riser to bring this material again through the bottom of the gasifier.

Bubbling fluidized bed gasifer is designed to keep the particles in suspension. This technology is based on a configuration where feedstock is added continuously to a bed of alumina, olivine (iron and magnesium orthosilicate) and/or other material serving as a heat carrier. This type of gasifier has high heat/mass transfer and conversion and is able to treat a large range of particles sizes and heterogeneous feedstock. Fluidized bed gasifier has the following advantages: flexibility and easiness for control and maintenance. Figure 1illustrates an overview of the process developed by Enerkem Inc based on a bubbling fluidized reactor.

Feeding is assured by conveyor, lock hoppers, rotary valves and cooled feeding screw. Oxygen and steam are added through the bed in order to fluidize the medium. Low severity conditions are used in the bed, i.e. pressure lower than 4 atm and temperature lower than 750°C (Chornet et al., 2010). The reactor itself is made from carbon steel with insulation and refractory.

Ash is collected at the cyclones and larger inert material exits at the bottom of the gasifier. Ash can be composed of silica, alumina, potassium, phosphorus, sodium, magnesium, ferric oxide, and smaller amounts of titanium oxide and sulfur compounds (Higman and van der Burgt, 2008).

CaO/MgO is also introduced in the gasifier for neutralising a part of HCl, H2S and COS formed. Literature reports that the use of calcined dolomite and NiO-loaded calcined dolomite helps in the reduction of tar and char formation (Corujo et al., 2010). This dolomite contained 24,6 % of Ca, 19,7 % of Mg and 0,035 % of Fe on a weight basis. Calcination of dolomite produces magnesium and calcium oxides. Corujo et al. have stated that the use of NiO-loaded calcined dolomite catalysts increased the total product gas volume by 30% and decreased the rate of char and tar formation leading to a higher gas energy yield. In fact, bed materials are selected for two criteria: attrition resistance and possible catalytic activity in hydrocarbon and tar reforming. Olivine was compared to silica sand, dolomite, ash, magnetite and iron ore. With the use of Olivine, higher conversion of toluene was observed during steam gasification at 900°C (Rauch et al., 2004). Comparison of Ni/olivine catalyst with olivine alone demonstrated a higher activity, a higher selectivity to hydrogen and carbon monoxide and a lower carbon deposition. These observations were explained by an enhanced steam adsorption on magnesium oxide facilitating surface carbon gasification rates and by nickel dilution effect, associated with Fe-Ni alloys, ensuring stable catalyst activity (Swierczynski et al., 2007)

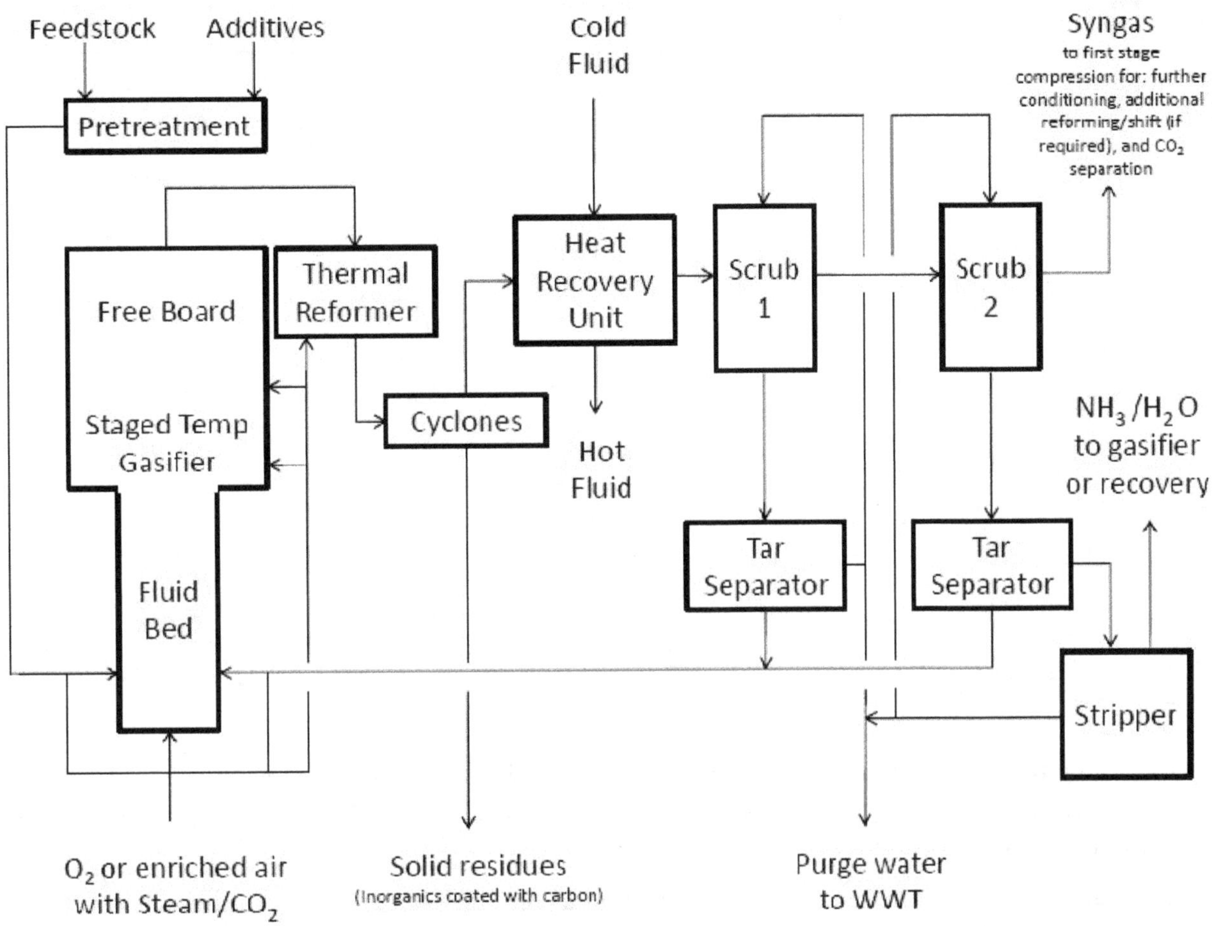

Figure 1. Schematic diagram of Enerkem's process

Leaving the freeboard of the gasifier, the syngas is composed of: H2, CO, CO_2 , H2O N2,hydrocarbons, tar, fines, char and contaminants. Tar is a complex blend of condensable hydrocarbons, which includes single to five-rings aromatic compounds along with oxygen-containing hydrocarbons and complex PAH (Yan et al., 2006). The free board is sized appropriately for disengagement of alumina and/or other solid fluidized bed materials. A reforming zone is used for converting hydrocarbons, tar and char to obtain more syngas (see Eq. 8,Eq. 9,Eq. 10andEq. 11). Since thermal decomposition of hydrocarbons requires activation and that reforming reactions are endothermic, heating the gas, by direct or indirect manner, is essential. For example, if oxygen is injected directly, it must be added carefully to avoid lowering the product gas heating value and cold gas efficiency. However, with just the right amount of the oxidizing gas, selective oxidation of methane, ethylene and propylene will occur (Villano et al., 2010). Target temperatures are in the 750°C to 1200°C range. Table 3 depicts general compositions of syngas produced out of some previously mentioned biomass.

Table 3. Solid, liquid and gas ratio as well as gas composition following gasification of common residual feedstocks.

Biomass	MSW	MSW	Sludges*	Res. Forest/Agriculture	Typical biomass
Conditions	700 °C, Fixed bed gasifier, Dolomite (Catalyst),	900 °C, Fixed bed gasifier, Dolomite (Catalyst), Steam (oxidizing agent)	829 °C Fluidized bed	900 °C, Fixed bed gasifier, Dolomite (Catalyst), Steam (oxidizing agent)	O_2 / Steam / CO_2 low severity gasification + Thermal cracking and reforming at higher severity
Char (%w/w)	19.15	12.65	n.d	n.d	n.d
Tar (%w/w)	12.94	2.62	n.d	n.d	n.d
Gas (%w/w)	94.52	145.23	n.d.	n.d.	n.d.
Gas composition	% mol	% mol	% mol	% mol	% mol
H_2	16.92	36.98	17.23	46.2	23-27
CO	20.33	27.37	26.17	33.2	21-23
CO_2	35.28	20.78	45.64	16.1	38-44
CH_4	21.44	9.94	7.96	4.4	6-8
C_2	6.03	4.93	3.00	0.1	-
C2H4	n.d	n.d	n.d	n.d	Traces
C2-C5	n.d	n.d	n.d	n.d	0.2-0.5
C5-C10	n.d.	n.d.	n.d.	n.d.	Traces
Reference	(He et al., 2009)	(He et al., 2009)	(Van der Drift et al., 2001)	(Corujo et al., 2010)	(Marie-Rose et al. 2010)

* Demolition wood + paper residue sludge
n.d. not determined

Increasing temperature cause the decomposition of char and tar by thermal cracking and steam reforming reactions. Moreover, this has a direct impact on the production and composition of the gas as seen fromtable 3(He et al., 2009)

Cyclones are used for entrained ash removal. The latter having typically a diameter greater than 10 microns. Efficiency of this cyclone system is from 90 to 95% and mostly recuperates char. Following the cyclone and recuperation of ash, the next unit operation is a heat exchanger. The latter cools the crude synthesis gas and allocates heat recovery for other application on the process using thermal oil or liquid water which turns to steam.

Note that the reforming step mentioned above could be by-passed directly to the cyclones if the feedstock contains inorganic materials that form eutectics with low melting point Indeed, high temperature in this situation would have as a consequence scaling and fouling of the walls of the system.

Quenching is required for further cooling and removal of condensable materials, tars and fines. In the following step, a wet venturi scrubber with alkaline water is used for neutralisation of acid gases as H2S and HCl. A coalescer/demister at the exit removes fines particles and mist.

A second scrubbing step involving neutral or slightly acidic water is used in order to capture ammonia, trace tars, residual fines and impurities such as chlorine and metals. After this cooling step, the cooled purified syngas is around 30 °C.

Water is recirculating after proper handling. The water loop includes knockout drums and separation of tars and fines particles with air flotation, decanters and/or centrifuges. The skimmed phase recovered (overflow), enclosing tars and fines particles, is emulsified and recycled to the gasifier. The underflow, containing heavier organics and particles, is also re-injected to the gasifier. A part of the recirculated water is withdrawn from the system to maintain the balance of water and contaminants. This purge is subjected to water treatment in the objective of meeting below the required environmental standards. Note that ammonia is separated from the water by steam stripping above 100°C and at about 1 to 3 atm.

The ammonia is sent back to the gasifier producing hydrogen and nitrogen.

Subsequently, an extra step can be implemented to convert light hydrocarbons into a higher proportion of CO and H_2, while also fine-tuning the molar ratio of H2/CO. Afterwards, the syngas undergoes filtration and adsorption processes to remove moisture, as well as eliminate any remaining chlorine, sulfur, and other impurities such metal carbonyls. An ultra-clean synthesis gas is produced, and then compression is used to eliminate carbon dioxide with the assistance of either amine-based methods or cold methanol.

3. SYNGAS CONVERSION INTO METHANOL

3.1. Thermodynamic consideration

Hydrogen and carbon monoxide, which are the primary constituents of synthesis gas, are commonly referred to as the fundamental elements of C1 chemistry. The conversion of syngas to liquid fuels, as well as the conversion rates, is directly influenced by the composition of the catalyst. Syngas can be effectively transformed into various compounds such as alcohols and aldehydes (Figure 2).

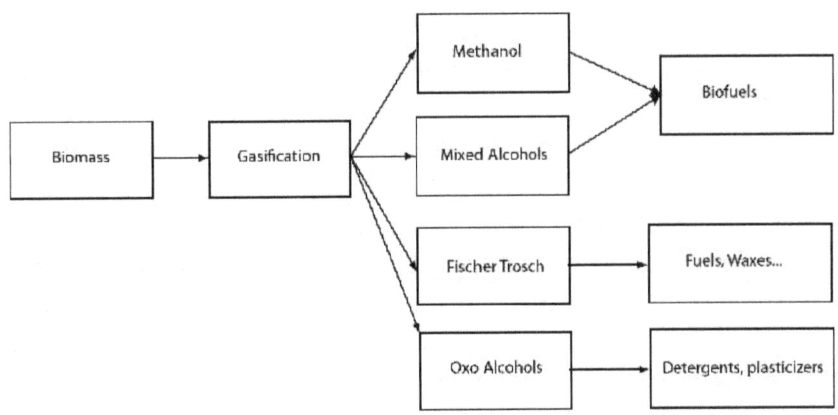

Figure 2. Examples of application for syngas produced from biomass (Higman and van der Burgt, 2008)

Although many routes are available, the most promising route at industrial level is the production of methanol since the synthesis yields are the highest. The conversion to synthetic liquids as methanol is strongly influenced by thermodynamic factors (Rostrup-Nielsen, 2000).

$$CO + 2H_2 = CH_3OH \quad -90.64\ KJ/mol \tag{12}$$

$$CO_2 + 3H_2 = CH_3OH + H_2O \quad -49.67\ KJ/mol \tag{13}$$

$$CO + H_2O = CO_2 + H_2 \quad -41.47\ KJ/mol \tag{14}$$

For methanol synthesis, a stoichiometric ratio defined as ($H2-CO_2$)/($CO+CO_2$), of about 2 is preferred, which implies that there should be just the stoichiometric amount of hydrogen needed for methanol synthesis. For kinetic reasons and in order to control by-products formation, a value slightly above 2 is normally preferred (Dybkjr and Christensen, 2001). Moreover, methanol synthesis is subjected to a thermodynamic equilibrium that limits the process to low conversion per pass and therefore implies a large recycle of unconverted gas. The reaction is strongly exothermic and consequently requires significant cooling duty.

Different phenomena exist at high and low pressure conditions. As example, when the pressure is relatively low, increasing temperature, CO conversion is not monotonic, and the trend is that of an increase followed by a decrease with the maximum conversion appearing near 250°C. This phenomenon is in agreement with many works in the literature (Li and Inui, 1996,Liaw and Chen, 2001,Wang et al., 2002). As the reaction temperature increases, the reaction rate gets higher and leads to the increase of CO conversion. However, methanol synthesis is an exothermic reaction and low temperature is more beneficial considering equilibrium. The conversion does not continue to increase due to the thermodynamic limitation and a decrease trend will even appear. When the system pressure is relative high, because of the relatively low CO conversion, the system is far from the thermodynamic equilibrium and under the control of reaction kinetics. In this case, CO conversion increases monotonically with an increase in temperature. From a mechanistic point of view for the methanol synthesis, two main reactions need to be in line with real world situation: CO hydrogenation and CO_2 hydrogenation.

3.1.1. The CO hydrogenation is the main reaction used for the production of methanol

In this case, the main reaction leading to methanol is the hydrogenation of CO.

$$CO + 2H_2 = CH_3OH \qquad -90.64\ KJ/mol \tag{15}$$

$$CO_2 + H_2 = CO + H_2O \qquad 41.14\ KJ/mol \tag{16}$$

According to equation (Eq. 15)methanol is predominantly synthesised via the direct hydrogenation of CO. The second reaction is the reverse of Water Gas Shift (RWGS). Experimental data involving typical syngas mixtures that contain 3 to 9% CO_2 show a decrease of its concentration in the reactor effluent stream (Sunggyu, 2007). It should be noted that the first reaction (methanol synthesis) is exothermic, whereas the second (RWGS) is endothermic. According to this depletion of carbon dioxide in the RWGS reaction, produce more reactant (CO) which overall boost the synthesis of methanol. Until 90s the role of CO_2 in the methanol synthesis was not clear. Deficiency of CO_2 in the feed composition can be extremely detrimental to the overall synthesis, very rapidly deactivating the catalysts and immediately lowering methanol productivity in the process. Typically, 2 to 4% of CO_2 is present in the syngas mixture for the vapor-phase synthesis of methanol, whereas this value is somewhat higher, with 4 to 9%, for liquid-phase synthesis (Cybulski, 1994).

The primary reaction for methanol production is the hydrogenation of CO2.

$$CO_2 + 3H_2 = CH_3OH + H_2O \qquad -49.67\ KJ/mol \tag{17}$$

$$CO + H_2O = CO_2 + H_2 \qquad -41.14\ KJ/mol \tag{18}$$

In this case, the main reaction leading to methanol is hydrogenation of CO_2 .

It should be noted that according to this reaction, the synthesis of methanol proceeds predominantly by direct hydrogenation of CO_2 and not CO. It should also be noted that the WGS reaction proceeds in the forward direction, consuming CO to produce the main reactants: CO_2 and H2, thus boosting the eventual methanol productivity. A number of authors (Cybulski, 1994,Lee et al., 1989) have tried a variety of reaction to elucidate the true reaction pathways or mechanistic pathways, including isotope labelling studies and kinetic studies involving complete absence of one of the syngas components.

3.2. Catalysts used

The first high-temperature, high-pressure ever used methanol synthesis catalysts were ZnO/Cr2O3and were operated at 350°C and 250-350 bar. Catalyst compositions contained 20-75 atom% Zn and these catalysts demonstrated high activity and selectivity for methanol synthesis and proved robust enough resist sulphur poisoning which is inherent when converting syngas from coal gasification. Over the years, as gas purification technologies improved, interest in the easily poisoned Cu catalysts for methanol synthesis was renewed. In 1966, ICI introduced a new, more active Cu/ZnO/Al2O3catalyst was the first of a new generation of methanol production using lower temperature (220-275°C) and lower pressure (50-100 bar) than the established ZnO/ Cr2O3catalysts. The last high temperature methanol synthesis plant was closed in the mid-1980s (Fiedler et al., 2003) and at the present, low temperature, low pressure processes based on Cu catalyst are used for all commercial production of methanol from syngas. The synthesis process has been optimised to the point that the modern methanol plants yield 1 kg of methanol/liter of catalyst/hr with >99.5% selectivity for methanol. Commercial methanol synthesis catalyst has lifetimes on the order of 3-5 years under normal operating conditions.

The Cu crystallites in methanol synthesis catalysts have been identified as the active catalytic sites although the actual state (oxide, metallic...) of the active Cu site is still being debated. The most active catalysts all have high Cu content, optimum about 60 wt% Cu on the catalyst that is limited by the need to have enough refractory oxide to prevent sintering of the Cu crystallites. Hindering agglomeration is why ZnO creates a high Cu metal surface area. ZnO also interacts with Al2O3to form a spinel that provides a robust catalyst support. Acidic materials like alumina, are known to catalyse methanol dehydration reactions to produce DME. By interacting with the Al_2O_3 support material, the ZnO effectively improves methanol selectivity by reducing the potential for DME formation. Catalysts are typically prepared by the co-precipitation of metal salts with a variety of precipitation agents. It is important to avoid contaminating methanol catalysts with metals that have FT (Fischer Tropsch) activity (Fe or Ni) during the synthesis. Incorporation of alkali metal in the catalyst formulation should also be avoided because they catalyse the increase of higher alcohols production.Table 4shows catalyst formulation from several commercial manufacturers.

Additional catalyst formulations have been presented in the literature with the purpose of improving per-pass methanol yields (Klier, 1982). The addition of Cs to Cu/ZnO mixtures has shown improved methanol synthesis yields. This only holds true for the heavier alkali metals, as the addition of K to methanol synthesis catalysts tends to enhance higher alcohols yields. The Cu/ThO_2intermetallic catalysts have also been investigated for methanol synthesis (Klier, 1982). These catalysts have demonstrated high activity for forming methanol from CO_2 -free syngas. Cu/Zr catalysts have proven active for methanol synthesis in CO-free syngas at 5 atm and 160-300°C (Herman, 1991). Supported Pd catalysts have also demonstrated methanol synthesis activity in CO_2 -free syngas at 5-110 atm and 260-350°C (Spath and Dayton, 2003).

Table 4. Commercial Methanol synthesis catalyst formulation (Spath and Dayton, 2003)

Manufacter	Cu (wt%)	Zn (wt%)	Al (wt%)	Other
IFP	45 – 70	15 – 35	4 – 20	Zr – 2-18
ICI	20 – 35	15 – 50	4 – 20	Mg
BASF	38.5	18.6	12.9	Rare Earth oxide – 5
Shell	71	24		
Sud Chemie	65	22	12	
Dupont	50	19	31	
Haldor Topsoe	"/>55	21 - 25	8 - 10	

3.3. Methanol synthesis reactors

Currently, most of the methanol is produced by synthesizing syngas through the steam reforming process of natural gas (SMR). The synthesis can be conducted using either the heat generated by a furnace situated near the tubular reactor, or using a combination of auto-thermal reforming (ATR) and steam reforming. After the natural gas is transformed, the resulting synthesis gas can be adjusted to achieve the desired H2/CO ratio and reduce the CO2 content to a few percentages, as previously indicated. The syngas is subsequently introduced into a reactor vessel, where it reacts with a catalyst to generate methanol and water vapor. The impure methanol, often including up to 18% water, as well as ethanol and other alcohols, is sent to a distillation facility comprising of two units: one for removing the volatile substances, and another for eliminating the water and higher alcohols. The unused syngas is returned to the methanol converter, leading to a total conversion efficiency of 99%. Figure 3 displays a standard process flow diagram (PFD) for methanol production.

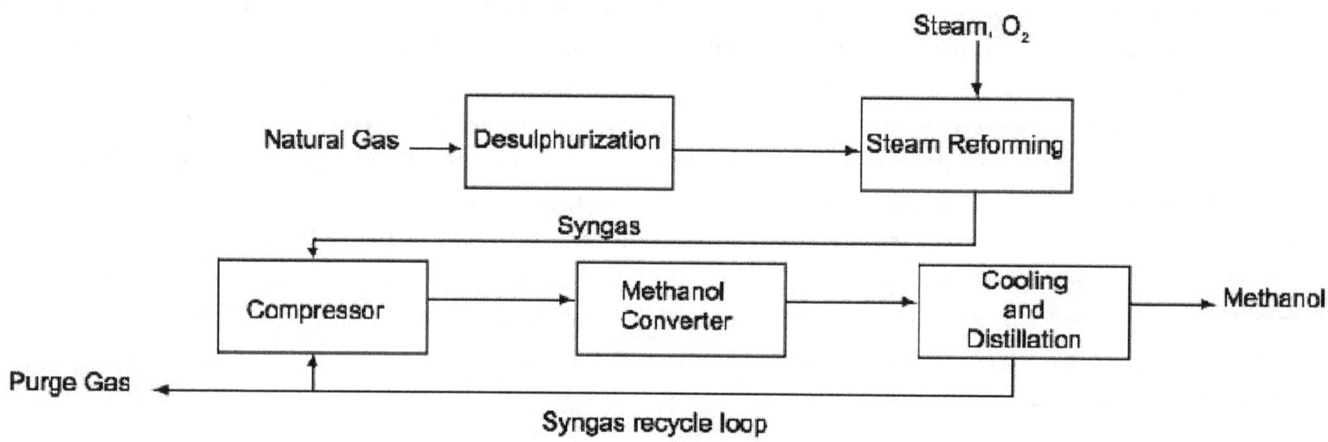

Figure 3. Simplified Methanol Synthesis PFD (Spath and Dayton, 2003).

As is the case with Fisher Tropsch synthesis, one of the challenges associated with commercial methanol synthesis is removing the large excess heat of reaction. Methanol synthesis increases at higher temperatures so does the chance for competing side reactions. Controlling and dissipating the heat of reaction and overcoming the equilibrium constraint to maximise the per-pass conversion efficiency are the two main process features that are considered when designing the methanol synthesis reactor, commonly referred to as a methanol converter. Numerous methanol converter designs have been commercialised over the years and these can be roughly separated into two categories: adiabatic or isothermal reactors.

Adiabatic reactors often imply multiple catalysts beds separated by gas cooling devices, either direct heat exchange or injection of cooled, fresh or recycled syngas. The isothermal reactors are designed to continuously remove the heat of the reaction so they operate essentially like a heat exchanger.

Haldor Topsoe low-pressure methanol synthesis process. This process is designed to produce methanol from natural gas or associated gas feedstocks, utilizing a two-step reforming process to generate a syngas mixture feed for the methanol synthesis (Sunggyu, 2007). Associated gas is natural gas produced with crude oil from the same reservoir. It is claimed that the total investment for this process is lower than with the conventional flow scheme based on straight steam reforming of natural gas by approximately 10%, even after considering an oxygen plant. As shown in figure 4, the two stage reforming is conducted by primary reforming in which a preheated mixture of natural gas and steam is reacted, followed by a secondary reforming which further converts the exit gas from the primary reformer with the aid of oxygen that is feed separately. The amount of oxygen required as well as the balance of conversion between the primary and the secondary reformers need to be properly adjusted so that a balanced syngas (in a stoichiometric ratio (2:1) of H2/CO) is obtained with a low inert content.

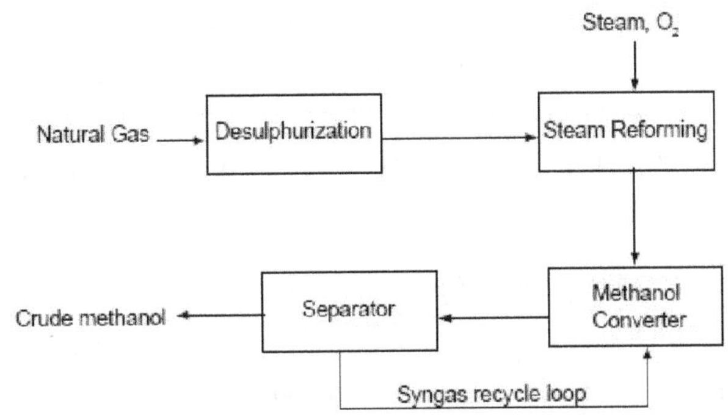

Figure 4. Haldor Topsoe methanol synthesis process (Sunggyu, 2007).

Liquid-Phase methanol process. The liquid phase methanol process was originally developed by Chem. Systems Inc in 1975 (Cybulski, 1994). The R&D of this process was sponsored by the U.S. Department of Energy and Electric Power Research Institute. Commercialised by Air Products and Chemicals Inc and Eastman Chemical Co. in the 1990s, the process is based on the low-pressure methanol synthesis using coal as the source of syngas. Recently, in Québec (Canada) Enerkem Inc. has developed a liquid phase methanol process using syngas produced from biomass. The chemical reaction is carried out in a slurry reactor using a Cu/ZnO/Al2O3catalyst at temperature ranging from 230 to 260 °C and 50 to 100 atm. The commercial reactor used a liquid entrained reactor in which fine grains of catalyst are slurried in an inert high-boiling oil typically white mineral oil. Pressurized gaseous reactants are dissolved in the oil and the dissolved molecular species are reacted on the catalytic surfaces of the grains present in a slurry. Thefigure 5shows the schematic of liquid phase methanol process of Enerkem Inc.

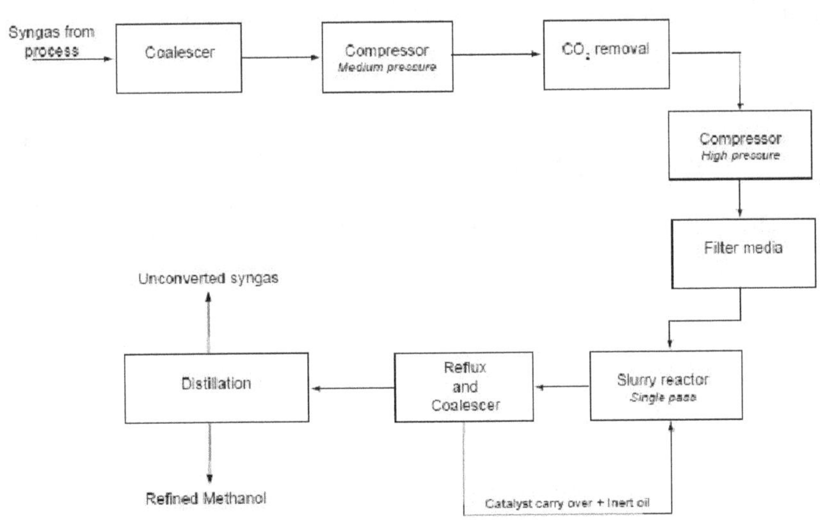

Figure 5. Enerkem Inc. liquid phase methanol process

4. SYNTHESIS OF ETHANOL

Ethanol can be easily generated by the process of fermenting simple sugars that have been hydrolyzed from starch crops. The feedstocks used for this fermentation process include corn, barley, potato, rice, and wheat (Cybulski, 1994). Grain ethanol is another term for sugar ethanol, while ethanol derived from cellulose biomass is referred to as cellulosic or second generation ethanol. Grain ethanol and cellulosic ethanol are both created by biological processes, but chemical ethanol is synthesized using chemical synthesis methods that do not entail fermentation. Indeed, the thermo-catalysis process approach presents itself as a viable alternative to the bioprocess path. In a recently documented procedure (Chornet et al., 2009), ethanol is synthesized by a two-part process. The first phase involves the conversion of methanol to methyl acetate through carbonylation. The second step involves the hydrogenolysis of methyl acetate. This process is represented in Figure 6.

4.1. Methanol carbonylation: methyl acetate synthesis

Methyl acetate is produced industrially by methanol carbonylation in the acetic acid production processes. Methanol carbonylation reaction is a well known reaction since Monsanto in 1970 performed it in liquid phase using a rhodium-base catalytic system. which was later acquired and optimized by Celanese. Such system had been preceded by a cobalt-based system developed by BASF, which suffered from significantly lower selectivity and the necessity for much harsher conditions of temperature and pressure. Although the rhodium catalysed system has much better activity and selectivity, the R&D has continued for new catalysts which improve efficiency even further. The strategies employed have involved either modifications to the rhodium-based system or insertion of another metal, eventually Iridium (Haynes, 2006). In the homogenous system, acetic acid was used as solvent, so esterification leads to substantial conversion into methyl acetate. Methyl acetate formation is facilitated by the methyl iodide used in carbonylation as a mediator. Heterogenous systems for methanol carbonylation have been suggested for several years. Research has been primarily focus on two possible catalysts for this reaction: 1) rhodium supported polymers (Acetica process as an example) or 2) zeolites and a variety of metals supported on activated carbon (Merenov and Abraham, 1998). The choice of the support seems to play an important role in the activity of the catalyst (Yashima et al., 1979). Another way to produce methyl acetate is by an esterification reaction. Eastman Chemical Company, produced methyl acetate by the esterification of methanol with acetic acid in a reactive distillation column.

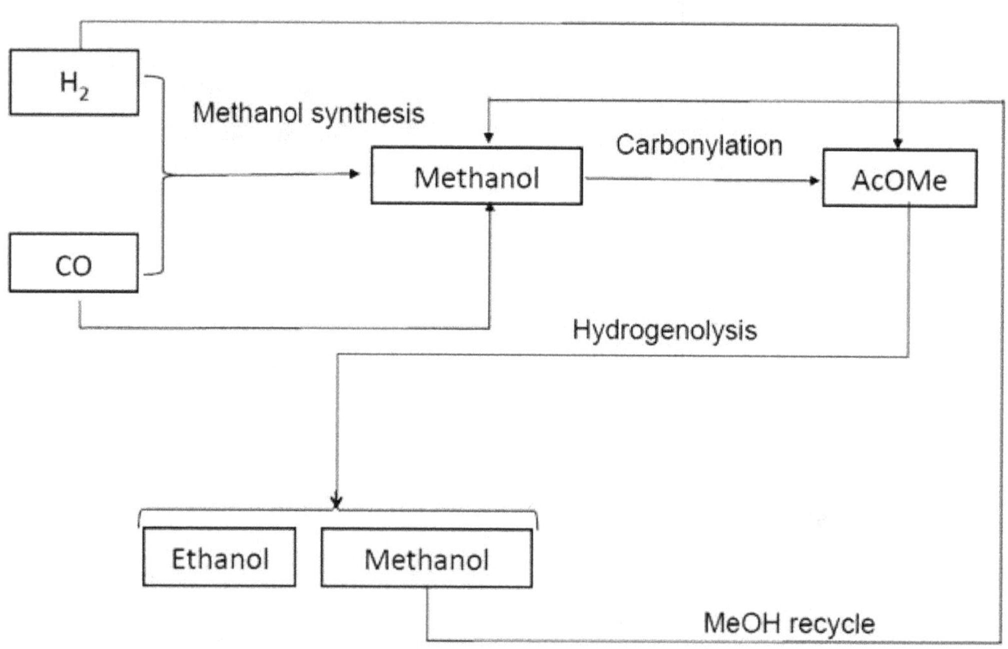

Figure 6. Enerkem Inc. ethanol pathway production.

This company claimed the first commercial plant of reactive distillation (Zoeller, 2009). As illustrated by figure 7, in the reactive distillation there are three sections in the column, water/methanol stripping zone in the bottom, a reaction zone in the center, and methyl acetate enriching at the top. Descriptively, methanol (b.p. 65°C) is added at the bottom of the reaction zone and acetic acid (b.p. 118°C) and acid catalyst are added at the top of the reaction zone.

As methanol ascends up the column it encounters a consistently richer acetic acid concentration, which drives the equilibrium toward the products and prevents methanol from leaving column. Likewise, as acetic acid descends down the column, it also encounters a continuously enriched stream of methanol, which also pushes the equilibrium toward products and prevents acetic acid reaching the bottoms of the column. However, the acetic acid also serves as extractant for water, breaking the various azeotropes. A short section is left at the base of the distillation column to strip all the methanol and a short section is added at the top of the column to enrich methyl acetate and prevent acetic acid from being entrained as the product is distilled overhead in the column. By properly balancing the addition rates, the reaction provides methyl acetate in 100% yield based on both acetic acid and methanol.

Enerkem Inc. has focused on the new gas phase approach to the carbonylation reaction. As acetic acid is not the desired end product of the carbonylation, the reaction conditions have been tailored to favour the methyl acetate directly according to the equation:

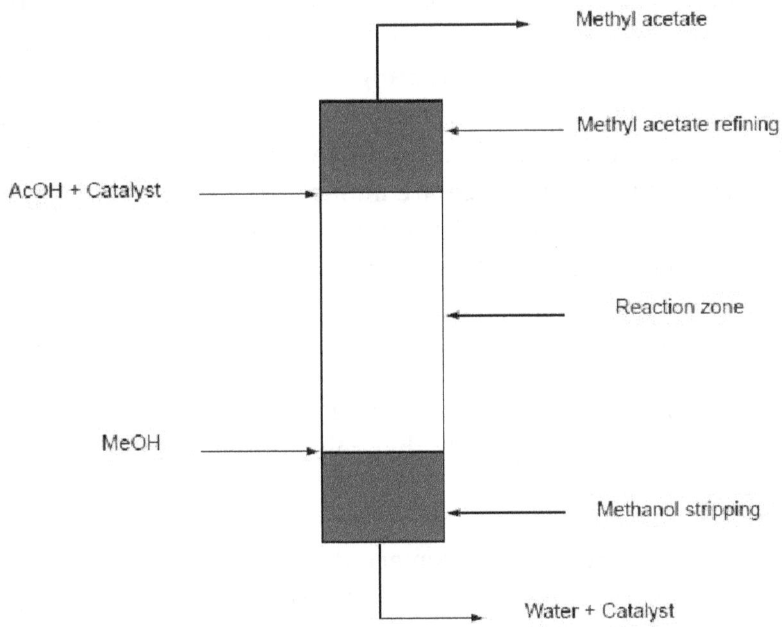

Figure 7. Reactive distillation of methanol (MeOH) and acetic acid (AcOH) to form methyl acetate.

$$2CH_3OH + CO = CH_3COOCH_3 + H_2O \quad -138.08\ KJ/mol \qquad (19)$$

This is beneficial for Enerkem Inc. as it streamlines their ethanol production process by eliminating one step. Promoting the conversion to methyl acetate can also be achieved by the utilization of a gas phase reactor. The methanol carbonylation process involves keeping methanol in the vapor phase using a fixed bed filled with a catalyst based on rhodium. Prior to entering the reactor, the methanol is evaporated under pressure and combined with the gas that has a high concentration of carbon monoxide. The molar ratio of methanol to CO falls within the range of 1 to 4. Methyl iodide, which serves as a co-catalyst and mediator, is introduced into the system at an appropriate molar ratio with respect to methanol. The operating conditions result in a variable GHSV (Gas Hourly Space Velocity) based on CO, ranging up to

2,000 h-1. When the temperature is between 170 to 300°C and the total pressure is from 10 to 50 atm, it has been observed that the conversion of CO to methanol reaches about 100% when the methanol:CO ratio is up to 2 (Chornet et al., 2009).

4.2. Methyl acetate hydrogenation

Esters, such as methyl acetate, are widely present in nature and have extensive industrial and commercial uses depending on the specific substituent groups. A significant category of industrial catalytic processes for reducing esters involves metal-catalyzed hydrogenolysis, which is used to produce alcohols such as ethanol (Adkins and Folkers, 1931, Adkins et al., 1933, Turek et al., 1994a, Xu and Xu, 2010). Recent research has discovered further uses for ester hydrogenolysis in the process of catalytic upgrading and conversion of renewable resources, including lignocellulosic biomass, into fuels or fine chemicals (Corma et al., 2007, Huber et al., 2006). Copper-based catalysts are commonly employed for the process of ester hydrogenolysis, as demonstrated by several studies (Adkins and Folkers, 1931; Cybulski et al., 2001; Thomas et al., 1992; Turek et al., 1994b; Turek et al., 1994a). These compounds exhibit excellent reactivity and specificity towards alcohols when subjected to elevated temperatures and hydrogen pressure (500-700K, 200-300 bar). Alloys containing precious metals, such as Rh-Sn, have the potential to reduce the harshness of the reaction conditions (by lowering hydrogen pressure to less than 100 bar). However, precious metals are expensive and tend to promote the synthesis of hydrocarbons. The methyl acetate that has been recovered is kept in a liquid state at a temperature of 20°C in our method. The substance is pressurized within the range of 10 to 50 atm and then sent through a heat exchanger where it is totally vaporized at temperatures ranging from 150 to 425°C. Hydrogen, which has been heated to the same temperature range, is combined with the methyl acetate vapor at the outlet of the heat exchanger. The hydrogen to methyl acetate molar ratio ranges from 4:11. The heated mixture is passed over a catalytic bed containing either CuO/Cr2O3 or CuO/ZnO/Al2O3 catalyst. Prior to the addition of methyl acetate, the CuO is subjected to reduction using H2/N2 combinations (Claus et al., 1991). The conversion of methyl acetate into ethanol and methanol is described by the following equation:

$$CH_3COOCH_3 + 2H_2 = CH_3OH + CH_3CH_2OH \quad -27\ KJ/mol \tag{20}$$

The GHSV (based on H2) of the reaction is comprised between 1,000 and 2,000 h-1and the methyl acetate conversion reached 95%, with a slightly higher selectivity towards ethanol.

5. CONCISE SUMMARY

Biomass, a renewable energy source, can be converted into biofuels as a means to decrease reliance on oil and minimize the carbon dioxide emissions associated with fossil fuels. This chapter has demonstrated the chemical processes required to transform heterogeneous biomass into bioethanol by gasification. The syngas generated undergoes catalytic conversion into biofuels, specifically methanol and ethanol. Enerkem employs a technique that involves using non-homogeneous residual biomass, particularly urban biomass, as feedstock for gasification. The syngas is purified before being utilized for the catalytic production of methanol. The methanol synthesis reactor employed is a three-phase reactor that operates on the Liquid-Phase methanol method. The methanol yield is approximately 1 kgMeOH/kgcat/h, whereas the reaction's selectivity towards methanol is around 99%. The purified methanol is subjected to carbonylation to yield methyl acetate. The production of methyl acetate occurs in a fixed bed reactor and involves the use of a halide component as a co-catalyst. The process of hydrogenolysis converts methyl acetate into equimolar amounts of ethanol and methanol. The methanol generated during the process is reused in the carbonylation unit.

REFERENCES

J.-L. Rastoin, (2008). Agrocarburant et sécurité alimentaire: l'alliance impossible, Vol.

M. J. Prins, K. J. Ptasinski and F. J. J. G. Janssen, (2005). Exergetic optimisation of a production process of Fischer-Tropsch fuels from biomass, Vol. 86,No. 4,pp. 375-389, 0378-3820

D. Ballerini and N. Alazard-Toux. (2006). Les biocarburants: état des lieux, perspectives et enjeux du développement, Technip, 9782710808695,

A. van der Drift, J. van Doorn and J. W. Vermeulen, (2001). Ten residual biomass fuels for circulating fluidized-bed gasification, Vol. 20,No. 1,pp. 45-56, 0961-9534

G. L. Borman and K. W. Ragland. (1998). Combustion engineering, McGraw-Hill, 9780070065673,

C. Higman and M. van der Burgt. (2008). Gasification (2nd Edition), Elsevier, 978-0-7506-8528-3,

J. P. Ciferno and J.J. Marano (2002). Benchmarking Biomass Gasification Technologies for Fuels, Chemicals and Hydrogen Production, report for US DOE.

S. A. Channiwala and P. P. Parikh, (2002). A unified correlation for estimating HHV of solid, liquid and gaseous fuels, Vol. 81,No. 8,pp. 1051-1063, 0016-2361

B. Wetherold, D. Orr, D. Maxwell (2000). A comparison of gasification and incineration Hazardous wastes. US DOE, DCN 99.803931.02

S. C. Marie-Rose, J.-M. Lavoie, D. Lynch and E. Chornet. (2011). From biomass-rich residues into fuels and green chemicals via gasification and catalytic synthesis, WIT Transaction on Ecoloy and the Environment, Vol. 143, pp. 123-132, 1743-3541

S. M. Villano, J. Hoffman, H.-H. Carstensen, A.M. Dean (2010). Selective Oxidation of Ethylene in a Model "Dirty Synthesis Gas Stream: An Experimental and Kinetic Modeling Study, ACS National meeting, San Francisco

T. Paes (2005). Modeling for control of biomass gasifier, Thesis of Endhoven University of Technology.

E. Chornet, B. Valsecchi, G. Drolet, M. Gagnon, B. Nguyen (2010). Production and conditioning of synthesis gas obtained from biomass. Patent Application number 20100051875.

A. Corujo, L. Yermán, B. Arizaga, M. Brusoni and J. Castiglioni, (2010). Improved yield parameters in catalytic steam gasification of forestry residue; optimizing biomass feed rate and catalyst type, Vol. 34,No. 12,pp. 1695-1702, 0961-9534

R. Rauch, H. Hofbauer, K. Bosch, I. Siefert, C. Aichernig, H. Tremmel, K. Voigtlaender, R. Koch, R. Lehner (2004). Steam Gasification of Biomass at CHP Plant in Guessing- Status of the Demonstration Plant. 2nd World Conference and Technology Exhibition on Bioamass for Energy, Italy.

D. Swierczynski, S. Libs, C. Courson and A. Kiennemann, (2007). Steam reforming of tar from a biomass gasification process over Ni/olivine catalyst using toluene as a model compound, Vol. 74,No. 3-4,pp. 211-222, 0926-3373

C. Yan, W. Yang, J. T. Riley and W.-P. Pan. (2006). A novel biomass air gasification process for producing tar-free higher heating value fuel gas, Elsevier, Amsterdam, PAYSBAS M. He, B. Xiao, Z. Hu, S. Liu, X. Guo and S. Luo, (2009). Syngas production from catalytic gasification of waste polyethylene: Influence of temperature on gas yield and composition, Vol. 34,No. 3,pp. 1342-1348, 0360-3199

J. R. Rostrup-Nielsen, (2000). New aspects of syngas production and use, Vol. 63,No. 2-4,pp. 159-164, 0920-5861

I. Dybkjr and T. S. Christensen. (2001). Syngas for large scale conversion of natural gas to liquid fuels, Studies in Surface Science and Catalysis, J. J. S. E. Iglesia and T. H. Fleisch,pp. 435-440, Elsevier, 0167-2991,

J. L. Li and T. Inui, (1996). Enhancement in methanol synthesis activity of a copper/zinc/aluminum oxide catalyst by ultrasonic treatment during the course of the preparation procedure, Vol. 139,No. 1-2,pp. 87-96, 0926-860X

B. J. Liaw and Y. Z. Chen, (2001). Liquid-phase synthesis of methanol from CO2/H2 over ultrafine CuB catalysts, Vol. 206,No. 2,pp. 245-256, 0926-860X

Y. W. Wang, Z. L. Wang, J. F. Wang and Y. Jin, (2002). Study on the methanol synthesis in slurry reactors, Vol. 31,No. 8,pp. 597-601, 10008144 (ISSN)

L. Sunggyu. (2007). Methanol Synthesis from Syngas, Handbook of Alternative Fuel Technology, Taylor & Francis Group, A. Cybulski, (1994). Liquid-phase methanol synthesis: Catalysts, mechanism, kinetics, chemical equilibria, vapor-liquid equilibria, and modeling - A review, Vol. 36,No. 4,pp. 557-615, 01614940 (ISSN)

S. Lee, V. R. Parameswaran, I. Wender and C. J. Kulik, (1989). Roles of carbon dioxide in methanol synthesis, Vol. 7,No. 8,pp. 1021-1057,

E. Fiedler, G. Grossmann, D. B. Kersebohm, G. Weiss and C. Witte. (2003). Ullmann's Encyclopedia of Industrial Chemistry Release 2003, Methanol, W.-V. V. G. C. KGaA,pp.

K. Klier. (1982). Methanol Synthesis, Advances in Catalysis, H. P. D.D. Eley and B. W. Paul,pp. 243-313, Academic Press, 0360-0564,

R. G. Herman. (1991). Chapter 7 Classical and Non-Classical Routes for Alcohol Synthesis, Studies in Surface Science and Catalysis, L. Guczi,pp. 265-349, Elsevier, 0167-2991, P. L. Spath, D.C. Dayton (2003). Preliminary Screening and Economy Assessment of Synthesis fuels and Cjemicals with Emphasis on the Potential for Biomass-Derived Syngas

A. Haynes. (2006). Acetic Acid Synthesis by Catalytic Carbonylation of Methanol, Catalytic Carbonylation Reactions, M. Beller,pp. 179-205, Springer Berlin / Heidelberg, A. S. Merenov and M. A. Abraham, (1998). Catalyzing the carbonylation of methanol using a heterogeneous vapor phase catalyst, Vol. 40,No. 4,pp. 397-404, 0920-5861

T. Yashima, Y. Orikasa, N. Takahashi and N. Hara, (1979). Vapor phase carbonylation of methanol over RhY zeolite, Vol. 59,No. 1,pp. 53-60, 0021-9517

J. R. Zoeller, (2009). Eastman Chemical Company's "Chemicals from Coal" program: The first quarter century, Vol. 140,No. 3-4,pp. 118-126, 0920-5861

H. Adkins and K. Folkers, (1931). The catalytic hydrogenation of Esters to alcohols, Vol. 53,No. 3,pp. 1095-1097, 0002-7863

E. Chornet, B. Valsecchi, Y. Avilla, B. Nguyen, J-M. Lavoie, (2009). Production of ethanol to methanol, patent Application, WO20090326080

H. Adkins, B. Wojcik and L. W. Covert, (1933). The Catalytic Hydrogenation of Esters to Alcohols. III, Vol. 55,No. 4,pp. 1669-1676, 0002-7863

T. Turek, D. L. Trimm and N. W. Cant, (1994a). The Catalytic Hydrogenolysis of Esters to Alcohols, Vol. 36,No. 4,pp. 645-683, 0161-4940

L. Xu and Y. Xu, (2010). Activation of methyl acetate on Pd(111), Vol. 604,No. 11-12,pp. 887- 892, 0039-6028

A. Corma, S. Iborra and A. Velty, (2007). Chemical Routes for the Transformation of Biomass into Chemicals, Vol. 107,No. 6,pp. 2411-2502, 0009-2665.

G. W. Huber, S. Iborra and A. Corma, (2006). Synthesis of Transportation Fuels from Biomass:,Äâ Chemistry, Catalysts, and Engineering, Vol. 106,No. 9,pp. 4044-4098, 0009-2665

A. Cybulski, J. Chrzszcz and M. V. Twigg, (2001). Hydrogenation of dimethyl succinate over monolithic catalysts, Vol. 69,No. 1-4,pp. 241-245, 0920-5861

D. J. Thomas, J. T. Wehrli, M. S. Wainwright, D. L. Trimm and N. W. Cant, (1992). Hydrogenolysis of diethyl oxalate over copper-based catalysts, Vol. 86,No. 2,pp. 101-114, 0926-860X

T. Turek, D. L. Trimm, D. S. Black and N. W. Cant, (1994b). Hydrogenolysis of dimethyl succinate on copperbased catalysts, Vol. 116,No. 1-2,pp. 137-150, 0926-860X

P. Claus, M. Lucas, B. Lücke, T. Berndt and P. Birke, (1991). Selective hydrogenolysis of methyl and ethyl acetate in the gas phase on copper and supported Group VIII metal catalysts, Vol. 79,No. 1,pp. 1-18, 0926-860X

www.ingramcontent.com/pod-product-compliance
Lightning Source LLC
LaVergne TN
LVHW062144200926
848484LV00007B/1857